Nature-Based Solutions-2nd Edition

Nature-Based Solutions-2nd Edition

Topic Editors

Panayiotis G. Dimitrakopoulos
Miklas Scholz
Mario A. Pagnotta
Arshiya Noorani

MDPI

Basel • Beijing • Wuhan • Barcelona • Belgrade • Novi Sad • Cluj • Manchester

Topic Editors
Panayiotis G. Dimitrakopoulos
Department of Environment
University of the Aegean
Mytilene
Greece

Miklas Scholz
Department of Civil Engineering Science
University of Johannesburg
Johannesburg
South Africa

Mario A. Pagnotta
Department of Agricultural and Forest Sciences
Tuscia University
Viterbo
Italy

Arshiya Noorani
Plant Production and Protection Division
FAO
Rome
Italy

Editorial Office
MDPI AG
Grosspeteranlage 5
4052 Basel, Switzerland

This is a reprint of the Topic, published open access by the journals *Forests* (ISSN 1999-4907), *Land* (ISSN 2073-445X) and *Sustainability* (ISSN 2071-1050), freely accessible at: https://www.mdpi.com/topics/MI3F953C5E.

For citation purposes, cite each article independently as indicated on the article page online and as indicated below:

Lastname, A.A.; Lastname, B.B. Article Title. *Journal Name* **Year**, *Volume Number*, Page Range.

ISBN 978-3-7258-7927-4 (Hbk)
ISBN 978-3-7258-7928-1 (PDF)
https://doi.org/10.3390/books978-3-7258-7928-1

Contents

About the Editors

Panayiotis G. Dimitrakopoulos

Panayiotis G. Dimitrakopoulos is Professor of Functional Ecology in the Department of Environment of the University of the Aegean, Greece. He obtained his Ph.D. from the University of the Aegean in 2001. He carried out his postdoctoral research at the University of Zurich, Switzerland, funded by the European Science Foundation (LINKECOL program). His research focuses on functional plant ecology, community ecology, biodiversity conservation, and conservation policy. He has served as President of the Hellenic Ecological Society (2012–2014), a member of the "Natura 2000" National Committee (2010–2020), and Honorary Visiting Senior Fellow at Anglia Ruskin University, Cambridge (2019–2022). He serves as the Section Editor-in-Chief of Ecology in *Biology* (Basel) and is a member of the Editorial Board of eight journals and a referee for more than 70 journals. He has published more than 100 papers in peer-reviewed journals.

Miklas Scholz

Miklas Scholz, cand ing, BEng (equiv), PgC, MSc, PhD, DSc, CWEM, CEnv, CSci, CEng, FHEA, FIEMA, FCIWEM, FICE, Fellow of IWA, Fellow of IETI, is a professional senior engineer and a Distinguished Professor of Civil Engineering Science at Johannesburg University, South Africa. He has published 11 books and 361 journal articles. Prof. Dr. Scholz has 20,900 citations, resulting in an h-index of 71, an i10-Index of 267, and an i100-Index of even 42. His top paper alone received 2306 citations. Miklas ranks among the top five wetland engineers worldwide. He belongs to the top 2% of academics regarding the i10-index. Miklas also belongs to the World's Top 2% Scientists by Stanford University. He was awarded €1.52M for the JPI Water 2018 project RAINSOLUTIONS and € 7M for the EU H2020 REA project WATERAGRI. According to People Ai, his relative social fame wealth (not income) as a brand name is estimated to be 493k million US$. Miklas's areas of expertise are water and environmental engineering; water resources engineering; agricultural water management; nature-based solutions; pollution control; biological filtration; wastewater treatment; decision support systems; artificial intelligence; integrated constructed wetlands; storm water management; sustainable drainage systems; and climate change adaptation.

Mario A. Pagnotta

Mario A. Pagnotta is Associate Professor in Agricultural Genetics at the "Dipartimento di Scienze Agrarie e Forestali" (Department of Agricultural and Forestry scieNcEs) (DAFNE) University of Tuscia, Viterbo, Italy. He obtained his Ph.D at Reading University (UK) in 1991 with the Thesis "The ecology and ecological genetics of pasture Legumes in Syria". He has been Research Associate at ICARDA (International Center for Agricultural Research in the Dry Areas), Aleppo, Syria. Research activities have focused mainly on the analysis of genetic variation in natural populations of several species of agricultural interest (*Triticum*, Emmer wheat, Globe Artichoke, *Capparis spinosa*, *Poplar*, forage legumes, Pistachio, Orchid), and the relationship between genetic variation and edaphic factors. These include biodiversity conservation, adaptation, and tools for variety identification. The analyses are conducted using traditional and advanced methodologies, based on different kinds of molecular markers. Particular attention is paid to the resistance to abiotic stress, especially drought stress, looking also at the distribution and identification of SNP markers for genes related to drought and salt tolerance, and investigating the role of root architecture in conferring tolerance.

Arshiya Noorani

Arshiya Noorani is an Agricultural Officer based at the FAO headquarters in Rome and is a member of the Seeds and Plant Genetic Resources Team, Plant Production and Protection Division (AGP). Arshiya has been with FAO since 2011 but has been working since 2000 in the conservation, assessment, and management of biodiversity, including protected areas. She holds a PhD in Plant Conservation Genetics, an MSc in Environmental Protection and Management, and BSc (Hons) in Zoology. Prior to joining FAO, she worked on *in situ* conservation and on-farm management of plant genetic resources for food and agriculture at Bioversity International. She previously taught ecological surveying and protected area management planning in both terrestrial and freshwater habitats at the University of Edinburgh.

At FAO, Arshiya works on developing and implementing projects, developing technical guidelines, and assessing data submitted by Members when monitoring the conservation and use of plant genetic resources. She supports Members in the implementation of Global Action for Plant Genetic Resources for Food and Agriculture and associated reporting, which feed into the preparation of global periodic assessments. She contributes to FAO's normative work, especially through the Commission on Genetic Resources for Food and Agriculture and the International Treaty on Plant Genetic Resources for Food and Agriculture.

Preface

This Reprint includes sixteen peer-reviewed papers that explore the evolving field of nature-based solutions (NBSs) within the broader context of sustainable land management, environmental planning, biodiversity management, and climate resilience.

The aim of this Reprint is to provide an articulate body of scientific work that advances both theoretical understanding and practical implementation of NBSs across diverse landscapes and levels, including urban, peri-urban, agricultural, and forested environments. It reflects the need to address pressing global challenges—such as climate change, land degradation, biodiversity loss, and resource scarcity—through solutions that are based on ecological integrity while remaining socio-economically viable.

This Reprint is addressed to researchers, practitioners, policy makers, and stakeholders engaged in sustainability transitions. It provides a structured overview of recent developments, methodological approaches, and real-world applications, while also offering insights that can inform future research on the design and implementation of nature-based strategies across multiple spatial and governance scales.

Panayiotis G. Dimitrakopoulos, Miklas Scholz, Mario A. Pagnotta, and Arshiya Noorani
Topic Editors

sustainability MDPI

Review

Phytoremediation of Mercury Contamination: Bibliometric Analysis

Lina Mosquera Chaverra [1], Diego Paredes Cuervo [2,*], Ana López Gutiérrez [3], Carlos A. Arias [4,5,*] and Pedro N. Carvalho [5,6]

1 Faculty of Engineering, Environmental Engineering, Universidad Tecnológica del Chocó, Quibdó 270002, Chocó, Colombia; d-lina.mosquera@utch.edu.co
2 Faculty of Basic Environmental Sciences, Universidad Tecnológica de Pereira, Pereira 660003, Risaralda, Colombia
3 Faculty of Agricultural Sciences and Agroindustry, Universidad Tecnológica de Pereira, Pereira 660003, Risaralda, Colombia; alopez@utp.edu.co
4 Department of Biology-Aquatic Biology, Aarhus University, Ole Worms Allé 1, Building 1135, 8000 Århus, Denmark
5 WATEC—Centre for Water Technology, Aarhus University, Ny Munkegade 120, 8000 Aarhus, Denmark; pedro.carvalho@envs.au.dk
6 Department of Environmental Science, Aarhus University, Frederiksborgvej 399, 4000 Roskilde, Denmark
* Correspondence: diparede@utp.edu.co (D.P.C.); carlos.arias@bio.au.dk (C.A.A.)

Abstract: Mercury is classified as one of the world's most toxic and dangerous pollutants as it tends to bioaccumulate and biomagnify within the trophic chain and is persistent. Various approaches are available to remediate Hg-affected sites including phytoremediation, which includes the use of plants to clean up contaminated environments. The phytoremediation of mercury contamination is attracting increasing attention because of its advantages: it is environmentally friendly, inexpensive, simple, and can improve soil fertility. In this report, VOSviewer (version 1.6.1) and Bibliometrix (version 4.16) software were used to analyze 457 and 697 documents published from 2000 to 2023, retrieved from the databases WoS and Scopus, respectively. China, India, the United States, and Spain were the top four most productive countries. The largest topic area was environmental sciences, and the Chinese Academy of Sciences was the organization that contributed the most to the overall number of publications. The keywords with the highest frequency excluding phytoremediation and mercury in WoS were heavy metals, accumulation, cadmium, soils, and phytoextraction. In Scopus, the most frequent keywords were bioremediation, heavy metals, soil pollution, bioaccumulation, biodegradation, and environmental. From the above analysis, we concluded that future research should focus on (1) finding native plants, (2) genetic engineering applications, (3) increasing remediation ability through assisted phytoremediation, and (4) the detoxification mechanism of mercury. This study provides insights into trending themes and serves as a reference for future research.

Keywords: Bibliometrix; bibliometric analysis; mercury; phytoremediation; VOSviewer

1. Introduction

Mercury (Hg) is a global pollutant because of its harmful effects on human and environmental health [1]. It is among the top 10 chemicals of concern and ranks third on the Agency for Toxic Substances and Disease Registry's priority list of substances [2]. It is poisonous, threatens health and ecosystems, and can accumulate to a high level, especially in fish, and cause serious diseases; it has carcinogenic, teratogenic, neuteratogenic, neurotoxic, and genotoxicity effects. Acute exposure to this metal also causes psychological problems such as anxiety, sleep disorders, and depression [3]; its exposure in soils has caused kidney damage and even death [4].

This pollutant exists in elemental, organic, and inorganic forms. In gaseous form, mercury is emitted through the atmosphere and reaches long distances before settling in the soil. According to the United Nations Environment Program, mercury emissions amount to around 2200 tons per year, of which artisanal and small-scale mining activities account for almost 38% of global emissions [5,6]. Volatile mercury is persistent and can be easily taken up by the plant system, accumulating in the food chain and entering the human body. The presence of reactive Hg in the soil under acidic conditions, low redox potential, and a high concentration of organic matter leads to the formation of methylmercury (MeHg) [7–9]. MeHg, the most common organic form, bioaccumulates efficiently in organisms and biomagnifies in the food chain. Its bioaccumulation within organisms severely distorts the biological food chain and eventually enters ecosystems [10]. Exposure to MeHg in fish is often the main concern associated with Hg contamination sites, making fish consumption the main route of mercury exposure in human beings [11,12].

Mercury does not degrade in the environment, and because of its bioaccumulative nature, it is necessary to look for alternatives for the remediation of soils contaminated with this metal so that it can be separated from the contaminating media, or otherwise transformed into less toxic species [13]. There are physical, chemical, and biological remediation techniques, of these, the biological remediation approach has gained the most attention due to its environmentally sustainable nature and ability to treat heavy metal-contaminated soil. It comprises two methods: bioremediation and phytoremediation as well as a combination of the two [14]. Bioremediation uses living organisms to remove and neutralize contaminants and involves the use of microorganisms (e.g., bacteria, microalgae, yeasts) to remove/immobilize, transform, or detoxify heavy metals from the environment [15–17]. Phytoremediation refers to the use of plants to extract, reduce, transform, or immobilize contaminants (organic and inorganic) contained in soils, sediments, and groundwater [5,18,19]. Plants have developed strategies to address toxic elements through chelation, extrusion, regulatory distribution, and vascular sequestration [20]. In addition, this technology involves translocation, accumulation, transport, transformation, and volatilization processes [21–23].

A large portion of the research on the phytoremediation of mercury has focused on finding mercury accumulator or hyperaccumulator plants because, despite their phytotoxic characteristics, there are plant communities that grow on mercury-contaminated sites [24–28]. Hyperaccumulator plants can accumulate large amounts of heavy metals at concentrations 10 to 100 times higher than non-hyperaccumulator plants can tolerate [29]. The limit of mercury hyperaccumulation in plant shoots is estimated to be 10 ppm, although this threshold needs to be more clearly established [30,31]. Therefore, the search for mercury hyperaccumulators and their application in practice has become a research hotspot [1].

In recent years, researchers have found remediating plants for mercury-contaminated soils, reporting on the mechanisms associated with Hg tolerance and the improvement of phytoremediation techniques by chemical and/or microorganism-assisted remediation. For example, *Cardamine violifolia* has a phytoremediation effect on soils contaminated with different mercury concentrations, reaching concentrations of 6000 μg/g [32]. The addition of sodium thiosulfate ($Na_2S_2O_3$) to mercury-contaminated pots showed that *Oxalis corniculata* could recover and remediate mercury-contaminated soils [33]. In the case of rhizobial associations, *Lupinus albus* L. plants were able to accumulate approximately 370 and 360 mg/kg Hg in the roots and nodules, respectively, and to maintain constant levels of photosynthetic pigments when inoculated with the mercury-tolerant *Bradyrhizobium canariense* L-7AH strain [34]. On the other hand, the PtABCC1 transporter gene from *Populus trichocarpa* was overexpressed in transgenic Arabidopsis and *Populus tomentosa* (poplar) plants, showing higher mercury tolerance than wild-type plants [35].

Bibliometric analysis identifies and summarizes the main research points relevant to expanding, publishing, and applying up-to-date knowledge on a subject. It was initially proposed by Pritchard (1969) to analyze books and other media [36]. It integrates mathematical, statistical, and bibliographic methods to quantitatively analyze the distribution structure and quantitative changes in the literature in a given field [37]. This popular and effective method is used to determine and suggest future research trends and essential problems based on historical publications [38,39]. Therefore, a comprehensive bibliometric review in the research field will help current and future researchers strategically identify gaps and leverage their studies [40].

In this paper, we used VOSviewer, and Bibliometrix to analyze publications related to the phytoremediation of mercury-contaminated sites in Scopus and WOS. We wanted to show: (1) the overlap between the two databases and the types of publications; (2) which countries had the highest number of publications and how countries cooperate; (3) which publications are the most cited and which subject area they belonged to; and (4) how the field of mercury contamination phytoremediation has advanced. The types of publication, citations, countries, thematic categories, and institutions were collected and analyzed. Document co-citation analysis was considered to establish the similarity between documents based on bibliographic references as well as the co-authorship and citation analysis of documents, giving information based on the number of documents in which they appeared together, the kinship among documents, and the number of times in which they cited each other. Finally, word co-occurrence analysis was considered, which measures the similarity of documents, relating them according to their degree of co-occurrence. Therefore, this study aimed to analyze the distribution of publications in the field of the phytoremediation of heavy metals, such as mercury, to report on their critical points and provide reference data for new research directions based on bibliometrics to analyze mapping and knowledge gaps.

2. Materials and Methods

A bibliometric analysis was performed to understand the trend of the phytoremediation of mercury contamination using papers published from January 2000 to June 2023 in the academic databases WoS and Scopus. Traditionally, WoS and Scopus have been the two most widely used databases for bibliometric analysis [41]. It is important to note that the selection of this period was made due to the low volume of publications (less than 20) before 2000.

In both databases, the search was performed in the subject field including article title, abstract, and keywords. The query sets used for the literature search were ("phytoremediation") AND ("mercury" OR "Hg") AND ("plant*" OR "phyto*" OR "vegetation" OR "biomass" OR "hyperaccumulator" OR *accumula*). All articles were individually screened to eliminate irrelevant documents (i.e., those dealing with heavy-metal phytoremediation in which mercury was not included). The search results indicated 515 articles for WoS and 813 for Scopus. After eliminating the documents that did not comply with the thematic work, it was reduced to 458 documents (88.9% of 515) for WoS and 698 (85.85% of 813) for Scopus. The documents included citation information, volume, year, authors, affiliations, abstracts and keywords, funding details, and other information such as conference information and references.

Different research methods have been proposed in the field of bibliometric analysis, with scientific mapping being a recent method, which not only reveals the laws of development but also graphically expresses the relationship and evolutionary laws of knowledge structure in related fields [42,43]. Currently, there are several tools for scientific mapping including knowledge graph analysis, which have advantages and characteristics of their own [44,45]. Bibliometrix is an open-source tool for quantitative research in scientometrics and bibliometrics developed by Massimo Aria and Corrado Cuccurullo. This tool includes the main bibliometric methods of analysis and allows for the construction of data matrices for co-citation, linkage, scientific collaboration analysis, and co-word analysis (https://www.bibliometrix.org) accessed on 31 August 2023. VoSviewer is software that is used to plot co-occurrence maps of authors, citations, keywords, and other data and was developed by Van Eck and Waltman [46,47]. In this article, these two tools (VoSviewer 1.6.16, Low Countries, and R-Studio's Bibliometrix R package version 4.1) were used to perform a global network map of collaborating countries, an author co-citation analysis was conducted to analyze the relationships between highly cited references, co-authorship for the analysis of the most productive authors, keyword or term co-occurrence analysis, and the strongest citation keywords. In turn, the "Analyze results" option was used in both databases to extract information on annual research, countries, journals, funders, and thematic areas (i.e., metric information on journals and documents). Before the diagrams were made, the data were processed manually in an Excel matrix to merge synonyms or similar terms and eliminate meaningless words to avoid affecting the analysis of the results.

3. Results and Discussion

3.1. Overlap Between WoS and Scopus Publications

A total of 1156 documents were found in the databases together, distributed between 458 documents for WoS and 698 for Scopus, of which 27% (313) were duplicates (i.e., they belonged to both databases) (Figure 1). Scopus is the largest database of abstracts and literature citations, with a broader coverage of journals compared with the WoS, but the latter is more selective [41].

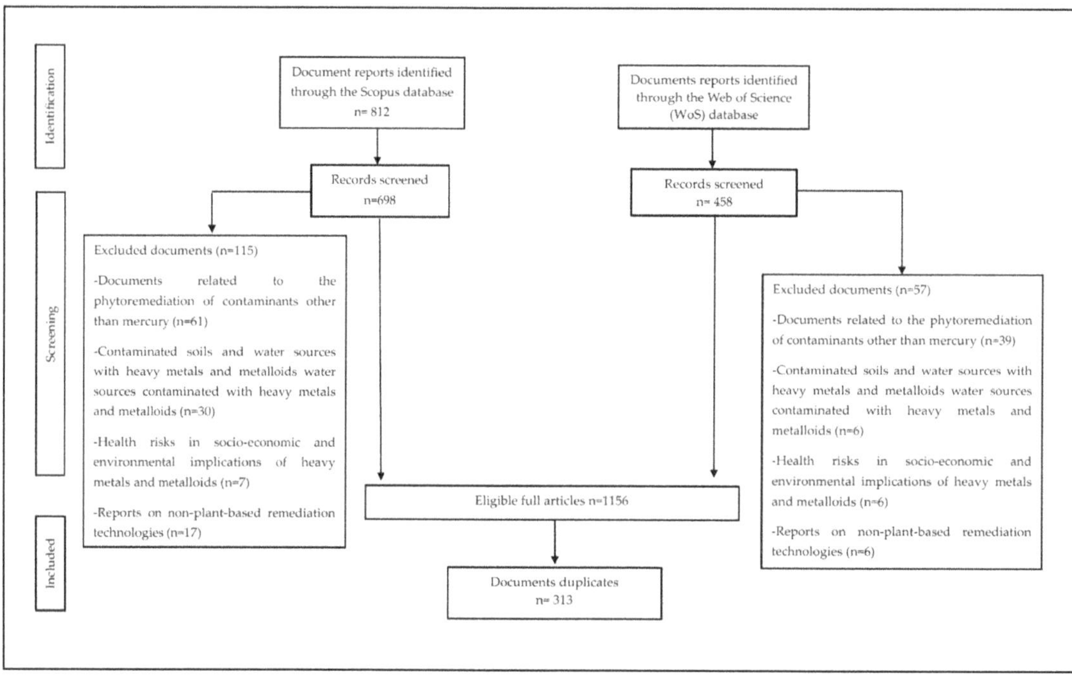

Figure 1. Prisma diagram.

3.2. Annual Trends

The types of scientific publications related to the phytoremediation of mercury contamination are presented in Table 1. The papers included research articles, review articles, books, conference papers, and short surveys. In both databases, research articles were the most published (468; 67%) for Scopus and (396; 87%) for WoS, followed by review papers (15 and 11%, respectively).

Table 1. Types of documents published in the period from 2000 to 2023.

Scopus			WoS		
Document Type	Documents	Percentage (%)	Document Type	Documents	Percentage (%)
Article	468	67.0	Article	396	86.5
Review article	108	15.5	Review article	51	11.1
Book chapter	76	10.9	Proceeding paper	5	1.1
Conference paper	35	5.0	Early access	5	1.1
Others	11	1.6	Others	1	0.2
Total	698	100	Total	458	100

On the other hand, the annual production of publications is shown in Figure 2. Published research on the phytoremediation of mercury contamination show an increasing trend in both databases. Minimizing mercury contamination in soil is of great significance for ensuring the quality and safety of crops and maintaining ecological safety [32]. As far as the Scopus database is concerned, there was a significant increase in publications from 2011. Between 2015 and 2023, the number of published papers (463) more than doubled. In 2022, 78 articles were published, 2.88 times more than in 2011. Similarly, in WoS, the same trend continued, with 360 documents published between 2015 and 2023, which was 3.34 times more than in 2011. More publications are expected in the future due to the importance of phytoremediation not only in remediating contaminated land and water bodies, but also to discover Hg hyperaccumulators and aided techniques for the improvement of this ecological remediation technology [5,48,49].

Figure 2. Number of annual publications on the phytoremediation of mercury contamination included in the Scopus and WoS databases.

3.3. Main Subject Areas

The analysis by category or subject area indicated that in both databases, most of the documents belonged to the category of Environmental Sciences, represented by 33.5% for Scopus and 58.6% for WoS. It should be mentioned that in the Scopus database, the six areas that followed in order were Agricultural and Biological Sciences (16.5%), Biochemistry, Genetics, and Molecular Biology (9.9%), Medicine (5.7%), Engineering (5.6%), Chemistry (5.1%), and Chemistry Engineering (5.0%) (see Figure S1A). In the case of WoS, Environmental Engineering (13.1%), Plant Sciences (10.3%), Water Resources (6.1%), Toxicology (5.0%), Biotechnology Applied Microbiology (4.8%), and Chemistry Multidisciplinary (4.6%) predominated (see Figure S1B). In [50], the authors reported that Scopus has an orientation toward biomedical research and natural sciences, while engineering seems to be represented to a greater extent in WoS.

The results of this study are similar to those reported by [37,51] through a scientometric study on phytoremediation and heavy-metal hyperaccumulation using the WoS database, which reported that Environmental Sciences, Engineering Environmental, Water Resources, Plant Sciences, Toxicology, Soil Science, and Biotechnology Applied Microbiology were the most important categories. On the other hand, [52], in their article titled "Scientometric study of treatment technologies of soil pollution: Present and future challenges" reported that Environmental Sciences covered the largest number of publications, with 47.4% and 43.0% for WoS and Scopus, respectively. In this work, the authors present current technologies and their combinations for the remediation of contaminated soils, and also discuss different techniques and mechanisms for soil remediation including physical, chemical, and biological methods. Among these, the biological technique of phytoremediation was the keyword with the highest number of occurrences in the WoS and Scopus databases.

3.4. Most Productive Countries and Organizations

The remediation of mercury-contaminated sites is a global concern because of its bio-accumulative nature in food chains and the environment; hence, the intensity of the investigation may vary between countries. Figure S2A,B shows the number of papers published by each country in the WoS and Scopus databases. The intensity of the blue color is related to the countries with the highest number of published papers. In contrast, the gray color is related to the countries where no papers related to this topic have been published. Regarding the number of documents published by each country in the WoS database, 80 countries have published research papers related to this topic. China, India, and Spain were the countries with the highest production, with 115, 58, and 40 papers, respectively.

Regarding the number of citations, this is defined as the number of publications that cite the articles published each year [36]. Similarly, China and India had the highest number of citations (2784 and 1551), while Spain as in fifth place (867). In Scopus, 77 countries were found to have published papers on this topic. India (145), China (131), and the United States (86) had the highest production of documents. The highest number of citations by country was in China (2784), followed by India (1551), and the USA (1222). Out of 80 countries, 36 had at least 5 papers in WoS, while out of 77 countries in Scopus, 41 had at least 5 papers. In both databases, Africa had the fewest publications, followed by most of South America (except Brazil) and some Asian nations.

The co-authorship network was constructed with more than five articles for each database (Figure 3A,B). In WoS, the countries with the highest co-authorship were China, India, USA, Saudi Arabia, and South Korea, with total link strength (TLS) values of 52, 33, 33, 33, 26, and 23, respectively. Eight groups were distinguished, the most relevant being the first group formed by India, Italy, Germany, and Poland. Group 2 consisted of China, Japan, Taiwan, and Pakistan, while Group 3 included Spain, Colombia, Ecuador, and England. The USA was related to most countries. As in WoS, the most cooperative countries in Scopus were China, India, USA, Saudi Arabia, and South Korea, with TLS values of 59, 58, 53, 31, and 24, respectively. Eight groups were detected, of which four were the most relevant. China appeared in the first group and was related to most other countries. The second group mostly comprised European countries such as the Czech Republic, Germany, Italy, Poland, and the United Kingdom. The third group included South American countries (Argentina, Colombia, Peru, Brazil, Ecuador) and Spain. Finally, the fourth group comprised the United States and South Africa.

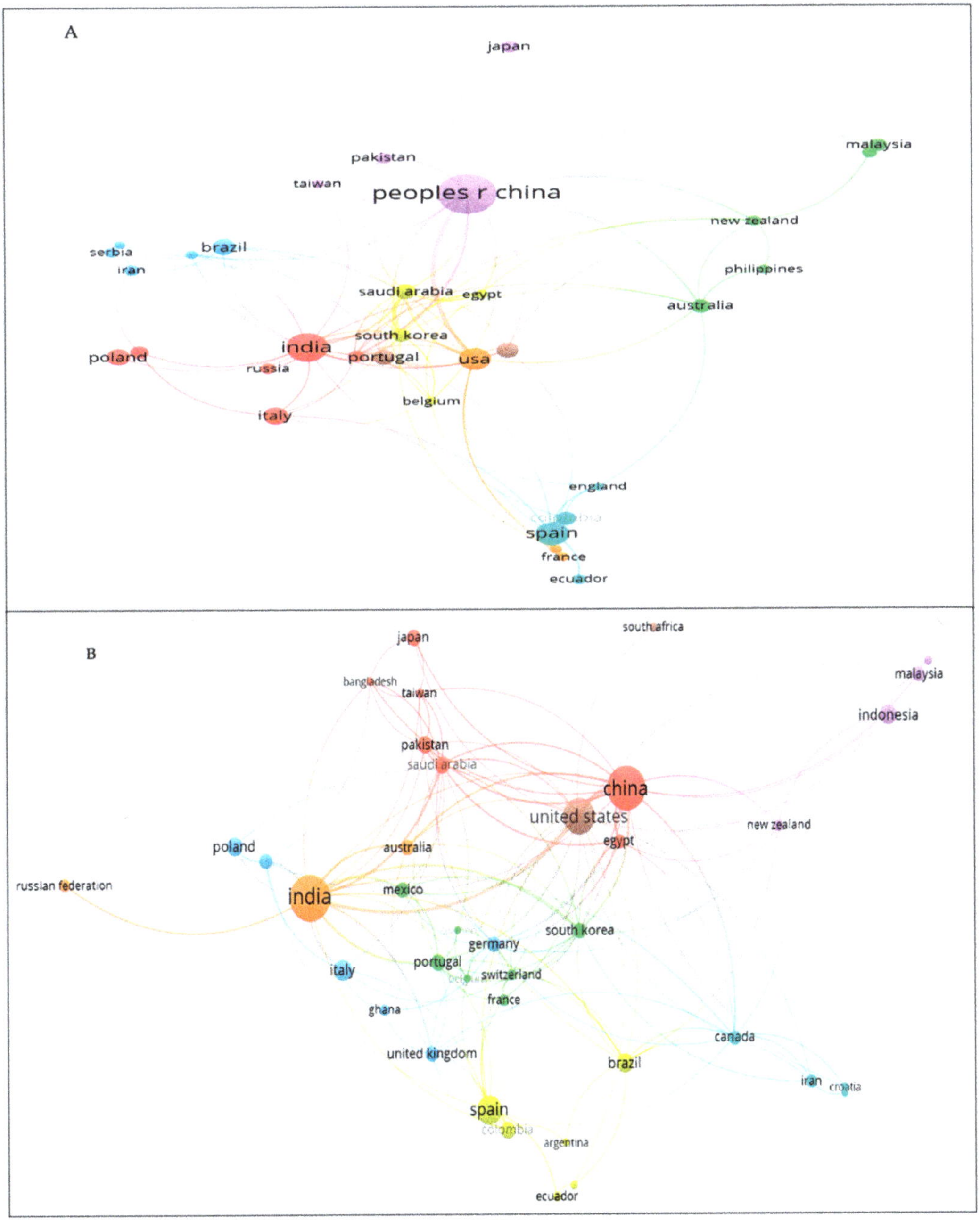

Figure 3. Map of country cooperation: (**A**) WoS and (**B**) Scopus.

In general, global emissions are mostly caused by human activity, the static burning of fossil fuels, and biomass, accounting for about 24% of emissions. However, small-scale artisanal gold mining represents nearly 38% of total global emissions and was the main source of emissions in South America and Sub-Saharan Africa [6]. In addition, mercury mines are one of the most persistent anthropogenic sources of environmental contamination. Historical mining and cinnabar ore retorting released many elemental Hg (Hg^0) and water-soluble Hg compounds into nearby surroundings [53,54] and generated numerous wastelands composed of mercury-enriched mine tailings (calcines) adjacent to abandoned retorts and smelters [5].

China was the country with the most publications related to the phytoremediation of mercury pollution. This could be attributed to the fact that, despite being rich in plant resources and preserving various ancient plants including more than 14,000 plant species, a significant number of mercury-enriched wastelands can be found in the Wanshan mining region. Mercury concentrations in calcines have been recorded as high as 4400 μg/g [36], continuously releasing mercury into surrounding ecosystems even after the mercury mines had been abandoned for several decades, emitting Hg^0 and secondary Hg compounds through natural weathering and runoff into the air, soil, and water, entering the biota [9,55,56]. Due to the large cinnabar ore reserves and production of elemental Hg, Wanshan was once called the "Mercury Capital" of China. Thus, the abundant plant resources and high mercury concentrations have provided the basis for the evolution of accumulator plants [26]. Furthermore, the Chinese government has been actively improving the environmental quality in recent years and has invested significant financial and human resources in remediating contaminated soils [57,58]. Furthermore, the most prolific institutions and authors came from this country (Tables 2 and S1). It is also worth noting that the three main sources of funding for research related to the phytoremediation of mercury contamination came from China: the National Natural Science Foundation of China, the Ministry of Science and Technology of the People's Republic of China, and the National Key Research and Development Program of China.

Table 2. Top 15 most prolific institutions in the phytoremediation of mercury contamination.

Rank	WoS			Scopus		
	Organization	**Documents**	**Country**	**Organizations**	**Documents**	**Country**
1	Chinese Academy of Sciences	29 (6.3%)	China	Chinese Academy of Sciences	39 (5.5%)	China
2	Guiyang Institute of Geochemistry Cas	15 (3.3%)	China	Institute of Geochemistry Chinese Academy of Sciences	20 (2.8%)	China
3	University of Chinese Academy of Sciences Cas	15 (3.3%)	China	University of Chinese Academy of Sciences	17 (2.4%)	China
4	Chongqing University	10 (2.2%)	China	Universidade de Aveiro	11 (1.6%)	Portugal
5	Consejo Superior de Investigaciones Cientificas Csic	10 (2.2%)	Spain	Massey University	10 (1.4%)	New Zealand
6	Universidade De Aveiro	10 (2.2%)	Portugal	University of Georgia	10 (1.4%)	United States
7	Centro de Investigaciones Energéticas Medioambientales Tecnológicas	8 (1.7%)	Spain	Ministry of Education of the People's Republic of China	9 (1.3%)	China
8	Massey University	8 (1.7%)	New Zealand	Chongqing University	9 (1.3%)	China

Table 2. *Cont.*

Rank	WoS			Scopus		
	Organization	Documents	Country	Organizations	Documents	Country
9	Yangtze Normal University	8 (1.7%)	China	Universidad de Castilla—La Mancha	8 (1.1%)	Spain
10	Council of Scientific Industrial Research Csir India	7 (1.7%)	India	Kitasato University	8 (1.1%)	Japan
11	Lodz University of Technology	7 (1.7%)	Poland	Banaras Hindu University	8 (1.1%)	India
12	Universidad de Córdoba	7 (1.7%)	Colombia	Indian Institute of Technology Indian School of Mines, Dhanbad	8 (1.1%)	India
13	Universidade Federal de Mato Grosso	7 (1.7%)	Brazil	Universidad de Córdoba, Monteria	8 (1.1%)	Colombia
14	University of Melbourne	7 (1.7%)	Australia	The State Key Laboratory of Coal Mine Disaster Dynamics and Control	8 (1.1%)	China
15	University of Oviedo	7 (1.7%)	Spain	Centro de Investigaciones Energéticas, Medioambientales y Tecnológicas	7 (1.0%)	Spain

Similarly, countries such as the United States and Spain have large-scale mercury mineralization zones distributed along the world's plate margins. For example, the New Almaden mine in the United States and the Spanish Almaden mine (the largest mercury mine in the world) with approximately 34,000 mg/kg of mercury, where the smelting process discharges a large amount of Hg into the environment [5,59]. Furthermore, the United States was the first country to actively participate in this field of study, indicating that it was a pioneer in this area [60,61].

It is also important to mention that in countries such as Colombia, Brazil, Southeast Asian countries, and some African countries, gold mining also causes severe mercury contamination [5,62,63].

Therefore, due to the remediation of large amounts of mercury-contaminated farmlands, phytoremediation, as an eco-friendly and efficient technology for the treatment of contaminated soils, has received significant attention and application in different countries.

On the other hand, 742 and 160 institutions for WoS and Scopus, respectively, contributed to the topic of the phytoremediation of mercury-contaminated sites. Institutions were ranked according to the total number of publications. In WoS and Scopus, six organizations had at least 10 papers. There as a greater diversity of institutions in WoS. In both databases, the Chinese Academy of Sciences contributed the most documents on this research topic with 29 for WoS and 39 for Scopus (Table 2).

3.5. Most Productive Authors

Determining the scientific production of the authors allowed us to analyze one of the most important structures that make up the scientific community on the phytoremediation of mercury-contaminated sites, reflecting the existing relationships among the members of this community. A total of 2476 authors had publications related to this topic in the Scopus database, whereas 1973 authors were registered in WoS. Table S1 shows the 10 authors with the highest number of publications related to the phytoremediation of mercury-contaminated sites, totaling 77 articles for WoS and 100 for Scopus.

Similarly, Figure 4 shows the authors who to date have contributed the largest number of publications on the phytoremediation of mercury contamination. Each circle in the network in Figure 4A,B symbolizes a researcher; the closer the circles are together, the more closely the researchers collaborate. The extent of the co-citation link's strength is shown by the size of the circle. Collaboration improves efficiency by making better use of available resources and transferring knowledge [64]. Seventeen authors produced at least five papers in WoS (Figure 4A). The top five authors were Liu Z. (10 publications), Feng X. (10 publications), Wang L. (9 publications), Wang J. (8), and Anderson C.W.N. (8). Three of the top writers were represented in cluster 1 of the six clusters (six colors). For Scopus, 41 authors met the requirement of at least 5 papers (Figure 4B). According to Scopus, the top five writers were Wang J. with 16 papers published, Feng X. (14), Kumar A. (11), Liu Z. (10), and Meagher R.B. (9). Three of the top authors were represented in group 1 of the five groups that were produced. From the Institute of Geochemistry, author Feng X. with 88 and 133 total links in WoS and Scopus, respectively, had the highest total link strength.

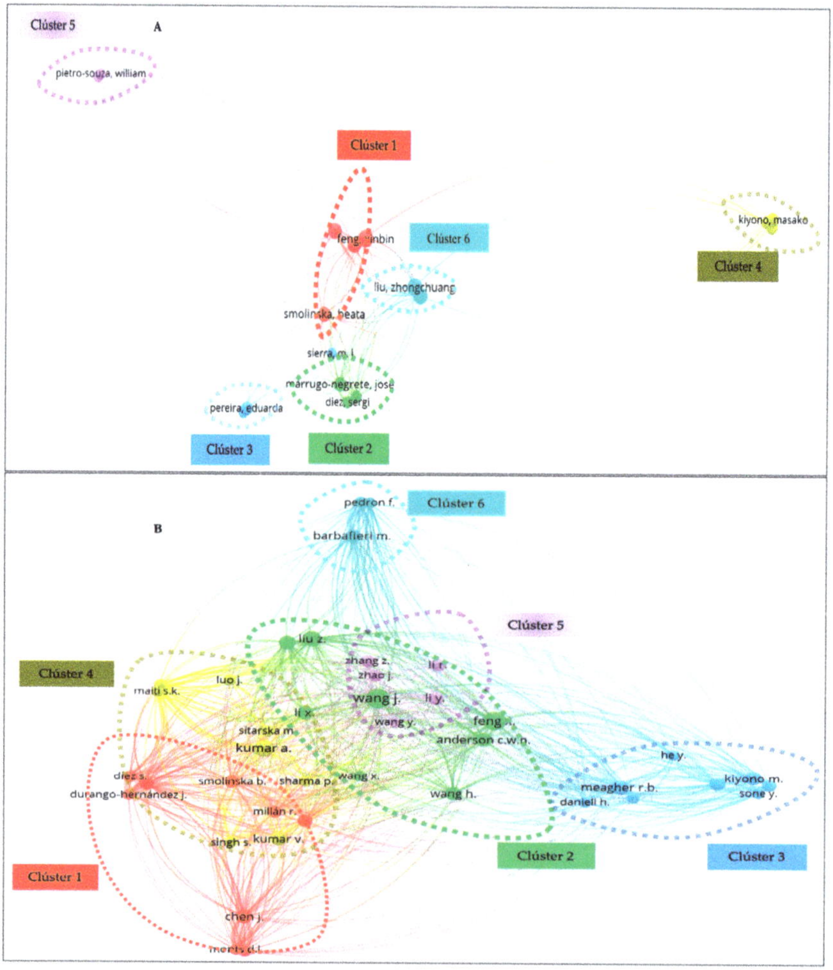

Figure 4. The main authors that contributed to the publication of scientific papers related to mercury phytoremediation: (**A**) WoS and (**B**) Scopus.

It should be emphasized that Feng X, Wang J, and Anderson C.W.N have together investigated *Brassica juncea* L. plants, evaluating mercury accumulation and its exposure to different chelating agents, in which ammonium thiosulfate and sodium thiosulfate, EDTA, among others, were used to improve the phytoextraction of this heavy metal in the plant as well as the leaching of bioavailable mercury caused by chelating agents [65,66].

In a recent publication, Feng X and Wang J et al. reported that the plant *Cadamine violifolia* could be considered as a mercury hyperaccumulator due to the high concentrations of mercury that accumulated in its shoots and roots, translocation factors higher than 1.0, and a bioconcentration between 1.8 and 223 [32].

3.6. Most Recognized Journals

Regarding the publications, the analysis of journals in each research area can help researchers to accurately understand the top journals in the field of that research; in addition, it can guide bibliographic queries, data collection, and article writing [67,68]. In WoS, 197 journals were found, 14 of which had 5 or more published papers (Figure 5A). However, in Scopus, 359 sources were found, but only 20 journals matched the requirement of 5 or more published articles (Figure 5B).

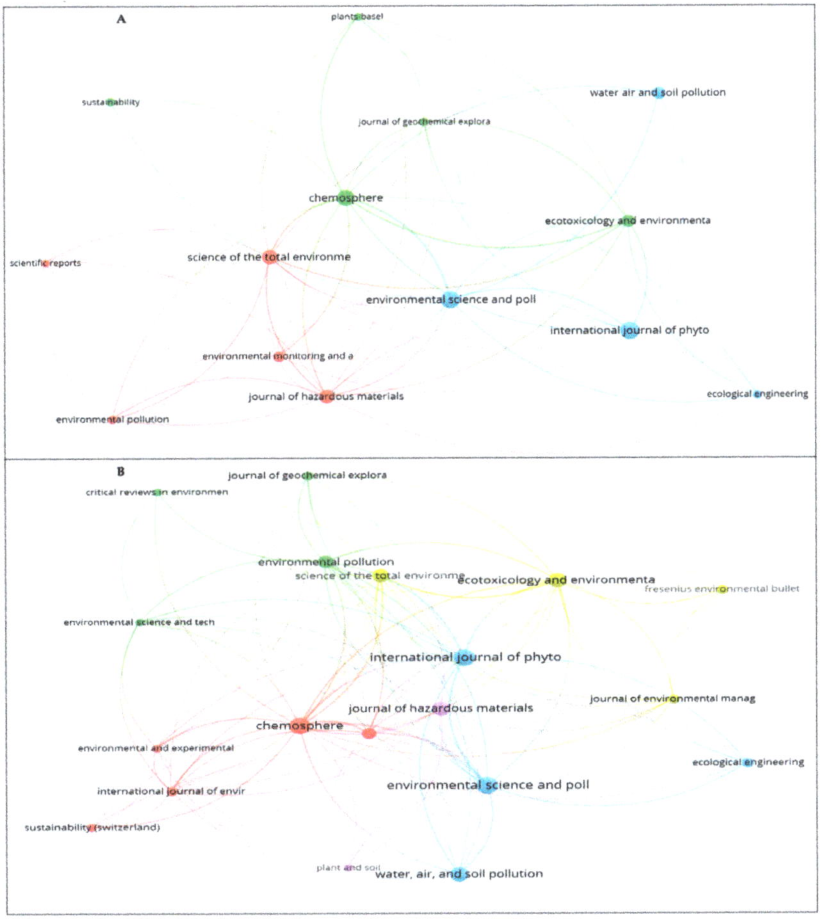

Figure 5. Research sources: (**A**) WoS and (**B**) Scopus.

The top 10 journals in WoS accounted for 35.4% of the total (Table S2). These journals included Environmental Science and Pollution Research with 26 papers (5.7% of publications), International Journal of Phytoremediation with 26 papers (5.7%), and Chemosphere with 23 (5.0%). As for Scopus, the top 10 journals accounted for 26.7%. The top three journals, with a combined 26 published papers (3.7%), were Chemosphere, Environmental Science and Pollution Research, and International Journal of Phytoremediation (Table S3).

Oliveira Fernandes et al. (2021), in a scientometric analysis of research on mercury in soil, found that the 10 most cited articles included four from the journal Science of the Total Environment and one from Chemosphere [7]. It should be mentioned that [52], in their bibliometric analysis on hyperaccumulators for potentially toxic elements in the WoS database, reported that International Journal of Phytoremediation ranked first, followed by Environmental Science and Pollution Research, and Chemosphere, coinciding with the information reported in this study. Meanwhile, [37], in their study on the phytoremediation of heavy-metal pollution, reported that the top three journals were Environmental Science and Pollution Research, International Journal of Phytoremediation, and Ecotoxicology and Environmental Safety.

The WoS database showed that Science of the Total Environment received the most citations (1551), followed by Chemosphere (875), Ecotoxicology and Environmental Safety, and other journals with citation counts between 875 and 1551. Five journals were in the top quartile (Q1), while six journals had an H-index of at least 150. Four journals are published in the Netherlands and three in the United Kingdom. The subject areas of Environmental Science, Medicine, and Agricultural and Biological Sciences were related (Table S2). The journal with the most citations in Scopus was Chemosphere (2031), followed by Science of the Total Environment (1582), and Ecotoxicology and Environmental Safety (1500). Seven journals had an H-index of 150 or more, and six of them were in the first quartile (Q1). Notably, four of the top ten most productive journals are published in the Netherlands, while three are published in the United Kingdom. Environmental Science, Medicine, Chemistry, and Agricultural and Biological Sciences were the subject areas associated with the phytoremediation of mercury exposure (Table S3). Both databases indexed all journals, which have been categorized under the Environmental Science category.

3.7. Most Cited Documents

In both databases, a document citation analysis was performed. The citation visualization network of documents is shown in Figure S3, with documents with at least 30 citations serving as the starting point. The papers "Remediation Techniques for Heavy Metal-Contaminated Soils: Principles and Applicability" [69], "The Phytochelatin Transporters AtABCC1 and AtABCC2 Mediate Tolerance to Cadmium and Mercury" [70], and "A Review on Heavy Metals Contamination in Soil: Effects, Sources, and Remediation Techniques" [71] had the most citations in WoS, with 814 (9.2%), 382 (4.3%), and 277 (3.1%), respectively. As for Scopus, the papers "Plant Responses to Abiotic Stresses: Heavy Metal-induced Oxidative Stress and Protection by Mycorrhization" [72] (1766; 5.7%), "Remediation Technologies for Metal-Contaminated Soils and Groundwater: An Evaluation" [73] (1235; 4.0%), and "Trace Elements in Agroecosystems and Impacts on the Environment" [74] (1171; 3.8%) were the most cited papers in the works considered for this study. In most of these papers, Hg was discussed with other toxic metals. Tables S4 and S5 list the 20 papers with the most citations in the WoS and Scopus databases.

The articles that were most frequently cited in WoS can be categorized into four groups. (1) Reviews of various remediation techniques for cleaning up heavy metal-contaminated sites. Among these, phytoremediation is a promising technique for cleaning up large sites with relatively low contaminant concentrations at shallow depths. It is considered cost-effective, ecologically beneficial, and has high public acceptance. (2) Studies that explore the mechanisms used by plants to accumulate, tolerate, and overexpress transporters associated with mercury detoxification. (3) Articles focused on the use of constructed wetlands to remove metals and metalloids from contaminated waters. (4) Studies investigating the health risks associated with heavy-metal contamination in soil, particularly mercury and cadmium. In Scopus, the most cited papers could be divided into six categories. (1) Molecular mechanisms associated with the role of antioxidants as protection against heavy-metal stress in roots, fungi, and mycorrhizae. (2) Selection of remediation methods for heavy-metal contaminated soils and sediments, which depend on the site characteristics, concentration, types of contaminants to be removed, and end use of the contaminated medium. (3) Improvements in the phytoremediation process such as plant-microorganism synergy. (4) Use of wetlands in their restoration processes. (5) Use of transgenic plants in the process of phytoremediation of heavy metals, in case mercury, transporters, and inclusion of mers-type genes are cited.

In WoS, more than 50% of the 20 highly cited papers came from the following five countries: China (15%), India (15%), USA (10%), Spain (10%), and Italy (10%). According to Table S4, reviews comprised 60% of the papers, whereas research articles comprised 40%. A total of 50% of the most cited papers were published in the journals with the highest scientific production (Table S2), predominantly Ecotoxicology and Environmental Safety, Chemosphere, and Science of the Total Environmental Safety. In contrast, the 20 most often cited reviews in Scopus were written by authors from the United States (25%), China (105), India (10%), Germany (10%), and the United Kingdom (10%) (Table S5). Major journals published only 25% of the manuscripts (Table S3), while China, the United States, and India worked closely together on both databases (Figure 3A,B). Information about a country was based on data provided by the corresponding author.

It should be noted that China, the United States, and India have focused on the accumulation of heavy metals by aquatic plants in aquatic matrices [75–77]. The potential to remove mercury from sediments was established by research on the aquatic plants *Carex heterostachya*, *C. brevicuspis*, *A. selengensis*, *Polygonum hydropiper*, and *G. longituba* [78]. According to Ulaganathan, *E. crassipes*, *A. cristata*, and *S. natans* are three more prospective candidate aquatic species for the phytoremediation of hazardous metals such as Pb, As, Hg, Cd, and Ni [77]. In China, mercury-contaminated croplands account for 1.6% of all croplands contaminated with heavy metals, resulting in mercury accumulation in the food chain. For example, rice (*Oryza sativa* L.) has been shown to absorb and accumulate mercury from the soil and transfer it to edible parts of plants, with mercury and methylmercury concentrations between 7.5 and 608 μg/kg and 7.37–62.3 μg/kg, respectively. Methylmercury absorbed by rice accounted for more than 30% of the total mercury. These results exceed the 20 μg/kg permitted limit for food safety [79,80]. Likewise, the effect of soil microorganisms on mercury uptake in rice plants was examined to determine possible soil phytoremediation agents, and it was observed that inoculations of arbuscular mycorrhizal fungi (AMF) significantly reduced the concentration of the metal in rice. Therefore, this study suggested the possibility of using microorganisms for the remediation of heavy-metal contamination in soil and provided valuable information on the reduction in human exposure to mercury through rice consumption [81].

On the other hand, recent studies in both databases on mercury phytoremediation have focused on the use of microorganisms to remediate mercury contamination, and good results have been obtained [82,83]. For example, the phytoremediation potential of *Boehmeria nivea* L. *Gaud* and the response of its rhizosphere soil microbiome to mercury contamination were evaluated, and *Proteobacteria*, *Actinobacteria*, *Gemmatimonadota*, *Latescibacterota*, and *NB1-j* were identified as potential mercury-tolerant taxa [84]. Furthermore, [85] demonstrated that the endophytic fungi *Westerdykella aquatica* P71 and *Pseudomonodictys pantanalensis nov.* A73 stimulated mercury uptake and promoted its accumulation in plant tissues, preferably in roots, mitigating the effects of metal phytotoxicity. Sharma et al. performed the phytoremediation of heavy metals using *Brevundimonas* sp. in the rhizosphere zone of *Saccharum munja* L, revealing that heavy metals such as Fe, Mn, Pb, Cd, Cr, Cu, Zn, Ni, As, and Hg were uptaken and translocated [86]. Bacterial isolates identified as *Jeotgalicococcus huakuii* and *Bacillus amyloliquefaciens* from *C. dactylon* and *E. indica* plants, respectively, showed the complete detoxification and bioaccumulation of mercury, in addition to indole acetic acid (IAA) production, ammonium production, and siderophore production [87]. On the other hand, many scholars have tried to find a hyperaccumulator of mercury, and more than 200 plant species have been used to test their ability to accumulate and transfer mercury [24,26]. *Cardamine violifolia* could be considered as a mercury hyperaccumulator as it accumulated concentrations of up to 6000 μg/g in its roots and upper parts [32].

In general, articles in recent years have focused on finding phytoremediation plants for mercury contamination from their natural habitats, on the improvement of crops for safer food through phytoremediation, the use of chelating ligands and plant–microorganism symbiotic relationships to strengthen the phytoremediation capacity of plants, the use of constructed wetlands to improve the phytoremediation of aquatic ecosystems as well as studies at the molecular level related to metal sequestration, detoxification, uptake and transport in plants, and omics approaches related to the analysis of proteins and genes expressed in plants.

3.8. Author Keyword Trend Analysis

Keywords are terms or short phrases that classify and address entries in indexing and information retrieval systems in databases for a particular manuscript or subject area. Keywords have become an essential bidirectional tool for people who write and people who search for information about related manuscripts or subject areas [52,53,88,89].

A criterion of selecting keywords with a minimum of five occurrences was applied. In WoS, the frequency of keywords in 457 documents was evaluated, and 188 keywords were found (Figure 6A). Meanwhile, 666 words were found in Scopus out of 697 analyzed (Figure 6B).

Excluding the terms phytoremediation and mercury in both databases, the most used keywords in WoS were heavy metals, with 242 (7.2%) occurrences, accumulation (196, 5.8%), cadmium (122; 3.6%), soils (115; 3.4%), and phytoextraction (70; 2.1%). In Scopus, excluding the words nonhuman and article, the keywords with the highest number of occurrences were bioremediation (333; 2.6%), heavy metals (299; 2.3%), soil pollution (277; 2.2%), bioaccumulation (198; 1.5%), and biodegradation and environmental (175; 1.4%). Even though they were not included in the descriptor, these keywords had a high co-occurrence strength, indicating that they were highly important and received a lot of attention.

In addition, Figure 6A displays the co-occurrence of keywords in the WoS database, categorizing them into eight clusters. Among the three major clusters, the red cluster, which used the term phytoremediation, was the largest. It connected phrases such as "oxidative stress", "detoxification", "tolerance", and "heavy metals such as cadmium and mercury" and contained the keywords antioxidant, growth, tolerance, resistance, glutathione, proteins, and phytotoxicity. Group 2 (green) was more related to soil contamination by heavy metals, their accumulation, and remediation and included the term heavy metals and heavy metals such as arsenic, chromium, lead, and zinc. It appeared in the cluster of agricultural soils, spatial distribution, and other remediation techniques such as electrokinetic remediation. All keywords in this cluster had a close conceptual link. Group 3 (blue cluster) contained different forms of mercury: atmospheric mercury, inorganic mercury, and methylmercury. Words such as exposure, extraction, fractionation, and speciation were part of this cluster. The Scopus database findings regarding the co-occurrence of terms were divided into six groups or clusters (Figure 6B), with the following being the most significant. The first cluster (red) focused on phytoremediation. The effectiveness of a plant with phytoremediation potential was defined by the terms bioaccumulation, translocation, bioconcentration factor, and translocation as well as bioremediation. The same was true for this cluster of connections between phytoremediation and aquatic settings including aquatic macrophytes, aquatic ecosystems and environments, wetlands, water pollution, and aquatic plants of the genus *Typha*. Group two (green color) involved genetic engineering associated with phytoremediation, gene and protein expression, genetically modified plants, mers genes, and other processes related to plant detoxification and tolerance to heavy metals such as mercury. In addition, this group contained several mercury salts and compounds including methylmercury and mercury chloride. Cluster three (blue) included environmental pollution by heavy metals and their removal processes. It also included phytoremediation processes such as phytoextraction, phytovolatilization, phytodegradation, phytostabilization, and hyperaccumulation, which define the effectiveness of phytoremediation. Heavy metals such as arsenic, cadmium, chromium, lead, and nickel, were found in this group. In general, the research topic can be expressed using keywords. Through these studies, it is possible to comprehend the key research areas and upcoming developments in the field [43].

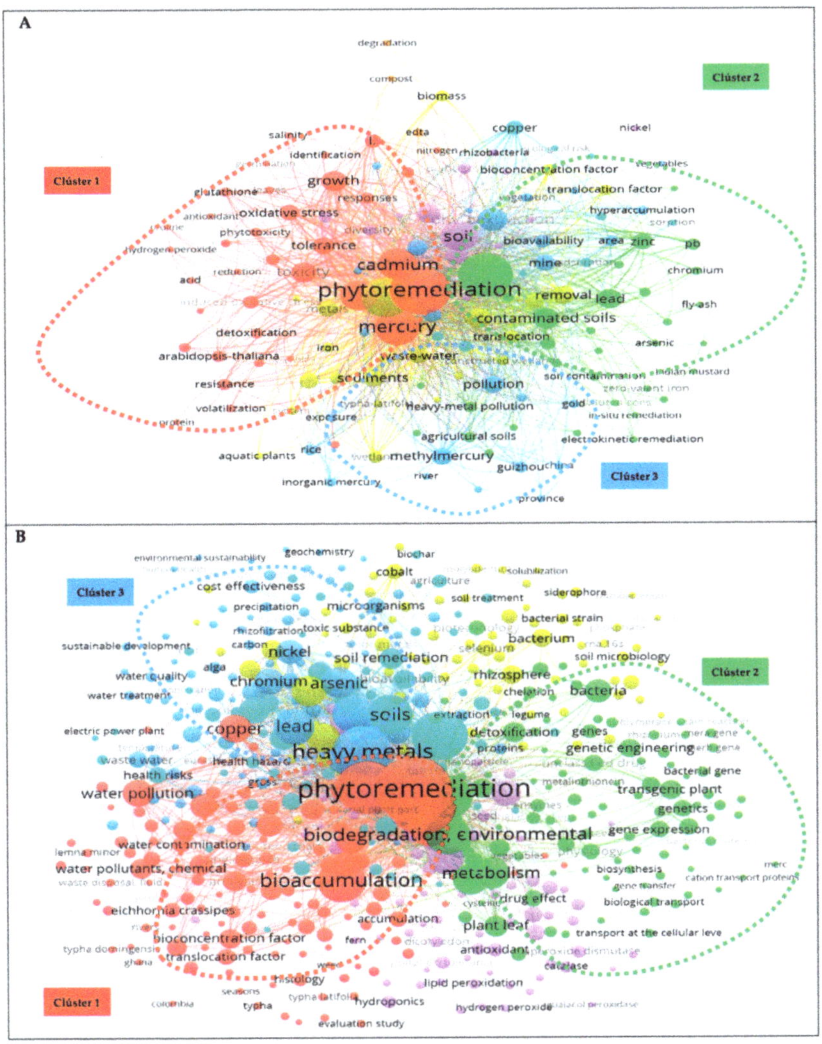

Figure 6. Keyword co-occurrence analysis: (**A**) WoS and (**B**) Scopus.

3.9. Keywords with the Strongest Citation Bursts

To obtain more information about the phytoremediation of mercury contamination, burst analysis in Bibliometrix software was used to perform a retrospective keyword analysis. Keyword analysis can be used to examine development trends and research frontiers. In the Bibliometrix software, the size of the circle indicates the strongest words, and the lines indicate the period. The program identified 28 and 18 main keywords with a minimum frequency of five for WoS and Scopus, respectively (Figure 7A,B, Supplementary Tables S6 and S7). In WoS, the keywords could be divided into the following main aspects:

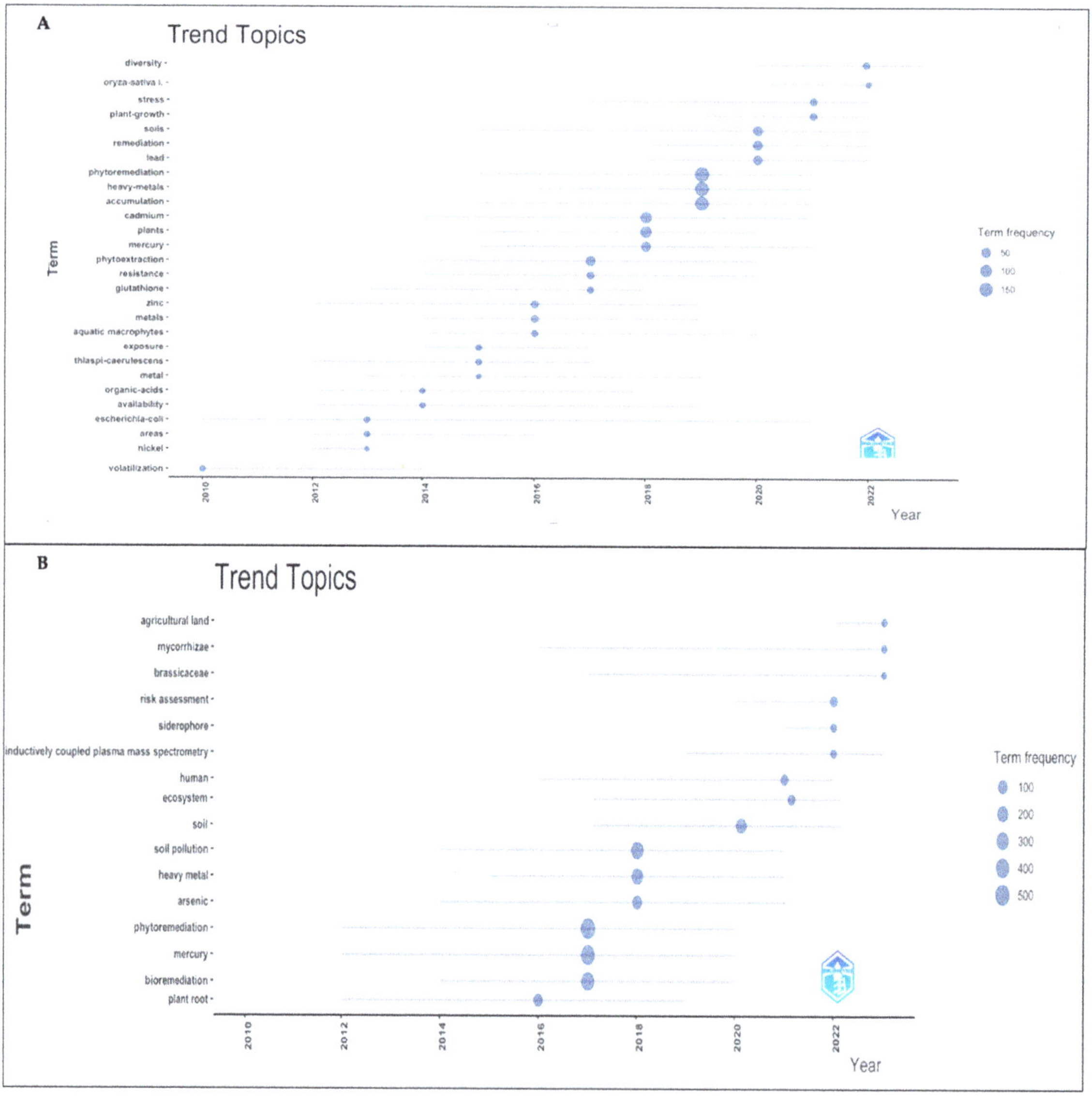

Figure 7. Top keywords with the strongest citation bursts: (**A**) WoS and (**B**) Scopus.

- Phytoremediation of heavy metals. The remediation of "heavy metals" (2016–2021), "metals" (2014–2019), and specific heavy metals such as "lead" (2018–2022), "cadmium" (2014–2021), "mercury" (2015–2021), "zinc" (2012–2019), and "nickel" (2012–2013) was investigated, with mercury and cadmium being the metals with the highest frequency. This group also included terms like "Remediation" (2018–2022), "Phytoremediation" (2015–2021), "Phytoextraction" (2014–2020), and "Volatilization" (2010–2014).

- Exploration of mechanisms by which plants accumulate metals. Exploring the mechanisms by which plants absorb, accumulate, resist, stress, and translocate metal helps in understanding the process of mercury detoxification in plants. This group included "Diversity" (2020–2023), "Plant-Growth" (2019–2022), "Stress" (2017–2022), "Accumulation" (2015–2021), "Resistance" (2014–2020), "Availability" (2012–2019), "Organic-Acids" (2012–2018), "Glutathione" (2013–2018), and "Exposure" (2014–2017). Among them, "Accumulation" showed the highest frequency, which could indicate ongoing research into how plants accumulate mercury, since although plants show a well-developed mechanism for the absorption, translocation, and detoxification of mercury, there are still knowledge gaps in the molecular understanding of these processes [1].
- Rice (*Oryza sativa*) (2020–2022) is one of the most important cereal crops, feeding nearly half of the world's population. Mercury concentrations in rice grains can reach up to 460 μg/kg when grown in Hg-contaminated soils [90]. Thus, the accumulation levels of mercury in rice have been studied [91]. Through genetic engineering, *E. coli* expressed four metallothioneins from rice (*Oryza sativa* L.), which conferred enhanced mercury tolerance, metal binding, and sequestration [92].
- Aquatic macrophytes (2014–2020) have been widely explored due to concern about the ecological water environment, and research has been conducted on the ability of several aquatic plants to absorb heavy metals [93,94].

Regarding the Scopus database, the keywords could be divided into the following aspects:

- Phytoremediation of heavy metals. The remediation of "heavy metals" (2015–2021) and individual heavy metals like "mercury" (2012–2020) and "arsenic" (2014–2021) were investigated. This group also included terms such as "Bioremediation" (2014–2020), "Phytoremediation" (2012–2020), and "Phytostabilization" (2015–2022).
- Search for strategies to improve phytoremediation efficiency. This group included "Siderophore" (2021–2022), "plant root" (2012–2019), and "Mycorrhizae", which began in 2016 and continues to date, indicating that it is a research hotspot.
- Identified hyperaccumulator families. There were 721 hyperaccumulator species divided into 52 families and 130 genera, ranging from annual grasses to perennial shrubs and trees, with the *Brassicaceae* family (2017–2023) being the largest [95–97].
- Risk assessment. This group included "Risk Assessment" (2020–2022), "Human" (2016–2022), "agricultural land" (2022–2023), "ecosystem" (2016–2022), and "soil pollution" (2014–2021). It is known that mercury is highly toxic to living organisms, and its presence in agricultural land poses another route of exposure. Therefore, phytoremediation combined with risk assessment is a topic that requires attention.
- China (2017–2022) is one of the top three countries publishing papers on the phytoremediation of mercury pollution and is a keyword with the highest citation frequency.

3.10. Points of Interest in the Research

Considering the keyword analysis, research on the phytoremediation of mercury contamination also informs on the degree of contamination in soils, sediments, and watersheds [24,98,99]. Thus, knowledge about local plant screening in mercury-contaminated sites will help to obtain a better effect on ecosystem remediation. In general, the keywords covered four research directions: (1) finding native plants, (2) genetic engineering applications (3) increasing remediation ability through assisted phytoremediation, and (4) the detoxification mechanism of Hg.

(1) Identification of mercury phytoremediation plants

Some plants have adapted to survive in mercury-contaminated sites thanks to evolution in molecular tolerance mechanisms, so they can behave as mercury accumulators or hyperaccumulators [98]. When the plant absorbs mercury, it accumulates in the vacuoles, which have low metabolic activity, keeping heavy metals away from important cellular metabolic processes. This may serve as a tolerance mechanism in plants [100]. More than 700 plant species have been identified with the ability to hyperaccumulate heavy metals, ranging from annual grasses to perennial shrubs and trees [96]. In terms of research uniqueness, researchers have reported different plant species identified in mercury-contaminated sites. For example, in China, Quian et al. found a total of 259 plants belonging to species in 29 families that grew on mercury-rich wastelands composed of cinnabar mine tailings (calcines) in the Wanshan region of southwest China, the third-largest Hg mining district in the world. Of these plant species, *Arthraxon hispidus*, *Eremochloa ciliaris*, *Fallopia multiflora*, and *Ixeris sonchifolia* exhibited higher concentrations of this metal in their roots and shoots compared to other species [26]. Similarly, Marrugo-Negrete et al., in a former mining district in Colombia, identified 24 plant species, grouped into 16 families, and the five species with the highest concentrations of the metal in their roots were *Jatropha curcas*, *Thalia geniculata*, *Piper marginathum*, *Cyperus ferax*, and *Ricinus communis* [24]. Xun et al., in the mountainous area of Dehua and Wanshan in China (highly mercury-contaminated sites), collected a total of 14 dominant wild plants, among which were *Cyrtomium macrohyllum*, *Woodwardia unigemmata*, *Dennstaedtia hirsute*, *Ageratum conyzoide*, and *Conyza canadensis*. All of these plants exhibited high resistance to mercury toxicity [25]. However, although some species are considered as potential mercury hyperaccumulators, no mercury hyperaccumulator has been reported to date. Therefore, discovering mercury hyperaccumulators is highly desirable [5,26,32].

It should be said that at present, many studies on the phytoremediation of mercury contamination are still in the laboratory stage [25,32], and very few studies have been conducted at contaminated sites [101]. These studies often overlook the complex dynamics of open fields such as climatic conditions, geographic factors, and the heterogeneity of mercury contamination (Table S8). Therefore, further research is required to verify the effectiveness of the phytoremediation process on a larger scale and develop site-specific protocols that account for the variability in plant behavior. Overall, the conditions of the contaminated site and the plant communities present must be considered to improve phytoremediation in mercury-affected sites. It is also necessary to investigate the mercury content in plant species that have not been studied yet [5]. Native plants are suitable for managing both contamination and the environment [37].

(2) Applications of Genetic Engineering (OK)

Currently, genetic engineering allows for the integration of genes from other organisms to enhance the capacity and accumulation of heavy metals in plants, thus improving the phytoremediation conditions. For example, genetically modified *Spartina alterniflora* for mercury phytoremediation is resistant to high concentrations of this metal [102]. Similarly, genetic engineering can also be used to change the storage location of heavy metals and reduce the damage to plants caused by stress to these pollutants [103]. Most genetically modified plants for mercury phytoremediation are based on the *mer* operon gene, and these genes are involved in the process of translocating mercury (Hg^{2+}) into the cell. They play a significant role in the biogeochemistry and bioremediation of mercury by converting reactive inorganic and organic forms of mercury into relatively inert, volatile, monoatomic forms [104]. For instance, transgenic rice (*Oryza sativa*) with the *merA* gene is able to withstand higher mercury concentrations compared to plants without the gene [105]. Additionally, the overexpression of *PtABCC1* in *Populus trichocarpa* has been shown to be effective in *A. thaliana* and *Populus tormentosa*, promoting growth at certain mercury concentrations and increasing mercury accumulation in both the roots and shoots compared to wild-type plants [35,106].

In general, genetically modified plants for mercury resistance have provided valuable insights, enabling significant steps toward achieving more successful phytoremediation methods [98]. Multi-omics methods such as transcriptomics, proteomics, and metabolomics are required to reveal the functional roles of genes, proteins, and associated metabolites in mercury accumulation.

(3) Assisted phytoremediation

- *Chelating agents*

  Several approaches have been widely used to improve the phytoremediation capacity in the field of heavy-metal pollution. Sometimes, the availability of metals in the soil is insufficient, and consequently, no active uptake of the metal by the root is generated. Mercury negatively interferes with plant growth as well as with their metabolic process, so some research has focused on the use of assisted enhancements to improve the efficiency of phytoremediation. It has been shown that sodium thiosulfate ($Na_2S_2O_3$), ammonium thiosulfate $(NH_4)_2S_2O_3$, ethylenediaminetetraacetic acid (EDTA), diethylaminapentaacetic acid (DTPA), $(NH_4)_2SO_4$, NH_4Cl, $NaNO_3$, $(NH_4)_2S_2O_3$, HCl, KI, and citric acid are related to the phytoremediation of mercury-contaminated soils, helping to improve their efficiency [66,69,107–109].

  Previous studies conducted on *Oxalis corniculata* and *Brassica juncea* L. found that chelating agents increased the concentration of mercury in the plant tissues of plants, achieving higher translocation and bioconcentration factors, which could be associated with the fact that the soluble complexes chelated by them can easily pass through the endodermis and enter the xylem [66,69].

  On the other hand, the use of organic amendments has also been used to increase the adsorption capacity of heavy metals. For example, the use of biochar (MB2) and attapulgite (MA2) showed that they can improve the growth capacity in *Solanum nigrum* L., a favorable feature for phytoremediation as well as remove the soil contaminant and increase the concentrations of metals in plant tissues [110]. Likewise, the addition of compost (derived from green waste) to the contaminated soil decreased the concentration of total mercury, where the higher the compost content, the lower the mercury contamination in soils [111].

  The application of selenium (Se) fertilizers has been used as a cost-effective strategy to ameliorate mercury accumulation and the noxious effects of mercury [112]. For example, the development of apoplastic barriers in the root endodermis of *Oryza sativa* crops in hydroponic solutions contaminated with mercury and methylmercury (MeHg) improved with the addition of selenium and also showed a remarkable decrease in Hg and MeHg concentrations [113].

  Therefore, the inclusion of these chelating agents or the development of new ones to improve the ability of plants to remove mercury from the soil is an important goal to consider.

- *Growth-promoting microorganisms*

  Growth-promoting rhizobacteria help plants cope with stressful environments in addition to forming beneficial host–plant relationships, which can increase the phytoavailability of metals while reducing their toxic effects. They stimulate the ability of the host plant to generate more biomass while storing significant concentrations of metals and decreasing toxicity [114–116]. Bacteria, fungi, and other microorganisms living in the rhizosphere can increase the availability of heavy metals by either changing the soil pH, releasing chelators, or having redox reactions with heavy-metal ions. Rhizosphere microorganisms in mercury-contaminated soils can reduce Hg^{2+} to less toxic volatile Hg^0 through enzymatic reduction [15]. There is a belief that beneficial fungi in the rhizosphere stimulate root proliferation and enhance plant response to soil chemistry through plant–microbe association [117,118]. It should be added that rhizobacterial fungi and plant growth-promoting rhizobacteria (PGPR) induce heavy-metal tolerance responses [119]. For example, the symbiotic association between *Lolium perenne* roots and arbuscular mycorrhizal fungi (*Glomus* sp., *Acaulospora* sp., *Entrophospora* sp., *Scutellosospora* sp. and *Glomus* sp., *Acaulospora* sp., *Entrophospora* sp., *Giaspora* sp.) showed that there was enhanced mercury uptake. Arbuscular mycorrhizae play an important role in heavy-metal tolerance and have several extracellular and intracellular defense mechanisms such as plant tissue growth mechanisms and biological mechanisms of nitrogen fixation [120]. Likewise, the high mercury accumulation in *Erato polymnioides* could be based on a core of highly infectious strains of arbuscular mycorrhizal fungi (AMF) that are capable of improving the nutritional status of the plant and making mercury bioavailable species [31]. Therefore, the identification and characterization of heavy-metal tolerant microorganisms in soil and the development of microbial agents is a key research objective.

(4) Mercury detoxification mechanisms

Plants that accumulate heavy metals have adopted strategies to cope with their harmful effects by carrying them from the source where they are found to their roots, either through the plasma membrane of root endodermal cells or by movement through the free space between the cell wall [121,122], where they can be stored or translocated to shoots through xylem vessels, where they are deposited in vacuoles. The travel of heavy metals to vacuoles is controlled and regulated by molecules involved in cross-membrane transport or their complexation and subsequent sequestration. ATP-binding cassette (ABC) transporters have been shown to transport As, Cd, and Hg [37]; it is also considered that Zn, Cu, or Fe transporters may be possible candidates to facilitate Hg influx [123].

It is worth mentioning that the formation of reactive oxygen species (ROS) is produced as by-products of several metabolic pathways that are located in different cellular compartments (chloroplasts, mitochondria). It is held that the Hg level in leaves interferes with mitochondrial activity and triggers ROS generation, malondialdehyde accumulation resulting from lipid peroxidation, and causes cell disruption [124]. Just as ROS can damage cells, they can also initiate responses related to gene overexpression. The accumulation of ROS can be counteracted by antioxidant enzymes including a variety of detoxifiers such as superoxide dismutase (SOD), ascorbate peroxidase (APX), glutathione peroxidase (GPX), glutathione-S-transferase (GST), catalase (CAT), and non-enzymatic low molecular weight enzymes such as ascorbic acid (ASH), glutathione (GSH), α-tocopherol, carotenoids, and flavonoids [125]. SOD, together with GSH, can scavenge reactive oxygen species and prevent lipid overoxidation in membranes [25]. SOD is considered to be the most efficient of the intracellular enzymes and is important in the plant's tolerance to all types of stresses because they provide the first line of defense against the effects of elevated ROS levels [125]. Increased SOD at low Hg concentrations can scavenge ROS and alleviate plant damage, while increased GSH at high Hg concentrations, aside from helping to scavenge ROS, can also induce the formation of phytochelatins and an increase in proline, an intracellular substance important for an osmotic adjustment that can regulate cellular osmotic homeostasis disturbed by heavy-metal stress and protect the structural and functional integrity of cells [25,126–128]. Thus, for example, *Cyrtomium macrophyllun*, exposed to concentrations up to 1000 mg kg^{-1} of Hg, presented high resistance to mercury stress due to the increase in activities such as SOD, GSH, and proline compared to the control. Phytotoxic effects such as chlorosis and necrosis were low [25]. The transcripts and activities of SOD, POD, and CAT were pronouncedly enhanced in transgenic Arabidopsis compared to the control when exposed to $HgCl_2$ [129].

Therefore, the synthesis of antioxidant and non-enzymatic enzymes in plant tissues may be an important mechanism for the establishment of the defense systems of reactive oxygen species under heavy-metal toxic stress [130].

3.11. Problems and Challenges

Significant progress has been made in addressing mercury contamination in soil through the use of emerging technologies such as phytoremediation, which remains a particularly attractive option. However, there are some issues to consider:

- The practice of phytoremediation can take years for the various contaminated fields, so the discovery of local species is necessary, as screening for locally tolerant species is becoming an alternative approach for soil restoration [121] including native species that have not been previously studied.
- Most studies have been conducted at the laboratory scale using nutrient cultures and pot experiments on soil contaminated with mercurial inorganic salts. These investigations should be transferred to field experiments because these experiments are not likely to be able to define a hyperaccumulator. In addition, the tests performed in the field or in the laboratory show significant differences. For example, van der Ent et al. pointed out that hyperaccumulators should be recorded from natural habitats and did not consider the extreme accumulation achieved by laboratory work with acidified and artificially modified soils or water to be hyperaccumulation [60].
- Plant growth and the bioavailability of metals are very important in the phytoremediation process. For plants, high biomass, environmental adaptability, response to Hg tolerance, root characteristics (roots with high root zone), and root protection, as it is the main organ affected by mercury, are factors to be considered. Most of the accumulator plants found so far have short cycles, with low biomass and present phytotoxic effects as the concentrations of the metal increase. Regarding the bioavailability of metals, the conditions, physicochemical properties, and characteristics of the medium (water, soil, mud, sediment, etc.) all have influences or impacts on the phytoremediation potential of plants.
- Phytoremediation assisted by chelating agents or microorganisms has been shown to enhance the phytoremediation process. Some microorganisms can promote plant growth and tolerate Hg, therefore, the rational screening of microorganisms should be carried out, and microbial taming techniques should be used to help plants ameliorate mercury-contaminated environments, so the mechanism of microbial symbiosis is worth exploring further. Studies conducted with chelating agents or amendments have shown an increase in the rate of accumulation, so there is an option to further investigate all soil amendments such as biochar, compost, and biosolids in the future.
- With the development of genetic engineering and molecular biology, improvements in Hg phytoremediation processes have been found. The inclusion of transporters and genes can improve the uptake, translocation, and transformation of Hg into less toxic forms. Researchers can use multi-omics methods to report on the different processes involved in gene overexpression, mercury oxidative stress, and metal translocation.
- After the phytoremediation process in mercury-accumulating plants, its removal should be considered. The question of how to treat plants with mercury after harvest should be studied. The release of atmospheric mercury from plants can also cause air pollution. Methods such as composting and pyrolysis have been used for the disposal of toxic plant residues after phytoremediation [131].

4. Conclusions and Perspectives

Conclusions

This study provided knowledge mapping of the phytoremediation of mercury contamination. Publications on this topic are increasing, with a total of 457 publications in WoS and 697 in Scopus, with higher growth in the last five years. In terms of research categories, Environmental Sciences was the most frequent subject area in both databases, followed by Agricultural and Biological Sciences in Scopus, and Environmental Engineering in WoS. China and India were the countries with the most publications in WoS and Scopus, respectively, followed by the United States, Spain, and Italy. Chinese organizations were the main drivers of research, with the Chinese Academy of Sciences being the most productive institution in this field of study, with these studies being mainly funded by the Natural Science Foundation of China. It also houses the most prolific authors.

The most prominent journals in this field were Chemosphere, International Journal of Phytoremediation, and Environmental Science and Pollution Research. The 20 most cited publications in Scopus were reviews, while in WoS, 60% were reviews and 40% were research articles. The most cited article in Scopus was from Germany, while the most cited in WoS was from the United States.

The co-occurrence and trending themes of the keywords made it possible to identify the critical points of the research and its evolution over time. For example, the terms with the highest number of occurrences were bioaccumulation and phytoextraction, which define the efficacy of a plant with phytoremediation potential. Genetic engineering, gene and protein expression, and genetically modified plants influence the process of assisted measures for improvements in the phytoremediation process. Similarly, more research should be conducted on the synergistic role between plants and microorganisms and on phytoremediation and chemical-assisted amendments. More research should also be conducted on aquatic plants for the remediation of ecosystems and aquatic environments contaminated with heavy metals.

Supplementary Materials: The following supporting information can be downloaded at: https://www.mdpi.com/article/10.3390/su16219408/s1, Figure S1. Documents by subject area: (A) Scopus and (B) WoS; Figure S2. Countries' scientific production: (A) WoS and (B) Scopus; Figure S3. Highly cited articles: (A) WoS and (B) Scopus; Table S1. Top 10 authors with the highest number of articles in WoS and Scopus on the phytoremediation of mercury contamination; Table S2. Journals with the greatest scientific production in WoS; Table S3. Journals with the greatest scientific production in Scopus; Table S4. Most cited papers on the phytoremediation of mercury pollution in WoS [69–71,132–147]; Table S5. Most cited papers on the phytoremediation of mercury pollution in Scopus [3,14,22,69–74,148–158]; Table S6. Top 31 keywords with the strongest citation bursts in WoS; Table S7. Top 24 keywords with the strongest citation bursts in Scopus; Table S8. Plants used in mercury phytoremediation [16,24,25,28,32,35,101,123,124,159–166].

Author Contributions: Conceptualization, L.M.C. and P.N.C.; Methodology; L.M.C. and D.P.C.; Writing—original draft preparation, L.M.C. and P.N.C.; Writing—review and editing, L.M.C., D.P.C., A.L.G., P.N.C. and C.A.A.; Supervision, D.P.C. and P.N.C. All authors have read and agreed to the published version of the manuscript.

Funding: This research received no external funding.

Acknowledgments: This work was supported by Universidad Tecnológica del Chocó, Colombia, Universidad Tecnológica de Pereira, Colombia, the Ministry of Science, and Technology and Innovation—MINCIENCIAS (Bicentennial Doctoral Excellence Scholarship Program), Colombia, and Aarhus University.

Conflicts of Interest: The authors declare no conflicts of interest.

References

1. Singh, A.; Khanna, K.; Kour, J.; Dhiman, S.; Bhardwaj, T.; Devi, K.; Sharma, N.; Kumar, P.; Kapoor, N.; Sharma, P.; et al. Critical review on biogeochemical dynamics of mercury (Hg) and its abatement strategies. *Chemosphere* **2023**, *319*, 137917. [CrossRef] [PubMed]
2. ATSDR. ATSDR Agency for Toxic Substances and Disease Registry. ATSDR's Substance Priority List. August 2022. Available online: https://www.atsdr.cdc.gov/spl/index.html (accessed on 25 July 2023).
3. Gall, J.E.; Boyd, R.S.; Rajakaruna, N. Transfer of heavy metals through terrestrial food webs: A review. *Environ. Monit. Assess.* **2015**, *187*, 201. [CrossRef] [PubMed]
4. Mahbub, K.R.; Krishnan, K.; Naidu, R.; Andrews, S.; Megharaj, M. Mercury toxicity to terrestrial biota. *Ecol. Indic.* **2017**, *74*, 451–462. [CrossRef]
5. Liu, Z.; Chen, B.; Wang, L.A.O.; Urbanovich, O.; Nagorskaya, L.; Li, X.; Tang, L. A review on phytoremediation of mercury-contaminated soils. *J. Hazard. Mater.* **2020**, *400*, 123138. [CrossRef]
6. The United Nations Environment Programme (UN Environment, 2019). Available online: https://www.unep.org (accessed on 25 July 2023).
7. Fernandes, I.O.; Gomes, L.F.; Monteiro, L.C.; Dórea, J.G.; Bernardi, J.V.E. A Scientometric Analysis of Research on World Mercury (Hg) in Soil (1991–2020). *Water Air Soil Pollut.* **2021**, *232*, 254. [CrossRef]
8. He, M.; Tian, L.; Braaten, H.F.V.; Wu, Q.; Luo, J.; Cai, L.M.; Lin, Y. Mercury–Organic Matter Interactions in Soils and Sediments: Angel or Devil? *Bull. Environ. Contam. Toxicol.* **2019**, *102*, 621–627. [CrossRef] [PubMed]

9. Tomiyasu, T.; Matsuyama, A.; Imura, R.; Kodamatani, H.; Miyamoto, J.; Kono, Y.; Kocman, D.; Kotnik, J.; Fajon, V.; Horvat, M. The distribution of total and methylmercury concentrations in soils near the Idrija mercury mine, Slovenia, and the dependence of the mercury concentrations on the chemical composition and organic carbon levels of the soil. *Environ. Earth Sci.* **2012**, *65*, 1309–1322. [CrossRef]
10. Cai, X.; Cai, B.; Zhang, H.; Chen, L.; Zheng, C.; Tong, P.; Lin, H.; Zhang, Q.; Liu, M.; Tong, Y.; et al. Establishment of High-Resolution Atmospheric Mercury Emission Inventories for Chinese Cement Plants Based on the Mass Balance Method. *Environ. Sci. Technol.* **2020**, *54*, 13399–13408. [CrossRef] [PubMed]
11. Eckley, C.S.; Gilmour, C.C.; Janssen, S.; Luxton, T.P.; Randall, P.M.; Whalin, L.; Austin, C. The assessment and remediation of mercury contaminated sites: A review of current approaches. *Sci. Total Environ.* **2020**, *707*, 136031. [CrossRef]
12. Gutiérrez-Mosquera, H.; Marrugo-Negrete, J.; Díez, S.; Morales-Mira, G.; Montoya-Jaramillo, L.J.; Jonathan, M.P. Distribution of chemical forms of mercury in sediments from abandoned ponds created during former gold mining operations in Colombia. *Chemosphere* **2020**, *258*, 127319. [CrossRef]
13. Mishra, B.; Chandra, M. Evaluation of phytoremediation potential of aromatic plants: A systematic review. *J. Appl. Res. Med. Aromat. Plants* **2022**, *31*, 100405. [CrossRef]
14. Ashraf, S.; Ali, Q.; Zahir, Z.A.; Ashraf, H.N.; Ashraf, S. Phytoremediation: Environmentally sustainable way for reclamation of heavy metal polluted soils. *Ecotoxicol. Environ. Saf.* **2019**, *174*, 714–727. [CrossRef]
15. Kumari, S.A.; Jamwal, R.; Mishra, N.; Singh, K. Recent developments in environmental mercury bioremediation and its toxicity: A review. *Environ. Nanotechnol. Monit. Manag.* **2020**, *13*, 100283. [CrossRef]
16. Cepoi, L.; Zinicovscaia, I.; Valuta, A.; Codreanu, L.; Rudi, L.; Chiriac, T.; Yushin, N.; Grozdov, D.; Peshkova, A. Bioremediation capacity of edaphic cyanobacteria Nostoc linckia for chromium in association with other heavy-metals-contaminated soils. *Environments* **2021**, *9*, 1. [CrossRef]
17. Raklami, A.; Meddich, A.; Oufdou, K.; Baslam, M. Plants—Microorganisms-Based Bioremediation for Heavy Metal Cleanup: Recent Developments, Phytoremediation Techniques, Regulation Mechanisms, and Molecular Responses. *Int. J. Mol. Sci.* **2022**, *23*, 5031. [CrossRef] [PubMed]
18. Pratush, A.; Kumar, A.; Hu, Z. Adverse effect of heavy metals (As, Pb, Hg, and Cr) on health and their bioremediation strategies: A review. *Int. Microbiol.* **2018**, *21*, 97–106. [CrossRef] [PubMed]
19. Ullah, A.; Heng, S.; Munis, M.F.H.; Fahad, S.; Yang, X. Phytoremediation of heavy metals assisted by plant growth promoting (PGP) bacteria: A review. *Environ. Exp. Bot.* **2015**, *117*, 28–40. [CrossRef]
20. Podar, D.; Maathuis, F.J.M. The role of roots and rhizosphere in providing tolerance to toxic metals and metalloids. *Plant Cell. Environ.* **2022**, *45*, 719–736. [CrossRef] [PubMed]
21. Shah, V.; Daverey, A. Phytoremediation: A multidisciplinary approach to clean up heavy metal contaminated soil. *Environ. Technol. Innov.* **2020**, *18*, 100774. [CrossRef]
22. Wang, J.; Feng, X.; Anderson, C.W.N.; Xing, Y.; Shang, L. Remediation of mercury contaminated sites—A review. *J. Hazard. Mater.* **2012**, *221*, 1–18. [CrossRef]
23. Wu, G.; Kang, H.; Zhang, X.; Shao, H.; Chu, L.; Ruan, C. A critical review on the bio-removal of hazardous heavy metals from contaminated soils: Issues, progress, eco-environmental concerns and opportunities. *J. Hazard. Mater.* **2010**, *174*, 1–8. [CrossRef] [PubMed]
24. Marrugo-Negrete, J.; Marrugo-Madrid, S.; Pinedo-Hernández, J.; Durango-Hernández, J.; Díez, S. Screening of native plant species for phytoremediation potential at a Hg-contaminated mining site. *Sci. Total Environ.* **2016**, *542*, 809–816. [CrossRef] [PubMed]
25. Xun, Y.; Feng, L.; Li, Y.; Dong, H. Mercury accumulation plant Cyrtomium macrophyllum and its potential for phytoremediation of mercury polluted sites. *Chemosphere* **2017**, *189*, 161–170. [CrossRef]
26. Qian, X.; Wu, Y.; Zhou, H.; Xu, X.; Xu, Z.; Shang, L.; Qiu, G. Total mercury and methylmercury accumulation in wild plants grown at wastelands composed of mine tailings: Insights into potential candidates for phytoremediation. *Environ. Pollut.* **2018**, *239*, 757–767. [CrossRef] [PubMed]
27. Petelka, J.; Abraham, J.; Bockreis, A.; Deikumah, J.P.; Zerbe, S. Soil Heavy Metal(loid) Pollution and Phytoremediation Potential of Native Plants on a Former Gold Mine in Ghana. *J. Water Air Soil Pollut.* **2019**, *230*, 267. [CrossRef]
28. Muryani, E.; Sajidan, S.; Budiastuti, M.T.S.; Pranoto, P. Diversity and potential of herbaceous plants as mercury (Hg) hyper-accumulators in small-scale gold mining sites in Pancurendang, Banyumas, Indonesia. *J. Biol. Divers.* **2023**, *24*, 3364–3372. [CrossRef]
29. Kukier, U.; Peters, C.A.; Chaney, R.L.; Angle, J.S.; Roseberg, R.J. The Effect of pH on Metal Accumulation in Two Alyssum Species. *J. Environ. Qual.* **2004**, *33*, 2090–2102. [CrossRef]
30. Santos, G.C.G.D.; Rodella, A.A.; Abreu, C.A.D.; Coscione, A.R. Vegetable species for phytoextraction of heavy metals Vegetable species for phytoextraction of boron, copper, lead, manganese and zinc from contaminated soil. *Sci. Agric.* **2010**, *67*, 713–719. [CrossRef]
31. Chamba-Eras, I.; Griffith, D.M.; Kalinhoff, C.; Ramírez, J.; Gázquez, M.J. Native Hyperaccumulator Plants with Differential Phytoremediation Potential in an Artisanal Gold Mine of the Ecuadorian Amazon. *Plants* **2022**, *11*, 1186. [CrossRef]
32. Cui, L.; Tian, X.; Xie, H.; Cong, X.; Cui, L.; Wu, H.; Wang, J.; Li, B.; Zhao, J.; Cui, Y.; et al. Cardamine violifolia as a potential Hg hyperaccumulator and the cellular responses. *Sci. Total Environ.* **2023**, *863*, 160940. [CrossRef]

33. Liu, Z.; Wang, L.; Ding, S.; Xiao, H. Enhancer assisted-phytoremediation of mercury-contaminated soils by *Oxalis corniculata* L., and rhizosphere microorganism distribution of *Oxalis corniculata* L. *Ecotoxicol. Environ. Saf.* **2018**, *160*, 171–177. [CrossRef] [PubMed]
34. Quiñones, M.A.; Ruiz-Díez, B.; Fajardo, S.; López-Berdonces, M.A.; Higueras, P.L.; Fernández-Pascual, M. Lupinus albus plants acquire mercury tolerance when inoculated with an Hg-resistant Bradyrhizobium strain. *Plant Physiol. Biochem.* **2013**, *73*, 168–175. [CrossRef] [PubMed]
35. Sun, L.; Ma, Y.; Wang, H.; Huang, W.; Wang, X.; Han, L.; Sun, W.; Han, E.; Wang, B. Overexpression of PtABCC1 contributes to mercury tolerance and accumulation in Arabidopsis and poplar. *Biochem. Biophys. Res. Commun.* **2018**, *497*, 997–1002. [CrossRef] [PubMed]
36. Mao, G.; Shi, T.; Zhang, S.; Crittenden, J.; Guo, S.; Du, H. Bibliometric analysis of insights into soil remediation. *J. Soils Sediments.* **2018**, *18*, 2520–2534. [CrossRef]
37. Yang, L.; Wang, J.; Yang, Y.; Li, S.; Wang, T.; Oleksak, P.; Chrienova, Z.; Wu, Q.; Nepovimova, E.; Zhang, X.; et al. Phytoremediation of heavy metal pollution: Hotspots and future prospects. *Ecotoxicol. Environ. Saf.* **2022**, *234*, 113403. [CrossRef]
38. Armah, F.A.; Obiri, S.; Yawson, D.O.; Onumah, E.E.; Yengoh, G.T.; Afrifa, E.K.A.; Odoi, J.O. Anthropogenic sources and environmentally relevant concentrations of heavy metals in surface water of a mining district in Ghana: A multivariate statistical approach. *J. Environ. Sci. Health Part A* **2010**, *45*, 1804–1813. [CrossRef]
39. Han, R.; Zhou, B.; Huang, Y.; Lu, X.; Li, S.; Li, N. Bibliometric overview of research trends on heavy metal health risks and impacts in 1989–2018. *J. Clean. Prod.* **2020**, *276*, 123249. [CrossRef]
40. Ubando, A.T.; Africa, A.D.M.; Maniquiz-Redillas, M.C.; Culaba, A.B.; Chen, W.-H.; Chang, J.-S. Microalgal biosorption of heavy metals: A comprehensive bibliometric review. *J. Hazard. Mater.* **2021**, *402*, 123431. [CrossRef]
41. Singh, V.K.; Singh, P.; Karmakar, M.; Leta, J.; Mayr, P. The journal coverage of Web of Science, Scopus and Dimensions: A comparative analysis. *Scientometrics* **2021**, *126*, 5113–5142. [CrossRef]
42. Chiu, W.-T.; Ho, Y.-S. Bibliometric analysis of tsunami research. *Scientometrics* **2007**, *73*, 3–17. [CrossRef]
43. Shi, D.; Xie, C.; Wang, J.; Xiong, L. Changes in the structures and directions of heavy metal-contaminated soil remediation research from 1999 to 2020: A bibliometric & scientometric study. *Environ. Res. Public Health* **2021**, *18*, 7358. [CrossRef]
44. Zhang, Y.; Huang, K.; Yu, Y.; Yang, B. Mapping of water footprint research: A bibliometric analysis during 2006–2015. *J. Clean Prod.* **2017**, *149*, 70–79. [CrossRef]
45. Hood, W.W.; Wilson, C.S. The Literature of Bibliometrics, Scientometrics, and Informetrics. *Scientometrics* **2001**, *52*, 291–314. [CrossRef]
46. van Eck, N.; Waltman, L. Software survey: VOSviewer, a computer program for bibliometric mapping. *Scientometrics* **2010**, *2*, 523–538. [CrossRef] [PubMed]
47. Yang, F.; Qiu, D. Exploring coal spontaneous combustion by bibliometric analysis. *Process Saf. Environ. Prot.* **2019**, *132*, 1–10. [CrossRef]
48. Ismael, M.A.; Elyamine, A.M.; Moussa, M.G.; Cai, M.; Zhao, X.; Hu, C. Cadmium in plants: Uptake, toxicity, and its interactions with selenium fertilizers. *Metallomics* **2019**, *11*, 255–277. [CrossRef]
49. Pasricha, S.; Mathur, V.; Garg, A.; Lenka, S.; Verma, K.; Agarwal, S. Molecular mechanisms underlying heavy metal uptake, translocation and tolerance in hyperaccumulators-an analysis: Heavy metal tolerance in hyperaccumulators. *Environ. Chall.* **2021**, *4*, 100197. [CrossRef]
50. Mongeon, P.; Paul-Hus, A. The journal coverage of Web of Science and Scopus: A comparative analysis. *Scientometrics* **2016**, *106*, 213–228. [CrossRef]
51. Zhang, D.; Dyck, M.; Filipović, L.; Filipović, V.; Lv, J.; He, H. Hyperaccumulators for potentially toxic elements: A scientometric analysis. *Agronomy* **2021**, *11*, 1729. [CrossRef]
52. Valdiviezo Gonzales, L.G.; Castañeda-Olivera, C.A.; Cabello-Torres, R.J.; García Ávila, F.F.; Munive Cerrón, R.V.; Alfaro, J. Scientometric study of treatment technologies of soil pollution: Present and future challenges. *Appl. Soil Ecol.* **2023**, *182*, 104695. [CrossRef]
53. Qiu, G.; Feng, X.; Wang, S.; Shang, L. Mercury and methylmercury in riparian soil, sediments, mine-waste calcines, and moss from abandoned Hg mines in east Guizhou province, southwestern China. *Appl. Geochem.* **2005**, *20*, 627–638. [CrossRef]
54. Qiu, G.; Feng, X.; Meng, B.; Zhang, C.; Gu, C.; Du, B.; Lin, Y. Environmental geochemistry of an abandoned mercury mine in Yanwuping, Guizhou Province, China. *Environ. Res.* **2013**, *125*, 124–130. [CrossRef] [PubMed]
55. Gosar, M.; Žibret, G. Mercury contents in the vertical profiles through alluvial sediments as a reflection of mining in Idrija (Slovenia). *J. Geochem. Explor.* **2011**, *110*, 81–91. [CrossRef]
56. Kocman, D.; Kanduč, T.; Ogrinc, N.; Horvat, M. Distribution and partitioning of mercury in a river catchment impacted by former mercury mining activity. *Biogeochemistry* **2011**, *104*, 183–201. [CrossRef]
57. Yang, Z.; Li, X.; Wang, Y.; Chang, J.; Liu, X. Trace element contamination in urban topsoil in China during 2000–2009 and 2010–2019: Pollution assessment and spatiotemporal analysis. *Sci. Total Environ.* **2021**, *758*, 143647. [CrossRef]
58. Yang, Q.; Li, Z.; Lu, X.; Duan, Q.; Huang, L.; Bi, J. A review of soil heavy metal pollution from industrial and agricultural regions in China: Pollution and risk assessment. *Sci. Total Environ.* **2018**, *642*, 690–700. [CrossRef]
59. Corella, J.P.; Valero-Garcés, B.L.; Wang, F.; Martínez-Cortizas, A.; Cuevas, C.A.; Saiz-Lopez, A. 700 years reconstruction of mercury and lead atmospheric deposition in the Pyrenees (NE Spain). *Atmos. Environ.* **2017**, *155*, 97–107. [CrossRef]

60. van der Ent, A.; Baker, A.J.M.; Reeves, R.D.; Pollard, A.J.; Schat, H. Hyperaccumulators of metal and metalloid trace elements: Facts and fiction. *Plant Soil.* **2013**, *362*, 319–334. [CrossRef]
61. Kumar, P.N.; Dushenkov, V.; Motto, H.; Raskin, I. Phytoextraction: The Use of Plants To Remove Heavy Metals from Soils. *Environ. Sci. Technol.* **1995**, *29*, 1232–1238. [CrossRef]
62. Palacios-Torres, Y.; Caballero-Gallardo, K.; Olivero-Verbel, J. Mercury pollution by gold mining in a global biodiversity hotspot, the Choco biogeographic region, Colombia. *Chemosphere* **2018**, *193*, 421–430. [CrossRef]
63. Chen, G.Q.; Li, J.S.; Chen, B.; Wen, C.; Yang, Q.; Alsaedi, A.; Hayat, T. An overview of mercury emissions by global fuel combustion: The impact of international trade. *Renew. Sustain. Energy Rev.* **2016**, *65*, 345–355. [CrossRef]
64. Gutiérrez, J.K.R.; Velasco, N.Y.G. Redes de coautoría como herramienta de evaluación de la producción científica de los grupos de investigación. *Rev. Gen. Inf. Doc.* **2017**, *27*, 279–297. [CrossRef]
65. Wang, J.; Xia, J.; Feng, X. Screening of chelating ligands to enhance mercury accumulation from historically mercury-contaminated soils for phytoextraction. *J. Environ. Manag.* **2017**, *186*, 233–239. [CrossRef]
66. Wang, J.; Anderson, C.W.N.; Xing, Y.; Fan, Y.; Xia, J.; Shaheen, S.M.; Rinklebe, J.; Feng, X. Thiosulphate-induced phytoextraction of mercury in *Brassica juncea*: Spectroscopic investigations to define a mechanism for Hg uptake. *Environ. Pollut.* **2018**, *242*, 86–993. [CrossRef]
67. Hu, L.; Zhang, L.; Wu, H. Experimental study of the effects of soil pH and ionic species on the electro-osmotic consolidation of kaolin. *J. Hazard. Mater.* **2019**, *368*, 885–893. [CrossRef]
68. Su, H.-N.; Lee, P.-C. Mapping knowledge structure by keyword co-occurrence: A first look at journal papers in Technology Foresight. *Scientometrics* **2010**, *85*, 65–79. [CrossRef]
69. Liu, L.; Li, W.; Song, W.; Guo, M. Remediation techniques for heavy metal-contaminated soils: Principles and applicability. *Sci. Total Environ.* **2018**, *633*, 206–219. [CrossRef]
70. Park, J.; Song, W.Y.; Ko, D.; Eom, Y.; Hansen, T.H.; Schiller, M.; Lee, Y. The phytochelatin transporters AtABCC1 and AtABCC2 mediate tolerance to cadmium and mercury. *Plant J.* **2012**, *69*, 278–288. [CrossRef] [PubMed]
71. Li, C.; Zhou, K.; Qin, W.; Tian, C.; Qi, M.; Yan, X.; Han, W. A review on heavy metals contamination in soil: Effects, sources, and remediation techniques. *Soil Sediment Contam. Int. J.* **2019**, *28*, 380–394. [CrossRef]
72. Schutzendubel, A.; Polle, A. Plant responses to abiotic stresses: Heavy metal-induced oxidative stress and protection by mycorrhization. *J. Exp. Bot.* **2002**, *53*, 1351–1365. [CrossRef]
73. Mulligan, C.N.; Yong, R.N.; Gibbs, B.F. Remediation technologies for metal-contaminated soils and groundwater: An evaluation. *Eng. Geol.* **2001**, *60*, 193–207. [CrossRef]
74. He, Z.L.; Yang, X.E.; Stoffella, P.J. Trace elements in agroecosystems and impacts on the environment. *J. Trace. Elem. Med. Biol.* **2005**, *19*, 125–140. [CrossRef] [PubMed]
75. Malar, S.; Sahi, S.V.; Favas, P.J.C.; Venkatachalam, P. Mercury heavy-metal-induced physiochemical changes and genotoxic alterations in water hyacinths [*Eichhornia crassipes* (Mart.)]. *Environ. Sci. Pollut. Res.* **2015**, *22*, 4597–4608. [CrossRef] [PubMed]
76. Gupta, S.; Sireesha, S.; Sreedhar, I.; Patel, C.M.; Anitha, K.L. Latest trends in heavy metal removal from wastewater by biochar based sorbents. *J. Water Process Eng.* **2020**, *38*, 101561. [CrossRef]
77. Ulaganathan, A.; Robinson, J.S.; Rajendran, S.; Geevaretnam, J.; Shanmugam, S.; Natarajan, A.; Abdulrahman, I.A.; Karthikeyan, P. Potentially toxic elements contamination and its removal by aquatic weeds in the riverine system: A comparative approach. *Environ. Res.* **2022**, *206*, 112613. [CrossRef]
78. Peng, D.; Chen, M.; Su, X.; Liu, C.; Zhang, Z.; Middleton, B.A.; Lei, T. Mercury accumulation potential of aquatic plant species in West Dongting Lake, China. *Environ. Pollut.* **2023**, *324*, 121313. [CrossRef]
79. Qiu, G.; Feng, X.; Li, P.; Wang, S.; Li, G.; Shang, L.; Fu, X. Methylmercury Accumulation in Rice (*Oryza sativa* L.) Grown at Abandoned Mercury Mines in Guizhou, China. *J. Agric. Food. Chem.* **2008**, *56*, 2465–2468. [CrossRef]
80. Feng, X.; Li, P.; Qiu, G.; Wang, S.; Li, G.; Shang, L.; Meng, B.; Jiang, H.; Bai, W.; Li, Z.; et al. Human Exposure To Methylmercury through Rice Intake in Mercury Mining Areas, Guizhou Province, China. *Environ. Sci. Technol.* **2008**, *42*, 326–332. [CrossRef]
81. Li, X.; Zhou, M.; Shi, F.; Meng, B.; Liu, J.; Mi, Y.; Dong, C.; Su, H.; Liu, X.; Wang, F.; et al. Influence of arbuscular mycorrhizal fungi on mercury accumulation in rice (*Oryza sativa* L.): From enriched isotope tracing perspective. *Ecotoxicol. Environ. Saf.* **2023**, *255*, 114776. [CrossRef]
82. Saldarriaga, J.F.; López, J.E.; Díaz-García, L.; Montoya-Ruiz, C. Changes in *Lolium perenne* L. rhizosphere microbiome during phytoremediation of Cd- and Hg-contaminated soils. *Environ. Sci. Pollut. Res.* **2023**, *30*, 49498–49511. [CrossRef]
83. Guo, Y.; Sommer, N.; Martin, K.; Rasche, F. Rhizophagus irregularis improves Hg tolerance of Medicago truncatula by upregulating the Zn transporter genes ZIP2 and ZIP6. *Mycorrhiza* **2023**, *33*, 23–32. [CrossRef] [PubMed]
84. Li, X.; Wang, X.; Zhao, L.; Wang, Z.; Teng, Y.; Luo, Y. Mercury Enrichment Characteristics and Rhizosphere Bacterial Community of Ramie (*Boehmeria nivea* L. Gaud.) in Mercury-Contaminated Soil. *Sustainability* **2023**, *15*, 6009. [CrossRef]
85. Senabio, J.A.; de Campos Pereira, F.; Pietro-Souza, W.; Sousa, T.F.; Silva, G.F.; Soares, M.A. Enhanced mercury phytoremediation by *Pseudomonodictys pantanalensis* sp. nov. A73 and Westerdykella aquatica P71. *Braz. J. Microbiol.* **2023**, *54*, 949–964. [CrossRef]
86. Sharma, P.; Chaturvedi, P.; Chandra, R.; Kumar, S. Identification of heavy metals tolerant *Brevundimonas* sp. from rhizospheric zone of *Saccharum munja* L. and their efficacy in in-situ phytoremediation. *Chemosphere* **2022**, *295*, 133823. [CrossRef]
87. Ustiatik, R.; Nuraini, Y.; Suharjono, S.; Jeyakumar, P.; Anderson, C.W.N.; Handayanto, E. Mercury resistance and plant growth promoting traits of endophytic bacteria isolated from mercury-contaminated soil. *Bioremediat. J.* **2022**, *26*, 208–227. [CrossRef]

88. Valdiviezo Gonzales, L.G.; García Ávila, F.F.; Cabello Torres, R.J.; Castañeda Olivera, C.A.; Alfaro Paredes, E. Scientometric study of drinking water treatments technologies: Present and future challenges. *Cogent Eng.* **2021**, *8*, 1929046. [CrossRef]
89. Zhu, Y.; Jiang, S.; Han, X.; Gao, X.; He, G.; Zhao, Y.; Li, H. A bibliometrics review of water footprint research in China: 2003-2018. *Sustainability* **2019**, *11*, 5082. [CrossRef]
90. Zhang, H.; Feng, X.; Larsen, T.; Shang, L.; Li, P. Bioaccumulation of Methylmercury versus Inorganic Mercury in Rice (*Oryza sativa* L.) Grain. *Environ. Sci. Technol.* **2010**, *44*, 4499–4504. [CrossRef] [PubMed]
91. Huang, Y.; Li, F.; Yi, J.; Yan, H.; He, Z.; Li, X. Transcriptomic and physio-biochemical features in rice (*Oryza sativa* L.) in response to mercury stress. *Chemosphere* **2022**, *309*, 136612. [CrossRef]
92. Shahpiri, A.; Mohammadzadeh, A. Mercury removal by engineered *Escherichia coli* cells expressing different rice metallothionein isoforms. *Ann. Microbiol.* **2018**, *68*, 145–152. [CrossRef]
93. Gomes, M.V.T.; de Souza, R.R.; Teles, V.S.; Mendes, É.A. Phytoremediation of water contaminated with mercury using Typha domingensis in constructed wetland. *Chemosphere* **2014**, *103*, 228–233. [CrossRef] [PubMed]
94. Marrugo-Negrete, J.; Enamorado-Montes, G.; Durango-Hernández, J.; Pinedo-Hernández, J.; Díez, S. Removal of mercury from gold mine effluents using Limnocharis flava in constructed wetlands. *Chemosphere 2017. 167*, 188–192. [CrossRef]
95. Dushenkov, S. Trends in phytoremediation of radionuclides. *Plant Soil.* **2003**, *249*, 167–175. [CrossRef]
96. Reeves, R.D.; Baker, A.J.M.; Jaffré, T.; Erskine, P.D.; Echevarria, G.; Ent, A. van der. A global database for plants that hyperaccumulate metal and metalloid trace elements. *New Phytol.* **2018**, *218*, 407–411. Available online: https://www.jstor.org/stable/90019919 (accessed on 23 September 2024). [CrossRef]
97. Salt, D.E.; Smith, R.D.; Raskin, I. Phytoremediation. *Annu. Rev. Plant. Physiol. Plant Mol. Biol.* **1998**, *49*, 643–668. [CrossRef]
98. Tiodar, E.D.; Văcar, C.L.; Podar, D. Phytoremediation and microorganisms-assisted phytoremediation of mercury-contaminated soils: Challenges and perspectives. *Int. J. Environ. Res. Public. Health.* **2021**, *18*, 2435. [CrossRef]
99. Gutiérrez-Mosquera, H.; Marrugo-Negrete, H.S.; Díez, S.; Morales–Mira, G.; Montoya-Jaramillo, L.J.; Jonathan, M.P. Mercury distribution in different environmental matrices in aquatic systems of abandoned gold mines, Western Colombia: Focus on human health. *J. Hazard. Mater.* **2012**, *4040*, 124080. [CrossRef] [PubMed]
100. Ai, H.; Khan, E.; Sajad, M.A. Phytoremediation of heavy metals—Concepts and applications. *Chemosphere* **2013**, *91*, 869–881. [CrossRef]
101. Marrugo–Madrid, S.; Turull, M.; Montes, G.E.; Pico, M.V.; Marrugo–Negrete, J.L.; Díez, S. Phytoremediation of mercury in soils impacted by gold mining: A case study of Colombia. *Bioremediation Environ. Sustain. Toxic. Mech. Contam. Degrad. Detoxif. Chall.* **2021**, *2021*, 145–160. [CrossRef]
102. Czako, M.; Feng, X.; He, Y.; Liang, D.; Marton, L. Transgenic Spartina alterniflora for phytoremediation. *Environ. Geochem. Health.* **2006**, *28*, 103–110. [CrossRef]
103. Yang, R.Y.; Li, T.T.; Yang, T.H.; Li, Y.P.; Liu, H.; Wang, L.; Lu, N.Q. Advances in enhanced phytoremediation by genetic engineering technology for heavy metal pollution in soil. *Res. Environ. Sci.* **2019**, *32*, 1294–1303.
104. Priyadarshanee, M.; Chatterjee, S.; Rath, S.; Dash, H.R.; Das, S. Cellular and genetic mechanism of bacterial mercury resistance and their role in biogeochemistry and bioremediation. *J. Hazard. Mater.* **2022**, *423*, 26985. [CrossRef] [PubMed]
105. Heaton, A.C.; Rugh, C.L.; Kim, T.; Wang, N.J.; Meagher, R.B. Toward detoxifying mercury-polluted aquatic sediments with rice genetically engineered for mercury resistance. *Environ. Toxicol. Chem.* **2003**, *22*, 2940–2947. [CrossRef]
106. Zhang, Y.; Liu, J.; Zhou, Y.; Gong, T.; Wang, J.; Ge, Y. Enhanced phytoremediation of mixed heavy metal (mercury)—Organic pollutants (trichloroethylene) with transgenic alfalfa co-expressing glutathione S-transferase and human P450 2E1. *J. Hazard. Mater.* **2013**, *260*, 1100–1107. [CrossRef]
107. Rodríguez, L.; Alonso-Azcárate, J.; Villaseñor, J.; Rodríguez-Castellanos, L. EDTA and hydrochloric acid effects on mercury accumulation by Lupinus albus. *Environ. Sci. Pollut. Res.* **2016**, *23*, 24739–24748. [CrossRef]
108. Elizalde-Velázquez, G.A.; Gómez-Oliván, L.M. Occurrence, toxic effects and removal of metformin in the aquatic environments in the world: Recent trends and perspectives. *Sci. Total Environ.* **2020**, *702*, 134924. [CrossRef] [PubMed]
109. Amir, W.; Farid, M.; Ishaq, H.K.; Farid, S.; Zubair, M.; Alharby, H.F.; Ali, S. Accumulation potential and tolerance response of *Typha latifolia* L. under citric acid assisted phytoextraction of lead and mercury. *Chemosphere* **2020**, *257*, 127247. [CrossRef]
110. Li, X.; Zhang, X.; Wang, X.; Cui, Z. Phytoremediation of multi-metal contaminated mine tailings with *Solanum nigrum* L. and biochar/attapulgite amendments. *Ecotoxicol. Environ. Saf.* **2019**, *180*, 517–525. [CrossRef]
111. Smolinska, B.; Leszczynska, J. Influence of combined use of iodide and compost on Hg accumulation by *Lepidium sativum* L. *J. Environ. Manage.* **2015**, *150*, 499–507. [CrossRef]
112. Wang, Y.; Dang, F.; Evans, R.D.; Zhong, H.; Zhao, J.; Zhou, D. Mechanistic understanding of MeHg-Se antagonism in soil-rice systems: The key role of antagonism in soil. *Sci. Rep.* **2016**, *6*, 19477. [CrossRef]
113. Wang, J.; Feng, X.; Anderson, C.W.; Wang, H.; Wang, L. Thiosulphate-induced mercury accumulation by plants: Metal uptake and transformation of mercury fractionation in soil—Results from a field study. *Plant Soil* **2014**, *375*, 21–33. [CrossRef]
114. Mhlongo, M.I.; Piater, L.A.; Madala, N.E.; Labuschagne, N.; Dubery, I.A. The chemistry of plant-microbe interactions in the rhizosphere and the potential for metabolomics to reveal signaling related to defense priming and induced systemic resistance. *Front. Plant Sci.* **2018**, *9*, 112. [CrossRef] [PubMed]

115. Ruley, J.A.; Tumuhairwe, J.B.; Amoding, A.; Westengen, O.T.; Vinje, H. Rhizobacteria Communities of Phytoremediation Plant Species in Petroleum Hydrocarbon Contaminated Soil of the Sudd Ecosystem, South Sudan. *Int. J. Microbiol.* **2020**, *2020*, 6639118. [CrossRef] [PubMed]
116. Sharma, P. Efficiency of bacteria and bacterial assisted phytoremediation of heavy metals: An update. *Bioresour. Technol.* **2021**, *328*, 124835. [CrossRef]
117. Gupta, D.K.; Huang, H.G.; Corpas, F.J. Lead tolerance in plants: Strategies for phytoremediation. *Environ. Sci. Pollut. Res.* **2013**, *20*, 2150–2161. [CrossRef]
118. Fasani, E.; Manara, A.; Martini, F.; Furini, A.; DalCorso, G. The potential of genetic engineering of plants for the remediation of soils contaminated with heavy metals. *Plant Cell Environ.* **2018**, *41*, 1201–1232. [CrossRef]
119. Ma, Y.; Prasad, M.N.V.; Rajkumar, M.; Freitas, H.J.B.A. Plant growth promoting rhizobacteria and endophytes accelerate phytoremediation of metalliferous soils. *Biotechnol. Adv.* **2011**, *29*, 248–258. [CrossRef]
120. Leudo, A.M.; Cruz, Y.; Montoya-Ruiz, C.; Delgado, M.d.P.; Saldarriaga, J.F. Mercury phytoremediation with Lolium perenne—Mycorrhizae in contaminated soils. *Sustainability* **2020**, *12*, 3795. [CrossRef]
121. Xiao, L.; Zhang, C.; Wang, G.; Guan, D.; Zhang, R.; Chen, Y.; Dai, J. Influencing pathways of soil microbial attributes on accumulation of heavy metals in brassica (*Brassica campestris* L. ssp. chinensis var. utilis Tsen et Lee) leaves. *Environ. Pollut.* **2020**, *262*, 114215. [CrossRef]
122. Thakur, S.; Singh, L.; Wahid, Z.A.; Siddiqui, M.F.; Atnaw, S.M.; Din, M.F.M. Plant-driven removal of heavy metals from soil: Uptake, translocation, tolerance mechanism, challenges, and future perspectives. *Environ. Monit. Assess.* **2016**, *188*, 206. [CrossRef]
123. Esteban, E.; Moreno, E.; Penalosa, J.; Cabrero, J.I.; Millán, R.; Zornoza, P. Short and long-term uptake of Hg in white lupin plants: Kinetics and stress indicators. *Environ. Exp. Bot.* **2008**, *62*, 316–322. [CrossRef]
124. Ibraheem, F.; Al-Hazmi, N.; El-Morsy, M.; Mosa, A. Ecological risk assessment of potential toxic elements in salt marshes on the east coast of the red sea: Differential physiological responses and adaptation capacities of dominant halophytes. *Sustainability* **2021**, *13*, 11282. [CrossRef]
125. Gill, S.S.; Tuteja, N. Reactive oxygen species and antioxidant machinery in abiotic stress tolerance in crop plants. *Plant Physiol. Biochem.* **2010**, *48*, 909–930. [CrossRef] [PubMed]
126. Yilmaz, D.D.; Parlak, K.U. Nickel-induced changes in lipid peroxidation, antioxidative enzymes, and metal accumulation in *Lemma gibba*. *Int. J. Phytorem.* **2011**, *13*, 805–817. [CrossRef]
127. Manara, A. Plant Responses to Heavy Metal Toxicity. In *Plants Heavy Metals*; Springer: Dordrecht, The Netherlands, 2012; pp. 27–53. [CrossRef]
128. Szalai, G.; Krantev, A.; Yordanova, R.; Popova, L.P.; Janda, T. Influence of salicylic acid on phytochelatin synthesis in Zea mays during Cd stress. *Turk. J. Bot.* **2013**, *37*, 708–714. [CrossRef]
129. Xu, J.; Bravo, A.G.; Lagerkvist, A.; Bertilsson, S.; Sjöblom, R.; Kumpiene, J. Sources and remediation techniques for mercury contaminated soil. *Environ. Int.* **2015**, *74*, 42–53. [CrossRef]
130. Sidhu, G.P.S.; Singh, H.P.; Batish, D.R.; Kohli, R.K. Tolerance and hyperaccumulation of cadmium by a wild, unpalatable herb *Coronopus didymus* (L.) Sm. (Brassicaceae). *Ecotoxicol. Environ. Saf.* **2017**, *135*, 209–215. [CrossRef]
131. Chen, Y.; Chen, Y.; Li, Y.; Wu, Y.; Zeng, Z.; Xu, R.; Zhang, J. Changes of heavy metal fractions during co-composting of agricultural waste and river sediment with inoculation of *Phanerochaete chrysosporium*. *J. Hazard. Mater.* **2019**, *378*, 120757. [CrossRef]
132. Marchand, L.; Mench, M.; Jacob, D.L.; Otte, M.L. Metal and metalloid removal in constructed wetlands, with emphasis on the importance of plants and standardized measurements: A review. *Environ. Pollut.* **2010**, *158*, 3447–3461. [CrossRef]
133. Xiao, R.; Wang, S.; Li, R.; Wang, J.J.; Zhang, Z. Soil heavy metal contamination and health risks associated with artisanal gold mining in Tongguan, Shaanxi, China. *Ecotoxicol. Environ. Saf.* **2017**, *141*, 17–24. [CrossRef]
134. Rahman, M.A.; Hasegawa, H. Aquatic arsenic: Phytoremediation using floating macrophytes. *Chemosphere* **2011**, *83*, 633–646. [CrossRef] [PubMed]
135. Khanam, R.; Kumar, A.; Nayak, A.K.; Shahid, M.; Tripathi, R.; Vijayakumar, S.; Pathak, H. Metal(loid)s (As, Hg, Se, Pb and Cd) in paddy soil: Bioavailability and potential risk to human health. *Sci. Total Environ.* **2020**, *699*, 134330. [CrossRef] [PubMed]
136. Hernández, L.E.; Sobrino-Plata, J.; Montero-Palmero, M.B.; Carrasco-Gil, S.; Flores-Cáceres, M.L.; Ortega-Villasante, C.; Escobar, C. Contribution of glutathione to the control of cellular redox homeostasis under toxic metal and metalloid stress. *J. Exp. Bot.* **2015**, *66*, 2901–2911. [CrossRef] [PubMed]
137. Wang, L.; Hou, D.; Cao, Y.; Ok, Y.S.; Tack, F.M.; Rinklebe, J.; O'Connor, D. Remediation of mercury-contaminated soil, water, and air: A review of emerging materials and innovative technologies. *Environ. Int.* **2020**, *134*, 105281. [CrossRef] [PubMed]
138. OB, A.; Muchie, M. Remediation of heavy metals in drinking water and wastewater treatment systems: Processes and applications. *Int. J. Phys. Sci.* **2010**, *5*, 1807–1817.
139. Arao, T.; Ishikawa, S.; Murakami, M.; Abe, K.; Maejima, Y.; Makino, T. Heavy metal contamination of agricultural soil and countermeasures in Japan. *Paddy Water Environ.* **2010**, *8*, 247–257. [CrossRef]
140. Gaur, N.; Flora, G.; Yadav, M.; Tiwari, A. A review with recent advancements on bioremediation-based abolition of heavy metals. *Environ. Sci. Process. Impacts* **2014**, *16*, 180–193. [CrossRef]
141. Singh, J.S.; Abhilash, P.C.; Singh, H.B.; Singh, R.P.; Singh, D.P. Genetically engineered bacteria: An emerging tool for environmental remediation and future research perspectives. *Gene* **2011**, *480*, 1–9. [CrossRef]

142. Bonanno, G. Comparative performance of trace element bioaccumulation and biomonitoring in the plant species Typha domingensis, Phragmites australis and Arundo donax. *Ecotoxicol. Environ. Saf.* **2013**, *97*, 124–130. [CrossRef]
143. Li, K.; Ramakrishna, W. Effect of multiple metal-resistant bacteria from contaminated lake sediments on metal accumulation and plant growth. *J. Hazard. Mater.* **2011**, *189*, 531–539. [CrossRef]
144. Monterroso, C.; Rodríguez, F.; Chaves, R.; Diez, J.; Becerra-Castro, C.; Kidd, P.S.; Macías, F. Heavy metal distribution in mine soils and plants growing in a Pb/Zn-mining area in NW Spain. *Appl. Geochem.* **2014**, *44*, 3–11. [CrossRef]
145. Mleczek, M.; Rutkowski, P.; Rissmann, I.; Kaczmarek, Z.; Golinski, P.; Szentner, K.; Stachowiak, A. Biomass productivity and phytoremediation potential of Salix alba and Salix viminalis. *Biomass Bioenergy* **2010**, *34*, 1410–1418. [CrossRef]
146. Bonanno, G.; Cirelli, G.L. Comparative analysis of element concentrations and translocation in three wetland congener plants: Typha domingensis, Typha latifolia and Typha angustifolia. *Ecotoxicol. Environ. Saf.* **2017**, *143*, 92–99. [CrossRef]
147. Li, W.C.; Tse, H.F. Health risk and significance of mercury in the environment. *Environ. Sci. Pollut. Res.* **2015**, *22*, 192–201. [CrossRef] [PubMed]
148. Tangahu, B.V.; Sheikh Abdullah, S.R.; Basri, H.; Idris, M.; Anuar, N.; Mukhlisin, M. A review on heavy metals (As, Pb, and Hg) uptake by plants through phytoremediation. *Int. J. Chem. Eng.* **2011**, 939161. [CrossRef]
149. McGrath, S.P.; Zhao, F.J. Phytoextraction of metals and metalloids from contaminated soils. *Curr. Opin. Biotechnol.* **2003**, *14*, 277–282. [CrossRef]
150. Wong, M.H. Ecological restoration of mine degraded soils, with emphasis on metal contaminated soils. *Chemosphere* **2003**, *14*, 277–281. [CrossRef]
151. Weis, J.S.; Weis, P. Metal uptake, transport and release by wetland plants: Implications for phytoremediation and restoration. *Environ. Int.* **2004**, *30*, 685–700. [CrossRef]
152. Ojuederie, O.B.; Babalola, O.O. Microbial and plant-assisted bioremediation of heavy metal polluted environments: A review. *Int. J. Environ. Res. Public Health* **2017**, *14*, 1504. [CrossRef]
153. Patra, M.; Sharma, A. Mercury toxicity in plants. *The Botanical Review.* **2000**, *66*, 379–422. [CrossRef]
154. McGrath, S.P.; Zhao, J.; Lombi, E. Phytoremediation of metals, metalloids, and radionuclides. *Adv. Agron.* **2002**, *75*, 1–56. [CrossRef]
155. Krämer, U. Phytoremediation: Novel approaches to cleaning up polluted soils. *Curr. Opin. Biotechnol.* **2005**, *16*, 133–141. [CrossRef] [PubMed]
156. Doty, S.L. Enhancing phytoremediation through the use of transgenics and endophytes. *New Phytol.* **2008**, *179*, 318–333. [CrossRef] [PubMed]
157. Eapen, S.; D'souza, S.F. Prospects of genetic engineering of plants for phytoremediation of toxic metals. *Biotechnol. Adv.* **2005**, *23*, 97–114. [CrossRef]
158. Lebeau, T.; Braud, A.; Jézéquel, K. Performance of bioaugmentation-assisted phytoextraction applied to metal contaminated soils: A review. *Environ. Pollut.* **2008**, *153*, 497–522. [CrossRef] [PubMed]
159. Zhao, R.; Cao, X.; Li, X.; Li, T.; Zhang, H.; Cui, X.; Cui, Z. Ecological toxicity of Cd, Pb, Zn, Hg and regulation mechanism in *Solanum nigrum* L. *Chemosphere* **2023**, *313*, 137447. [CrossRef]
160. Marrugo-Negrete, J.; Durango-Hernández, J.; Pinedo-Hernández, J.; Olivero-Verbel, J.; Díez, S. Phytoremediation of mercury-contaminated soils by *Jatropha curcas. Chemosphere* **2015**, *127*, 58–63. [CrossRef]
161. Marrugo-Negrete, J.; Durango-Hernández, J.; Pinedo-Hernández, J.; Enamorado-Montes, G.; Díez, S. Mercury uptake and effects on growth in *Jatropha curcas. Environ. Sci.* **2016**, *48*, 120–125. [CrossRef]
162. Durante-Yánez, E.V.; Martínez-Macea, M.A.; Enamorado-Montes, G.; Combatt Caballero, E.; Marrugo-Negrete, J. Phytoremediation of soils contaminated with heavy metals from gold mining activities using *Clidemia sericea* D. *Don. Plants.* **2022**, *11*, 597. [CrossRef]
163. Liu, Z.; Wang, L.A.; Xu, J.; Ding, S.; Feng, X.; Xiao, H. Effects of different concentrations of mercury on accumulation of mercury by five plant species. *Ecol. Eng.* **2017**, *106*, 273–278. [CrossRef]
164. Ding, W.; Zhang, J.; Wu, S.C.; Zhang, S.; Christie, P.; Liang, P. Responses of the grass Paspalum distichum L. to Hg stress: A proteomic study. *Ecotoxicol. Environ. Saf.* **2019**, *183*, 109549. [CrossRef] [PubMed]
165. Garcia-Ordiales, E.; Roqueñí, N.; Loredo, J. Mercury bioaccumulation by Juncus maritimus grown in a Hg contaminated salt marsh (northern Spain). *Mar. Chem.* **2020**, *226*, 103859. [CrossRef]
166. Shiyab, S.; Chen, J.; Han, F.X.; Monts, D.L.; Matta, F.B.; Gu, M.; Su, Y. Phytotoxicity of mercury in Indian mustard (*Brassica juncea* L.). *Ecotoxicol. Environ. Saf.* **2009**, *72*, 619–625. [CrossRef] [PubMed]

sustainability

MDPI

Article

How Administrative Traditions Shape Policy Experiments in European Nature-Based Solutions

Nick Kirsop-Taylor * and Duncan Russel

Department of Politics, The University of Exeter, Exeter EX4 4RJ, UK; d.j.russel@exeter.ac.uk
* Correspondence: n.a.kirsop-taylor@exeter.ac.uk

Abstract: Public administrators are empirical experimenters by nature. This paper argues that policy experiments are functions of administrative institutional settings. These administrative traditions influence bureaucrat-led policy experiments. This argument is explored through the case of nature-based solutions in the European Union; a field that is reporting increasing policy experimentation across diverse geographies and across four of the archetypal administrative traditions. Our review of this nascent literature revealed 19 sources across multiple different disciplinary sources. These revealed that the Nordic tradition is effective due to its culture of discretionary risk taking in experimentation; the Rechtsstaat tradition supports longevity in policy experiments; and the Westminster tradition facilitates broad and inclusive experimental spaces. This offers significant new contributions to the tradition and policy experimentalist literature. By drawing out the relevance of heterogeneous institutional administrative settings on experiments, it adds to the growing discourse evidencing the policy impact of administrative tradition literature and showing why nature-based solutions in the EU have significant empirical value to these policy and administration literatures.

Keywords: Napoleonic; Westminster; Nordic; Rechtsstaat; culture; governance

1. Summary

> *An idea is true if it 'works'—if it has desirable effect when tried. It places emphasis upon experience and is hence characterised by empiricism. Intelligent use of experience in testing for truth is an experiment; so 'experimental is a term frequently found in pragmatic writings.*Dwight Waldo—(1948, p. 85)

This article is based within the sociology of public administrative systems. It explores how these systems of rich and complex 'administrative traditions' [1,2] affect the policy-making process in and around European nature-based solutions (NBSs) [3]. These NBS approaches are increasingly understood as critical interventions to meet multiple normative sustainability agendas such as nature restoration, climate resilience, and pollution control [3,4]. NBS policy experiments are an increasingly key plank of European Union (EU) sustainability efforts and agendas [4]. As Sengers and colleagues' meta-analysis evidences [5], sustainability transitions necessitate policy and governance experiments for the purpose of trying new approaches, finding out what works, and situating new approaches within often crowded governance spaces. In addition, as Kooiman argues [6], the bureaucracy (e.g., public administration) remains, even in the age of new public management and plural governance, the linchpin of effective policy and governance for sustainability [4]. Indeed, these new NBS approaches are concomitantly having to manifest in congested

 https://doi.org/10.3390/su17156869

and complex governance spaces [4] that are subject to national/place-based administrative or bureaucratic institutional and sociological phenomena. This is described as the 'administrative traditions' that fundamentally shape and discipline the organic qualities of governance through bureaucracy [1,2]. Therefore, understanding why and how administrative traditions affect policy experiments is an important scholarly endeavour [2]. More specifically, how this dynamic affects sustainability transitions and European NBSs is an important and timely topic for this journal [7]. Based on and within the logic and rationale set out in the next three sections, this article reviews 19 sources from a diverse research literature and asks how archetypal administrative traditions are affecting European NBS policy experiments.

2. Introduction: Policy Experiments and Administrative Traditions

Dwight Waldo famously argued [8] that public administrators are driven primarily by a pragmatic need to find what works. This makes administrators rational empiricists by nature and pragmatic experimenters in practice. This is an argument evidenced in later works on the deep roots of empirical experimentalism in public administration literature and theory [9]. Recent decades have witnessed a proliferation of accounts of administrative experimentalism, what Hutiema et al. [10] describe as the 'experimentalist turn' that spans and bridges the fields of public policy and administration [11,12]. Recent decades have witnessed the emergence of a small but insightful intersectional experimentalist public policy and administration literature [13–15]. This literature has tended to focus on issues of learning [16,17], autonomy [18], and adaptivity [19] at the policy experimental/administrative tradition intersection. Often undertaken on decentralised scales, policy experiments offer opportunities for temporally and spatially limited trials with limited cost and risk profiles [20]. This literature argues that experimentation has been increasingly popular with policy makers in the interest of addressing complex and intractable policy problems [19], such as the provision of social welfare (see: [21]) or policy for natural resource allocations and trade-offs [16].

A challenge arises where this experimentalist literature tends towards monolithic conceptions of the institution of modern public administration(s). There are, of course, many good reasons for this—such as the homogenising forces of globalisation ([22,23], pp. 248–250), managerialism [24], and Europeanisation [25]. However, these homogenising forces fail to account for the institutionally 'sticky' [26] nature of public administrative systems that are as different as the states, peoples, polities, and cultures they represent ([27], pp. 52–66). Institutionalist and interpretivist accounts of public administrations consider them socially constructed sociopolitical constructs that vary depending on the institutional forces that historically and contemporarily shape them. This sociological notion of institutional variability in administrative forms is captured neatly in the theory and literature of administrative traditions. Whilst there have been several iterations of administrative tradition that accentuate different variables/cleavages (see: [13,28,29]), here, Painter and Peters' text *Tradition and Public Administration* ([30], pp. 3–30) is utilised. Painter and Peters [30] argue that administrative traditions are fundamentally sociological phenomena that represent and manifest heterogeneous approaches to bureaucracy that exist in different places and polities. They conceptualise these heterogeneous approaches along three key variables—(1) history, (2) culture, and (3) governance experiences. By utilising Painter and Peters' three-part conceptualisation, we can address the experimentalist policy literature's deficit in accounting for different institutionally constructed administrative settings and, conversely, the administrative tradition literature's deficit in considering how comparative traditions impact policy experiments. This paper bridges and addresses these lacunae.

This argument is contextualised and explored through dynamic and innovative literature exploring policy experiments for leveraging nature and natural processes to address the global climate crises [31,32], that is, the 'actions to protect, sustainably manage and restore natural or modified ecosystems, (to) address societal challenges (e.g., climate change) effectively and adaptively, while simultaneously providing human well-being and biodiversity benefits' ([32], p. 7). These nature-based solutions (NBSs) are particularly attractive where they tend to offer a range of additional co-benefits beyond simply their stated objectives [33,34], as well as where NBSs are perceived as lower-cost than the engineered solutions of previous decades [35].

The past ten years have seen particular growth in the literature on the potential role of NBSs in the European Union (EU), catalysed by the adoption of NBSs as a priority area for research by EU Research and Innovation in 2015 [36]. Part of the promise of NBSs in EU settings is their potential to meet international climate commitments, build urban resilience through adaptation, and prioritise a host of co-benefits [36] through a lower cost suite of solutions that have the added benefits of being more open to citizen participation [37], equity [38], and justice [39]. For example, NBS policy experiments might include public funding and planning support for retrofitting urban green roofs for multiple ecological and social co-benefits [40] or mandated agricultural set-aside margins for biodiversity, soil health, and carbon sequestration.

The EU is comprised of states and bureaucracies with archetypal administrative traditions—the 'common European legacy', as described by Painter and Peters [30]. Europe has historically been a source of significant historical administrative innovation and is comprised of an intersecting and overlapping constellation of administrative traditions [30]. This heterogeneity offers diverse and rich comparatives for exploring the relationships between policy experimentation under administrative traditions.

This paper shows how and why policy experiments are contingent upon the administrative tradition of the bureaucrats leading, participating in, and shaping the experiments. In addition, it shows that the history, culture, and governance experiences of public administrative systems, as well as their deep and sticky institutional legacies [30], play a significant role in shaping these policy processes. This speaks directly to administrative tradition literature and scholars by offering a new application of theory in policy practice. It evidences the critical importance of the institutional administrative dynamics of contemporary policy. It also speaks to policymakers considering the scaling up of NBS policy experiments and challenges them to critically consider how administration will affect how experiments translate into programmes and legislation.

3. EU Nature-Based Policy Experiments Under Four Administrative Traditions

Of the nine core families of administrative tradition that Painter and Peters [30] offer, four of these are core to what they describe as the 'Western' archetypal forms and can be used to conceptualise the principal European traditions. These are the Westminster, Napoleonic, Rechtsstaat, and Nordic archetypal traditions. Naturally, these are ideal types of administrative traditions. However, the majority of EU states can be characterised in terms of hybridised forms [41]. Similarly, when considering these four ideal European types, it is important to note their appropriative histories from the older East Asian (especially Confucian Chinese bureaucracy) and Islamic administrative traditions ([27], pp. 52–54). This paper adopts a normative perspective on the role and functioning of the four ideal types of European administrative traditions as a tool to illuminate their power in shaping NBS policy experiments.

As noted, Painter and Peters' [30] interpretivist and institutionalist conception of administrative traditions is based around the three intersecting dynamics—history, culture, and governance. Historical institutionalism offers a clear and cogent series of arguments for how institutions emerge from and are embedded in concrete temporal processes [42]. It tends to orient analysis around temporal concepts such as critical junctures, path dependencies, intercurrence, and modes of institutional change [43]. Painter and Peters [30] build on these understandings by positioning temporal influences as a determinant of traditions in public administrations. Similarly, they note how different cultures have varying affinities towards bureaucracy. For example, Waldo [8] uses culture as a tool to describe the emergence of US bureaucratic character in the 19th and early 20th centuries. Others highlight how the culture of society places expectations, establishes norms, and influences the frames of bureaucrats within traditions [44]. Painter and Peters ([30], p.5) describe the role of culture in shaping tradition:

> *"societies that stress egalitarian or individualistic values are much less likely to find the formality of a Weberian bureaucracy an acceptable mechanism for organization, whether in the public or the private sector. Likewise, a more hierarchical society, and its administrative system, may find adjusting to many of the changes in contemporary management that stress participation and involvement of lower-echelon workers and clients to be inconsistent with their ideas of good management."*

Finally, Painter and Peters' concept of administrative traditions [30] is rooted in the patterns of governance; that is, administrative traditions are mere components of wider state traditions, and the administrative tradition is, in fact, a function of formal and informal interactions between various state and non-state actors.

These three connected perspectives of administrative tradition help elucidate how administrative traditions are institutionally constructed, replicated, and maintained. Although this paper is oriented towards interpretivist conceptualisations of administrative tradition, as outlined above, for comparative purposes, it distils specific definitions of four archetypal European administrative traditions to avoid what Russel and Serban ([45], p. 3) describe as a problematic mixing of interpretivist and positivist perspectives of administrative tradition. The questions presented in this paper are rooted in an interpretivist framing of administrative traditions and explored comparatively through somewhat stricter definitions of what comprises each tradition; see Table 1 below.

Table 1. The four 'European' administrative traditions.

	Westminster	Napoleonic	Rechtsstaat	Nordic
Legal basis (for the state)	No	Yes	Yes	Yes
State and society	Pluralist	Interventionist	Organicist	Organicist; welfare; open government
Organisation of government	Limited government; weak local self-government	Hierarchical; centralised	Integrated; cooperative federalism	Decentralised administration
Civil service	Quite high status; unified; increasing politicisation of appointments; permanent; generalist	Very high status; permanent; specialised elite training; segmented	Very high status; permanent; legal training; openly partisan; permanent (at upper ranks)	High status; professional; non-politicised

Table 1. *Cont.*

	Westminster	Napoleonic	Rechtsstaat	Nordic
History	Imperial	Imperial; Jacobin	Prussian; state building; federalist	Mixed
Culture	Plural; politics over bureaucracy	Liberty; high trust; high legitimacy	Hierarchical; statist; legalist	Plural
Governance	Anglo-Saxon common law; pluralist state–civil society relations	Roman civil law; unitary state	Civil law; prescribed	

Derived from Painter and Peters [30] with author additions.

In Table 1, each of the four archetypal traditions is characterised first in terms of the legal basis for the state, the relationship between state and society, the organisation of government, and their form of civil service. Secondly, Table 1 compares each tradition in terms of Painter and Peters' core variables of history, culture, and governance.

The Westminster administrative tradition, from a European perspective, is the dominant tradition in the United Kingdom (UK) and the Republic of Ireland. The idea of the Westminster tradition is contested, and as noted by Rhodes et al. [46], it is perhaps best articulated by a series of common but blurred intersecting beliefs about constitutional arrangements, executive power, and party systems. Within a normative view of the Westminster tradition, the civil service is professional, high status, and dominated by permanent bureaucratic generalists. Furthermore, there is only a limited constitutional relationship between the state and the citizen. It tends towards limited government overall but relatively centralised administrative and policymaking functions. Accountability is held through political departmental Ministers [47], and power is dispersed throughout apolitical delivery agencies. The Westminster administrative tradition is a function of English history as far back as the English Civil War, the court of Henry VIII, the fourteenth century, or even further ([27], pp. 246–261). However, Weller and Haddon [48] and many others define the Northcote–Trevelyan report (1853) as the inception point of the Westminster administrative tradition. The development of the Westminster administrative tradition was also influenced by the Indian Civil Service and bureaucracy's role as administrators of the British Empire [49]. Certainly, this tradition underwent a significant period of change following the Thatcher Administration's New Public Management (NPM) civil service reforms of the 1980s. However, even these were not enough to substantively change the core institutional nature of the Westminster tradition, and as Cooper [50] suggests, it lives on through the institutional memories of permanent secretaries. The Westminster tradition can also be understood through the weak conception of the central state held culturally in the society of the UK. The culture instead emphasises governance as a shared endeavour between a limited state, markets, and civil society partners [51]. Arguably, since the rise of NPM, there has been a political–cultural aversion to an unelected elite bureaucratic cadre. Under NPM, there has been a shift in the pattern of governance towards a more explicit focus on public service provision [29].

The Napoleonic tradition is the dominant administrative tradition in France and has enjoyed wide global proliferation due to French imperial and colonial history, including in Spain, Portugal, and Greece [41] (e.g., hybrids). In France, it operates within well-articulated and codified state–citizen relationships and views its legal system as 'an instrument of the state for intervening in society, rather than serving as a means of conflict resolution between different societal actors' ([25], p. 65). It is an étatist tradition that ascribes to an organic state-citizen relationship with all manner of relations and interactions described and accounted for through its system of civil law, derived from the Justinian system by Emperor Napoleon I. This civil code offers a detailed prescription for the function and role of the bureaucracy in civic life, checked by a robust system of judicial review ([30], p. 21).

Fukuyama ([52], pp. 337–354) argues that the chronic inefficiencies and injustices of the pre-revolutionary preceding ancien réegime bureaucracy necessitated the creation of the 'new' administrative tradition based on orientalist sentiments for the values of bureaucratic meritocracy, professionalism, and state-building. The Jacobin state has endowed France with a unitary, inclusive, and étatist political culture that emphasises the role of an active and interventionist bureaucracy in the lives and decisions of civil society, though less so in the everyday lives of citizens [41]. Its historical Jacobin cultural orientation towards professional meritocracy has influenced the emergence of an executive administrative class, selected and trained at elite educational institutions (grandes écoles). The clear codification of political and administrative roles in the Napoleonic code facilitates movement between the administration and politics, and therefore, there is a degree of politicisation in its administrative tradition.

The Rechtsstaat tradition is the predominant bureaucratic tradition in Austria, Switzerland, the Netherlands, and, most notably, Germany. This tradition is based on a statist and legalist conception of state–citizen relationships. Though the functioning of the state apparatus is highly prescribed through a civil law code, much like the Napoleonic tradition, it normalises and legitimises a heavy presence of the state in the everyday lives of citizens (unlike France) [30]. Fukuyama ([27], pp. 77–80) details the historical roots of the Rechtsstaat administrative tradition in Prussian state building in the eighteenth century. For example, Gorski [53] notes that when the Hohenzollern family converted to Calvinism (mid-1600s), it put them in conflict with their Lutheran nobility, which precipitated three major effects on the Prussian bureaucracy ([27], p. 69). First, to counteract the influence of their nobility, the Hohenzollerns staffed their new Prussian central bureaucracy with Dutch and Huguenot coreligionists, which imbued the bureaucracy with a culture of autonomy from, and superiority over, society at large. Second, it gave them cultural alignment towards puritanical values of personal austerity, thrift, and a hatred of corruption. It also precipitated the creation of a host of new Calvinist social institutions: schools, poor relief houses, parish record keeping, etc. These were, in time, institutionalised into the Prussian (and then German) state, laying the foundations for the social-administrative reforms of Chancellor Bismarck. This history and culture have gifted the Rechtsstaat bureaucracy an elite and hierarchical nature based within a values-based and ethical framework. As noted by Knudsen, it comprised:

> *"a highly bureaucratic, rule-bound and hierarchic structure designed to promote objective decision making on the basis of appropriate laws or regulations in obedience to the current government, whatever its political complexion"*([54], p. 37)

Rechtsstaat bureaucrats are said to enjoy a monopoly on expertise and are the prime intercession between state and citizens [30]. Governance under the Rechtsstaat tradition is both hierarchical and elite but also displays elements of plurality and collaboration derived from, as Painter and Peters [30] argue, its federalist polity, in which federal states enjoy significant power. This has a tendency to lead to horizontal administrative replication of departments and functions, though this is offset by the Calvinist values. For example, under the German Rechtsstaat system, many non-state corporations, organisations, and governance collaborations are given protected legal status (and state funding in places), leading to an institutionalisation of collaborative governance forms and offering the potential for checking, balancing, and regulating elite administrative interests.

The Nordic administrative tradition nominally accounts for administrative systems in Denmark, Sweden, Norway, and Finland (and hybrids in Iceland, the Faroes, and the Baltic states). Whilst there are many commonalities that bind these state traditions together into this Nordic family [30], there are also many structural differences that differentiate them [55]. The Nordic administrative traditions combine an étatist and organicist inheritance with a strong focus on social welfarism (with similarities to the Rechtsstaat tradition) and a strong state tradition for consensual democracy through local government ([54], p. 96). The administrative traditions in these states share a strong sense of Rawlsian social egalitarianism [56] and institutional rights linked to citizenship [51]. The strong state foundation of this tradition adopts an interventionist and 'directing' administrative–state positionality in welfare and industry ([57], p. 9), with the aim of maximising employment opportunities and redistributive welfare policy in partnership with strong trade unions and civil society engagement in administration and governance. It has been argued that historical preferences for communitarianism and local democracies paved the way for the contemporary political and administrative Nordic culture ([54], p. 12). Greve et al.'s work [55,56] has found that the administrative culture across the Nordic states has a distinctly pragmatic perspective on governance and that the administrative culture generally encourages reform through layering and a general culture for administrative innovation, and as noted by Knudsen [54], relatively low levels of bureaucratic corruption. These studies concluded that the Nordic administrative tradition is inherently agile and adaptive, which in part contributes to the overall value of the 'Nordic model' and makes the Nordic tradition particularly well disposed to adaptive governance [58,59]. There are, however, also notable differences between the different Nordic states, which challenge the notion of a unified administrative tradition [30]. For example, the Swedish administrative system as set out in Ansell et al. [60] and Pierre [61], and arguably the 'East Nordic (Finnish) tradition' [62] and the Danish systems [63].

4. Expectations from Theory

Based on the characterisation of each of the four principal European administrative traditions, we can see that each is constructed from the intersection of their history, culture, and governance. The next task involves exploring what extant theory can reveal about how each tradition might influence NBS policy experiments.

The Westminster tradition has a diminutive view of the potential of the state and the capacity, competence, or function of public administrators in intervening in the innovation and operationalisation of policy experiments [30]. Certainly, within this tradition, there is an apparent culture of openness and inclusivity in governance for innovation and experimentation. However, simultaneously, it is relatively politically centralised, with limited decentralised power for risk-taking. The historic preferences are for governance through minimal state intervention. Instead, it seeks to crowd in voluntary and private sector partners and remove state actors as quickly as possible. This perception of legitimacy is based on private sector actors being subject to the laws of economics, which, under the economically orientated Westminster tradition, is culturally considered a higher-order driver of innovation than public administrators. Its focus (especially in recent decades) on 'service delivery' over 'administrative structure and hierarchy' suggests that it might offer increased agility across departmental silos for operationalising interdisciplinary NBS policy experiments. Similarly, theorising about the Westminster tradition [47] suggests that ministerial accountability should embolden administrators to take risks and innovate, which, when coupled with the service delivery ethos, could be beneficial for NBS policy experiments. However, this still raises the question of who starts and brings the early support (e.g., funding and partnership building) required for the operationalisation of policy experiments—it might be argued that Westminster public quangos with an interest in NBS might play this role. Congruent with the significant agentification of UK regulatory governance [3], this role might conceivably be filled by public agencies facilitating pre-packaged partnerships led by larger voluntary organisations. Critical to this, though, is that budgetary control under the Westminster system is problematic, as the state's centralising tendencies, a function of national political culture and the history of the 'United' Kingdom, restrict local government discretionary funding to support new policy experiments, such as NBS policy experiments.

NBS policy experiments undertaken under the Napoleonic administrative tradition are likely to benefit from its culture of openness and inclusivity in the practices of governance. The state and bureaucrats at all levels are likely to adopt a strong leadership positionality in framing, shaping, and supporting policy experiments, yet nevertheless remain open to citizens and civil society actors both procedurally (in law) and substantively (in culture and governance). French bureaucrats, we might expect, enjoy substantial discretionary powers within the confines of the civil code to support and promote policy experiments. Moreover, their 'elite' position in French civic culture means that public administrators should be the integral components of NBS policy partnerships. The Napoleonic administrative–civic culture of interventionism should mean that public administrators are often integral to decision-making but in balance with the ideas and interests of citizens. Similarly, the relative budgetary autonomy of Napoleonic administrative regions should provide the scope for discretionary funding support, so long as partnerships for NBS experiments effectively engage public administrators. However, 'engagement' with the public sphere is a near certainty under the heavily prescribed and relatively centralised Napoleonic tradition. That said, this prescribed nature of governance tends towards administrative siloing, which has the potential to stymie effective experiments or hinder their translation into policy. Conceivably, the heavy siloing in the French system could also hinder effective funding and interdisciplinary partnership building, though these challenges of working across silos might be circumvented by advocacy or activist individuals moving between local/politics and administration and vice versa.

NBS policy experiments under the Rechtsstaat tradition are likely to be relatively prescribed and closed. That is not to suggest that they are elite; rather, the compositions will be more prescribed by the governance culture of prescribed and legitimate non-state actors. These should still fulfil the needs for inclusivity where the range of prescribed partners is inclusive of major stakeholder groups, but there is less scope for openness. Elite public administrators will likely play a central role in bringing together and supporting NBS experiment partnerships and exercise significant bureaucratic discretion (at more senior grades) in ensuring their longevity, crossing political–electoral timescales (a problem in the Westminster and, to a lesser extent, the Napoleonic tradition). Discretionary behaviours should be freer compared to the Napoleonic tradition; whilst both have civil law systems that mandate state–citizen relations, the Rechtsstaat tradition is relatively less prescriptive. The politics of NBS should be less pronounced under the Rechtsstaat tradition, where senior public administrators are more accountable to law and professional standards than politicians, which should make administrators more effective institutional champions for NBS policy experiments. However, the Rechtsstaat tradition is famous for its history and culture of strict hierarchies and departmental segregations, which could hamper adaptivity and agility in seeking to operationalise interdisciplinary NBS experiments. In addition, whilst the Rechtsstaat tradition does not enjoy as much of the advantage of activist administrator–politicians as the Napoleonic system, it does retain the potential for these to act as boundary actors in the service of working across administrative silos.

Despite the relative heterogeneity between the composite parts of the 'Nordic tradition', we can derive broad-brush expectations about how NBS experiments might fare under this tradition. The communitarian history and culture of the Nordic countries [57] suggest that NBS partnerships ought to be structurally open and inclusive. Nominally, under this tradition, it would be expected that the somewhat interventionist state and its administrators would play an active role in NBS experiments. However, the recent view of Bergholm and Bieler [64] cannot be ignored; as a tradition, Nordic administrations are increasingly mirroring the Westminster tradition with regard to positionality and the interventionist role of the state. Therefore, it might be expected that the state plays a less central role in shaping, framing, and supporting NBS policy experiments than imagined. Whilst there is evidence of departmental siloing in the Nordic tradition [65], these are not a constitutional or legal artefact of Nordic states, and so there are likely better cross-departmental policy experiments than in either the Napoleonic or Rechtsstaat traditions. The relatively high levels of public trust in the state and administration [58], coupled with higher levels of bureaucratic discretion based on professionalism and expertise, means that administrators have the potential to play key roles in the framing and delivery of NBS experiments. The relative budgetary autonomy at scales enjoyed under the Nordic tradition (ibid.), coupled with high levels of discretion, means that there is significant scope for administrators supporting experimentation for NBSs.

5. Method and Findings

This paper reviewed the emergent NBS literature to better understand how archetypal administrative traditions are affecting European NBS policy experiments across the EU. The emergent nature of this research literature across multiple disciplines suggests that a more ad-hoc snowballing literature review was likely to yield meaningful insights to address the core research conundrum/question articulated in the paper. The search engines Google Scholar and ScienceDirect were initially utilised, with search words related to 'Europe/EU, 'nature-based solutions', 'policy', 'administration', 'bureaucracy', and some others. All returned sources had their abstracts read to determine whether they were in or out of scope. Potential sources were subjectively deemed in scope where they were saying something substantive at the intersection of NBS and taking a qualitative-interpretivist perspective on the bureaucracy/public administration of policy experiments. This led to 19 sources being discerned, as per Appendix A. This comprised 9 sources speaking strictly to NBS policy experiments under the Rechtsstaat tradition, 3 for the Nordic tradition, 2 for the Napoleonic, and 2 for the Westminster tradition.

These 19 sources were analysed considering the research question, and the comparative dynamics of different traditions were highlighted. It is important to note that these sources only offered empirical insights into the experimental phase of EU NBS policy experiments and made no reference to how different traditions might offer different advantages when experimental-phase policy turns into national-scale policy (as is the EU agenda). Furthermore, only some of these sources [33,39,66–73] offered insights about strict, if oblique, iterations of four administrative traditions (e.g., cases from Germany, Sweden, and the UK), with the remainder offering insights from within hybrid traditions (Spain, Netherlands, and Belgium) [74].

From these 19 sources, we can explore and, to a degree, verify the expectations (given in Section 4) about the influence of administrative traditions on NBS policy experiments. The major sources that we can draw upon to validate expectations about NBS policy experiments under the Westminster tradition are from works by Frantzeskaki and colleagues [66,70]. These drew upon observations of experiments taking place in the cities of London, Brighton [66], and Glasgow [70]. In their attempt to create a framework for assessing and implementing the co-benefits of NBSs in the EU, Raymond et al. [33] offered some insights into the development of urban NBS experiments in London (UK). The early stages of establishing the ideas and partnership for developing the NBS policy experiments featured a strong local government presence, and the initial stages of partnership building were open and plural. Public partners appeared keen to draw in local small- to medium-scale enterprises (private sector), though there was a quick transition after setting up from substantive public involvement to the state (represented by the Greater London Authority) withdrawing from the experiment completely. Frantzeskaki et al. [70] reported on the case of urban NBSs in the Scottish city of Glasgow (UK). Arguably, the Scottish administrative tradition varies slightly from that of the UK, but not so much as to classify it as a hybrid variant. Similar to the London case, they reported an initial setup spurred by local government, coupled with a keenness and openness to engaging private sector partners early in the process of experimenting with NBSs. However, the process of successfully scaling up the experiment was reportedly hampered by structural administrative siloing, exacerbated by a lack of competencies and capacities in local government, which brought about ten years of political austerity. Frantzeskaki et al. [70] further reported a differential between a symbolic culture of openness and plurality, undermined by a structural governance path dependency for centralised government power. This meant that 'substantive partnership building' was difficult. Congruent with Painter and Peters [30] and others, these cases evidenced a strong locus of central government power coupled with low levels of confidence in the competency of public administrators [75]. The low level of administrative interventionalism was borne out where the public administrators were key to the inception of new NBS experiments but quickly exited partnerships and relinquished control to major private and voluntary organisations. Whilst there was an evident appetite for local-scale administrators to innovate and take on risks, they were constrained by their recent history of public austerity, which has hollowed out the UK public sector at all levels and reduced its ability to engage in governance activities [76].

Suleiman et al. [40] reported on the case of Barcelona, which operates administratively under a hybrid variant of the Napoleonic tradition. The three cases they report on were all initiated by authoritative public administrators, though subsequently funded through a range of commercial, citizen-based, and public means. Whilst other authors report on NBS policy experiments in France, none of these offered insights into how the administrative–institutional setting affected the experiments.

There is a relatively higher number of case-based reports on the administrative dynamics of NBS experiments in Germany [40,66,67,69,77,78] and the hybrid Rechtsstaat administrative states of Belgium [66,70] and the Netherlands [33,68]. In their reporting on urban NBS experiments in Dresden, Frantzeskaki et al. [70] discussed how German cultural norms for openness and inclusivity in governance and the structural 'governance spaces' that are made available for bottom-up civil society innovation were key to the success of their NBS experiments. Frantzeskaki et al. [66] commented on the German administrative culture for bottom-up initiatives striving to attract the attention and patronage of authoritative and influential public administrators, who held power over funding and support for longevity and prioritisation. Duskova and Haas [69], reporting the progress of NBS policy experiments in the city of Leipzig, found that many of the projects were initiated by either city administrators or civil society actors and groups and that such projects were funded by different public, institutional, and NGO sources. In exploring administrative funding mechanisms for NBS experiments across Germany, Droste et al. [64] discerned that funding mechanisms were well prescribed, with functions delegated from higher (elite) administrative actors to lower scales. This meant low levels of fiscal autonomy to support NBS in German municipalities, and NBS experiments often had to innovate around seeking funding and support for new NBS experiments or partnering with elite and influential public actors [79].

In the case of the Nordic tradition, the NBS policy experiment literature predominantly reports on Swedish cases in Stockholm [40,66] and Malmö [39]. This is problematic due to noted differences between different states within the Nordic administrative tradition and between 'Sweden and the others' [58] in particular. Thus, these cases offer a partial reflection about the Swedish administrative sub-tradition that we can use to, in part, draw conclusions about the wider Nordic tradition. Suleiman et al. [40] reported the experiment with urban rainwater run-off on Hornsgatan street in Stockholm. This experiment was led by the city council, which exercised significant discretionary powers in taking risks with new approaches to NBSs. This aligns with Ansell et al. [60] and Capobianco et al. [80], who speak to the Swedish administrative culture for discretionary risk-taking by bureaucrats. Frantzeskaki et al. [66] reported on how the administrative and governance cultures in Stockholm played an important facilitative role in the development of NBS experiments. Similar to Suleiman et al. [40], they reported that many projects begin with the strong support (financial and organisational) or that new community-originated projects exist within an institutional governance setting that encourages their engagement with public actors. Sweden is described as a relatively open and plural experimental space for NBS and one where decisions about the role of state engagement are decided through discourse and negotiation.

6. Discussion

The studies reviewed in this analysis revealed that in many (but not all) cases, policy experimentation for NBSs was contingent upon, and proceeded according to, the prevailing administrative institutional settings of place, polity, and culture. Overall, Painter and Peters' [30] conceptualisation of administrative traditions, comprised of history, culture, and governance, offered a broadly effective analytical scope to capture the comparative nuances between the underlying institutional drivers of bureaucratic–administrative behaviour in these cases. Other conceptualisations within this field, such as Knill's 'instrumental' vs. 'autonomous' cleavage [13] or Page and Wright's 'public authority' vs. 'service provision' [29] would not, on reflection, have offered the scope to account for historical institutionalist forces that shape tradition ([27], pp. 52–66) or accounted for institutionalist pressures of previous bureaucratic decision-making in governance to create path dependencies on contemporary behaviour and decision-making in policy experiments. Thus, Painter and Peters' [30] interpretivist and institutionalist conceptualisation of administrative traditions has clear analytical value in accounting for administrative behaviours in the exercise of policy experiments. However, the value of this approach likely has declining usefulness in hybrid administrative tradition settings, in which it is harder to specifically triangulate upon the key dynamics of the institutional administrative dynamics, as established in Table 1. For example, how possible was it to attribute bureaucratic decision-making around NBS policy experiments in the city of Rotterdam [65] to a function of its hybrid Rechtsstaat tradition, using an idealised form of the Rechtsstaat tradition (as per Table 1)? Trying to draw conclusions about the influences of hybrid administrative institutional drivers on policy processes appears, in this case, more tenuous.

The EU NBS literature was found to have variable usefulness in drawing out multiple cases of NBS policy experimentation under the four European archetypal administrative traditions. There was a heterogeneous spread of cases, with nine occurring under the Rechtsstaat tradition (or hybrid variants thereof), and only two in hybrid forms of the Napoleonic tradition [78]. Therefore, whilst this policy experimental literature brought value to this paper in terms of its dynamic and contemporary sources, the disparity in the spread of sources was problematic. This meant that the confirmation of administrative tradition theory about the Napoleonic tradition through NBS cases was lacking. This is not to suggest that other policy experimental literature might not have cases that could be used, only that the NBS literature offered little in this case. However, for those cases that were revealed through the review, there were some connections between the theory-derived expectations about the effects of administrative tradition on policy experiments and the case study of NBS policy experiments. Overall, therefore, it can be considered that the selection of the NBS literature had value in evidencing the relationship between administrative tradition and policy experiments. However, there is clearly scope for additional research to validate these findings, and especially relating to how the Napoleonic tradition influences policy experiments (and indeed other policy processes).

The sources reviewed evidenced how the Nordic administrative tradition shaped and facilitated smaller-scale (e.g., municipality-scale) policy experimentation. This was due to its culture of consensual democracy [54,61] and risk-taking by administrators, along with its history of an open and inclusive approach to governance. Many of the reported NBS policy experiments displayed the underlying administrative institutional forces of what Greve et al. [58,59] describe as a 'pragmatic perspective on governance' [39]. These were coupled to what Greve describes as the high levels of bureaucratic autonomy [55] within a wider administrative culture framed by Nordic high-trust polities, as well as adaptiveness in policy experimentation [60]. There were similarities in terms of openness of governance in the service of policy experimentation in accounts of NBS policy experiments in Rechtsstaat administrative settings. What Painter and Peters describe as the 'monopoly of expertise' enjoyed by public servants in acting between the state (and in particular state funding streams) and bottom-up policy experiments was borne out in the cases of Droste et al. [67], Frantzeskaki et al. [66], and Duskova and Haas [69]. However, this literature also highlighted how in Rechtsstaat administrative settings, there was a high (and competitive) threshold for institutional administrative patronage for policy experiments. However, once patronage was bestowed, policy experiments had significant institutional support longitudinally. The cases offered tantalising evidence of how institutional dynamics within the Westminster tradition influence NBS policy experiments. In these cases, such as those reported in Frantzeskaki et al. [66], in which in London Docklands, the NBS policy experiment was initiated by a partnership that included public actors, but these quickly extricated themselves to crowd in private-sector partners; this is congruent with Grube and Howard's [51] perspective on the Westminster culture for governance. This case, and that of Glasgow reported in Raymond et al. [33], was congruent with the Westminster history and culture for non-interventionalism.

7. Conclusions

Waldo argued that bureaucrats are, in their fundamental decision orientations, pragmatic empiricists focused on 'what works' ([5], pp. 83–85) and so are naturally inclined towards experimentation. However, bureaucracies are not homogenous, and correspondingly, this paper set itself the task of making the case for how and why administrative traditions affect bureaucratic policy experimentation through an exploration of European NBSs. Our core findings and conclusions, therefore, must be that what we might expect from NBS policy experimentation under each of the archetypal traditions was broadly borne out in the various research sources analysed. This goes some way to answering the research question by showing us how NBS policy experiments in different European locales were being affected or operationalised in formats congruent with the administrative traditions in which they were occurring. Indeed, the rapidly evolving European NBS experimental policy environment proved an insightful body of research for this purpose because of the multiple accounts of NBS policy experimentation taking place under the four archetypal administrative traditions, rich with their own cultures, histories, and governing modalities. However, critically, the particular dynamics of place-based NBS policy experimentation by bureaucrats vary according to their institutional–administrative tradition [14]. This paper offers an initial perspective on this phenomenon in response to the research question.

There are limitations to this study. There were markedly fewer contributions in the NBS policy experimentation literature pertaining to the influences beyond the Rechtsstaat tradition on bureaucratic policy experimentation. Most EU-facing literature spoke about Germany (as a country) as opposed to the suite of states that operate under/within what might be called the Nordic tradition, though the relative paucity of Westminster tradition-facing sources (UK and Ireland) and the strict Napoleonic tradition (France) was harder to account for. Moreover, whilst this initial review of the literature offers a sense that the expectations from administrative traditions theory are being borne out in policy experimental practice, there is still a great deal further we could go in terms of validating what these results suggest. The limits of the research literature precluded (at the time of writing) a more systematic review of the literature, which could also be an important future step as and when the literature continues to grow with more and richer accounts. The relatively immature and emergent body of thought and theory that comprises 'administrative traditions' as a sociocultural lens offering thick accounts of place-based bureaucracies was also a limitation. The archetypes used here are crude, and any effort to assign positivist or deterministic interpretations should be avoided.

This paper's primary contributions are to the individual administrative traditions literature [2,22,30,45], the experimentalist policy sciences literature [9–11], and the NBS policy and governance literature [3,4,33,36,37,39,40,74,75,78–81]. By presenting an argument about how normative conceptualisations of administrative traditions influence contemporary and highly salient policy agendas [14,63], such as NBS policy experiments, it highlights for the administrative tradition literature and scholarly community the impact of applying their theoretical insights to contemporary policy challenges. Similarly, it also speaks to policymakers and practitioners [3] considering the scaling-up of policy experiments [4] and challenges them to critically consider how sticky institutional forces that scaffold public administrations influence the delivery and translation of policy experiments into programmes and legislation. For the policy sciences scholarly community and literature, it evidences how and why experimentation is contingent upon sticky institutional administrative forces. The key message for policymakers and urban planners is that the sociocultural phenomena of administrative traditions need to be accounted for in the construction of future policy experiments for NBS, if not designed into them. We should also conclude that because bureaucratic institutions are place-based and socioculturally constituted, they will continue to exert influence on and shape all stages of the NBS policy-cycle across the EU from experiments to scaling-up.

We might also conclude that the sociocultural phenomena of administrative traditions affect all stages of the national policy cycle writ large. In addition, as such, this has wider implications for national policy and governance efforts towards meeting the United Nations 2030 Sustainable Development agenda. There is an opportunity for public administration and policy scholars to engage with the implications of this conclusion through explorations in other policy spheres or administrative settings. For example, it might rightly be asked how sustainability-facing policy experimentation is influenced by the Confucian administrative tradition in China and its hybrid forms (e.g., South Korea or Vietnam), or what different administrative traditions in comparative settings might tell us about the effectiveness, impact, or role of policy experiments on national public policy.

Author Contributions: We acknowledge that this paper was substantively conceived, undertaken, and written by N.K.-T., with conceptual and proofreading support from D.R. All authors have read and agreed to the published version of the manuscript.

Funding: This research received no external funding.

Data Availability Statement: The academic sources that comprised the data for this study (given in Appendix A) are all publicly available through standard academic search engines.

Conflicts of Interest: The authors declare no conflict of interest.

Appendix A. Sources for Literature Review

Authors	Archetypal Tradition	Year of Publication
Zingraff-Hamed et al. [71]	Reechstaat	2021
Scolobig et al. [74]	Reechstaat	2023
Tafel et al. [72]	Reechstaat	2022
Albert et al. [78]	Reechstaat	2021
Brillinger et al. [79]	Reechstaat	2020
Droste et al. [67]	Reechstaat	2017
Dunn et al. [68]	Reechstaat	2017
Raymond et al. [33]	Reechstaat	2017
Duskova & Hass [69]	Reechstaat	2020
Duffaut et al. [81]	Napoleonic	2022
Lepesant [82]	Napoleonic	2024
Olsson et al. [39]	Nordic	2020
van Post et al. [73]	Nordic	2023
Capobianco et al. [80]	Nordic	2024
Kirsop-Taylor & Russel [3]	Westminster	2020
Dick et al. [75]	Westminster	2020
Suleiman et al. [40]	Reechstaat, Nordic & Napoleonic	2020
Raymond et al. [33]	Reechstaat, Nordic & Westminster	2017
Frantzeskaki et al. [70]	Reechstaat & Westminster	2020

References

1. Peters, G. *The Politics of Bureaucracy*, 1st ed.; Routledge: London, UK, 2010.
2. Peters, G. *Administrative Traditions: Understanding the Roots of Contemporary Administrative Behaviour*; Oxford University Press: Oxford, UK, 2018.
3. Kirsop-Taylor, N.; Russell, D. Agencies navigating the political at the science-to-policy interface for nature-based solutions. *Environ. Sci. Policy* **2020**, *127*, 303–310. [CrossRef]
4. Petersen, C.J.; Russel, D.; Jensen, A.; Pedersen, A.B.; Kirsop-Taylor, N. Evaluation of innovative policy for nature-based solutions: Analysis from three European city case studies. *Discov. Cities* **2025**, *2*, 48. [CrossRef]
5. Sengers, F.; Wieczorek, A.J.; Raven, R. Experimenting for sustainability transitions: A systematic literature review. *Technol. Forecast. Soc. Change* **2019**, *145*, 153–164. [CrossRef]
6. Kooiman, J. *Governing and Governance*, 1st ed.; Sage: London, UK, 2003.
7. Dorst, H.; van der Jagt, A.; Toxopeus, H.; Tazer, L.; Raven, R.; Runhaar, H. What's behind the barriers? Uncovering structural conditions working against urban nature-based solutions. *Landsc. Urban Plan.* **2022**, *220*, 104335. [CrossRef]
8. Waldo, D. *The Administrative State: A Study of Political Theory of American Public Administration*, 1st ed.; Ronald Press Company: New York, NY, USA, 1949.
9. Bauwman, R.; Grimmelikhuijsen, S. Experimental public administration from 1992 to 2014: A systematic literature review and ways forward. *Int. J. Public Sect. Manag.* **2016**, *29*, 10–31. [CrossRef]
10. Huitema, D.; Jordan, A.; Munaretto, S.; Hilden, M. Policy experimentation: Core concepts, political dynamics, governance and impacts. *Policy Sci.* **2018**, *51*, 143–159. [CrossRef]

11. Bozeman, B. Experimental Design in Public Policy and Management Research: Introduction. *J. Public Adm. Res. Theory* **1992**, *2*, 289–291. [CrossRef]
12. Ewert, B.; Kathrin, L.; Thomann, E. Beyond nudge: Advancing the state-of-the-art of behavioural public policy and administration. *Policy Politics* **2020**, *49*, 3–23. [CrossRef]
13. Knill, C. Implementing European Policies: The Impact of National Administrative Traditions. *J. Public Policy* **1998**, *18*, 1–28. [CrossRef]
14. Capano, G. Administrative Traditions and Policy Change: When Policy Paradigms Matter. The Case of Italian Administrative Reform During the 1990s. *Public Adm.* **2003**, *81*, 781–804. [CrossRef]
15. Raudla, R.; Douglas, J.W.; Windett, J.H. Comparative Analysis of Financial Bureaucracies of Europe: The Influence of Policy Sector vs Administrative Tradition on the Attitudes of Finance Officials. *J. Comp. Policy Anal. Res. Pract.* **2020**, *23*, 393–415. [CrossRef]
16. McFadgen, B.; Huitema, D. Are all experiments created equal? A framework for analysis of the learning potential of policy experiments in environmental governance. *J. Environ. Plan. Manag.* **2016**, *60*, 1765–1784. [CrossRef]
17. McFadgen, B.; Huitema, D. Stimulating Learning through Policy Experimentation: A Multi-Case Analysis of How Design Influences Policy Learning Outcomes in Experiments for Climate Adaptation. *Water* **2017**, *9*, 648. [CrossRef]
18. Bandiera, O.; Prat, A.; Valetti, T. Active and Passive Waste in Government Spending: Evidence from a Policy Experiment. *Am. Econ. Rev.* **2009**, *99*, 1278–1308. [CrossRef]
19. Haasnoot, M.; Kwakkel, J.H.; Walker, W.E.; ter Maat, J. Dynamic adaptive policy pathways: A method for crafting robust decisions for a deeply uncertain world. *Glob. Environ. Change* **2013**, *23*, 485–498. [CrossRef]
20. Mei, C.; Wang, X. Wire-Walking: Risk Management and Policy Experiments in China from a Comparative Perspective. *J. Comp. Policy Anal. Res. Pract.* **2020**, *22*, 360–382. [CrossRef]
21. Ettelt, S.; Mays, N.; Allen, P. Policy experiments: Investigating effectiveness or confirming direction. *Evaluation* **2015**, *21*, 292–307. [CrossRef]
22. Chandler, J.A. *Comparative Public Administration*; Routledge: London, UK, 2003.
23. Heichel, S.; Pape, J.; Sommerer, T. Is there convergence in convergence research? An overview of empirical studies on policy convergence. *J. Eur. Public Policy* **2005**, *12*, 817–840. [CrossRef]
24. Rouillard, C.; Giroux, D. Public Administration and the Managerialist Fervour for Values and Ethics: Of Collective Confusion in Control Societies. *Adm. Theory Prax.* **2005**, *27*, 330–357. [CrossRef]
25. Knill, C. *The Europeanisation of National Administrations: Patterns of Institutional Change and Persistence*; Cambridge University Press: Cambridge, UK, 2001.
26. Boettke, P.; Coyne, C.; Leeson, P. Institutional stickiness and the new development economics. *Am. J. Econ. Sociol.* **2008**, *67*, 331–358. [CrossRef]
27. Fukuyama, F. *Political Order and Political Decay: From the Industrial Revolution to the Present Day*; Profile Books: London, UK, 2014.
28. Halligan, J. The Fate of Administrative Tradition in Anglophone Countries during the Reform Era. In *Tradition and Public Administration*; Painter, M., Peters, G., Eds.; Palgrave Macmillan: London, UK, 2010.
29. Page, E.; Wright, V. *Bureaucratic Elites in Western European States: A Comparative Analysis of Top Officials*; Oxford University Press: Oxford, UK, 1999.
30. Painter, M.; Peters, G. *Tradition and Public Administration*; Palgrave Macmillan: London, UK, 2010.
31. Cohen-Shacham, E.; Walters, G.; Janzen, C.; Maginnis, S. (Eds.) *Nature-Based Solutions to Address Global Societal Challenges*; IUCN Press: Gland, Switzerland, 2016.
32. Seddon, N.; Chausson, A.; Berry, P.; Girardin, C.; Smith, A.; Turner, B. Understanding the value and limits of nature-based solutions to climate change and other global challenges. *Philos. Trans. R. Soc. B* **2020**, *375*, 20190120. [CrossRef]
33. Raymond, C.M.; Frantzeskaki, N.; Kabisch, N.; Berry, P.; Breil, M.; Nita, M.R.; Geneletti, D.; Calfapietra, C. A framework for assessing and implementing the co-benefits of nature-based solutions in urban areas. *Environ. Sci. Policy* **2017**, *77*, 15–24. [CrossRef]
34. Giordano, R.; Pluchinotta, I.; Pagano, A.; Scrieciu, A.; Nanu, F. Enhancing nature-based solutions acceptance through stakeholders' engagement in co-benefits identification and trade-offs analysis. *Sci. Total Environ.* **2020**, *713*, 136552. [CrossRef] [PubMed]
35. Keesstra, S.; Nunes, J.; Novara, A.; Finger, D.; Avelar, D.; Kalantari, Z.; Cerdà, A. The superior effect of nature based solutions in land management for enhancing ecosystem services. *Sci. Total Environ.* **2018**, *610–611*, 997–1009. [CrossRef]
36. Faivre, N.; Fritz, M.; Freitas, T.; de Boissezon, B.; Vandewoestijne, S. Nature-Based Solutions in the EU: Innovating with nature to address social, economic and environmental challenges. *Environ. Res.* **2017**, *159*, 509–518. [CrossRef] [PubMed]
37. Pagano, A.; Pluchinotta, I.; Pengal, P.; Cokan, B.; Giordano, R. Engaging stakeholders in the assessment of NBS effectiveness in flood risk reduction: A participatory System Dynamics Model for benefits and co-benefits evaluation. *Sci. Total Environ.* **2019**, *690*, 543–555. [CrossRef]

38. La Rosa, D.; Pappalardo, V. Planning for spatial equity—A performance-based approach for sustainable urban drainage systems. *Sustain. Cities Soc.* **2020**, *53*, 101885. [CrossRef]
39. Olsson, J.A.; Brunner, J.; Nordin, A.; Hanson, H.I. A just urban ecosystem service governance at the neighbourhood level-perspectives from Sofielund, Malmö, Sweden. *Environ. Sci. Policy* **2020**, *112*, 305–313. [CrossRef]
40. Suleiman, L.; Olofsson, B.; Sauri, D.; Palau-Rof, L. A breakthrough in urban rain-harvesting schemes through planning for urban greening: Case studies from Stockholm and Barcelona. *Urban For. Urban Green.* **2020**, *51*, 126678. [CrossRef]
41. Ongaro, E. The Napoleonic Administrative Tradition and Public Management Reform in France, Greece, Italy, Portugal and Spain. In *Tradition and Public Administration*; Painter, M., Peters, G., Eds.; Palgrave Macmillan: London, UK, 2010.
42. Thelen, K. Historical institutionalism in comparative politics. *Annu. Rev. Political Sci.* **1991**, *2*, 369–404. [CrossRef]
43. Fiortos, O.; Falleti, T.G.; Sheingate, A. (Eds.) Historical Institutionalism in Political Science. In *The Oxford Handbook of Historic Institutionalism*; Oxford University Press: Oxford, UK, 2016.
44. Hofstede, G.; Neuijn, B.; Ohayv, D.; Sanders, G. Measuring Organisational Culture: A Qualitative and Quantitative Study Across Twenty Cases. *Adm. Sci. Q.* **1990**, *35*, 286–316. [CrossRef]
45. Russell, M.; Serban, R. The Muddle of the 'Westminster Model': A Concept Stretched Beyond Repair. *Gov. Oppos.* **2020**, *56*, 1–21. [CrossRef]
46. Rhodes, R.A.W.; Wanna, J.; Weller, P. *Comparing Westminster*; Oxford University Press: Oxford, UK, 2009.
47. Hood, C. Public Administration and Public Policy: Intellectual Challenges for the 1990s. *Aust. J. Public Adm.* **1989**, *48*, 346–358. [CrossRef]
48. Weller, P.; Haddon, C. Westminster Traditions: Continuity and Change. *Governance* **2016**, *29*, 483–498. [CrossRef]
49. Grube, D.C. Back to the future: Rediscovering the lost arts of the Victorian mandarin. *Parliam. Aff.* **2016**, *69*, 708–728. [CrossRef]
50. Cooper, C.A. Politics and the permanency of permanent secretaries: Testing the vitality of the Westminster administrative tradition, 1949–2014. *Br. Politics* **2019**, *15*, 311–325. [CrossRef]
51. Grube, D.C.; Howard, C. Is the Westminster System Broken Beyond Repair? *Governance* **2016**, *29*, 467–481. [CrossRef]
52. Fukuyama, F. *The Origins of Political Order: From Pre-Human Times to the French Revolution*; Profile Books: London, UK, 2011.
53. Gorski, P.S. *The Disciplinary Revolution: Calvinism and the Rise of the State in Early Modern Europe*; University of Chicago Press: Chicago, IL, USA, 2003.
54. Knudsen, T. State Building in Scandinavia: Denmark in a Nordic Context. In *Welfare Administration in Denmark*; Knudsen, T., Ed.; Ministry of Finance: Copenhagen, Denmark, 1991.
55. Lægreid, P. Nordic administrative traditions. In *Handbook of Scandinavian Politics*, 1st ed.; Nedergaard, P., Wivel, A., Eds.; Routledge: London, UK, 2017.
56. Fllesdal, A. Rawls in the Nordic Countries. *European Journal of Political Theory.* **2002**, *1*, 2, 181–198. [CrossRef]
57. Pollitt, C. Politics, Administration and Performance: A continuing search, but no one best way? In *Statskontoret Den Effektiva Staten: En Antologi Fran Statskontoret*; Statskontoret: Stockholm, Sweden, 2012; pp. 117–160.
58. Greve, C.; Lægreid, P.; Rykkja, L.H. Introduction: The Nordic Model in Transition. In *Nordic Administrative Reforms: Lessons for Public Management*; Greve, C., Lægreid, P., Rykkja, L.H., Eds.; Palgrave MacMillan: London, UK, 2016.
59. Greve, C.; Ejersbo, N.; Lægreid, P.; Rykkja, L.H. Unpacking Nordic Administrative Reforms: Agile and Adaptive Governments. *Int. J. Public Adm.* **2020**, *43*, 697–710. [CrossRef]
60. Ansell, C.; Trondal, J.; Øgård, M. (Eds.) *Governance in Turbulent Times*; Oxford University Press: Oxford, UK, 2017.
61. Pierre, J. Administrative Reform in Sweden: The Resilience of Administrative Tradition? In *Tradition and Public Administration*; Painter, M., Peters, G., Eds.; Palgrave Macmillan: London, UK, 2010.
62. Johansen, J.E.; Johnsen, A.; Pekkola, E.; Affleck Reid, S. Strategic Management in Finnish and Norwegian Government Agencies. *Adm. Sci.* **2019**, *9*, 80. [CrossRef]
63. Rhodes, R.A.W. Traditions and Public Sector Reform: Comparing Britain and Denmark. *Scand. Political Stud.* **1999**, *22*, 341–370. [CrossRef]
64. Bergholm, T.; Bieler, A. Globalization and the erosion of the Nordic model: A Swedish–Finnish comparison. *Eur. J. Ind. Relat.* **2013**, *19*, 55–70. [CrossRef]
65. Leiren, M.D.; Jacobsen, J.K.S. Silos as barriers to public sector climate adaptation and preparedness: Insights from road closures in Norway. *Local Gov. Stud.* **2018**, *44*, 492–511. [CrossRef]
66. Frantzeeskai, N.; Borgström, S.; Gorissen, L.; Egermann, M.; Ehner, F. Nature-Based Solutions Accelerating Urban Sustainability Transitions in Cities: Lessons from Dresden, Genk and Stockholm Cities. In *Nature Based Solutions to Climate Change Adaptation in Urban Areas*; Kabisch, N., Korn, H., Stadler, J., Bonn, A., Eds.; Springer International Publishing: Cham, Switzerland, 2017. Available online: https://link.springer.com/book/10.1007/978-3-319-56091-5 (accessed on 1 March 2024).
67. Droste, N.; Schröter-Schlaack, C.; Hansjürgens, B.; Zimmermann, H. Implementing Nature-Based Solutions in Urban Areas: Financing and Governance Aspects. In *Nature Based Solutions to Climate Change Adaptation in Urban Areas*; Kabisch, N., Korn, H.,

Stadler, J., Bonn, A., Eds.; Springer International Publishing: Cham, Switzerland, 2017. Available online: https://link.springer.com/book/10.1007/978-3-319-56091-5 (accessed on 1 March 2024).
68. Dunn, G.; Brown, R.R.; Bos, J.J.; Bakker, K. The role of science-policy interface in sustainable urban water transitions: Lessons from Rotterdam. *Environ. Sci. Policy* **2017**, *73*, 71–79. [CrossRef]
69. Duskova, D.; Haas, D. Not Simply Green: Nature-Based Solutions as a Concept and Practical Approach for Sustainability Studies and Planning Agendas in Cities. *Land* **2020**, *9*, 19. [CrossRef]
70. Frantzeskaki, N.; Vandergert, P.; Connop, S.; Schipper, K.; Zwierzchowska, I.; Collier, M.; Lodder, M. Examining the policy needs for implementing nature-based solutions in cities: Findings from city-wide transdisciplinary experiences in Glasgow (UK), Genk (Belgium) and Poznań (Poland). *Landuse Policy* **2020**, *96*, 104688. [CrossRef]
71. Zingraff-Hamed, A.; Hüesker, F.; Albert, C.; Brillinger, M.; Huang, J.; Lupp, G.; Scheuer, S.; Schlätel, M.; Schröter, B. Governance models for nature-based solutions: Seventeen cases from Germany. *Ambio* **2021**, *50*, 610–1627. [CrossRef] [PubMed]
72. Tafel, L.; Ott, E.; Brillinger, M.; Schulze, C.; Schröter, B. Attitudes of administrative decision-makers towards nature-based solutions for flood risk management in Germany. *Sustain. Sci.* **2022**, *17*, 135–139. [CrossRef]
73. van Post, M.; Knaggård, A.; Alkan Olsson, J.; Olsson, O.; Sofie Persson, A.; Ekroos, J. The Swedish green infrastructure policy as a policy assemblage: What does it do for biodiversity conservation? *People Nat.* **2023**, *5*, 839–851. [CrossRef]
74. Scolobig, A.; Linnerooth-Bayer, J.; Pelling, M.; Martin, J.G.C.; Deubelli, T.M.; Liu, W.; Oen, A. Transformative adaptation through nature-based solutions: A comparative case study analysis in China, Italy, and Germany. *Reg. Environ. Change* **2023**, *23*, 69. [CrossRef]
75. Dick, J.; Carruthers-Jones, J.; Carver, S.; Dobel, A.J.; Miller, J.D. How are nature-based solutions contributing to priority societal challenges surrounding human well-being in the United Kingdom: A systematic map. *Environ. Evid.* **2020**, *9*, 25. [CrossRef]
76. Kirsop-Taylor, N.; Russel, D.; Winter, M. The contours of state retreat from collaborative environmental governance under austerity. In: Governance for sustainable development in troubled times. *Sustainability* **2020**, *12*, 2761. [CrossRef]
77. International Union for Conservation of Nature. Nature Based Solutions to Address Climate Change. 2016. Available online: https://uicn.fr/wp-content/uploads/2016/09/Plaquette-Solutions-EN-07.2016.web1_.pdf (accessed on 1 March 2024).
78. Albert, C.; Brillinger, M.; Guerrero, P.; Gottwald, S.; Henze, J.; Schmidt, S.; Ott, E.; Schroeder, B. Planning nature-based solutions: Principles, steps, and insights. *Ambio* **2021**, *50*, 1446–1461. [CrossRef]
79. Brillinger, M.; Dehnhardt, A.; Schwarze, R.; Albert, C. Exploring the uptake of nature-based measures in flood risk management: Evidence from German federal states. *Environ. Sci. Policy* **2020**, *110*, 14–23. [CrossRef]
80. Capobianco, V.; Palau, R.M.; Solheim, A.; Gisnås, K.; Gilbert, G.; Danielsson, P.; van der Keur, P. The potential use of nature-based solutions as natural hazard mitigation measure for linear infrastructure in the Nordic Countries. *Geoenvironmental Disasters* **2024**, *11*, 1–33. [CrossRef]
81. Duffaut, C.; Frascaria-Lacoste, N.; Versini, P.A. Barriers and Levers for the Implantation of Sustainable Nature-Based Solutions in Cities: Insights from France. *Sustainability* **2022**, *14*, 9975. [CrossRef]
82. Lepesant, G. Between hard protection measures, nature-based solutions, and managed retreat: Adapting coastal areas to sea level rise in the Netherlands and France. *Z. Vgl. Polit. Wiss.* **2024**, *18*, 141–156. [CrossRef]

sustainability

MDPI

Article

A Selection Model of Compositions and Proportions of Additive Lime Mortars for Restoration of Ancient Chinese Buildings Based on TOPSIS

Xiaolu Long [1,*], Lizhi Liu [2] and Qi Liu [1]

[1] School of Architecture and Design, Hunan University of Science and Technology, Xiangtan 411201, China; qi.liu_hnust@hotmail.com
[2] Xiangtan Economic and Technological Development Zone Management Committee, Xiangtan 411201, China; lizhi.liu_hnust@hotmail.com
* Correspondence: 1100129@hnust.edu.cn

Abstract: To improve the accuracy of choosing restoration materials for repairing ancient Chinese buildings and to mitigate the risk of decision-making, this paper establishes a novel selection model of compositions and proportions of additive lime mortars for the restoration of ancient Chinese buildings. The selection process is influenced by multi-criteria and determined by a group of experts through comprehensive judgment. Thus, it is a multi-criteria group decision-making (MCGDM) problem. Firstly, considering subjective and objective criteria simultaneously, establish a selection index system for compositions and proportions of additive lime mortars in the restoration of ancient Chinese buildings. Secondly, applying a neutrosophic set to characterize experts' evaluation information and quantify the evaluation information. Thirdly, the best–worst method (BWM) is implemented to obtain criteria weights, and the entropy weight method is utilized to obtain index weights. Finally, obtaining the priority of each alternative solution by using the TOPSIS (Technique for Order Preference by Similarity to Ideal Solution) ranking technique. The practicality of the proposed model was demonstrated through a specific case of the selection of repair materials for a decorative window in one ancient Chinese building. The comparative analysis was carried out to verify the reliability and validity of the model.

Keywords: restoration of ancient buildings; additive lime mortars; multi-attribute group decision making; entropy weight; TOPSIS method

1. Introduction

Lime usage has a long history in China [1]. The advantages of lime mortar added with glutinous rice slurry are high strength, reasonable toughness, and sound anti-seepage performance. This method is widely used in the construction of ancient tombs, water conservancy projects, buildings, bridges, and so on. However, since the early 19th century, the invention and widespread use of cement have greatly reduced the demand for lime. In the field of cultural heritage protection, cement has side effects such as excessive strength, low porosity, incompatibility with ancient architectural materials, and the introduction of soluble salts during use, which was gradually pointed out by cultural heritage protection experts. Given the outstanding advantages of traditional lime mortar technology in durability, strength, and harmony with ancient buildings, using traditional lime mortars instead of modern cement materials to restore ancient Chinese buildings is of great significance to ensure the sustainability of ancient building protection and restoration.

The selection of repair materials is a crucial step in the restoration process of ancient Chinese buildings [2,3]. Therefore, it is necessary to develop an appropriate method to select the most suitable compositions and proportions of additive lime mortars based on the actual situation of ancient Chinese buildings restoration. The selection of compositions and proportions of additive lime mortars is influenced by factors such as the basic situation of the repaired components, the performance of the additive lime mortars, and the possible repair results. It is usually determined by a group of experts through comprehensive judgment. Thus, it is a multi-criteria group decision-making (MCGDM) problem.

Although some progress has been made in existing research, there are still some issues that require further consideration. Firstly, the selection of materials for the restoration of ancient buildings is complex, and many factors need to be considered. However, the evaluation information tends to be partial, and the standardization research on the selection process of repair materials is insufficient [4]. Secondly, in practice, experts usually select compositions and proportions of additive lime mortars based on experience, and the comprehensiveness of information judgment is limited by the expert's personal experience [5]. In this case, there exists relatively little research on using quantitative methods to determine the decision-making process of compositions and proportions of additive lime mortars for the restoration of ancient Chinese buildings.

To sum up, the research methods of this study are as follows:

- Regarding the complexity of evaluating information, the evaluation of compositions and proportions of additive lime mortars includes both subjective and objective criteria. Some studies have used the objective criteria to quantitatively evaluate the restoration of ancient Chinese buildings [6,7]. Subjective criteria, such as possible repair outcomes, have also been mentioned in Lourenço, P.B.'s research [8]. However, in the index system of the proposed model, subjective criteria are combined with objective ones.
- To address the issue of partial use of evaluation information, it is appropriate to use a neutrosophic set to describe evaluation information. Smarandache, F. [9] introduced neutrosophy, whose main argument is that every concept contains a certain degree of truth, as well as a certain degree of falsehood and uncertainty, all of which need to be considered independently of each other [10]. Zhang et al. [11,12] believed that neutrosophic sets can effectively evaluate fuzzy information. Thus, it is necessary to convert the evaluation results into neutrosophic numbers. For example, if the aesthetics of the evaluation repair result are "high", then the corresponding value of the membership degree of truth is 0.8, the membership degree of falsehood is 0.2, and the membership degree of uncertainty is 0.15.
- In order to prioritize the compositions and proportions of additive lime mortars based on MCGDM, the TOPSIS method could be introduced. According to relevant scholars' research, TOPSIS, as a practical technique for ranking and selecting many externally determined alternative solutions through distance measurement, is related to multi-criteria group decision-making (MCGDM) [13,14].

The expected results of this study might be (1) establishing an index system for the selection of compositions and proportions of additive lime mortars to repair ancient Chinese buildings; (2) using multi-granularity language terms to represent experts' information, convert them into single-valued neutrosophic sets, and quantify such information to improve the comprehensiveness and reliability of the evaluation results; (3) combining BWM method and entropy weight method to obtain the criteria and index weights; (4) utilizing the TOPSIS method to determine the ranking of available solutions; and (5) demonstrating the process of the proposed model through empirical research by analyzing a particular case.

Based on the above, this paper aims to propose a new model for selecting compositions and proportions of additive lime mortars in the restoration of ancient Chinese buildings, which can improve the accuracy and standardization of material selection in the restoration process of ancient Chinese buildings, and protect the heritage of ancient Chinese buildings from a decision-making perspective. The content of this paper is organized as follows. A brief review of previous research is provided in Section 2. Section 3 introduces some basic concepts and definitions of neutrosophic sets. In Section 4, compositions and proportions of additive lime mortars selection model for the restoration of ancient Chinese buildings based on MCGDM are constructed. Subsequently, a case study is presented in Section 5, aiming to verify the proposed model through comparative analysis. In the end, Section 6 briefly summarizes the content of this paper and proposes some practical directions for further research.

2. Literature Review

2.1. Research on Lime Mortars

In recent years, many scholars have studied the effects of different additives on the performance of traditional lime mortars and explored the application of lime mortars with different additives in the restoration of ancient buildings [15–17]. These achievements have promoted the development of heritage conservation practices and provided more choices for materials used in the restoration of ancient Chinese buildings. Research on lime mortars has been continuously developed. For example, the comparative analysis of the performance evaluation of additive lime mortars provides an experimental basis for the selection of restoration materials for ancient Chinese buildings. Usually, mass spectrometry thermogravimetric analysis, X-ray powder diffraction (XRD), and scanning electron microscopy (SEM) are used to test the mechanical properties and microstructure of additive lime mortars [18,19]. Maria Apostolopoulou et al. [20] used artificial neural networks (ANN) to draw a development chart of mortar characteristics, which can help reveal the impact of parameter settings on each mortar property. B.A. Silva et al. [21] evaluated that different maintenance conditions will have an impact on the carbonation process of lime-based mortars and slurry. Maria Apostolopoulou et al. [22] proved that different proportions of binder and aggregate can significantly alter the mechanical and physical properties of mortars. The most important direction of research on lime mortars is the changes in the properties of lime mortars with different additives and their comparative analysis, especially the improvement of lime mortar properties by organic additives. Hee-Young Hwang et al. [23] conducted a comparative analysis of the performance improvement of lime mortars using three traditional natural organic additives commonly used in East Asia: starch, glutinous rice, and seaweed slurry. Yang, F. et al. [1] conducted a systematic study on the technology of glutinous rice lime mortar to help determine appropriate action plans for restoring ancient buildings. Other lime mortar additives, such as paraffin [24], biological additives [25,26], pozzolan [27], etc., have also been tested for their effects on various properties of lime mortars. Despite the continuous deepening of research on lime mortar materials, the research results have not yet been reasonably applied in practical engineering. In actual projects, experts need to consider not only the performance of lime mortar materials but also the basic situation of the ancient buildings that need to be restored and the possible restoration results. The principles of authenticity, recoverability, and minimal impact should be followed when selecting restoration materials for ancient Chinese buildings. If inappropriate compositions and proportions of additive lime mortars are used, it will lead to the failure of the restoration, even further damaging the ancient buildings.

2.2. Research on Related Algorithm Models

At present, relevant factors that affect the selection of repair materials for traditional building heritages have been considered. For example, some studies explore the current situation of ancient buildings that need to be restored. The existing theory suggests that the best way to protect and maintain ancient buildings is to combine active and passive maintenance strategies [28]. Gao et al. [29] introduced a fuzzy element theory based on asymmetric proximity improvement for efficient evaluation of the health of timber structures. The performance of compositions and proportions of lime mortars is also a factor that affects the selection of repair materials, mainly the improvement of durability, crack resistance, freeze–thaw resistance, corrosion resistance, adhesion, etc. [16]. There have also been studies analyzing the results of the restoration of ancient buildings. Lourenço [8] believes that the continuation of the original structural and material features of ancient buildings is of great significance for the value of architectural restoration and the authenticity of the restoration results. In the research on factors mentioned above, there are subjective judgments about the influencing factors of materials used in the restoration of ancient buildings, as well as objective judgments based on experiments. However, there is a lack of consideration for both subjective and objective standards, making it impossible to comprehensively evaluate the selection of repair material types. However, in practical cases, the factors that affect the selection of compositions and proportions of additive lime mortars are both subjective and objective [30]. Therefore, when constructing the index system, it is necessary to comprehensively consider the subjective and objective criteria of the basic conditions of the components, the performance of the lime mortars, and the repair effect.

Usually, we use subjective weighting and objective weighting methods to determine weights. Rezaei [31] proposed the best–worst method (BWM) for calculating subjective weights to solve multi-criteria decision-making problems. Compared to other methods, such as AHP, that determine subjective weights, the BWM method excels in having fewer data comparisons and higher consistency. However, if only subjective weights are used, the objectivity of the evaluation results cannot be guaranteed, which will bring difficulty and the possibility for decision-makers to make mistakes in their analysis. Therefore, the entropy weight method [32,33] will be used to calculate objective weights. Entropy value helps determine the degree of dispersion of the index by displaying their randomness and disorder. Hence, in order to make the results of index weights more reliable, it is necessary to combine the BWM method with the entropy weight method to calculate the weights of the index, not only the intrinsic characteristics of the index data are considered, but also the experience of experts is taken into account [34].

The neutrosophic set has a certain degree of similarity with human thinking, as it reflects the incompleteness of knowledge, the possibility of acquiring incorrect knowledge, or the uncertainty caused by random guessing [9]. Neutrosophic sets have also been recommended in a medical diagnosis system, as shown in the study of Abdel et al. [35]. In the decision-making process of selecting compositions and proportions of additive lime mortars for ancient building heritage restoration, due to the complexity of actual restoration projects and the multifaceted performance of additive lime mortars, expert evaluation information is often difficult to be comprehensive, and evaluation information often represents experiential preferences, making it difficult to express evaluation information in an accurate data format [36]. Hence, introducing neutrosophic numbers to describe the evaluation results is a suitable strategy.

In summary, it can be seen that fuzzy group decision theory and quantitative analysis have not yet been applied to solve the problem of materials selection in the restoration of ancient Chinese buildings. Therefore, this paper aims to design a model to assist experts in selecting compositions and proportions of additive lime mortars in the restoration process of ancient Chinese architecture. The relevancy between the main literature and the algorithm of the proposed model is shown in Table 1.

Table 1. The relevancy between the main literature and algorithm of the proposed model.

Literature Source	Research Findings	Relevancy
Naciri, K.; Jayasingh, S. et al. [15–27]	Introduce the research progress of additive lime mortars, mainly including performance testing, composition and proportion comparison, application directions, etc.	This is the research background of this study, providing experts with multiple possible alternative solutions.
Lourenço, P.B.; Gao, Z.; Jayasingh, S. et al. [8,16,29,30]	Research on the factors affecting the selection of repair materials for ancient buildings restoration mainly includes the current state of the building, material properties, and possible restoration outcomes.	Providing reference for the establishment of an index system that combines subjective and objective criteria in this study.
Rezaei, J. [31]	Proposed the best–worst method (BWM) for calculating subjective weights to solve multi-criteria decision-making problems.	Applying the BMW method to calculate criteria weights.

Table 1. *Cont.*

Literature Source	Research Findings	Relevancy
Majumdar, P. et al. [32,33]	Suggest using entropy weight method to calculate objective weights.	Utilizing the entropy weight method to calculate index weights.
Smarandache, F.; Abdel-Basset, M. [9,35]	Introduced neutrosophy, whose elementary sentence is that every concept contains a certain degree of truth, a certain degree of falsehood and uncertainty, all of which need to be considered independently of each other.	Using neutrosophic numbers to represent expert evaluation information.
Afshar, A.; Shih, H.S. et al. [13,14]	TOPSIS, as a practical technique for ranking and selecting many externally determined alternative solutions through distance measurement, is related to multi-criteria group decision-making (MCGDM).	Ranking alternative solutions using TOPSIS method based on MCGDM.

3. Preliminaries

In this section, some concepts and definitions that contribute to the development of compositions and proportions of additive lime mortar selection models are introduced.

Definition 1. *Neutrosophic set [37]. Let X be a space of points (objects), with a generic element in X denoted by x. Then, a neutrosophic set A in X is characterized by three membership functions, including a truth membership function* T_A*, an indeterminacy membership function* I_A*, and a falsity membership* F_A *and is defined as* $A = \{<x, T_A(x), I_A(x), F_A(x)> | x \in X\}$*, where* $T_A(x)$*,* $I_A(x)$ *and* $F_A(x)$ *are real standard or nonstandard subsets of* $]0^-, 1^+[$*, that is,* $T_A(x){:}X \to]0^-, 1^+[$*,* $I_A(x){:}X \to]0^-, 1^+[$*,* $F_A(x){:}X \to]0^-, 1^+[$*, and satisfying* $0^- \leq T_A(x) + I_A(x) + F_A(x) \leq 3^+$*.*

Definition 2. *Single-valued neutrosophic set [38]. Let X be a space of point (object), with a generic element in X denoted by x. A single-valued neutrosophic set (SVNS) A in X is characterized by truth membership function* T_A*, indeterminacy membership function* I_A*, and falsity membership function* F_A *with* $T_A, I_A, F_A \in [0, 1]$ *for all x in X. The sum of three memberships of an SVNS A, for all* $x \in X$*,* $0 \leq T_A(x) + I_A(x) + F_A(x) \leq 3$*.*

Definition 3 ([39]). *Let* $A = \{(x_1 | <T_A(x_1), I_A(x_1), F_A(x_1)>), \ldots, (x_n | <T_A(x_n), I_A(x_n), F_A(x_n)>)\}$*, and* $B = \{(x_1 | <T_B(x_1), I_B(x_1), F_B(x_1)>), \ldots, (x_n | <T_B(x_n), I_B(x_n), F_B(x_n)>)\}$ *be two SVNSs for* $x_i \in X$ $(i = 1, 2, \ldots, n)$*. Then, the normalized Euclidean distance between A and B can be defined as follows:*

$$D(A,B) = \sqrt{\frac{1}{3n}\sum_{i=1}^{n}\left\{(T_A(x_i) - T_B(x_i))^2 + (I_A(x_i) - I_B(x_i))^2 + (F_A(x_i) - F_B(x_i))^2\right\}} \quad (1)$$

Definition 4 ([32]). *Through the research of Majumdar et al., for single-valued neutrosophic set* $A = \{<x, T_A(x), I_A(x), F_A(x)> | x \in X\}$*, an entropy on neutrosophic set A can be calculated using the following equation:*

$$E(A) = 1 - \frac{1}{n}\sum_{x_i}(T_A(x_i) + F_A(x_i)) \otimes |I_A(x_i) - I_{A_C}(x_i)| \quad (2)$$

Definition 5 ([30]). *According to the study by Tan et al., the entropy weight of a neutrosophic set is computed as follows:*

$$W_j = (1 - E(x_j)) / \sum_j^n (1 - E(x_j)) \quad (3)$$

Definition 6. *In the study by Biswas et al. [40], fuzzification of SVNS* $\widetilde{N} = \{(x \mid <T_{\widetilde{N}}(x), I_{\widetilde{N}}(x), F_{\widetilde{N}}(x)> \mid x \in X\}$ *can be shown as follows:*

$$\mu_{\widetilde{F}}(x) = 1 - \sqrt{\frac{1}{3}\left\{(1 - T_{\widetilde{N}}(x))^2 + I_{\widetilde{N}}(x)^2 + F_{\widetilde{N}}(x)^2\right\}} \quad (4)$$

Definition 7 ([41])**.** *According to Ye's research, the single-valued neutrosophic weighted averaging (SVNWA) aggregation operator is defined as follows:*

$$F_{A_i} = \psi_1 A_1 \oplus \psi_2 A_2 \oplus \ldots \oplus \psi_n A_n = \langle 1 - \prod_{i=1}^{n} (1 - T_{A_i})^{\psi_i}, \prod_{k=1}^{n} (I_{A_i})^{\psi_i}, \prod_{k=1}^{n} (F_{A_i})^{\psi_i} \rangle \quad (5)$$

In which $\Psi = (\Psi_1, \Psi_2, \ldots, \Psi_n)$ is the weight vector of A_i $(i = 1, 2, \ldots, n)$, $\Psi_i \in [0, 1]$, and satisfying $\sum_{i=1}^{n} \psi_i = 1$.

4. Methodology

In the case of ancient Chinese buildings restoration, the selection of compositions and proportions of additive lime mortars should be determined by a panel of experts and is the result of collective decision-making. In the decision-making process, there are many factors to consider, and many indices cannot be evaluated with accurate values. Therefore, a single-valued neutrosophic set is adopted to express the characteristics of each index. Figure 1 shows the detailed process of the proposed model. The rest of this section will explain the detailed information of the model. The process of the proposed model and its corresponding algorithm and principle of TOPSIS method are shown in Figure 2.

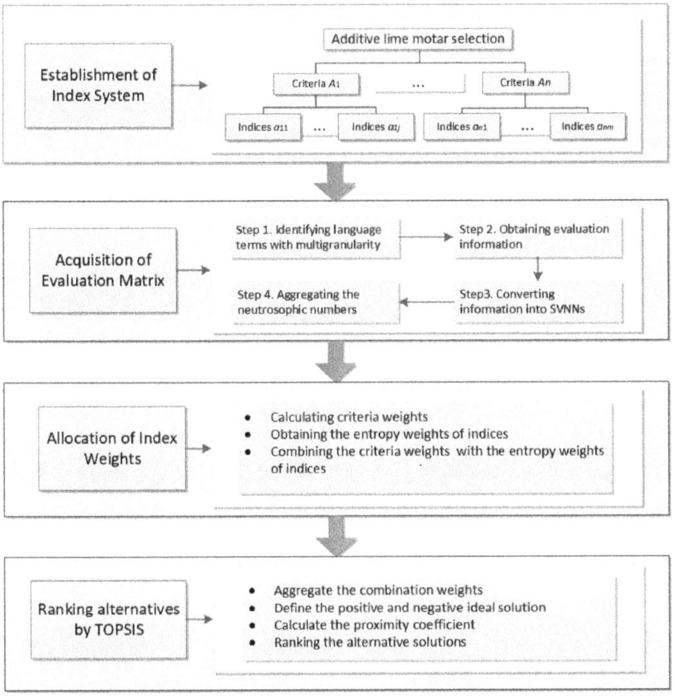

Figure 1. The framework of the proposed model.

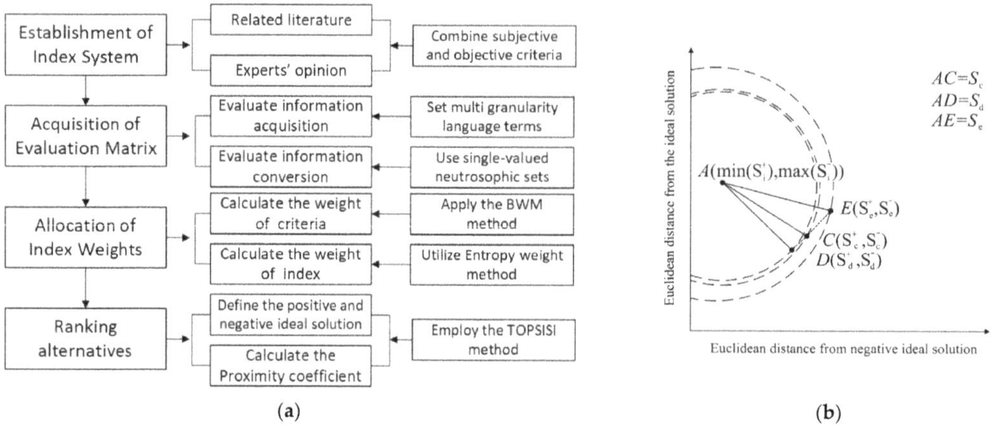

Figure 2. The schematic diagram of the proposed model algorithm: (**a**) the process of the proposed model and its corresponding algorithms; and (**b**) schematic diagram of the principle of TOPSIS method.

4.1. The Establishment of Index System

The selection of compositions and proportions of additive lime mortars in the restoration of ancient Chinese buildings is generally determined by a panel of k experts. According to the literature analysis, experimental results, and expert opinions, the determination of compositions and proportions of additive lime mortars can be described by three categories, represented by three criteria A_i ($i = 1, 2, 3$): basic condition of component (A_1), performance of different compositions and proportions of additive lime mortars (A_2), and restoration outcomes (A_3). In addition, there are several subcriteria under each criterion that affect the selection of lime mortar materials. Thus, we establish an index system, as shown in Table 2.

Table 2. Index system of compositions and proportions of additive lime mortars selection for ancient Chinese buildings restoration.

Criteria	Index	Definition	Index Type
Basic condition of component (A_1)	Contact surface material (a_{11})	The degree to which the contact surface material is suitable for the mortar material	Benefit
	Contact surface morphology (a_{12})	The degree to which the contact surface morphology is suitable for the mortar material	Benefit
	Component structure (a_{13})	The degree to which the component structure is suitable for the mortar material	Benefit
	Degree of damage (a_{14})	The degree of damage to the component is suitable for the mortar material	Benefit
Performance of different compositions and proportions of additive lime mortars (A_2)	Hardness (a_{21})	The hardness of the mortar material is suitable for repairing the component	Benefit
	Compressive strength (a_{22})	The compressive strength of the mortar material is suitable for repairing component to a certain extent	Benefit
	Freeze–thaw resistance (a_{23})	The degree to which the freeze–thaw resistance of the mortar material is suitable for repairing component	Benefit
	Shrinkage (a_{24})	The degree to which the shrinkage of the mortar material is suitable for repairing component	Benefit

Table 2. *Cont.*

Criteria	Index	Definition	Index Type
Restoration outcomes (A_3)	The degree of preservation of historical information (a_{31})	The degree of preservation of historical information by the repaired component using the mortar material	Benefit
	The possibility of irreversible impact (a_{32})	The possibility of irreversible effects caused by the repair of component using the mortar material	Cost
	Durability (a_{33})	The durability of the repaired component with the mortar material	Benefit
	Aesthetics (a_{34})	The aesthetic appearance of the repaired component with the mortar material	Benefit
	Maintenance difficulty (a_{35})	The difficulty of maintenance after repairing the component with the mortar material	Cost

4.2. The Acquisition of Evaluation Matrix

The selection of compositions and proportions of additive lime mortars to repair ancient Chinese buildings requires consideration of actual situation on site, experts, building components to be restored, and various other factors. Experts may provide different evaluation results based on their knowledge structure, experimental data, work experience, different evaluation criteria, and performance of different compositions and proportions of additive lime mortars. Due to the diversity, ambiguity, and complexity of information sources and types, the evaluation values of index are often difficult to represent accurately with numbers. To address this issue, decision-makers typically use language terms such as fuzzy numbers with multiple granularities to evaluate index. Therefore, we convert the evaluation results acquired from experts into single-valued neutrosophic numbers to obtain the evaluation matrix $R = (r_{ij})$. The detailed process in this section can be shown as follows:

Step 1. Identifying language terms with multi-granularity. Experts usually use language terms according to their own habits, which may result in different experts having different evaluation values due to semantic differences in language terms. In order to unify the evaluation values of index provided by expert groups in the form of multi-granularity language terms [42], it is necessary to preset a language term set containing ordered language terms. Language term sets can characterize different membership function features. For example, the language term set $\{n_0, n_1, n_2, n_3, n_4, n_5, n_6, n_7, n_8\}$ can be expressed as {extremely bad, very bad, bad, medium bad, medium, medium good, good, very good, extremely good}, while the language term set $\{l_0, l_1, l_2, l_3, l_4, l_5, l_6\}$ can be donated as {very low, low, medium low, medium, medium high, high, very high} in linguistic terms. Obviously, different experts may have varying interpretations of the degree to which these terms are expressed. Therefore, it is necessary for experts to provide evaluation values for different compositions and proportions of additive lime mortar selection schemes based on a predetermined language term set.

Step 2. Based on the symmetrical language evaluation scale, convert the evaluation information from expert's questionnaire into single-valued neutrosophic numbers. The language term set used in this research is {extremely high, extremely high, high, medium high, medium low, low, extremely low}, which includes nine granularity language terms, as shown in reference [43]. Table 3 defines language terms and SVNNs to evaluate each alternative based on each index.

Table 3. Language terms for rating alternative solutions using SVNNs.

Language Terms	SVNNs
Extremely high	<1.00, 0.00, 0.00>
Very high	<0.90, 0.10, 0.05>
High	<0.80, 0.20, 0.15>

Table 3. *Cont.*

Language Terms	SVNNs
Medium high	<0.65, 0.35, 0.30>
Medium	<0.50, 0.50, 0.45>
Medium low	<0.35, 0.65, 0.60>
Low	<0.20, 0.75, 0.80>
Very low	<0.10, 0.85, 0.90>
Extremely low	<0.05, 0.90, 0.95>

Step 3. Calculating experts' weights. Different experts have different backgrounds and experiences; therefore, it is necessary to determine the importance of experts in the decision-making process. In this article, the decision-making ability of experts can be represented by a set of language terms {very important (VI), important (I), medium (M), unimportant (UI), very unimportant (VUI)}. Table 4 shows language terms and SVNNs to evaluate the importance of experts. Let the weight of expert be represented as e_k (k = 1, 2, . . ., n). Based on Equation (4), we can calculate the weights of experts using Equation (6) [40]:

$$e_k = u_k / \sum_{k=1}^{l} u_k = \frac{1 - \sqrt{\left\{(1 - T^k)^2 + (I^k)^2 + (F^k)^2\right\}/3}}{\sum_{k=1}^{l}\left(1 - \sqrt{\left\{(1 - T^k)^2 + (I^k)^2 + (F^k)^2\right\}/3}\right)} \quad (6)$$

Table 4. Language terms for rating the importance of experts with SVNNs.

Language Terms	SVNNs
VI	<0.90, 0.10, 0.05>
I	<0.80, 0.20, 0.15>
M	<0.50, 0.40, 0.45>
UI	<0.35, 0.60, 0.70>
VUI	<0.10, 0.80, 0.90>

Step 4. Aggregating the neutrosophic numbers. The single-valued neutrosophic weighted average (SVNWA) aggregation operator shown in Equation (5) expressed in Section 2 is used to aggregate neutrosophic numbers. Hence, the evaluation matrix $R = (r_{ij})$ shall be obtained.

4.3. *The Allocation of Index Weights*

In this section, a combination of subjective and objective methods is utilized to determine index weights. Firstly, we use an effective BWM method to compute the criteria subjective weights. Then, the entropy weight of each index is calculated based on the principle of objectivity. Finally, according to subjective–objective method, the comprehensive weight of each index is determined. The specific steps are as follows.

4.3.1. Calculating Criteria Weights

Raize proposed BWM method to solve multi-criteria decision-making problems [44]. Defining a set of decision criteria $\{A_1, A_2, \ldots, An\}$, then, the BWM method is applied it to calculate the three criteria weights in this section [45].

Step 1. The best (e.g., the most ideal and important) and worst (e.g., the least ideal and least important) criteria are determined by decision-makers. Based on the BWM questionnaire, experts should choose a number between 1 and 9 to determine the preference of the best criteria relative to all other criteria. The resulting best-to-others vector would be $A_B = (a_{B1}, a_{B2}, \ldots, a_{Bn})$, among them, a_{Bj} indicates the preference of the best criteria B over criteria j. This can be obtained that $a_{BB} = 1$. Then, we can compute the preference of all the criteria over the worst criteria by choosing a number between 1 and 9. Then, we can obtain the resulting others-to-worst vector $A_W = (a_{1W}, a_{2W}, \ldots, a_{nW})$, in which a_{jW} represents the preference of the criteria j over the worst criteria W, satisfying $a_{WW} = 1$.

Step 2. Obtaining the optimal weights $\{w_1{}^*, w_2{}^*, \ldots, w_n{}^*\}$ by solving (7). The optimal weight for the criteria is the one where for each pair of w_B/w_j and w_j/w_W, we have $w_B/w_j = a_{Bj}$ and $w_j/w_W = a_{jW}$. In order to meet these requirements for j, we shall find a method to meet the maximum absolute differences $\left|\frac{w_B}{w_j} - a_{Bj}\right|$ and $\left|\frac{w_j}{w_w} - a_{jw}\right|$ for all j are minimized. Given that the nonnegativity and condition for the weights, the following problem is the result:

$$\begin{aligned} &\min \max_j \left\{ \left|\frac{w_B}{w_j} - a_{Bj}\right|, \left|\frac{w_j}{w_w} - a_{jw}\right| \right\} \\ &s.t. \\ &\sum_j w_j = 1 \\ &wj \geq 0,\ for\ all\ j. \end{aligned} \tag{7}$$

Equation (7) is equivalent to Equation (8):

$$\begin{aligned} &\min \xi \\ &s.t. \\ &\left|\frac{w_B}{w_j} - a_{Bj}\right| \leq \xi,\ for\ all\ j \\ &\left|\frac{w_j}{w_w} - a_{jw}\right| \leq \xi,\ for\ all\ j \\ &\sum_j w_j = 1 \\ &w_j \geq 0,\ for\ all\ j. \end{aligned} \tag{8}$$

The optimal weights $(w_1{}^*, w_2{}^*, \ldots, w_n{}^*)$ and ξ^* (consistency index) can be obtained through Equation (8). Then, we can calculate the consistency ratio. When consistency ratio is between [0, 1] and the value is close to 0, it indicates better consistency; conversely, if the value is close to 1, it means less consistency.

Hence, the weight vector $\{w_1, w_2, w_3\}$ of basic condition of component A_1, performance of different compositions and proportions of additive lime mortars A_2, restoration outcomes A_3 is obtained.

4.3.2. Obtaining Index Weights

We calculate the weights of index based on the principle of combining subjective and objective factors in this section. First of all, the entropy weight method is used to calculate each index entropy weight. Then, the criteria weights are combined with entropy weights to obtain the comprehensive weights of the index. The specific process is as follows:

Step 1. Normalizing the evaluation index. We obtain the normalization matrix $R = (\beta_{ij})$ using Equation (9), an equation for normalizing [46] the neutrosophic numbers.

$$r_{ij} = \begin{cases} T_{ij}, I_{ij}, F_{ij} & if\ j\ is\ benefit\ index \\ 1 - T_{ij}, 1 - I_{ij}, 1 - F_{ij} & if\ j\ is\ cost\ index \end{cases} \tag{9}$$

Step 2. Computing the entropy weights of indices. We can calculate the entropy value of each index $E(x_j)$ using Equations (2) and (3) mentioned in Section 2; thus, the entropy weight w_j of each index can be acquired.

Step 3. Acquiring combination weights. The weights of evaluation index of selecting compositions and proportions of additive lime mortars are crucial, as they can affect the correctness of the evaluation results. The subjective weight method mainly depends on expert experience, which often leads to arbitrary evaluation results. However, the objective weighting method can compensate for this drawback, as it is based on objective reality and can greatly reduce subjective arbitrariness. Therefore, combining subjective and objective methods is an appropriate approach. According to the explanation of BWM in Section 4.3.1, the optimal weight vector of criteria is w_i*. Then, we calculate a comprehensive index weight H_{ij} based on the following Equation (10) [47].

$$H_{ij} = \frac{w_i^* \times w_j}{\sum_{i=1}^{n} w_i^* \times w_j} \tag{10}$$

4.4. The Decision Process Based on TOPSIS

Hwang and Yoon [48] proposed TOPSIS method to deal with multicriteria decision-making problems. It obtains the best alternative by calculating the choice that is as close as possible to the optimal solution. The basic operation process of TOPSIS method is introduced as the following [49]:

Step 1. Aggregate the combination weights to the decision matrix by making $V_{ij} = H_{ij}r_{ij}$.

Step 2. Define the positive ideal solution (PIS), v_j^+, and the negative ideal solution (NIS), v_j^-, for each criterion.

$$\begin{aligned} v_j^+ &= <T_j^+, I_j^+, F_j^+> = <\max T_{ij}^{wj}, \min I_{ij}^{wj}, \min F_{ij}^{wj}>, i = 1, 2, \ldots, m, j = 1, 2, \ldots, n \\ v_j^- &= <T_j^-, I_j^-, F_j^-> = <\min T_{ij}^{wj}, \max I_{ij}^{wj}, \max F_{ij}^{wj}>, i = 1, 2, \ldots, m, j = 1, 2, \ldots, n \end{aligned} \tag{11}$$

Step 3. Calculate the distance between the alternative solution and the positive ideal solution and the distance between the alternative solution and the negative ideal solution according to Equations (12) and (13).

$$S_i^+ = \sqrt{\sum_{j=1}^{n} \left(v_j^+ - v_{ij}\right)^2}, \quad i = 1, 2, \ldots, m \tag{12}$$

$$S_i^- = \sqrt{\sum_{j=1}^{n} \left(v_j^- - v_{ij}\right)^2}, \quad i = 1, 2, \ldots, m \tag{13}$$

Step 4. Calculate the proximity coefficient between each alternative solution and the ideal solution according to Equation (14).

$$CC_i = \frac{S_i^-}{S_i^- + S_i^+} \tag{14}$$

Step 5. Rank alternative solutions based on CC_i. The bigger CC_i is, the better the alternative A_i will be.

5. Empirical Study

In this section, we apply the proposed model to solve the decision problem of repairing materials for ancient Chinese buildings. A decorative exterior window of an ancient building called "Dafudi" located in the central region of Hunan, China, needs to be restored. Dafudi is a national cultural heritage site built during the Qing Dynasty over 150 years ago. The exterior windows of Dafudi are made of cedar wood skeleton, which is decorated with mortar and depicted in different shapes, with strong decorative aesthetics and high historical, cultural, and artistic value (as shown in Figure 3). According to local residents, the decorative mortar for the exterior windows is made by mixing lime and glutinous rice. The management agency of Dafudi entrusted company R (Loudi, China) to repair it. A decorative exterior window to be repaired is donated by W_1. There are four experts, donated by DM_1, DM_2, DM_3, and DM_4, who conducted on-site inspections to prepare for obtaining the composition of lime mortar. The experts extracted exterior window decorative mortar material for XRD analysis on site. Through XRD (Instrument model: Bruker AXS X-ray diffractometer D8 Advance (Bruker, Billerica, MA, USA)), it was found that the proportion of calcium carbonate in the exterior window mortar was 69.12%, and there were also compounds of calcium (as shown in Figure 4). Then, considering the improvement of mortar material properties by additives, experts provided four possible repair materials, including non-added glutinous rice lime mortar, 3% tung oil glutinous rice lime mortar, 3% paper-reinforced glutinous rice lime mortar, and 6% aluminum sulfate glutinous rice lime mortar, which are donated by X_1, X_2, X_3, and X_4. In order to assist the experts in making more scientific decisions, the company R used experimental methods to produce samples of glutinous rice mortar with four different additives and observed the microstructure of the repair materials using SEM (Instrument model: Thermo Scientific TM Apreo SEM (Thermo Scientific, Waltham, MA, USA)) (shown in Figure 5). Experts evaluated the four types of repair materials through several evaluation indices (as shown in Figure 2) to determine their compatibility with the building components to be repaired.

(a)

(b)

Figure 3. Exterior windows decorated with mortar in ancient architecture: (**a**) north facade exterior window; and (**b**) east facade exterior window.

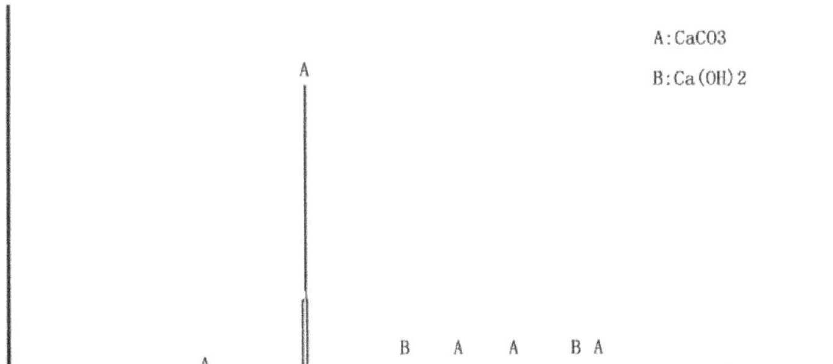

Figure 4. XRD pattern of exterior window decorative mortar extract.

According to the decision model proposed in Section 3, firstly, the evaluation matrix through the basic condition of components, performance of different compositions and proportions of additive lime mortars, and restoration outcomes can be obtained. Then, we calculate the index weights by combining criteria weights and entropy weights. Finally, the TOPSIS method can be used to obtain the ranking order of the four repair materials. In addition, the effectiveness and reliability of the proposed model are verified through a comparative analysis.

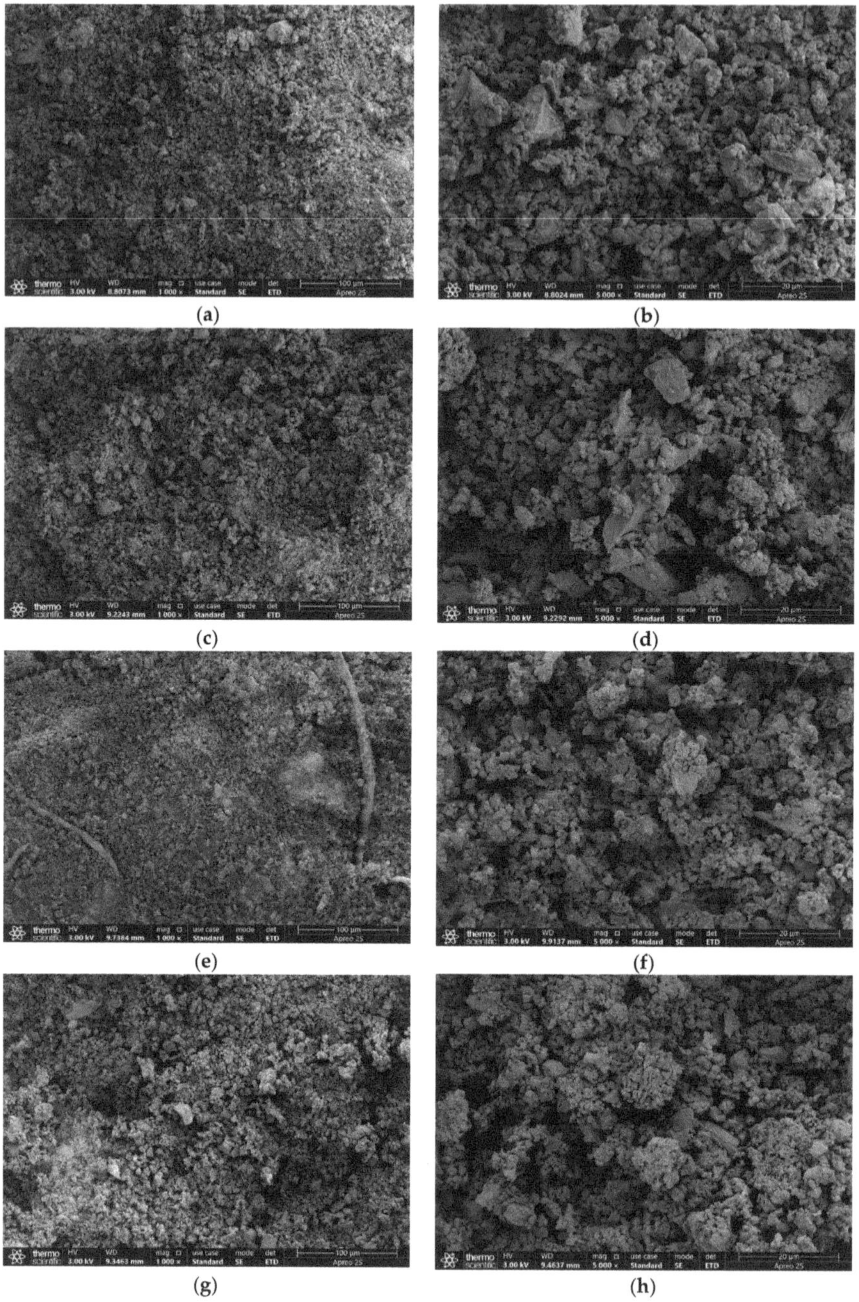

Figure 5. SEM images of two types of precision for different compositions and proportions of additive lime mortars. (**a**) SEM photo of 1000 times of non-added glutinous rice lime mortar. (**b**) SEM photo of 5000 times non-added glutinous rice lime mortar. (**c**) SEM photo of 1000 times 3% tung oil glutinous rice lime mortar. (**d**) SEM photo of 5000 times 3% tung oil glutinous rice lime mortar. (**e**) SEM photo of 1000 times 3% paper-reinforced glutinous rice lime mortar. (**f**) SEM photo of 5000 times of 3% paper-reinforced glutinous rice lime mortar. (**g**) SEM photo of 1000 times 6% aluminum sulfate glutinous rice lime mortar. (**h**) SEM photo of 5000 times 6% aluminum sulfate glutinous rice lime mortar.

5.1. The Acquisition of Evaluation Matrix

Obtain an evaluation matrix from experts based on the evaluation index system. According to the evaluation method mentioned in Section 4.2, experts evaluate using the index evaluation values, which are represented by language terms (LT) with multi-granularity to evaluate the four repair materials. According to Table 3, we convert the language terms into single-valued neutrosophic numbers (SVNNs). Then, by assessing the project experience and educational background of the experts, the importance of the four experts is shown in Table 5. The importance of each expert was determined using the language terms and their related SVNNs in Table 4, as shown in Table 5. In light of Equation (6), the weights of experts (e_k) are calculated separately: $e_1 = 0.187$, $e_2 = 0.187$, $e_3 = 0.295$, $e_4 = 0.331$. Next, it is necessary to aggregate the evaluation values of these indices. According to Equation (4), the single-value neutrosophic weighted averaging (SVNWA) aggregation operator should be used to aggregate these indices. Finally, based on the single-valued neutrosophic sets, we obtained the evaluation matrix $R = (r_{ij})$, outlined in Table 6.

Table 5. Importance of experts expressed with SVNNs.

	DM_1	DM_2	DM_3	DM_4
LT	M	M	I	VI
SVNS	<0.5, 0.5, 0.45>	<0.5, 0.5, 0.45>	<0.8, 0.2, 0.15>	<0.9, 0.1, 0.05>

Table 6. The aggregated SVNNs decision matrix.

Criteria	Index	Alternatives			
		X_1	**X_2**	**X_3**	**X_4**
A_1	a_{11}	<0.504, 0.496, 0.445>	<0.484, 0.516, 0.463>	<0.759, 0.241, 0.189>	<0.606, 0.394, 0.34>
	a_{12}	<0.411, 0.589, 0.539>	<0.436, 0.554, 0.522>	<0.759, 0.241, 0.189>	<0.65, 0.35, 0.297>
	a_{13}	<0.689, 0.311, 0.061>	<0.532, 0.468, 0.417>	<0.77, 0.23, 0.171>	<0.733, 0.267, 0.215>
	a_{14}	<0.625, 0.368, 0.072>	<0.546, 0.448, 0.406>	<0.791, 0.209, 0.15>	<0.65, 0.35, 0.3>
A_2	a_{21}	<0.236, 0.732, 0.512>	<0.341, 0.639, 0.627>	<0.733, 0.267, 0.215>	<0.857, 0.143, 0.088>
	a_{22}	<0.126, 0.824, 0.746>	<0.236, 0.732, 0.744>	<0.886, 0.114, 0.061>	<0.404, 0.596, 0.546>
	a_{23}	<0.125, 0.825, 0.795>	<0.125, 0.825, 0.875>	<1, 0, 0>	<0.309, 0.678, 0.653>
	a_{24}	<0.182, 0.768, 0.691>	<0.182, 0.768, 0.818>	<0.475, 0.525, 0.475>	<0.778, 0.222, 0.171>
A_3	a_{31}	<0.824, 0.176, 0.023>	<0.527, 0.473, 0.421>	<0.55, 0.45, 0.399>	<0.504, 0.496, 0.445>
	a_{32}	<0.23, 0.73, 0.632>	<0.509, 0.491, 0.44>	<0.404, 0.596, 0.546>	<0.532, 0.468, 0.417>
	a_{33}	<0.168, 0.782, 0.751>	<0.231, 0.736, 0.751>	<0.716, 0.284, 0.232>	<0.556, 0.444, 0.394>
	a_{34}	<0.2, 0.75, 0.667>	<0.375, 0.618, 0.582>	<0.764, 0.236, 0.184>	<0.663, 0.337, 0.284>
	a_{35}	<0.54, 0.46, 0.189>	<0.397, 0.597, 0.558>	<0.315, 0.65, 0.666>	<0.267, 0.695, 0.718>

5.2. The Calculation of Index Weights

In this part, we compute the subjective weights of the three criteria using the BWM method described in Section 4.3.1. Firstly, according to the expert's opinion, the criteria in Table 2, denoted by A_1 to A_3, the performance of different compositions and proportions of additive lime mortars (A_2) is considered the most important criteria, while the restoration outcomes (A_3) is considered the least important criteria. Then, we can obtain the pairwise comparison vectors for the best and the worst criteria, as shown in Tables 7 and 8. Table 7 represents that the preference values of the best criteria (A_2) over criteria (A_1) and criteria (A_3) are 3 and 5. The preference values of criteria (A_1) and criteria (A_2) over criteria (A_3) are 3 and 5, respectively. Finally, the weight vector of criteria $w^* = \{0.218, 0.652, 0.13\}$ is calculated by Equation (8).

Table 7. Pairwise comparison vector of the most important criteria.

Best Criteria (A_1 or A_2 or A_3)	A_1	A_2	A_3
A_2	3	1	5

Table 8. Pairwise comparison vector of the least important criteria.

Worst Criteria (A_1 or A_2 or A_3)	A_1	A_2	A_3
A_3	3	5	1

Afterward, we can compute the weights of the index according to the entropy weight method introduced in Section 4.3.2. Firstly, the evaluation index can be normalized by using Equation (9), we obtain the normalized evaluation matrix $R = (\beta_{ij})$, as shown in Table 9. Then, we can obtain the entropy weight of the index according to the normalized evaluation matrix.

Table 9. The normalized evaluation matrix.

Criteria	Index	Alternatives			
		X_1	X_2	X_3	X_4
A_1	a_{11}	<0.504, 0.496, 0.445>	<0.484, 0.516, 0.463>	<0.759, 0.241, 0.189>	<0.606, 0.394, 0.34>
	a_{12}	<0.411, 0.589, 0.539>	<0.436, 0.554, 0.522>	<0.759, 0.241, 0.189>	<0.65, 0.35, 0.297>
	a_{13}	<0.689, 0.311, 0.061>	<0.532, 0.468, 0.417>	<0.77, 0.23, 0.171>	<0.733, 0.267, 0.215>
	a_{14}	<0.625, 0.368, 0.072>	<0.546, 0.448, 0.406>	<0.791, 0.209, 0.15>	<0.65, 0.35, 0.3>
A_2	a_{21}	<0.236, 0.732, 0.512>	<0.341, 0.639, 0.627>	<0.733, 0.267, 0.215>	<0.857, 0.143, 0.088>
	a_{22}	<0.126, 0.824, 0.746>	<0.236, 0.732, 0.744>	<0.886, 0.114, 0.061>	<0.404, 0.596, 0.546>
	a_{23}	<0.125, 0.825, 0.795>	<0.125, 0.825, 0.875>	<1, 0, 0>	<0.309, 0.678, 0.653>
	a_{24}	<0.182, 0.768, 0.691>	<0.182, 0.768, 0.818>	<0.475, 0.525, 0.475>	<0.778, 0.222, 0.171>
A_3	a_{31}	<0.824, 0.176, 0.023>	<0.527, 0.473, 0.421>	<0.55, 0.45, 0.399>	<0.504, 0.496, 0.445>
	a_{32}	<0.77, 0.27, 0.368>	<0.491, 0.509, 0.56>	<0.596, 0.404, 0.454>	<0.468, 0.532, 0.0.583>
	a_{33}	<0.168, 0.782, 0.751>	<0.231, 0.736, 0.751>	<0.716, 0.284, 0.232>	<0.556, 0.444, 0.394>
	a_{34}	<0.2, 0.75, 0.667>	<0.375, 0.618, 0.582>	<0.764, 0.236, 0.184>	<0.663, 0.337, 0.284>
	a_{35}	<0.46, 0.54, 0.0.811>	<0.603, 0.403, 0.442>	<0.685, 0.35, 0.334>	<0.733, 0.305, 0.282>

Finally, through a combination of the subjective weights of criteria and the entropy weights of the index, index weights can be acquired, which is shown in Table 10.

Table 10. The weights of criteria and index for repair materials selection.

Criteria	Weight	Index	Weight
A_1	0.218	a_{11}	0.051
		a_{12}	0.046
		a_{13}	0.081
		a_{14}	0.071
A_2	0.652	a_{21}	0.156
		a_{22}	0.166
		a_{23}	0.211
		a_{24}	0.132
A_3	0.13	a_{31}	0.021
		a_{32}	0.009
		a_{33}	0.022
		a_{34}	0.019
		a_{35}	0.018

5.3. *The Sequence of Repair Materials*

We prefer the TOPSIS method based on MCGDM expressed to solve the repair materials selection problem according to Section 4.4. First, determine the positive ideal solution v_j^+ and negative ideal solution v_j^- according to Equation (11). Then, calculate the distances S_i^+ and S_i^- of each alternative solution to v_j^+ and v_j^- according to Equations (12) and (13). Finally, calculate the proximity coefficient CC_i of each alternative solution (X_i) to the ideal solution according to Equation (14), the result is obtained as in Table 11. Hence, as $CC_3 > CC_4 > CC_1 > CC_2$, the most appropriate compositions and proportions of additive lime mortars for W_1 is "X_3", namely, 3% paper-reinforced glutinous rice mortar.

Table 11. Ranking orders of compositions and proportions of additive lime mortars.

Alternatives	S_i^+	S_i^-	CC_i	Ranking Orders
X_1	0.1209	0.0154	0.113	3
X_2	0.121	0.0097	0.0744	4
X_3	0.0227	0.1172	0.8379	1
X_4	0.0821	0.0674	0.4511	2

5.4. Comparison Analysis

To verify whether the model proposed in Section 4 can be effectively and practically used to determine the most suitable compositions and proportions of additive lime mortars for a specific ancient Chinese building, considering the basic condition of the component to be repaired, the performance of different compositions and proportions of additive lime mortars and restoration outcomes, we employ the TODIM [50,51] method to make comparative analysis of the above empirical study. When $\theta = 2.25$, the final dominance (ε_i) of the four alternatives X_1, X_2, X_3, and X_4 are 0.081, 0, 1, and 0.685, respectively, the ranking of the proposed model computed by the TODIM method is $X_3 > X_4 > X_1 > X_2$ (shown in Table 12), which is consistent with the result computed by the proposed model. That is to say, the effectiveness of the proposed model can be verified. Compared with existing methods for selecting compositions and proportions of additive lime mortars for the restoration of ancient Chinese buildings, the advantages of the model proposed in this study can be summarized as follows:

- The proposed model combines fuzzy set theory and quantitative analysis and comprehensively considers subjective and objective criteria to develop an index system, which makes the selection of compositions and proportions of additive lime mortars for ancient Chinese buildings restoration more in line with the actual situation.
- The evaluation information is evaluated using a single-valued neutrosophic set that has a certain degree of similarity to human thinking [9], effectively solving the problems of diversity, ambiguity, and complexity in the sources and types of evaluation information, making compositions and proportions of additive lime mortars selection more accurate and reliable.
- The TOPSIS method is used to determine the superiority of alternative restoration materials, which are more flexible and operable in solving multi-criteria group decision-making problems [49]. Therefore, the proposed model can select the most suitable compositions and proportions of additive lime mortars for ancient Chinese building restoration.

Table 12. Ranking comparison with TODIM.

Alternatives	Values ε_i	Ranking Orders
X_1	0.081	3
X_2	0	4
X_3	1	1
X_4	0.685	2

6. Conclusions

This thesis establishes a selection model for compositions and proportions of additive lime mortars in the restoration of ancient Chinese buildings, which facilitates professionals to choose suitable restoration materials. When establishing the index system, both subjective and objective criteria were used. Furthermore, using single-valued neutrosophic sets represents evaluation information provided by experts using multi-granularity language terms, in which such information can be quantified. Moreover, by combining the BWM method and entropy weight method, the index weights can be obtained. In the end, the ranking of alternative solutions was determined using the TOPSIS method. Empirical research has been conducted on the proposed model in the selection of compositions and proportions of additive lime mortars for Dafudi to determine the type of mortar materials selection for a specific decorative exterior window, whose process was elaborated in detail. In addition, comparative analysis has demonstrated the effectiveness and reliability of the proposed model.

In summary, the model proposed in this study promotes the existing methods in the field of material selection for ancient Chinese architectural restoration and provides reasonable quantitative support for experts in the decision-making process. Therefore, this study embodies both theoretical and practical value. The contributions of this paper can be summarized as follows: (1) firstly, the proposed model fully considers the complexity of material selection in the restoration process of ancient building heritages, as well as the limitations of experts' decision-making based on experience, which may come across practical issues and challenges; (2) the proposed model enriches the quantitative research methods for the sustainability of ancient building heritages protection. By optimizing the selection of restoration material, we believe more efficient and effective protection of antithetical heritages can be realized; and (3) this model could play a supporting role in the decision-making, which might be conducive to assisting experts in making the most appropriate decisions, improving the efficiency of ancient buildings restoration, avoiding potential damages, and promoting the sustainable protection of ancient building heritages.

However, the proposed model still shows some limitations. Firstly, the proposed model is aimed at the selection of mortar materials in the restoration of ancient buildings. However, in fact, there are various types of materials available in the restoration of ancient buildings, and surely adjustments are needed according to the actual situation in specific tasks. Secondly, there is a wide variety of compositions and proportions of lime mortars, but the limited use of samples during the research process makes it difficult to fully compare the differences between different types of lime mortar material. Thirdly, the proposed model should also adopt different methods for evaluating information processing based on the different varieties of data obtained in specific cases. This study can be further improved from the following aspects. First, the proposed model could consider the selection information of more types of restoration materials and further improve the index system with the development of related studies in future research. Second, decision methods such as TODIM, VIKOR, and other algorithms [52] can also be considered in materials selection when the project situation and data type are appropriate. Third, the model could be more applicable to the selection of compositions and proportions of additive lime mortars and could extend to the selection of other types of restoration materials for ancient Chinese buildings to ensure the sustainability of ancient buildings' protection and restoration.

Author Contributions: Conceptualization, X.L. and L.L.; methodology, X.L.; software, L.L.; validation, X.L. and L.L.; formal analysis, Q.L.; investigation, Q.L.; data curation, X.L.; writing—original draft preparation, X.L.; writing—review and editing, X.L. and L.L. All authors have read and agreed to the published version of the manuscript.

Funding: This research was funded by the Project of the Hunan Provincial Social Science Achievement Review Committee (XSP2023YSC010).

Institutional Review Board Statement: There are no human subjects in this article and informed consent is not applicable.

Informed Consent Statement: Not applicable.

Data Availability Statement: The data used to support the findings of this study are included within the article.

Acknowledgments: The authors would like to thank the experts who participated in the survey.

Conflicts of Interest: Author Lizhi Liu was employed by the company Xiangtan Economic and Technological Development Zone Management Committee. The remaining authors declare that the research was conducted in the absence of any commercial or financial relationships that could be construed as a potential conflict of interest.

Abbreviations

The following abbreviations are used in this manuscript:

MCGDM	Multi-Criteria Group Decision-Making
BWM	The Best–Worst Method
TOPSIS	Technique for Order Preference by Similarity to Ideal Solution
SVNWA	Single-valued Neutrosophic Weighted Averaging
SVNN	Single-valued Neutrosophic Number
XRD	X-ray powder Diffraction
SEM	Scanning Electron Microscopy
TODIM	An acronym in Portuguese for interactive multi-criteria decision making

References

1. Yang, F.; Zhang, B.; Ma, Q. Study of Sticky Rice-Lime Mortar Technology for the Restoration of Historical Masonry Construction. *Acc. Chem. Res.* **2010**, *43*, 936–944. [CrossRef] [PubMed]
2. Peng, Y.; Wang, C.; Zhang, X.; Zheng, W.; Yu, Y. Inorganic-accelerated aging method: An efficient and simple strategy to obtain antique Chinese fir wood for the restoration of ancient wooden architecture. *J. Build. Eng.* **2024**, *84*, 108372. [CrossRef]
3. Xu, S.; Ma, Q.; Xu, S. Fuzzy comprehensive evaluation of the compatibility of restoration materials—Case study in the rammed earth restoration of the M2 Han tomb in Dingtao, Shandong Province. *J. Cult. Herit.* **2022**, *57*, 131–141. [CrossRef]
4. Fregonese, L.; Barbieri, G.; Biolzi, L.; Bocciarelli, M.; Frigeri, A.; Taffurelli, L. Surveying and Monitoring for Vulnerability Assessment of an Ancient Building. *Sensors* **2013**, *13*, 9747–9773. [CrossRef]
5. Sodangi, M.; Kazmi, Z.A. Integrated Evaluation of the Impediments to the Adoption of Coconut Palm Wood as a Sustainable Material for Building Construction. *Sustainability* **2020**, *12*, 7676. [CrossRef]
6. Qing, C.; Hui, J.; Yuande, W.; Yidan, H.; Zhendong, M.; Qi, C.; Shiyu, M. Quantitative evaluation method of structural safety status of ancient Chinese three-hole stone arch bridges—A case study of the Yongchang Bridge. *J. Asian Archit. Build. Eng.* **2023**, *22*, 2143–2160. [CrossRef]
7. Chen, S.; Hu, Q.; Wang, S.; Yang, H. A Virtual Restoration Approach for Ancient Plank Road Using Mechanical Analysis with Precision 3D Data of Heritage Site. *Remote Sens.* **2016**, *8*, 828. [CrossRef]
8. Lourenço, P.B. Recommendations for restoration of ancient buildings and the survival of a masonry chimney. *Constr. Build. Mater.* **2006**, *20*, 239–251. [CrossRef]
9. Smarandache, F. *A Unifying Field in Logics: Neutrosophic Logic. Neutrosophy, Neutrosophic Set, Neutrosophic Probability*; American Research Press: Rehoboth, NM, USA, 1999.
10. Luo, S.; Cheng, P.; Wang, J.; Huang, Y. Selecting Project Delivery Systems Based on Simplified Neutrosophic Linguistic Preference Relations. *Symmetry* **2017**, *9*, 151. [CrossRef]
11. Zhang, H.; Ji, P.; Wang, J.; Chen, X. A novel decision support model for satisfactory restaurants utilizing social information: A case study of TripAdvisor.com. *Tour. Manag.* **2017**, *59*, 281–297. [CrossRef]
12. Li, D.; He, J.; Cheng, P.; Wang, J.; Zhang, H. A Novel Selection Model of Surgical Treatments for Early Gastric Cancer Patients Based on Heterogeneous Multicriteria Group Decision-Making. *Symmetry* **2018**, *10*, 223. [CrossRef]
13. Afshar, A.; Mariño, M.A.; Saadatpour, M.; Afshar, A. Fuzzy TOPSIS multi-criteria decision analysis applied to Karun Reservoirs System. *Water Resour. Manag.* **2011**, *25*, 545–563. [CrossRef]
14. Shih, H.S.; Shyur, H.J.; Lee, E.S. An extension of TOPSIS for group decision making. *Math. Comput. Model.* **2007**, *45*, 801–813. [CrossRef]
15. Naciri, K.; Aalil, I.; Chaaba, A. Eco-friendly gypsum-lime mortar with the incorporation of recycled waste brick. *Constr. Build. Mater.* **2022**, *325*, 126770. [CrossRef]
16. Jayasingh, S.; Baby, J. Influence of organic addition on strength and durability of lime mortar prepared with clay aggregate. *Mater. Today Proc.* **2022**, *64*, 1006–1013. [CrossRef]
17. Manoharan, A.; Umarani, C. Lime Mortar, a Boon to the Environment: Characterization Case Study and Overview. *Sustainability* **2022**, *14*, 6481. [CrossRef]
18. Frankeová, D.; Koudelková, V. Influence of ageing conditions on the mineralogical micro-character of natural hydraulic lime mortars. *Constr. Build. Mater.* **2020**, *264*, 120205. [CrossRef]
19. Branco, F.G.; Belgas, M.D.; Mendes, C.; Pereira, L.; Ortega, J.M. Mechanical Performance of Lime Mortar Coatings for Rehabilitation of Masonry Elements in Old and Historical Buildings. *Sustainability* **2021**, *13*, 3281. [CrossRef]
20. Apostolopoulou, M.; Asteris, P.G.; Armaghani, D.J.; Douvika, M.G.; Lourenço, P.B.; Cavaleri, L.; Bakolas, A.; Moropoulou, A.I. Mapping and holistic design of natural hydraulic lime mortars. *Cem. Concr. Res.* **2020**, *136*, 106167. [CrossRef]
21. Silva, B.A.; Ferreira Pinto, A.P.; Gomes, A.; Candeias, A. Effects of natural and accelerated carbonation on the properties of lime-based materials. *J. CO_2 Util.* **2021**, *49*, 101552. [CrossRef]
22. Apostolopoulou, M.; Bakolas, A.; Kotsainas, M. Mechanical and physical performance of natural hydraulic lime mortars. *Constr. Build. Mater.* **2021**, *290*, 123272. [CrossRef]
23. Hwang, H.; Kwon, Y.; Hong, S.; Kang, S. Comparative study of effects of natural organic additives and cellulose ether on properties of lime-clay mortars. *J. Build. Eng.* **2022**, *48*, 103972. [CrossRef]

24. Feijoo-Conde, J.; Álvarez-Feijoo, M.Á.; Fort, R.; Arce, E.; Ergenç, D. Effects of Paraffin Additives, as a Phase Change Materials, on the Behaviour of a Traditional Lime Mortar. *SSRN Electron. J.* **2022**, *361*, 129734. [CrossRef]
25. Yadav, M.; Pal, S. Experimental study to check the efficacy of adding organic additives on engineering properties of lime mortar. *Mater. Today Proc.* **2023**, *93*, 538–544. [CrossRef]
26. Manoharan, A.; Umarani, C. Properties of Air Lime Mortar with Bio-Additives. *Sustainability* **2022**, *14*, 8355. [CrossRef]
27. Brzyski, P. The effect of pozzolan addition on the physical and mechanical properties of lime mortar. *E3S Web Conf.* **2018**, *49*, 9. [CrossRef]
28. Forster, A.M.; Kayan, B.A. Maintenance for historic buildings: A current perspective. *Struct. Surv.* **2009**, *27*, 210–229. [CrossRef]
29. Gao, Z.; Ma, D.; Wang, W.; Guo, X.; Ge, Q. Development and Application of Ancient Timber Buildings Structural Condition Assessment Model Based on a Fuzzy Matter-Element Model that Includes Asymmetric Proximity. *Math. Probl. Eng.* **2018**, *2018*, 7426915. [CrossRef]
30. Tan, R.P.; Zhang, W.D.; Chen, L.L. Study on emergency group decision making method based on VIKOR with single valued neutrosophic sets. *J. Saf. Sci. Technol.* **2017**, *13*, 79–84.
31. Rezaei, J. Best-worst multi-criteria decision-making method. *Omega-Int. J. Manag. Sci.* **2015**, *53*, 49–57. [CrossRef]
32. Majumdar, P.; Samanta, S.K. On similarity and entropy of neutrosophic sets. *J. Intell. Fuzzy Syst.* **2013**, *26*, 1245–1252. [CrossRef]
33. Güzelci, O.Z.; Şener, S.M. An Entropy-Based Design Evaluation Model for Architectural Competitions through Multiple Factors. *Entropy* **2019**, *21*, 1064. [CrossRef]
34. You, X.; Chen, T.; Yang, Q. Approach to Multi-Criteria Group Decision-Making Problems Based on the Best-Worst-Method and ELECTRE Method. *Symmetry* **2016**, *8*, 95. [CrossRef]
35. Abdel-Basset, M.; Gamal, A.; Manogaran, G.; Son, L.H.; Long, H.V. A novel group decision making model based on neutrosophic sets for heart disease diagnosis. *Multimed. Tools Appl.* **2019**, *79*, 9977–10002. [CrossRef]
36. Tian, Z.; Nie, R.; Wang, J. Probabilistic linguistic multi-criteria decision-making based on evidential reasoning and combined ranking methods considering decision-makers' psychological preferences. *J. Oper. Res. Soc.* **2019**, *71*, 700–717. [CrossRef]
37. Liu, P.; Cheng, S.; Zhang, Y. An Extended Multi-criteria Group Decision-Making PROMETHEE Method Based on Probability Multi-valued Neutrosophic Sets. *Int. J. Fuzzy Syst.* **2018**, *21*, 388–406. [CrossRef]
38. Kharal, A. A Neutrosophic Multi-Criteria Decision Making Method. *New Math. Nat. Comput.* **2014**, *10*, 143–162. [CrossRef]
39. Wang, H.; Smarandache, F.; Zhang, Y.Q.; Sunderraman, R. Single valued neutrosophic sets. *Multisp. Multistruct.* **2010**, *4*, 410–413.
40. Biswas, P.; Pramanik, S.; Giri, B.C. TOPSIS method for multi-attribute group decision-making under single-valued neutrosophic environment. *Neural Comput. Appl.* **2016**, *27*, 727–737. [CrossRef]
41. Ye, J. A multicriteria decision-making method using aggregation operators for simplified neutrosophic sets. *J. Intell. Fuzzy Syst.* **2014**, *26*, 2459–2466. [CrossRef]
42. Zhang, X.; Zhang, H.; Wang, J. Discussing incomplete 2-tuple fuzzy linguistic preference relations in multi-granular linguistic MCGDM with unknown weight information. *Soft Comput.* **2019**, *23*, 2015–2032. [CrossRef]
43. Song, C.; Wang, X.; Cheng, P.; Wang, J.; Li, L. SACPC: A framework based on probabilistic linguistic terms for short text sentiment analysis. *Knowl. Based Syst.* **2020**, *194*, 105572. [CrossRef]
44. Rezaei, J. Best-worst multi-criteria decision-making method: Some properties and a linear model. *Omega-Int. J. Manag. Sci.* **2016**, *64*, 126–130. [CrossRef]
45. Mahdiraji, H.A.; Arzaghi, S.; Stauskis, G.; Zavadskas, E.K. A hybrid fuzzy BWM-COPRAS method for analyzing key factors of sustainable architecture. *Sustainability* **2018**, *10*, 1626. [CrossRef]
46. Peng, J.J.; Wang, J.Q.; Zhang, H.Y.; Chen, X.H. An outranking approach for multi-criteria decision-making problems with simplified neutrosophic sets. *Appl. Soft Comput.* **2014**, *25*, 336–346. [CrossRef]
47. Yu, F.; Fang, G.; Shen, R. Study on comprehensive early warning of drinking water sources for the Gucheng Lake in China. *Environ. Earth Sci.* **2014**, *72*, 3401–3408. [CrossRef]
48. Hwang, C.L.; Yoon, K.P. *Multiple Attributes Decision Making: Methods and Applications*; Springer: New York, NY, USA, 1981.
49. Lourenzutti, R.; Krohling, R.A. A generalized TOPSIS method for group decision making with heterogeneous information in a dynamic environment. *Inf. Sci.* **2016**, *330*, 1–18. [CrossRef]
50. Hu, J.; Yang, Y.; Chen, X. A Novel TODIM Method-Based Three-Way Decision Model for Medical Treatment Selection. *Int. J. Fuzzy Syst.* **2017**, *20*, 1240–1255. [CrossRef]
51. Long, X.; Liu, L.; Xiao, C.; Cheng, P.; Fu, C. Restoration Methods Selection for Wood Components of Chinese Ancient Architectures Based on TODIM with Single-Valued Neutrosophic Sets. *Math. Probl. Eng.* **2020**, *2020*, 5049360. [CrossRef]
52. Zhang, X.; Ding, Y.; Zhao, H.; Yi, L.; Guo, T.; Li, A.; Zou, Y. Mixed Skewness Probability Modeling and Extreme Value Predicting for Physical System Input–Output Based on Full Bayesian Generalized Maximum-Likelihood Estimation. *IEEE Trans. Instrum. Meas.* **2024**, *73*, 2504516. [CrossRef]

sustainability

MDPI

Article

Co-Developing Business Plans for Upscaled Coastal Nature-Based Solutions Restoration: An Application to the Venice Lagoon (Italy)

Umberto Pernice [1,*], Francesca Coccon [2], Fabienne Horneman [3,4], Caterina Dabalà [2], Silvia Torresan [3,4] and Laura Puertolas [5]

[1] Pernice Collaborative Innovation Management, Viale Michelangelo 2315, 90145 Palermo, Italy
[2] Consortium for Coordination of Research Activities Concerning the Venice Lagoon System (CORILA), San Polo 19, 30125 Venezia, Italy; coccon@corila.it (F.C.); dabala@corila.it (C.D.)
[3] CMCC Foundation-Euro-Mediterranean Center on Climate Change, 30175 Venice, Italy; fabienne.horneman@cmcc.it (F.H.); silvia.torresan@cmcc.it (S.T.)
[4] Department of Environmental Sciences, Informatics and Statistics, Venice Ca' Foscari University, 30170 Venice, Italy
[5] Albirem Sustainability, Olzinelles 70 B2, 08014 Barcelona, Spain; lpuertolas@albirem.com
* Correspondence: collaborate@umbertopernice.com; Tel.: +39-3282837781

Abstract: Upscaling coastal nature-based solutions (NbSs), from the river to sea scale, essential to the ultimate success of restoration, requires larger investments, greater socio-economic engagement and a transformative shift in governance. Consequently, the ability to catalyze blended finance from public and private investors through business model innovation and accessibly written long-range business plans is imperative. This research proposes a methodology to systematically co-design, through iterative stakeholder engagement, suitable and effective business models for upscaling NbSs coastal restoration, as well as a framework to develop business plans adaptable to different investors needs and financial criteria. This method was tested and validated in a saltmarsh restoration project coordinated by public agencies in the Venice Lagoon. It delivered an advanced business model canvas, carefully assessed through a stakeholder questionnaire and a related business case centered on ecotourism complemented by a willingness-to-pay analysis. The resulting business plan lays out the governance, management, business and financial strategies for successful restoration, addressing specific interconnected issues, improving five main ecosystem services and delivering NbS multifunctional environmental, economic, social and cultural benefits. The methodology has proven to be applicable beyond the Venice Lagoon saltmarsh case and types of restoration projects.

Keywords: ecosystem services; nature-based solutions; coastal restoration upscaling; business plan; business model; stakeholder co-development; eco-tourism

1. Introduction

It is widely scientifically acknowledged that climate change, notably an exacerbation of natural hazards and ecosystem degradation, will define our future [1]. This progressive degradation of ecosystems adversely impacts the provisioning of ecosystem services (ESs), and thus constitutes a systematic risk to societies, economies and companies [2,3]. The global community has outlined ambitious ecosystem restoration targets [4], including the political push to integrate nature-based solutions (NbSs) in adaptation efforts. Cohen-Shacham et al. [5] defines NbSs as "measures that aim to protect, sustainably manage and restore ecosystems, thereby addressing societal challenges while simultaneously providing human well-being and biodiversity benefits". These NbSs utilize natural features and processes, are resource-efficient and are tailored to address social, environmental and economic challenges in diverse environments [6]. The acceptance and application of

NbSs is on the rise due to their ability to simultaneously address looming challenges, offer additional ESs, as well as benefits to multiple stakeholders [7]. Moreover, NbSs are firmly encouraged in the EU policy agenda, including proposed actions on flood management, climate change adaptation, water retention, disaster management [8], the recent EU Restoration law, the EU biodiversity strategy, the EU Green deal, as well as the Urban Agenda 2030, the UN Climate Action Summit 2019 and the Convention of Biological Diversity, advocating for the use of natural solutions over grey or engineered, physical infrastructure [9]. While NbSs are considered to be more cost-effective and efficient than grey infrastructure, especially in attenuating hazards [10], NbS implementation remains slow, partially limited by precarious financial funding [4,9].

NbSs generate wide-spread public benefits, but do not produce immediate revenue streams and, as such, must rely on public funding, often subject to political and governmental restrictions [4,9]. Public financing, however, most often is insufficient to tackle larger international restoration targets [3]. A strategic NbS has the potential to align environment goals with private sector economic interests, stimulating private investments from banks, institutions and venture capitalists as the most promising alternative source of key funding [11]. To attract private investments, a supportive environment that harmonizes financial incentives with sustainability objectives must be created [12]. This includes providing more reliable and granular data to estimate up-front and running costs, as well as the short and long-term merits of restoration projects. Investors and other project stakeholders will be focused on potential investment returns, unfolding of private–public partnerships and realistic assessment of financial risk/reward ratios [13]. Cost-benefit and cost-effectiveness analyses equip policymakers with quantitative tools to make decisions regarding the implementation of NbS projects. However, stakeholders in the private sector are looking more to enhanced profits as a standard for willingness to participate, including the economic evaluation of ES improvements generated by restoration activities [14,15]. Underlining the necessity to conduct a comprehensive analysis of the costs and benefits associated with NbSs, it is crucial to integrate both monetary valuation and non-monetary impact assessments to address social, environmental and cultural effects [16]. Another difficult task of the NbS is in capturing the value of intangible rewards, such as better health or increased biodiversity, for the private investors deciding where to make the greatest difference in alignment with their own financial priorities and sense of responsibility to the public [17].

Barriers to NbS implementation are complex and interwoven, spanning technical, financial, economic and governing domains [18,19]. To overcome the barriers and bridge the gap, a keener knowledge and awareness of NbS business models is needed to stimulate the involvement of private stakeholders in financing NbS upscaling [20]. As such, a shift is occurring, so is diving deeper into the mechanisms that unlock private resources for NbSs from institutional investors, businesses and citizens [9]. The inclusion of various stakeholders is deemed a suitable and sustainable way towards the mainstreaming of NbS implementation; allowing for new synergistic alliances to create more flexible and innovative milieus, sustainable perspectives and reduced dependencies upon temporal financing [21].

The current literature depicting business models reveals an important knowledge gap concerning the design and implementation of value creation and value capture processes across organizations [22,23]. This vacuum creates a pressing need to facilitate the development of business models and case study examples that identify the necessary elements for achieving value [20]—capable of evaluating and integrating environmental, economic, social and cultural value propositions—arising from NbSs restoration efforts spanning different time periods. An innovative concept, although still in its infancy, refers to regenerative business models that aim to contribute to ecological restoration and societal well-being, generating value across multiple stakeholder levels, including nature, societies, customers, suppliers, shareholders and employees [24].

A business model (BM) represents the "design or architecture of the value creation, delivery, and capture mechanisms" of a firm [25]. Following this, a business plan (BP) formulates specific management, business and financial strategies aimed at achieving revenues and funding that mirror the objectives of NbSs restoration. In the context of NbSs restoration, the BM reveals how an organization generates, delivers and captures environmental, economic and social value through restoration [26]. One of the most important forms of BM innovation nowadays is the move from selling products to selling outcome-based services [27–30], where the provider assumes responsibility for the performance outcomes of the services offered and penalties in case of fault [30]. In the NbSs interventions context, the performance outcomes are linked to the expected ESs and biodiversity improvements, which are difficult to measure, with a crucial need for governance support and social acceptance. Drawing from the BM Canvas [26], business modelling tools have evolved to address various aspects such as sustainability [31–33], circular economy principles [34] and in recent years natural assurance schemes and NbSs [20,21,35].

A high-risk BM innovation strategy, infused with holistic thinking that embraces entrepreneurship, ESs, NbSs, life cycle thinking, sustainability and circular economy, must be adopted [21]. A clear assessment of the underlying reasons for the value attributed to NbSs is expected to help with the prioritization and uptake of NbS projects by managing expectations and providing a basis for a more transparent decision-making process [1]. Such an evaluation should consider critical research domains including the challenges associated with integrating various financial mechanisms, the complexities in evaluating multiple value propositions (including environmental, economic, social and cultural elements) from different stakeholders and the difficulties in transitioning from identifying the overall social impact of NbSs restoration to establishing effective value capture mechanisms. Therefore, this work addresses the complexity of the topic by proposing a systematic methodology for co-developing business plans based on innovative business models for upscaling NbSs restoration, capable of addressing NbSs multifunctionality while incorporating participatory stakeholder engagement. The methodology has been developed, tested and validated using the Venice Lagoon (Italy) as one of the pilot sites of the broader collaborative research project, REST-COAST, the large-scale RESToration of COASTal ecosystems through rivers-to-sea connectivity.

2. Materials and Methods

This section includes three parts: (i) exploring scientific and grey literature on business model innovation and NbSs restoration complemented by a scientometric analysis; (ii) systematizing the co-development of a BP based on BM innovative stakeholder participation; (iii) characterizing the case study.

2.1. *Unveiling Business Models and Business Plans in the Current Literature through a Scientometric Analysis*

The framework design for innovative business models relies on the extensive knowledge in the scientific and grey literature related to key principles of NbSs restoration. The analysis and quantification of suitable business models and plans for upscaling coastal restoration using a scientometric approach has been included in this article. A scientometric analysis assists in identifying structures and capturing the development of a discipline to generate a comprehensive and objective portrait of the state of knowledge [36]. This quantitative review, through the open-source Bibliometric Package, builds on the metadata—i.e., citation and bibliographic information, as well as the abstracts and keywords—of the publications identified through Scopus.

Scopus is considered the largest curated bibliographic abstract and citation database, containing over 76 million records [37], and has been selected as the main source of information for this review due to its comprehensive and accurate coverage of academic content. A total of 15 queries were made in Scopus on the 5 August 2024, combining relevant keywords such as "business model", "business plan", "business model canvas", "finance", "nature-based solution" and "restoration" in various combinations through Boolean operators, i.e., OR, AND. This resulted in the identification of 1236 publications, published between 1992 and 2024, primarily written in English (Appendix A), revealing a lack of consideration about the role that business plans can play in NbSs restoration, complementing the strong focus of the literature on financial aspects for restoration.

This analysis was complemented with knowledge from the past and ongoing EU projects, selected through the Research and Innovation community platform of the European Commission, CORDIS. Emphasis was given to the recent EU projects which are part of the NetworkNature European initiative, including the recent efforts from its Task Force 3 about financing and business models for NbSs in a nature-positive economy (Blessing and Barrientos, 2023), thereby highlighting a focus on BM and evaluation methods, while limited attention is paid to BPs. Through this scientometric analysis, it becomes clear that there is a niche for NbSs in coastal restoration business plans. This article pursues this opportunity through the introduction of a methodology to systematically co-develop business plans.

2.2. Hybrid Approach to Systematically Co-Develop a Business Plan for NbSs Restoration

The methodology for co-developing the BP is based on five main steps (Figure 1), as well as an iterative stakeholder engagement and coaching process by economic consultants and professionals in business management, tested and validated at the Venice Lagoon. This resulted in the (i) stakeholder (STK) analysis and interest–influence map (step 1); (ii) NbS-ES Business Model Canvas (step 2-3-4); (iii) NbS-ES Business Plan (step 5). The three main categories of STK identified in the Venice Lagoon were: (i) the core team of STKs, composed of experts representing different scientific disciplines, owning a deep knowledge of NbSs restoration in the Venice Lagoon and its stakeholders and beneficiaries; (ii) the broad STKs group, consisting of an extensive ensemble of stakeholders potentially impacted by/or impacting NbSs restorations; (iii) the tourists and visitors, representing potential beneficiaries and customers.

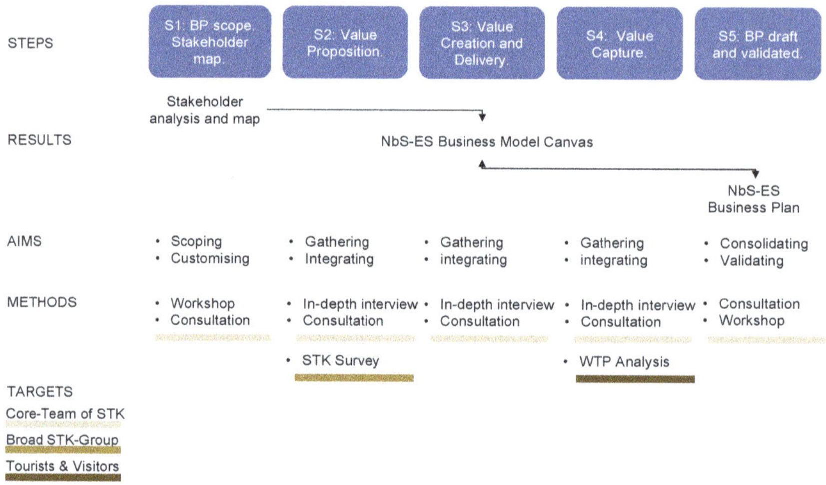

Figure 1. Methodological approach for co-developing the NbS-ES Business Plan, showing steps, results, aims, methods used and targets involved.

The core team of STKs was constantly engaged through in presence and remote workshops with the aim of scoping the BP, as the development required site-specific knowledge, customization, flexibility and contextual adaptation. Subsequently, the core team, nested within the broad STKs group, acted as the main interface with the two other STK categories, facilitating their engagement throughout the iterative process. Step 1, the stakeholder analysis and mapping, is based on information provided by the core team regarding two factors: interest and influence scaled from 1 (lowest) to 5 (highest), commonly used in participatory processes [38]. The ensuing map plots the position of stakeholders into four different quadrants and foresees different types of engagement: i.e., actively engaged, keep informed, keep satisfied and monitor. Therefore, it allows to analyze, categorize and map all stakeholders to be involved in the process, confirming the core team scoping for the BP and integrating the stakeholder' perspectives and needs in the co-development of the NbS-ES Business Model Canvas [39].

The NbS-ES Business Model Canvas (NbS-ES BMC) is at the core of the formulation of the BP, including the co-developed value proposition (step 2), creation/delivery (step 3) and capture (step 4). The NbS-ES BMC developed in this research builds on former business model approaches, emphasizing the contribution of NbSs restoration to the improvement of ESs and biodiversity. Quantitative and qualitative data were gathered through in-depth interviews and consultations with open-ended questions with the core team of STKs. The co-development process was complemented with (i) an online survey to the broad STK group (Appendix B) in order to co-define multiple value propositions depending on the type of environmental, economic, social and cultural benefit, assessed by stakeholders between 12 August 2024 and 10 September 2024; and (ii) a willingness-to-pay (WTP) study targeting tourists and visitors conducted between 12 August 2024 and 26 August 2024 (Appendix C), to strengthen the analysis of value capture using an ecotourism business case based on a selection of benefits and value propositions. In step 5, the information included in the BMC were used for the co-development and validation of the BP with the core team of STKs, to better align value creation and value capture [40] previously defined in the BM and deepened through the business case.

The template illustrated in Figure 2, helps to frame the process and all elements involved in the provision of coastal restoration through NbS interventions and the BM for upscaling restoration. This latter contemplates information gathered through in-depth consultation into three main clusters: (i) the value proposition that NbSs restoration interventions aim to offer to stakeholders and beneficiaries through multiple benefits—environmental, economic, social and cultural; (ii) the value creation/delivery, consisting of the specific restoration activities, ruled and impacted by regulation and governance, executed by key partners and other implementers, the resources needed, the delivery channels and customer relationship foreseen to deliver the value proposition to stakeholders, beneficiaries and customers; (iii) the value capture, comprising costs, revenues and additional financial sources to financially sustain the NbS restoration investments; (iv) horizontal aspects, including technological, governance and financial barriers and enablers, impact indicators, risks and mitigation solutions affecting the business model development and implementation.

The different elements of analysis indicated in the ES-NbS BMC, grouped into 20 main categories, represent the basis for the NbS-ES Business Plan, elaborated through additional analysis and consultation. The innovative BP structure was designed by adapting the mainstream template of the business plan to peculiarities of NbS investments and financial requirements from investors. It comprises five main chapters: the value proposition shapes the NbS restoration objectives (I) and the multifunctional NbS offering (II), the latter calibrated through the market analysis; the value creation and delivery from one side contributes to assessing the feasibility of the NbS's multifunctional offering (I) and from the other allows to define the management, business and financial strategies (III) comprising the organizational model supporting the execution of the activities, and the intertwined strategies for business development generating revenues and gathering funding; the value capture complements with quantitative information about costs, revenue, cash flow and financial gap; the business and financial strategies (III) are temporally estimated into economics and financial projections (IV); the horizontal aspects allow the risk and contingency plan to be defined (V), considering the impact of barriers, the contribution from enablers and establishing thresholds for impact indicators.

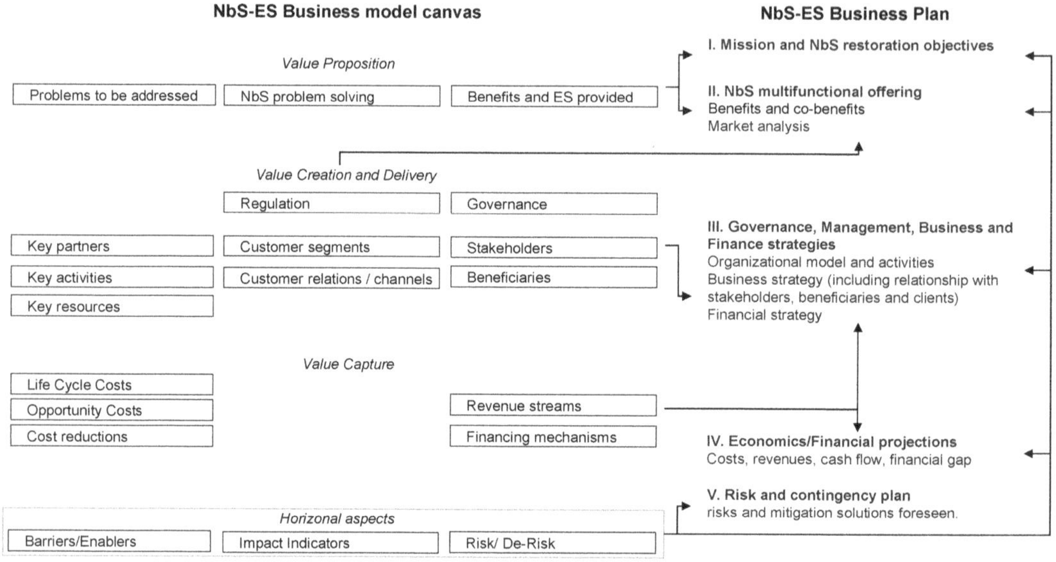

Figure 2. Connection flow diagram highlighting the contribution of the four categories of elements of the NbS-ES Business Model canvas to the five chapters of the NbS-ES Business Plan.

2.3. *Characterizing the Venice Lagoon as a Case Study for Restoration Upscaling*

The Venice Lagoon (Italy) is the largest lagoon in the Mediterranean Sea, located in the north-western Adriatic between the current mouths of the Brenta and Sile rivers, spanning roughly 550 km^2 [41,42]. The morphology within the Venice Lagoon includes different habitats whose distribution depends on the hydrodynamic regime [43], creating a complex mosaic of coastal habitats including 6800 ha of saltmarshes and mudflats, 6000 ha of seagrasses and islands [44–46], all of which are interwoven with cultural heritage and urban centers.

Venice and its lagoon have been a UNESCO World Heritage site since 1987, and therefore require the preservation of the landscape itself. Since its foundation, the Venetian Republic considered Venice and its lagoon as a single entity that should be carefully preserved in all its components. The safeguarding of Venice and its lagoon is dictated by a set of precise laws (e.g., L. 171/73, L. 798/84, L. 360/91, and L. 139/92). The Venice Water Authority (i.e., Provveditorato Interregionale per le Opere Pubbliche per il Veneto, Trentino Alto Adige e Friuli Venezia Giulia, hereafter Provv. OO. PP.), as the peripheral body of the Italian Ministry of Transport and Infrastructure (MIT), is responsible for safeguarding of Venice and its lagoon. In particular, the legal framework of the Venice Lagoon's governance is the result of overlapping three legal systems: (1) ordinary, applying EU Directives on various subjects including habitats, species, water quality, ports, mobility and use of sediments; regulatory and policy instruments of national, regional, provincial and municipal competence; (2) special, based on the national interest of safeguarding the city of Venice and its lagoon; (3) commissarial, dealing with different types of socio-economic, environmental and hydraulic emergencies with different commissioners, each for a specific emergency. The Venice Lagoon's governing structure is composed of nine municipalities, two provinces (Venice and Padua), the Metropolitan City of Venice which is responsible for urban restoration and social vitalization, the Veneto Region which looks after the de-pollution of the drainage basin, the Italian State which takes charge for the physical safeguard from sea and rivers and port functioning, and the Comitatone, including ministries and local administration, which oversees the coordination among the main actors. Altogether, the regulations contribute to defining a "lagoon law", with origins in the institution of the "Magistrato alle Acque di Venezia" (Venice Water Authority), now Provv. OO. PP, during the Republic of Venice in 1501.

The Provv. OO. PP. has been responsible for the planning, implementation and maintenance of the works in the lagoon since 1992, implementing a variety of interventions in order to counter the negative morphological evolution of the lagoon, as well as the disappearance and degradation of the lagoon's unique habitats. One of the key habitats that has been negatively impacted by human activities in the 20th century are saltmarshes [42], with the habitat's extent in the Venice Lagoon diminished by over 76% in the past two centuries [44]. In the past three decades, 1600 ha of artificial saltmarshes and mudflats have been created by the Provv. OO. PP. with the aim to limit the loss of sediment, protect internal marshes from wave energy and thus erosion, preserve the ecological functioning of the saltmarshes and help re-establish the transitional environment. In line with these objectives, 138 ha of saltmarshes are currently under "active" maintenance in the central lagoon.

In particular, the current NbS interventions consist of two phases, spanning ± 5 years, including partially concurrent planning, implementing and monitoring activities, subsequently followed by the naturalization of the marsh, that allows for the provisioning of ESs. The first phase, initiated by the Provv. OO. PP., outlines the restoration of the degraded edges and prepares the site for the upcoming second phase. The latter is executed by the Venice Port Authority (VPA) as part of controlling and maintaining port operationality and refers to the nourishment of the saltmarshes using sediments dredged from the channels to keep the lagoon navigable. Through nature-based processes, saltmarshes play an important role in food supply, nutrient cycling, contaminant filtering, sediment storage and flood control at the local scale, while at a larger scale the restored marshes help to regulate regional climate and provide habitat of conservation importance [47]. Various legislative and planning documents highlight the saltmarshes' capacity for ES provisioning, including reduction of erosion and flood risk (RCF), water quality purification (WP), carbon sequestration/CCR), food provisioning (FP), as well as biodiversity (BdV) (these ecosystem services are aligned with those studied in the REST-COAST project and are associated with various societal problems faced by coastal communities including flooding, degradation of coastal landscapes due to erosion, water quality degradation and eutrophication), and as such the importance of restoring these habitats is crystal clear.

3. Results

3.1. Stakeholder Analysis and Interest–Influence Map

The core team of STKs supported the identification of the broad STKs group starting from a list of over 70 stakeholders (many of these stakeholders also participated in the Venetian REST-COAST project participatory table, known as CORE-PLAT (Coastal Restoration PLATform), established as a platform for dialogue and collaboration to foster the exchange of knowledge, and ultimately develop a shared vision for the future of the lagoon. Although the recommendations made during CORE-PLAT meetings are not binding for public administrations, the platform aims to play a crucial role in shaping discussions around restoration activities), arising from the "Contratto per la laguna Nord" which is a voluntary tool for strategic and negotiated planning on a participatory basis, with the objective of safeguarding the northern part of the lagoon. These 70 stakeholders are impacted or impact NbSs restoration upscaling in the Venice Lagoon. From this list, 38 stakeholders have been contacted based on their alignment and past engagement with natural and artificial saltmarsh restoration. These 38 stakeholders can be categorized into two broader classes, i.e., the public sector (n = 13) and the private sector (n = 25), and eight categories, i.e., Governments, NGOs, Education and Research institutions, Environmental associations, Economic associations, Companies, Local Communities and Other. Subsequently, the stakeholders have been mapped (Figure 3), by the core team of STKs, according to their influence and interest in restoration upscaling to provide key insights on their role and expectations for developing and advancing the business plan.

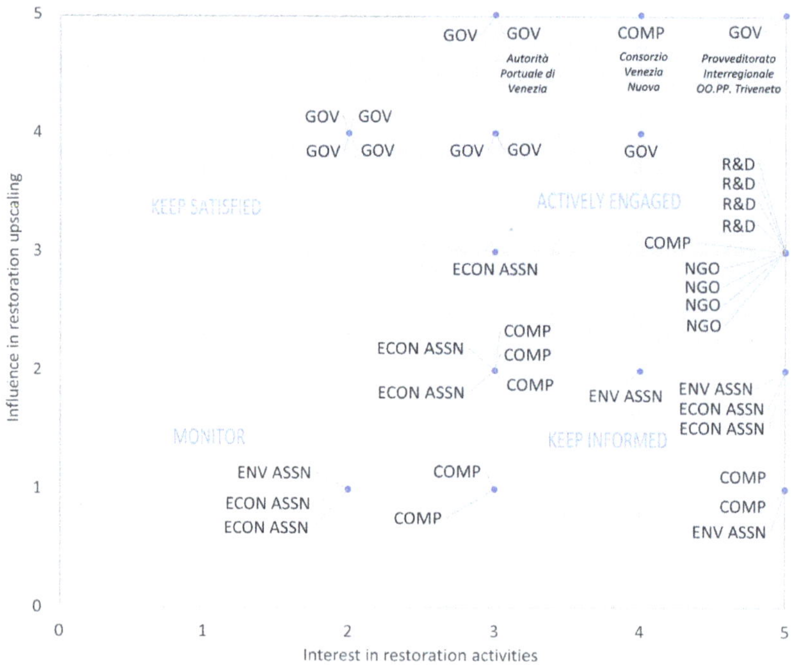

Figure 3. Interest–influence map of stakeholders' category for restoration upscaling on the Venice Lagoon Pilot. The map only mentions the name of the two initiators of the upscaling restoration work (Provv. OO. PP. and VPA) supported by the concessionary company Consorzio Venezia Nuova in implementing the work.

The upper-right quadrant shows two governmental organizations with a very high score, i.e., Provv. OO. PP. (5,5) and VPA (3,5), who are the initiators of the current and upscaling restoration activities in the Venice Lagoon and are responsible for safeguarding Venice and its lagoon (the former) and keeping the channels navigable (the latter). These are followed by a company, Consorzio Venezia Nuova (5,4), who is the concessionaire of Provv. OO. PP. for the implementation of the interventions to protect Venice and its lagoon, including the MOSE system (the MOSE consists in a series of mobile barriers located at the lagoon inlets to protect the city of Venice from high tides). Indeed, among the multiple competent authorities and stakeholders, the only one legally empowered to decide and implement restoration activities is the Provv. OO. PP. The initiators are followed, in terms of score, by the four research organizations involved in the REST-COAST project (the Consortium for the coordination of research activities concerning the Venice Lagoon system (CORILA), University Ca'Foscari of Venice (UNIVE), Euro-Mediterranean Center on Climate Change (CMCC) and University of Padova (UNIPD)), all NGOs and one economic association. The engagement strategy was to actively collaborate in co-shaping the BP through dedicated meetings, workshops and events.

3.2. Framing the NbS-ES Business Model Canva through Stakeholder Co-Development

The conceptualization of the BMC and its content is the result of a strong iterative process with the core team of STKs spanning 10 months. This was started from the value proposition, built-up by assessing the relevance of the problems impacting the Venice Lagoon, and was addressed through restoration of the saltmarshes, the relevance of ESs and biodiversity (BdV) improvements derived by NbSs restoration activities and the potential rewards. The value proposition was nourished with stakeholder participation through the survey, which highlighted the importance of saltmarshes and their benefits for Venetian STKs. From the 38 stakeholders in the sample, 28 responses have been received, coming from the different STK categories: Environmental associations, Companies, Education and Research institutions, and Other with five responses each, followed by Government and Environmental NGOs, with three responses each, and, finally, Economic associations and Local Community represented by one answer, respectively.

Approximately one third of the respondent STKs are personally involved in the Venice Lagoon including scientific research (36.4%) and/or recreational activities (27.3%), while political and administrative work was around one fifth (18.2%). Finally, a minority of the respondents took part in hunting (9.1%), professional fishing (6.1%) and/or navigation for passenger transport (3.0%). On average, respondents indicated that they take part in three activities related to saltmarsh restoration, with a predominant role of scientific research (27.9%), very close to recreational activities such as rowing and birdwatching (22.1%), policymaking and administration (18.6%), hunting (10.5%), professional fishing (9.3%), passenger navigation (7.0%) and freight transport (4.7%). Thereafter, the respondents were introduced by various problems affecting the lagoon, and were asked to rank the degree of importance of each of the problems, i.e., erosion, biodiversity loss, habitat loss, sediment loss, water pollution, subsidence, flood damage and soil contamination, showing that all the problems were considered more than neutral on average (Figure 4A), confirming the assessment of these problems from the core team. Considering the remediating influence of saltmarsh restoration, assessed with the next question, there is coherence between the most relevant problems and the highest influence from restoration, showing as the highest influential on erosion (78%), biodiversity loss (75%), habitat loss (71%), sediment loss (54%), flood damage (39%), water pollution (7%), subsidence (7%) and soil contamination (4%) (Figure 4B–I). This confirms the perception from stakeholders of the capability of saltmarshes restoration to positively impact the most important problems.

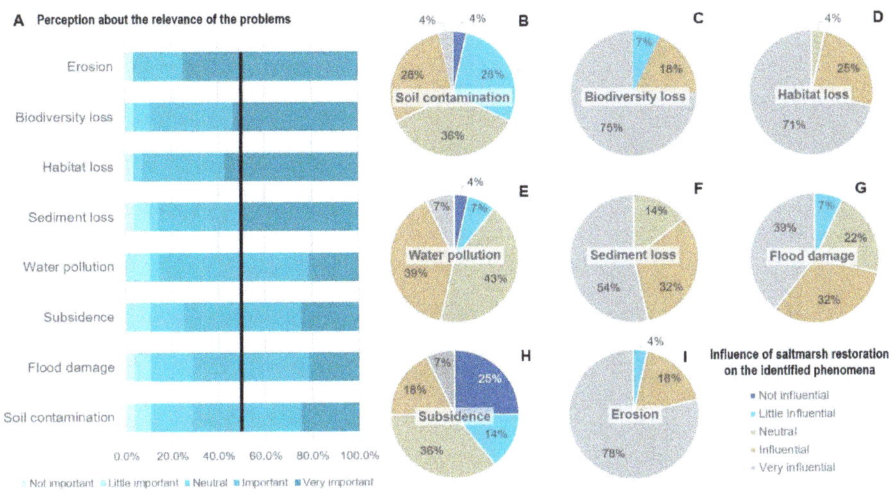

Figure 4. The figure shows two main graphs illustrating, respectively, the importance of the phenomena affecting the Venice Lagoon (**A**) and the positive influence of saltmarsh restoration on these problems (**B**–**I**).

Additional problems were identified by six STKs, such as wave motion, limited speed control of motorboats and loss of territorial identity.

In order to assess the value proposition, ESs and BdV were introduced, and the STKs were asked to indicate the importance of restoration for ES provisioning, highlighting that most respondents think restoration is important or very important for biodiversity [BdV] (96.4%), reduction of coastal erosion [RCE] (92.9%), carbon sequestration [CCR] (82.1%), reduction of flood risk [RCF] (75.0%), food provisioning [FP] (64.3%) and water quality purification [WP] (60.7%). Moreover, the STKs were presented with environmental, economic, social and cultural benefits associated with the restored saltmarshes occurring alongside the provisioning of ES and BdV improvements, indicating an overall agreement with the identified benefits. Overall, among the seven environmental benefits, the STKs primarily indicated a very high and high agreement, highlighting their knowledge of restoration impact such as naturalization of saltmarsh dynamics (89.3%), availability of habitat and improvement of biodiversity (92.9%), carbon capture and storage (82.1%), improved water quality purification (82.1%), provisioning of protection and nursery for juveniles (89.3%) and availability of new areas colonizable by native vegetation (89.3%).

The majority of the 20 economic benefits associated with the ESs and BdV received a large, varied distribution of responses. Focusing on the economic benefits only derived by biodiversity (considering the environmental benefit with highest importance), the survey revealed among the most important economic benefits ecotourism activities, birdwatching (both 82.1%), opportunities for developing educational activities (78.6%), corporate social responsibility activities related to NbSs (64.3%) and activities arising from circular economy (46.4%). The same dispersion pattern can also be observed for the six social benefits, as well as the three cultural benefits (see the Supplementary Material).

Finally, since the objectives of the restoration plan are correlated with the value proposition (Section 2.2), the participants were asked to indicate their agreement with the inclusion of nine different objectives in the BP in relation to the protection and restoration of saltmarshes (Figure 5). Consistent with the previous answers, on the perception of the problems and the influence of restoration to solve them, the last question shows an overall agreement from STKs with the majority of the objectives.

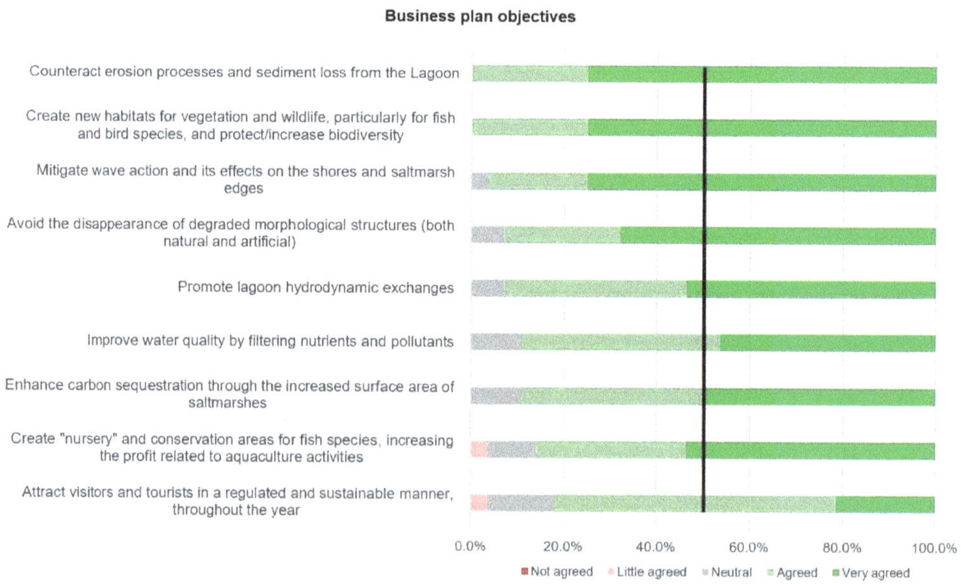

Figure 5. Level of agreement to include specific restoration objectives in the business plan. The legend illustrates the percentage of answers for the five potential options in the Liekert scale, from "Not agreed" to "Very agreed" although none of the stakeholders had chosen the lowest level of agreement (depicted in red).

Concerning the value capture, to complement the information collected from the core team of STKs and broad STK group, a questionnaire was circulated among tourists and visitors. Of the 91 respondents, 84 identified as visitor/tourist and eight identified as resident. In total, 58% of the respondents identify as male, 35% as female, 1% as non-binary and 2% prefer not to disclose this information (Figure 6A,B). Overall, the respondents indicated 190 reasons to visit the saltmarshes (Figure 6C), principally visiting the saltmarshes to observe nature (28.9%) and to enjoy the beauty of the area (26.3%). Subsequently, each of the respondents was presented with a detailed description of three visitation scenarios including (i) enjoying the beauty of nature; (ii) a birdwatching trip; (iii) a recreational fishing trip. These scenarios were populated with information about the Venetian saltmarshes, the flora and fauna that might be observed and the benefits that could be enjoyed during a visit in said scenario. Then, the respondents were asked what they would be willing to pay for such a visit (Figure 6D), and most of the respondents indicated a willingness to pay up to EUR 25 for each of the scenarios, 49.5%, 45.1% and 18.7% for enjoying the beauty of nature, birdwatching and recreational fishing, respectively.

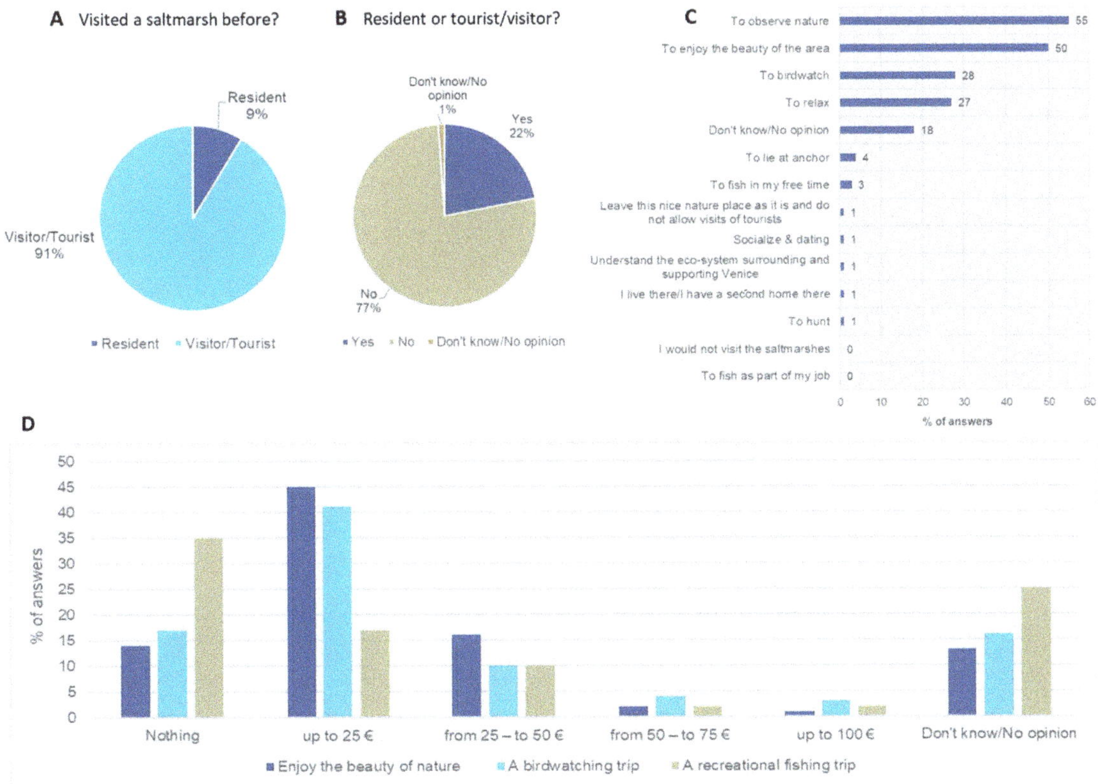

Figure 6. Displaying the results from the WTP analysis, including (**A**) whether respondents already visited saltmarshes in the Venice Lagoon, (**B**) the type of respondent, (**C**) the reasons for the visits to the saltmarshes and (**D**) the willingness to pay.

The results of the iterative process, including the collaboration with the core team of STKs, the broad STKs questionnaire and the WTP questionnaire to the tourists and visitors is synthesized in the NbS-ES Business Model Canvas in Figure 7.

VALUE PROPOSITION		
Problems to be addressed -Erosion -Damaged from flooding -Subsidence -Loss of sediments -Wave motion -Water pollution and solid waste -Loss of habitats and biodiversity	**Environmental benefits** -Natural (re)vegetation dynamics [RCF] -Habitat and biodiversity [RCE] -CO_2 capture and reduction [CCR] -Higher self-depuration capacity [WP] -Protection and nursery [FP] -New areas colonised by autochthonous vegetation species; habitats for feeding, roosting and breeding birds [BdV]	**Economic benefits** -Ecotourism including birdwatching [RCF, RCE, BdV] -Reduction of damage costs from flooding, coastal erosion [RCF, RCE] and sediment transport to landfill -Fish yield and recreational fishing [FP] -Eco-labeled goods [BdV] -CSR and philanthropy [RCF, RCE, BdV] -Restoration works [RCE, BdV] -Circular-economy activities, including artistic activities [BdV] -Educational activities [BdV]
Restoration problem-solving -Avoiding disappearance of degraded morphological structures -Contrasting erosion and sediments loss -Attenuating wave motion -Reducing water pollution and solid waste -Creating new habitats for wildlife	**Social benefits** -Safer life [RCF] -Space of quality, higher sense of attachment, higher human health [RCE] -Traditional fishing [FP] -Better water quality for social purposes [WP]	**Cultural benefits** -Traditional fishing techniques [FP] -Maintain cultural landmarks [RCF, RCE, BdV] -Human-nature experience [BdV] -Educational activities [BdV]
VALUE CREATION/DELIVERY		
Key partners Provv OO. PP. and Port Authority of Venice (initiators), supported, through procurement arrangement, by several implementers of scientific (CORILA, CMCC, UNIVE and UNIPD) and technical competence.	**Regulation** -UNESCO World heritage site -3 Legal systems: ordinary, special, and commissarial -3 Policy levels: EU, National, Regional -Authority for the Lagoon of Venice replacing Provv OO. PP.	**Governance** Multiple administrative levels: n.9 municipalities, n.2 provinces (Venice and Padua), Metropolitan City of Venice, Veneto Region, Italian State, n.1 Ministry and Local Administration Committee ("Comitatone")
Key Resources -Scientific and technical expertise (director of works, multidisciplinary team, RUP, CSP, CSE) -Technical means (equipment) -Tools (e.g., monitoring instruments)	**Customers segments** No direct customers -Ecotourists -Students for educational activities -Entrepreneurs contributing to restoration activities	**Stakeholders** 38 stakeholders (25 private and 13 public) split into 7 categories: 10 Gov org., 4 Research org., 8 Companies, 4 NGOs, 7 Economic associations, and 4 Environmental associations
Key activities -Phase 1. Interventions to protect the edges of the saltmarshes (5 yrs) -Phase 2: Nourishment of the salt marshes with new sediments (1 yr)	**Customer relations & channels** -Wide range of means and social media -Series of attractions reward NbS-friendly actions and services -Annual advanced training course -Basic e-learning courses -Events, workshops, seminars, etc.	**Beneficiaries** -tourists and visitors -(social) entrepreneurs -companies -Artists and cultural organisations -Schools -Citizens and local communities -Policy makers
VALUE CAPTURE		
Costs -Costs for phase 1 (362 €/LM), phase 2 (19 €/m^3 of sediment nourished) -Costs for naturalisation (2,900 €/ha of saltmarsh) -Saving costs for sediments (by water and land) transport to landfill (53€/ha of saltmarsh)	**Revenue streams** None directly to the initiators, but to other stakeholders from NbS interventions, ecotourism, eco-labelling, education.	**Financing and funding** Taxation, EU projects, Public grants (Superintendence of Archaeology, Fine Arts and Landscape for the Municipality of Venice and Lagoon, Coldiretti, Veneto Agricoltura), private investors and companies.
HORIZONTAL ASPECTS		
Barriers/Enablers -Limited stakeholder engagement, communication, collaboration, bureaucratic rigidity -Improved transparency	**Impact Indicators** -Impact on ES, biodiversity -Impact on human capital and produced capital	**Risks/De-Risk** -Emissions for sediment transport -Delay for changes in responsibilities -Additional funding availability

Figure 7. Overview of the NbS-ES Business Model Canvas, indicating the value proposition in light green, the value creation/delivery in green and the value capture in dark green.

3.3. From Business Modelling to Business Plan

Following the methodological framework explained in Section 2.2, this work allows for the development of a BP tailored to the peculiarities of NbSs in the Venice Lagoon. The BP delineates as a NbS upscaling plan the active maintenance of the already existing 1600 hectares of artificial saltmarshes and mudflats, by utilizing the results of the ongoing restoration (i.e., 138 ha of restored artificial marshes in the central lagoon) to model the upscaling. The BP describes the governance, management, business and financial strategies for upscaling saltmarshes restoration, considering a temporal horizon of five years, building on the results achieved through the NbS-ES business model canvas. The main findings are presented in the Sections 3.3.1–3.3.5.

3.3.1. Mission and NbS Restoration Objectives

Two public bodies, Provv. OO. PP and VPA, have jointly advanced, up to now, the recovery and maintenance of the lagoon of Venice habitats and morphology, according to their institutional mission. The Provv. OO. PP. is responsible for the safeguarding and management of the Venice Lagoon, through ordinary and extraordinary maintenance works contracted through its concessionaire CVN or private companies that secure contracts relative to specific projects; the VPA is accountable for planning, coordinating, promoting and controlling port operations, by dredging the lagoon canals to guarantee the safety of entering ships. Soon, Provv. OO. PP. will be replaced by the Authority for the Lagoon of Venice, whose stated mission, addressing mainly public services, drives the BP for upscaling restoration and its objectives. The nine main objectives of the upscaling restoration plan (Figure 6), endorsed by the stakeholders, include (i) avoiding the disappearance of the degraded natural and artificial morphological structures, (ii) contrasting the erosion process and the negative trend of loss of sediments in the lagoon, (iii) attenuating the wave motion and reducing its effects on the shores, (iv) favoring hydrodynamic exchanges within the lagoon, (v) filtering nutrients and pollutants resulting in an improvement of water quality, (vi) creating new habitats for wildlife, in particular for birds, thus improving and safeguarding the biodiversity, (vii) creating nursery, protection and provision areas for fish and increasing yields of fisheries, (viii) attracting, in a regulated and sustainable manner, visitors and tourists all year round and finally (ix) capturing carbon (CO_2) and sequestering it in the soil.

3.3.2. NbSs Multifunctional Offering

The upscaled restoration plan delineates different interrelated problems (Section 3.1, Figure 4) allowing the improvement of five ESs as well as BdV, that will produce a large variety of environmental, economic, social and cultural benefits to multiple stakeholders and beneficiaries, assessed through the STK questionnaire (Section 3.2). These benefits are listed in the NbS-ES Business Model Canvas (Figure 7) and concurrently appease multiple stakeholders alongside the two initiators of the restoration upscaling. Through the kaleidoscopic composition of the NbS offering, economic gains have been analyzed more closely, since they may generate potential profits for several stakeholders positively impacted by saltmarshes restoration that can contribute to financially sustaining NbS restoration investments as a form of compensation for their own economic gains.

- Restoration works [RCE, BdV]: Economic associations and companies being paid for delivering goods and services in NbS restoration and environmental monitoring. Environmental associations and NGOs being paid for services related to protecting the lagoon. Social cooperatives being paid for delivering goods and services that could also employ vulnerable people, contributing to social inclusion. Research institutions and education being paid to support the initiators in the restoration interventions;
- Ecotourism [RFC, RCE, BdV]; birdwatching [BdV]; recreational fishing [BdV, FP]; traditional fishing [FP]: Companies, economic and environmental associations, players in the eco-touristic value chain benefiting from the increase of visitors and tourists who will pay for different services (e.g., recreational fishing, birdwatching, etc.);

- Increase of fish yield [FP]: Fishermen and valley owners/managers being paid for a higher quantity of fish due to the improved water quality and habitats;
- Hunting [FP, BdV]: Companies and associations providing hunting services benefiting from an improved lagoon able to attract more birds, some of which are huntable, leading to a higher number of hunters willing to pay for this service;
- Eco-labeled goods [BdV]: Companies and economic associations being paid for producing goods in an eco-sustainable manner and also with the use of native vegetation species of the lagoon for the production of typically Venetian goods, saltmarsh honey, artisanal liqueur, seaweed fabrics, natural cosmetics, herbal teas, etc.;
- Educational activities [BdV]: Research institutions and environmental associations being paid for providing educational activities and training about NbSs restoration;
- Corporate social responsibility [RCF, RCE, BdV, CCR] and philanthropic activities [RCF, RCE, BdV, CCR]: Companies from the Venice area, and elsewhere, may invest to create shared value in the territory receiving back a positive reputational impact, and can benefit from offsetting, i.e., from carbon emissions, water pollution, etc. Similarly, philanthropists may want to have a positive impact in the Venice area;
- Circular economy activities [BdV] and artistic activities [BdV]: The removal of waste from the saltmarshes could contribute to the creation of a circular economy model by valorizing the materials and turning them into products to be reused in artistic products, testing the opportunity to reuse them for the maintenance of the saltmarshes, or creating Venetian souvenirs that help in raising awareness on waste reduction in the lagoon (e.g., starting a collaboration with a waste management operator);
- Cost reduction for solid waste management [WP]: The restoration works in the saltmarshes, including waste removal, offer a strategic opportunity to reduce costs. Typically, the removal of waste from these areas involves significant expenses. Aside from some occasional voluntary clean-up efforts organized by local associations, the costs of waste removal would need to be fully borne by the competent authorities. By integrating waste removal into large-scale restoration activities, it is possible to avoid duplicating efforts and expenses, such as those related to transportation. Additionally, the restored and cleaned environment brings broader benefits, benefiting both the environment and the community;
- Reduction of damage costs from flooding, coastal erosion [RCF, RCE]: Governments responsible for environmental risk management can benefit from cost saving due to the mitigation effect that a saltmarsh can have on flooding and coastal erosion;
- Cost reduction for water purification [WP]: Governments and agencies responsible for water resource management can benefit from cost saving due to an improved self-depuration capacity of the lagoon system thanks to saltmarsh restoration.

The market analysis pinpoints four main and strongly connected market sectors: (i) ecotourism; (ii) economic activities for organizations contributing to implementing NbS interventions; (iii) economic activities linked to eco-labelled products; (iv) educational activities on nature restoration. In particular, according to the International Ecotourism Society, ecotourism can be considered a subcategory of sustainable tourism restricted to the natural environment, and is driven by an unspecified need to experience a natural habitat that is largely unaffected and untouched [48]. It addresses several market niches (agro, rural, enogastronomy, wildlife, heritage, cycling, kayaking, etc.), and can deliver products or services in a synergistic manner to other market sectors, while protecting nature and bringing prosperity to local communities. Ecotourists also spend more than the average tourists [49]; many ecotourists place more value on the learning experience and environmental education, and are more satisfied if they have a knowledgeable guide [50]. Similar to the ecotourist, the birdwatcher is likely a highly educated and high-income tourist, and spends more than the average tourist; they generally prefer local guides, and because of this, they are more likely to converse with and integrate into the local community [51].

The NbSs multifunctional offering, however, must take into account the other type of benefits assessed with stakeholders, including the following:

- Environmental benefits: Natural (re)vegetation dynamics [RCF], habitat and enhanced biodiversity [RCE], carbon capture, CO_2 reduction in the air [CCR], higher self-depuration capacity [WP], offering protection and nursery for fish communities [FP], new areas colonized by autochthonous vegetation species and habitats for feeding, roosting and breeding birds [BdV];
- Social benefits: Safer life [RCF], space of quality, higher sense of attachment [RCE], traditional fishing [FP] and better water quality for social purposes [WP];
- Cultural benefits: Traditional fishing techniques [FP], maintain cultural landmarks [RCF, RCE, BdV], human–nature experience [BdV] and educational activities [BdV].

The majority of these benefits are difficult to monetize, requiring complex evaluation methods and data availability, but their potential positive result has to be judged by investors, as these elements contribute in creating shared value, allowing activities that are critical for the local community.

3.3.3. Governance, Management, Business and Finance Strategies

The BP describes four strategies, i.e., governance, management, business and finance that are strongly interconnected. The governance strategy compounds the regulatory framework and governance structure for the restoration in the Venice Lagoon described in Section 2.3. The new Authority for the Lagoon of Venice will be established soon, constituting a centralized entity for all the decisions related to Venice and its lagoon, favoring and streamlining the planning and management processes of the interventions at the lagoon scale. The authority manages public lands and waters within the lagoon, issues relevant permits and coordinates with other agencies on infrastructure and environmental projects. The regulatory laws frame the objectives, responsibilities, instruments, measures and financial resources for carrying out future restoration activities.

The management strategy defines the organizational model through which "key partners" and other implementers achieve the work, including the necessary monetary and non-monetary resources. The initiators of the restoration works, i.e., the Provv. OO. PP. and VPA, will arrange procurements with several implementers, taking decisions leveraging the knowledge of the research institutions for activities including planning, data analysis and monitoring. The core team of STKs, together with other relevant stakeholders, including those from the CORE-PLAT, are willing to participate in the planning process for upscaling restoration, by identifying, among others, NbS interventions to be implemented and the most suitable locations in the lagoon.

The NbS interventions consist of the maintenance of the 1600 ha of saltmarshes in the Venice Lagoon and will be conducted in two phases, as previously described (see Section 2.3), spanning five years, including partially concurrent planning and implementing and monitoring activities. The two phases are the precursors for the naturalization of the saltmarshes allowing for the provisioning of ESs.

The "key resources" for the restoration activities include (i) a multidisciplinary scientific and technical team with in-depth knowledge and experience in the lagoon; a director of the works to coordinate the implementation of the interventions supported by a project manager (RUP), a Safety Planning Coordinator (CSP) and a Safety Coordinator of the Execution phase (CSE); (ii) technical means and tools, e.g., equipment required by the contracting companies, multiparametric probes for evaluating water quality, soil sampling cores, sediment trap, (kaoline) markers for sediment compaction, biological monitoring means, including drones to confirm the naturalization process continues as expected. The monitoring verifies that the ecosystem is moving in the desired direction. Moreover, a detailed map of the status of the saltmarshes in the lagoon is crucial for assessing the restoration costs and setting out the works on site.

The co-development of the value proposition with stakeholders helps to discern the financial rewards, and therefore to assess the NbSs multifunctional offering, that the business strategy has in view to land in the market, spatially and temporally generating revenues from economic activities executed by different stakeholders, including:

- Companies and economic associations engaged in the ecotourism sector;
- Companies and economic associations (farming, fishing, hunting, etc.) benefitting from economic activities closely related to biodiversity and/or selling eco-labeled products;
- Research organizations, companies, NGOs and environmental associations engaged in educational activities;
- Companies, NGOs, economic and environmental associations engaged as implementers and subcontractors in restoration works.

Except for the implementation of the restoration works, assigned by the initiators through procurement arrangements to companies and other stakeholders, none of the described economic activities are performed by the initiators. Therefore, there are no direct clients and no revenues for them that the business strategy can directly take into account. Customers mainly pertain to the other categories of stakeholders, generating economic transactions that can result in indirect value capture streams, e.g., from taxations. Additionally, the business strategy shall consider potential environmental, social and cultural benefits even if these cannot immediately generate direct revenues, but instead play a crucial role to sustain human activities in the area, even when their negative externalities are not measured and threaten the carrying capacity of the Venice Lagoon ecosystems.

The category of "customer relationship activities" do not directly refer to the initiators, rather mainly relate to stakeholders for goods and services offered to customers and obtained by the improvements of ESs/BdV. The initiators can promote the generation of economic transactions between stakeholders to strengthen the endorsement of the upscaling restoration plan by the whole business community and civil society. This can have an impact on customer relationship activities, e.g., the initiators can stimulate the activation of a series of attractions that reward climate-friendly actions and services where tourists and residents may redeem rewards by showing proof of green actions; for instance, by participating in clean-up efforts in the lagoon.

In terms of "distribution channels", NbSs restoration interventions are not intended to transfer commercial products or services to the market by the initiators, and therefore distribution channels should be reinterpreted in terms of communication strategies and channels necessary to engage a broad set of stakeholders and beneficiaries, including the general public. Several stakeholders, such as research institutes and universities, may organize advanced courses on key aspects of environmental protection and restoration, targeting master or PhD students and practitioners, as well as a basic training course configured as "knowledge pills" on primary concepts of environmental protection and restoration; ad hoc events and debates are key to raising awareness among the general public.

The finance strategy looks at attracting funding to close the financial gap for upscaling restoration interventions, including research and testing of more sustainable NbS techniques, otherwise too expensive, and executing a widespread and continuous recovery plan of the lagoon morphology and habitats, that would prevent the need to spend very large sums to repair the significant damages that occur if the ordinary maintenance is not carried out. Moreover, funding will be needed to increase biodiversity through dedicated projects seeking the recovery of specific endangered or declining target species. The finance strategy will pursue blended finance, considering public and private investments. Public investors could comprise those with a high interest in preserving the lagoon landscape, such as the Superintendence of Archaeology, Fine Arts and Landscape for the Municipality of Venice and Lagoon. Similarly, agencies whose core business is to improve the competitiveness of businesses and production chains in the agricultural, agri-food, forestry and fishing sectors could see benefits from an upscaling plan for restoration in the Venice Lagoon as a means for improving the productivity of the area. Another source of public funding may stem from EU projects, already used for ongoing restoration interventions. Private investors are not yet identified. Investors could curry favor with headquarters, storefronts in the city or might be private companies steeped in tourism that could be encouraged to invest if they foresee economic returns in terms of an increased number of eco-tourists. Private investors can also include companies that try to offset the impact of their activities and projects.

3.3.4. Economics and Financial Projections

The economics of the BP, projected in a temporal range of 5 years, include costs and revenues. The cost structure analyzed three main components, i.e., life cycle costs, opportunity costs and costs reduction. Life cycle costs include all costs to be afforded for all three stages of saltmarshes restoration, i.e., planning, implementing and monitoring. Their estimation was based on the analysis of the metric calculations of the saltmarshes' executive projects included in REST-COAST for a total of 138.43 ha (Appendix D). The costs, validated by Provv. OO. PP. and summarized in Table 1, are described for each restoration stage:

- Planning: Surveys, assessments and investigations to support the project design (e.g., expeditious monitoring and survey on of the morphological/ecological state of the saltmarshes, chemical analysis costs and technical checks), supervision of the works, coordination for safety during the design, execution and testing phase of the works;
- Implementation: Interventions to protect the edges (first phase) and nourishment of the saltmarshes (second phase) (This estimation is based on the regional price list for 2023. To these values, 20% must be added to consider the increased cost of materials registered in 2024 (see https://prezziario.regione.veneto.it/, accessed on 18 June 2024));
- Monitoring: Multiparametric probes, top-bathymetric surveys, monitoring of terrestrial vegetation and sedimentation, monitoring BdV (i.e., birds and seagrasses).

Table 1. Cost structure of the NbS restoration upscaling plan.

Restoration Activities	Total Duration	Planning	Implementation	Monitoring	Total Costs
First Phase—Intervention to protect the edges	5 years	2 years	1 year	At least 1 year	EUR 362/LM
Second Phase—Nourishment of the saltmarshes	5 years	2 years	1 year	At least 1 year	EUR 19/m^3 of sediment nourished
Monitoring/Control Phase [1]	Up to 10 years		Between 2 months and a year	At least 1 year	EUR 2900/ha of saltmarsh

[1] It also follows the naturalization process that reaches the optimum condition for biodiversity after 3 to 10 years.

No opportunity costs can be assessed since no other alternatives to saltmarshes restoration and creation are contemplated/planned in the lagoon of Venice. Concerning the potential cost reductions, four main considerations can be made: firstly, reuse of good quality sediments dredged from the lagoon channels for saltmarsh nourishment, besides being a form of circular economy, may reduce the costs for the VPA compared to its transfer to a landfill (once moved, the sediments are considered hazardous waste), leading to a significant environmental impact due to the emissions linked to the sediments' transport by water and land, thus a reduced carbon footprint. Since in the coming years VPA plans to extract circa 3 million m^3 of sediments, this represents an important opportunity for the restoration upscaling plan. Secondly, reduction of damage costs from erosion and flooding are difficult to measure since there is a lack of precise information regarding the investments already in place to fight erosion, and the non-proximity of the saltmarshes to the old city of Venice with assets at risk of flooding. Thirdly, a reduction of costs for water purification is difficult to assess since there is no association between saltmarsh restoration and reusing water for commercial use (e.g., drinking, irrigation). Finally, reducing costs for solid waste management is difficult to quantify, as no regular service for cleaning the saltmarshes of the Venice Lagoon has been implemented yet. However, based on data collected during voluntary clean-up initiatives in March 2023 and April 2024 (Legambiente Venezia, pers. comm.), we can estimate that the waste density in the saltmarshes is approximately 0.22 tons per hectare. Waste removal from these areas implies significant costs, including labor, mechanical equipment, suitable boats for transportation and disposal in legal landfills, with the latter costing EUR 240 per ton (April 2024). All these expenses would need to be fully covered by the competent authorities. Integrating waste removal into large-scale restoration activities helps avoid duplicating efforts and related costs, such as transportation. By combining restoration with waste removal, resources can be optimized, leading to a reduction in overall costs compared to addressing these tasks separately.

Concerning revenues, these can be generated through direct and indirect value capture streams. In the NbS-ES BP, direct value capture streams are linked to ES and BdV improvements that can be monetized and/or translated to economic activities that are based on said improvements and stimulate economic transactions by trading goods and/or services. However, at present, the initiators do not conduct economic activities for which these economic transactions are generated. On the contrary, other stakeholders are involved in such economic transactions, delivering goods and services as described in the NbSs offering and business strategy. As a consequence, revenues cannot be assessed in a conventional manner and they do not directly contribute to the compensation for the restoration costs borne by the initiators. On the other hand, the economic transactions will generate indirect value capture streams from taxations (e.g., value-added tax), and positively contribute to a broad range of economic activities and sectors impacted by restoration, including the green economy and bioeconomy, with positive fallout in terms of employment, GDP, investments. In particular, leveraging the well-consolidated touristic sector that characterizes the city of Venice and its lagoon, and the challenge to sponge it and find a better balance between the growing demand and the need for sustainability, the business strategy can select ecotourism, considering its various interconnections with the other sectors identified in the NbS multifunctional offering.

Financial information usually includes the estimation of the financial gap assessed as a difference between all costs (including costs reduction) and revenues. Since neither the overall costs are available nor are these directly counterbalanced by revenues, the financial gap cannot be calculated. Moreover, the absence of specific revenues directly associated with the initiators of the upscaling restoration activities, and the lack of information about the temporal distribution for funding, does not allow for the calculation of cash flow.

An overview of the business model framework is provided in Figure 8. It portrays four types of arrangements and potential transactions [11]: (i) granting, provided by capital suppliers to the initiators of the upscaling NbSs restoration interventions with the expectation of receiving non-monetary rewards; (ii) financing, provided by financiers interested in making productive use of their capital by securing future monetary returns; (iii) procurement, provided as form of payments for services outsourced by the initiators to implementers or subcontractors; (iv) value capture, provided as forms of payment to goods and services paid by clients to stakeholders, but not directly to the initiators.

3.3.5. Risk and Contingency Plans

Governance and regulatory risks relate to, e.g., the DM 86/2023, i.e., the "Regulation containing provisions for the issuing of authorizations for the movement, in sea areas located within the lagoon border of Venice, of the sediments resulting from the excavation of the seabed of the lagoon border", which rules sediment reuse. Indeed, the sediments now have to be characterized not only in the place of dredging but also in the place where they will be released before the refilling phase can start. This will limit the general sediment availability for refilling the saltmarshes. The VPA will continue to dredge the channels to make them navigable and this could lead to the risk that the VPA will sustain additional expenses for the transport of the dredged sediments to the landfill, leading in this way to a further environmental impact due to the emissions linked to the sediments transport by water and by land. Another risk may arise with the appointment of the new Authority for the Lagoon of Venice that could lead to some changes in the definition of responsibilities, causing delays in the full operation of lagoon management, as well as in the administration of works (e.g., definition of sites to be restored and timing).

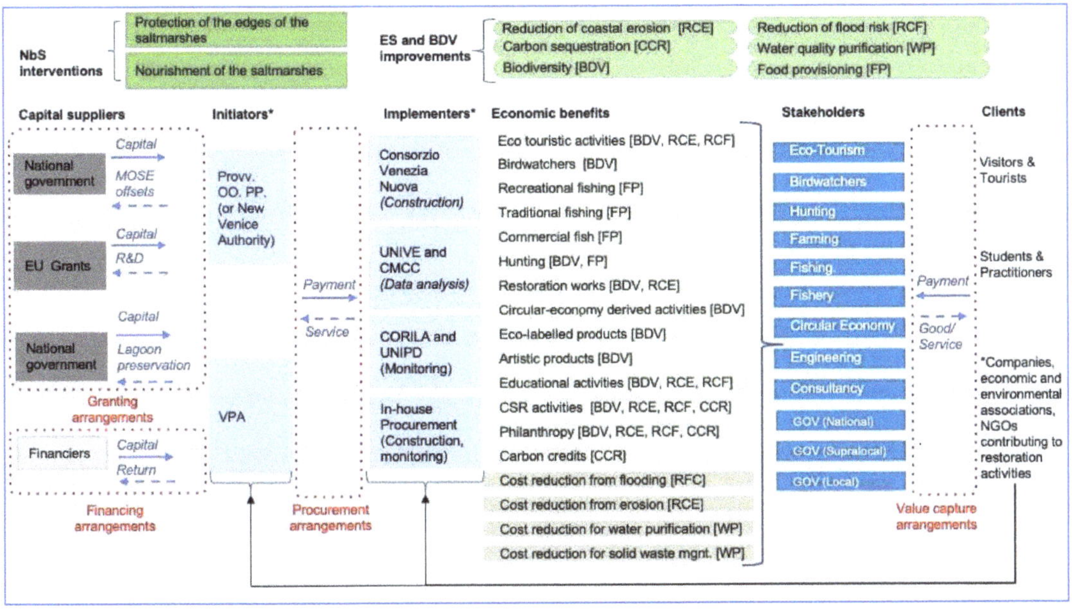

Figure 8. Advanced business model framework for the Venice Lagoon. The asterisk shows that companies, economic and environmental associations and NGOs may be paid by the implementers and/or the initiators of the restoration works through procurement arrangements.

Although of good chemical and ecotoxicological quality, a biophysical risk may be linked to the type of sediments used for the restoration; in particular, for the refilling of saltmarshes; sediments are generally mainly constituted of sand, and this could adversely affect the colonization of the saltmarsh by typical vegetation species. Particular attention should also be paid to evaluate the altimetry of the sediments nourished as this may create an opportunity for the invasive species (e.g., *Baccharis halimifolia*) to displace native species such as *Spartina*, *Salicornia*, *Limonium* spp. Additionally, some risks originating from climate change and extreme meteorological conditions may hamper the success of restoration or alter the natural revegetation dynamics.

Funding and market risks could stem from the challenge of significant upfront investment and securing funding, as well as the demand for NbSs restoration interventions which may be affected by changes in political priorities and economic conditions.

4. Discussion

Coastal NbSs can be more effective if designed for a river-to-sea scale, considering the adjacent landscape and the socio-economic context of stakeholders and beneficiaries [52]. Innovative NbS techniques, enhanced funding, socio-economic engagement and a transformative shift in governance can act as enablers for upscaling restoration, overcoming technical, financial, economic and governance barriers, thus contributing to filling the present implementation gap [17,19]. In this case study, a limitation for restoration upscaling proved to be the lack of a precise map and spatial plan of the current status of the saltmarshes in the Venice Lagoon (i.e., level of degradation, morphological and ecological status) which curtailed the proper estimation of restoration activities and related costs. Moreover, it has been challenging to apply the necessary economic evaluation methods for properly quantifying the various economic benefits, considering the temporal horizon of the present research and the available data that are mostly not owned or shared by the initiators.

Despite the above hindrances, decision-makers and initiating actors of NbS projects are starting to demand user-friendly tools that offer guidance at least for the selection of suitable business models for NbSs [35]. The findings of the scientometric analysis have confirmed a gap in the available tools to better connect business planning with coastal restoration projects. The methodology for co-developing the BP, based on enabling stakeholders' participation at the key stages and capturing, through surveys and economic evaluation methods (e.g., WTP), their insights, perceptions and willingness to pay, shows evidence that it is crucial to frame the business plan within the local context, reducing the complexity of the process. The methodological approach faced the challenge of cultivating a large variety of relevant representative stakeholders in the Venice Lagoon, but the survey achieved a 73.7% of the sample, which is considered a high representation, substantiating the inputs collected in the research. This adheres to the need for creating legitimate participation processes, highlighted as key to the general acceptance of NbS initiatives [35]. In this line, the results of this research revealed a strong alignment between the stakeholders' perception about the importance of the problems impacting the Venice Lagoon and the perception about the potential influence of saltmarsh restoration to solve them. Similarly, their endorsement on all the objectives proposed in the business plan (Section 3.2). Moreover, the coaching component empowered the team in charge of business planning to further customize the business plan according to a variety of investor needs and financial criteria. In this sense, it has proven to be applicable beyond the specific case and type of restoration activities currently ongoing. The implementation of NbS restoration projects faces difficulties to properly consider NbS ES provisioning multi-functionality for developing suitable business and/or investment cases that can stimulate adequate fundings [21,53]. In this case, the NbS multifunctionality has revealed a large amount of potential economic gains but not directly for the initiators, often spread between different stakeholders potentially competing among them. Some examples of these circumstances have been identified for the restoration works, ecotourism, educational activities, etc. Additionally, the potential cost savings arising (flooding, erosion, waste management, etc.) are difficult to measure and benefit. Similar to findings from other authors in the field [35], this study encountered several types of environmental, social and cultural benefits presenting major difficulties to be quantified (e.g., space of quality, human health, higher sense of attachment and maintaining cultural landmarks). This requires carefully co-designed win–win strategies that may unlock synergies and cooperation between stakeholders that will enable them to capture all the value created in a holistic manner [21] as well as governance models and financial mechanisms enabling the consideration of revenues generated by other stakeholders as contributing to support the investments in the multi-party restoration plan.

Managing the interconnections between the four strategies of the BP is particularly challenging [35], and requires taking into account several aspects, as follows: (i) The presence of a clear governance strategy ready to be implemented. The "special" legal system sets different responsibilities of the different levels of administration (national and local) and requires their cooperation. After nearly 50 years since the passage of the first law, it remains necessary and urgent to define a more comprehensive planning and programming framework. This duly affects the management strategy as it may enable or impede the design and implementation of restoration, even from a business and financial innovation perspective. (ii) The border between business and finance is fluid, because innovative financial mechanisms, e.g., green bonds and environmental impact bonds, green loans and credit lines and green microfinance, not yet fully exploited in the Venice Lagoon, can trigger unexpected business opportunities as sources of financial creation [54] but also financial criteria from investors can shape the value proposition and the business strategy.

The WTP analysis confirmed, with a percentage of 49.5% of visitors and tourists willing to pay up to EUR 25, the key role that ecotourism and its market niche birdwatching, with

45.1%, can play for stimulating indirect value capture mechanisms (including taxation and private investments). This is aligned with the findings from the previous LIFE project VIMINE, which assessed the importance of saltmarshes for the people [55] who visit/live in or near the Venice Lagoon, and their preferences about the management of the landscape. Consequently, there is a potential to position ecotourism at the core of the business strategy for upscaling coast restoration.

This may transform the overexploitation of tourism in Venice into a NbS-driven tourism offer, more sustainable and decompressed, taking into consideration how to manage the social–ecological system in a way that the flow of multiple ESs is provided without eroding the ESs capacity [56]. This approach also raises the possibility of delivering value in a synergetic manner to multiple sectors under the umbrella of bioeconomy and green economy. Part of the revenues generated by ecotourism can be channeled directly back into conserving the ecosystem, and can be a vital tool to make sure these protected areas receive the funds they need [57]. However, it is crucial not to exceed carrying capacity, and utilize breaks and/or off-seasons, since overcrowding negatively impacts ESs and biodiversity that people pay to see [56,58,59].

Therefore, the results of this research demonstrate that it is possible to demystify the belief that business plans just pursue the commodification of nature, and instead steer us toward the emerging concept of a nature positive approach in which businesses, governments, public institutions and the general public take collective action at scale to reduce and remove the drivers and pressures fueling the degradation of nature, and work to actively improve the state of nature and the ESs it provides.

5. Conclusions

This study has pioneered the co-development of a framework bridging business modelling innovation tools with business planning in the NbSs context for restoration upscaling advocating multi-stakeholder engagement. The innovative process combines business management consulting skills with a holistic knowledge on ESs provisioning through NBSs restoration projects, from a multidisciplinary scientific perspective, as well as an executive approach. Additionally, this research produced an innovative business plan conceptualized for the public sector, able to connect public and private services, across multiple economic sectors. This link is considered of special significance since public bodies usually act as initiators in restoration projects, accessing mainly public funds (granting arrangements). Indeed, the business plan reveals how increasing costs for upscaling will call for a financial strategy that stretches beyond public funding, by spawning private investments through creative financing and value capture arrangements stimulated by economic advances in ecotourism, eco-labelled products, complemented by educational activities with financial rewards and compensated by appreciable amounts of cost savings. This process has empowered the Venice Lagoon core team while raising awareness about the multifunctional merits of NbSs among all stakeholders.

The application of the framework contributes to initiating an enabling environment at a scale that can attract investors and sustain the financing of landscapes in the near future. Future research steps should assess extending the adoption of methods for quantifying non-market values in monetary equivalent terms, identifying governance and financial mechanisms that can allow part of revenue generation to compensate the restoration costs afforded by the initiators, and putting the upscaling framework into practice in different geographic areas for outscaling.

Supplementary Materials: The following supporting information can be downloaded at: https://www.mdpi.com/article/10.3390/su16208835/s1, Figure S1: Graphs from the results of the Stakeholder survey.

Author Contributions: Conceptualization, U.P. and L.P.; methodology, U.P. and L.P.; formal analysis, U.P., L.P., F.C. and F.H.; funding acquisition, U.P., L.P., F.C., C.D. and S.T., investigation, U.P., L.P., F.C., F.H., S.T. and C.D.; writing—original draft preparation, U.P., L.P. and F.H.; writing—review and editing, F.C. and C.D.; visualization, U.P., L.P. and F.H. All authors have read and agreed to the published version of the manuscript.

Funding: This project has received funding from the European Union's Horizon 2020 research and innovation programme under grant agreement No. 101037097 (REST-COAST project).

Data Availability Statement: The data in this study are available on request from the corresponding author.

Acknowledgments: The authors of this work would like to express their sincere appreciation to the stakeholders involved in the REST-COAST platform for their invaluable support and collaboration, particularly in sharing ideas and insights, as well as participating in the surveys. They are especially grateful to VENIS S.p.A.—Venezia Informatica e Sistemi and Vela S.p.A. for their crucial collaboration in disseminating the WTP questionnaire through their institutional tourism promotion channels. Special thanks need to be expressed to E. Pagan from CORILA for her assistance in reaching out to stakeholders for market analysis purposes. Lastly, the authors wish to express our warmest gratitude to the Provveditorato Interregionale per le Opere Pubbliche for Veneto, Trentino Alto Adige, and Friuli Venezia Giulia and in particular to G. Assalone for his valuable technical support, availability and collaboration throughout the study.

Conflicts of Interest: Author Umberto Pernice was employed by the company Pernice Collaborative Innovation Management; Authors Francesca Coccon and Caterina Dabalà were employed by the company Consortium for Coordination of Research Activities Concerning the Venice Lagoon System (CORILA); Authors Fabienne Horneman and Silvia Torresan were employed by the company CMCC Foundation—Euro-Mediterranean Center on Climate Change; Author Laura Puertolas was employed by the company Albirem Sustainability.

Appendix A

The Scopus search was conducted on the 5 August 2024 using 15 queries (Table A1), combining relevant keywords such as "business model", "business plan", "business model canvas", "finance", "nature-based solution" and "restoration" in various combinations through Boolean operators, i.e., OR, AND.

Table A1. Queries used to support the scientometric analysis.

Query	Date of Search	Number of Publications
("Business plan") AND (Restoration)	5 August 2024	16
("Business plan") AND (NBS OR "Nature-based solution" OR "Nature based solution")	5 August 2024	0
("Business plan") AND (Restoration) AND (NBS OR "Nature-based solution" OR "Nature based solution")	5 August 2024	0
("Business plan") AND (Restoration OR NBS OR "Nature-based solution" OR "Nature based solution")	5 August 2024	16
("Business model") AND (restoration)	5 August 2024	82
("Business model") AND (NBS OR "Nature-based solution" OR "Nature based solution")	5 August 2024	28
("business model") AND (NbS OR "Nature-based solution" OR "Nature based solution") AND (restoration)	5 August 2024	3
("Business model") AND (NBS OR "Nature-based solution" OR "Nature based solution" OR restoration)	5 August 2024	107
("Business model canva") AND (restoration)	5 August 2024	0
("Business model canva") AND (NBS OR "Nature-based solution" OR "Nature based solution")	5 August 2024	0
("Business model canva") AND (NBS OR "Nature-based solution" OR "Nature based solution") AND (restoration)	5 August 2024	0
(Finance) AND (restoration)	5 August 2024	540
(Finance) AND (NBS OR "Nature-based solution" OR "Nature based solution")	5 August 2024	78

Table A1. *Cont.*

Query	Date of Search	Number of Publications
(Finance) AND (NBS OR "Nature-based solution" OR "Nature based solution") AND (restoration)	5 August 2024	14
(Finance) AND (NBS OR "Nature-based solution" OR "Nature based solution" OR restoration)	5 August 2024	604

This resulted in the identification of 1236 publications, published between 1992 and 2024 (Figure A1) primarily written in English. It should be noted that queries combining "business plan" AND "nature-based solutions" resulted in zero publications, while those including the "finance" term show a higher number of publications, in particular in relation to "restoration", while less information is available in combination with "nature-based solutions". This reveals a lack of consideration about the role that business plans can play in NBS restoration, complementing the strong focus of literature on financial aspects for restoration. This is one of the aims that this research addresses.

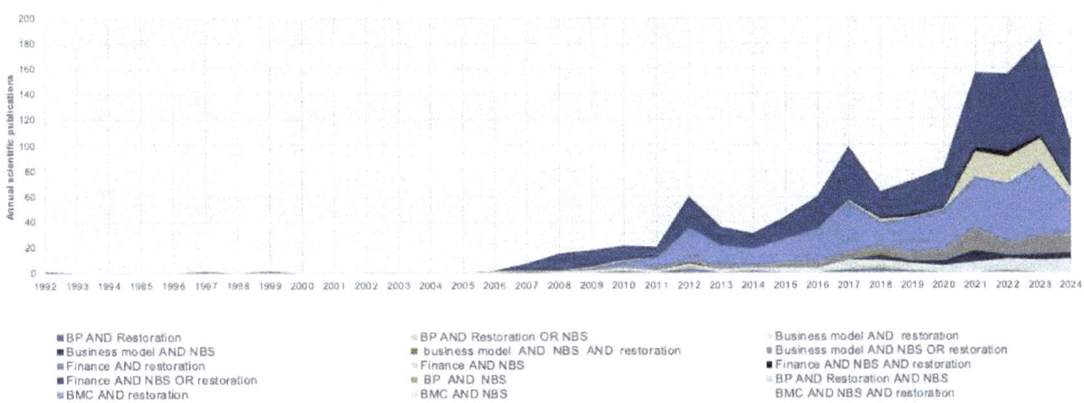

Figure A1. Scientific annual production of the different queries, highlighting an increasing trend of publications in the period between 1992 and 2024.

This bibliometric analysis also highlights that in relation to "business plan", emerging terms include "stakeholders" and "decision-making", showing the multifunctionality of NbSs for multiple stakeholders and the role that a business plan can play as a tool for decision-makers. These terms are repeated when it comes to "business model", and complemented with terms such as "cost-effectiveness", "cost-benefit" and "sustainable business development". These terms can be attributed to the business model's attribution of value creation, delivery and mechanisms, and associated evaluation methods able to capture socio-economic value frequently mentioned by other authors. Moreover, in relation to "nature-based solutions", central terms included "ecosystems", "biodiversity" and "climate change". This is rather fitting with the definition of the IUCN that NBS should contribute to preserving biodiversity and managing or restoring ecosystems sustainably while delivering a range of ESs [60]. NbSs are a way to leverage ecosystem services, where NbSs are the physical solutions that utilize the ecosystem's nature and natural capital to adapt to climate change and other stressors [61,62].

Appendix B. Broad STK Questionnaire: A Business Plan for the Protection and Restoration of the Saltmarshes of the Venice Lagoon, Only the English Version Is Reported Here for Readers' Convenience

Introduction

As part of the Horizon 2020 REST-COAST project, CORILA (Consortium for the Coordination of Research on the Venice Lagoon System), in collaboration with other project partners, is conducting a survey to understand the importance of the saltmarshes in the Venice Lagoon and to assess how to financially support the conservation and restoration of these environments. The aim of this questionnaire, developed within the European Horizon 2020 REST-COAST project, is to explore how to financially support the protection and morphological restoration of the typical environments of the Venice Lagoon, primarily mudflats and saltmarshes. Your contributions will help us understand how important the restoration activities of the lagoon saltmarshes are to you and what benefits may arise from them.

Privacy disclaimer for participation

We are committed to protecting your privacy. The information collected through this questionnaire will be used exclusively for research purposes and will be kept strictly confidential. Your responses will be anonymized and will not be linked to any personally identifiable information. The data will be processed in accordance with the General Data Protection Regulation (GDPR) [63] to ensure your privacy and data security. The survey takes less than 7 min to complete. We thank you in advance for your time and participation.

Section 1

1. **Please indicate the type of organization you belong to:**
   - ○ Government entity
   - ○ Non-Governmental Organization (NGO) of Environmental nature
   - ○ Other environmental association (Non-NGO)
   - ○ Non-Governmental Organisation (NGO) of Social nature
   - ○ Local community
   - ○ Economic association
   - ○ Company/Business operating in the Venice Lagoon
   - ○ Educational and research institution
   - ○ Other
2. Indicate the legal nature of the organization.
   - ○ Private
   - ○ Public
   - ○ Other
3. What activities do you carry out in the lagoon? (You can answer both in a professional capacity and as a regular citizen).
   - ○ Political or administrative functions
   - ○ Recreational activities in nature (rowing, birdwatching, etc.)
   - ○ Scientific research
   - ○ Navigation for freight transport
   - ○ Navigation for passenger transport
   - ○ Recreational boating
   - ○ Professional fishing
   - ○ Amateur/recreational fishing
   - ○ Hunting
   - ○ Other...
4. And which activities, in your opinion, are connected to the protection and restoration of saltmarshes?
   - ○ Political or administrative functions
   - ○ Recreational activities in nature (rowing, birdwatching, etc.)

- ○ Scientific research
- ○ Navigation for freight transport
- ○ Navigation for passenger transport
- ○ Recreational boating
- ○ Professional fishing
- ○ Amateur/recreational fishing
- ○ Hunting
- ○ Other...

Section 2: Perception of the problems affecting the Venice Lagoon.

5. In your opinion, which of the following problems affecting the Venice Lagoon are most relevant? For each problem, please indicate the degree of importance.

	Not Important	Little Importance	Neutral	Important	Very Important
Erosion	o	o	o	o	o
Subsidence	o	o	o	o	o
Damage caused by flooding	o	o	o	o	o
Sediment loss	o	o	o	o	o
Water pollution	o	o	o	o	o
Habitat loss	o	o	o	o	o
Biodiversity loss	o	o	o	o	o
Sediment contamination	o	o	o	o	o

6. Please add any other problems not listed, indicating their degree of importance.
7. Among the problems listed below, which could be positively influenced by the protection and restoration of saltmarshes? For each problem, indicate the degree of influence or contribution to solving the problem.

	Not Influential	Little Influential	Neutral	Influential	Very Influential
Erosion	o	o	o	o	o
Subsidence	o	o	o	o	o
Damage caused by flooding	o	o	o	o	o
Sediment loss	o	o	o	o	o
Water pollution	o	o	o	o	o
Habitat loss	o	o	o	o	o
Biodiversity loss	o	o	o	o	o
Sediment contamination	o	o	o	o	o

Section 3: Perception of the ability of saltmarsh restoration to contribute to solving the problems listed above.

8. How important do you consider the improvements that conservation and restoration activities for saltmarshes could bring in terms of providing ecosystem services and increasing biodiversity? (Ecosystem services are the benefits and services that nature provides to humans free of charge).

	Not Important	Little Importance	Neutral	Important	Very Important
Biodiversity	o	o	o	o	o
Food supply (fish resources)	o	o	o	o	o
Water purification	o	o	o	o	o
Reduction of coastal erosion risk	o	o	o	o	o
Carbon capture/sequestration and climate regulation	o	o	o	o	o
Reduction of coastal flood risk	o	o	o	o	o

9. Please add any other ecosystem services not listed, indicating their degree of importance.

Section 4: Restoration of saltmarshes and expected benefits.

10. With reference to the ecosystem service 'reduction of coastal flood risk', please indicate your level of agreement regarding the expected benefits, categorized into four groups: Environmental (Env.); Economic (Econ.); Social (Soc.); Cultural (Cult.).

	Not Agreed	Little Agreed	Neutral	Agreed	Very Agreed
Promote the renaturalization dynamics of saltmarshes (Env.)	o	o	o	o	o
Benefits in terms of reducing costs caused by flood damage (Econ.)	o	o	o	o	o
Opportunities for developing educational activities (Econ.)	o	o	o	o	o
Promote ecotourism activities (guided tours) (Econ.)	o	o	o	o	o
Support corporate social responsibility activities related to nature-based solutions (Econ.)	o	o	o	o	o
Support philanthropic activities (Econ.)	o	o	o	o	o
Provide benefits in terms of quality of life (Soc.)	o	o	o	o	o
Support the maintenance of a landscape and cultural identity (Cult.)	o	o	o	o	o

11. With reference to the ecosystem service 'reduction of coastal erosion risk', please indicate your level of agreement regarding the expected benefits, categorized into four groups: Environmental (Env.); Economic (Econ.); Social (Soc.); Cultural (Cult.).

	Not Agreed	Little Agreed	Neutral	Agreed	Very Agreed
Availability of habitats and improvement of biodiversity (Env.)	o	o	o	o	o
Reduction of costs caused by erosion damage (Econ.)	o	o	o	o	o
Opportunities for developing educational activities (Econ.)	o	o	o	o	o
Ecotourism activities (e.g., guided tours) (Econ.)	o	o	o	o	o
Corporate social responsibility activities related to nature-based solutions (Econ.)	o	o	o	o	o
Philanthropic activities (Econ.)	o	o	o	o	o
Increased sense of belonging or identity (Soc.)	o	o	o	o	o
Maintenance of the cultural value of the landscape (Cult.)	o	o	o	o	o

12. With reference to the ecosystem service 'carbon sequestration and climate regulation', please indicate your level of agreement regarding the expected benefits, categorized into four groups: Environmental (Env.); Economic (Econ.); Social (Soc.); Cultural (Cult.).

	Not Agreed	Little Agreed	Neutral	Agreed	Very Agreed
Carbon capture and sequestration in sediment (Env.)	o	o	o	o	o
Carbon credits (CO_2) for commercialization (Econ.)	o	o	o	o	o

13. With reference to the ecosystem service 'water purification', please indicate your level of agreement regarding the expected benefits, categorized into four groups: Environmental (Env.); Economic (Econ.); Social (Soc.); Cultural (Cult.).

	Not Agreed	Little Agreed	Neutral	Agreed	Very Agreed
Increased water self-purification capacity (Env.)	o	o	o	o	o
Higher survival rate of commercial and non-commercial fish species (Econ.)	o	o	o	o	o
Improved water use in agriculture (Econ.)	o	o	o	o	o
Opportunities to benefit from an improved lagoon environment in terms of water quality for social reasons (Soc.)	o	o	o	o	o

14. With reference to the ecosystem service 'food provisioning (fish resources)', please indicate your level of agreement regarding the expected benefits, categorized into four groups: Environmental (Env.); Economic (Econ.); Social (Soc.); Cultural (Cult.).

	Not Agreed	Little Agreed	Neutral	Agreed	Very Agreed
Provision of protection and nursery for juveniles (Env.)	o	o	o	o	o
Increase in the availability of economically valuable fish resources (Econ.)	o	o	o	o	o
Practice of traditional fishing (Econ.)	o	o	o	o	o
Restoration of traditional fishing practice for local and island communities (Soc.)	o	o	o	o	o
Social inclusion through fishing-related activities (Soc.)	o	o	o	o	o
Preservation of traditional fishing techniques (Cult.)	o	o	o	o	o

15. With reference to biodiversity, please indicate your level of agreement regarding the expected benefits, categorized into four groups: Environmental (Env.); Economic (Econ.); Social (Soc.); Cultural (Cult.).

	Not Agreed	Little Agreed	Neutral	Agreed	Very Agreed
Availability of new areas for colonization by native vegetation (Env.)	o	o	o	o	o
New habitats for breeding and feeding birds (Env.)	o	o	o	o	o
Ecotourism activities (e.g., guided tours) (Econ.)	o	o	o	o	o
Birdwatching (Econ.)	o	o	o	o	o
Circular economy activities (e.g., reuse of waste in artistic objects, reduction of costs for waste recovery and treatment) (Econ.)	o	o	o	o	o
Opportunities for developing educational activities (Econ.)	o	o	o	o	o
Corporate social responsibility activities related to nature-based solutions (Econ.)	o	o	o	o	o
Increased sense of cultural attachment/identity (Soc.)	o	o	o	o	o
Improvement of human health (Soc.)	o	o	o	o	o

16. Please add any other benefits not listed above, indicating their degree of importance.

Section 5: Investigation of the Objectives for Saltmarsh Restoration in the Venice Lagoon.

17. How much do you agree with including the following objectives in a business plan aimed at funding conservation and restoration activities for saltmarshes in the Venice Lagoon? Please indicate your level of agreement.

	Not Agreed	Little Agreed	Neutral	Agreed	Very Agreed
Prevent the disappearance of degraded morphological structures (both natural and artificial)	o	o	o	o	o
Counteract erosion processes and loss of sediment from the lagoon	o	o	o	o	o
Mitigate wave action and its effects on shores and saltmarsh edges	o	o	o	o	o
Promote lagoon hydrodynamic exchanges	o	o	o	o	o
Improve water quality by filtering nutrients and pollutants	o	o	o	o	o
Create new habitats for vegetation and wildlife, particularly for fish and bird species, and protect/increase biodiversity	o	o	o	o	o
Create "nursery" and conservation areas for fish species, increasing profit related to aquaculture activities	o	o	o	o	o
Attract visitors and tourists in a regulated and sustainable manner throughout the year	o	o	o	o	o
Promote carbon sequestration through increased saltmarsh area	o	o	o	o	o

Please add any other objectives not listed above, indicating their degree of importance.

Appendix C. Survey on the Importance of Saltmarshes in the Venice Lagoon

Introduction

Within the framework of the Horizon 2020 project REST-COAST, CORILA (Consorzio per il coordinamento delle ricerche inerenti al sistema lagunare di Venezia), in collaboration with other project partners, is carrying out a survey to understand the importance of saltmarshes in the lagoon of Venice and to explore how to financially support the upscaling of saltmarshes restoration

What Is Rest-Coast

The overall goal of REST-COAST (*Large-scale RESToration of COASTal ecosystems through rivers to sea connectivity*) is to address, with effective and innovative tools, the key challenges faced by coastal ecosystem restoration across Europe. The project will deliver improved techniques and practices for hands-on ecosystem restoration, supported by the co-design of innovative governance and financial arrangements for planning and implementing coastal restoration activities with the engagement of local stakeholders and communities

Why This Survey?

This survey will allow the importance and attractiveness of saltmarshes in the lagoon of Venice to be understood for visitors and tourists, and in particular, their willingness to pay for experiencing eco-touristic trips to these precious environments. The survey will take you less than 5 min to fill it in. We thank you in advance for your time and collaboration! Do you have questions or want to learn more about REST-COAST? Please do not hesitate to contact Caterina Dabalà (dabala@corila.it) and Francesca Coccon (coccon@corila.it).

Privacy Participation Disclaimer.

We are committed to protecting your privacy. The information collected in this questionnaire will be used solely for research purposes and will be kept strictly confidential. Your responses will be anonymized and will not be linked to any personally identifiable information. Your data will be processed in accordance with the General Data Protection Regulation (GDPR) (EU) 2016/679 and other relevant privacy law to ensure your privacy and data security.

What Are Saltmarshes?

The saltmarshes (in Italian *barene*), both natural and artificial, are unique environments that characterize the lagoon of Venice. They are low-lying areas that are periodically submerged by the tide, characterized by a unique vegetation specially adapted to survive in salty and waterlogged conditions. The saltmarshes hold significant conservation importance due to their high productivity and richness in biodiversity, in particular birds, that use these areas for nesting, feeding, and resting during migration periods.

View of the saltmarshes in front of the island of Burano, in the northern lagoon of Venice.

(1) Have you already visited a saltmarsh in the Venice lagoon?

- ○ Yes
- ○ No
- ○ Don't know/no opinion

(2) Why have you visited (or would you visit) the saltmarshes and the environment surrounding them? multiple choice.

- ○ To relax
- ○ To fish in my free time
- ○ To fish as part of my job
- ○ To birdwatching
- ○ To hunt
- ○ To observe nature
- ○ To lie at anchor
- ○ To enjoy the beauty of the area
- ○ I live there/I have a second home there
- ○ Don't know/no opinion
- ○ Other

Willingness to pay for a trip to the saltmarshes to enjoy the beauty of nature, including birdwatching and recreational fishing.

The Venice Lagoon, besides hosting the historical city of Venice, is characterized by a mosaic of coastal wetland habitats, primarily saltmarshes, which host a particular rich biodiversity in terms of flora and fauna. Visiting theses environments, whether by land or water, allows you to deeply connect with the surrounding nature, appreciating the scents and colors of the typical lagoon vegetation, which between July and August turns violet thanks the blooming of *Limonium*. You will be able to admire the numerous species of birds that use these areas for nesting, feeding and resting during migration periods, enjoy a recreational fishing experience or an aperitif accompanied by the sounds of the lagoon and gently rocked by the typical lagoon boats. On the way to the saltmarshes, you will have the chance to visit some of the wonderful and characteristic islands, where you can disembark and enjoy the local products and flavors in one of the many charming traditional venues. Stunning emotions and breathtaking photos are guaranteed ingredients in this visit to the Venice saltmarshes.

(3) How much would you be willing to pay for **a trip to the saltmarshes to enjoy the beauty of nature** and the above-mentioned activities.

- ○ Nothing
- ○ Up to EUR 25
- ○ From EUR 25–50
- ○ From EUR 50–75
- ○ Up to EUR 100
- ○ Don't know/no opinion

Willingness to pay for a trip to the saltmarshes to birdwatching.

The lagoon of Venice is the largest Important Bird Area (IBA) on a national scale, counting the highest number of species of community importance to be preserved. Every year, hundreds of thousands of birds choose the lagoon of Venice for wintering, while a total of 140 species breed in the area, representing around 55% of the known species for Italy. Many of these species are threatened, according to the Red List of IUCN, because of the restricted areal distribution linked to wetlands, or because of their small populations. The visit will allow you to admire and take some pictures of the numerous species of birds that use these areas for nesting, feeding and resting during migration periods ("stop over sites").

(4) How much would you be willing to pay for **birdwatching trip** as that above-described?

- ○ Nothing
- ○ Up to EUR 25
- ○ From EUR 25–50
- ○ From EUR 50–75
- ○ Up to EUR 100
- ○ Don't know/no opinion

Willingness to pay for a trip to the saltmarshes to recreational fishing

The Venice Lagoon is characterized by a very large number of fish species of which marine stragglers are the most abundant category (55 species), followed by marine migrants (16 species) and lagoon residents (15 species). You will have the opportunity to experience fishing activities in accordance with current legislation, get to know different local fish species and practice artisanal techniques. This will allow you to enjoy and connect with the surrounding nature, while also appreciating ways to contribute to the preservation of natural and local resources.

(5) How much would you be willing to pay for **recreational fishing trip** as that above described?

- ○ Nothing
- ○ Up to EUR 25
- ○ From EUR 25–50
- ○ From EUR 50–75

○ Up to EUR 100
○ Don't know/no opinion

Thank you for your time in taking this survey. If you have any comment, you can share it with us here.

Getting to know venetian tourists/visitors.

We aim to collect some general and anonymous information about your profile to better understand the main touristic trends in the Venice Lagoon.

Are you a resident or a visitor/tourist?

○ Resident
○ Visitor/Tourist

What Country are you coming from?

Gender: How do you identify?

○ Man
○ Woman
○ Non-binary
○ Prefer not to say

What age group do you belong to?

○ Under 18
○ 18–29
○ 30–39
○ 40–49
○ 50–59
○ Over 60

What is your occupation?

○ Student/in training
○ Employed
○ Freelance
○ Unemployed
○ Retired
○ Other

What is your level of education?

○ Elementary school
○ Junior high school
○ High school
○ University: Bachelor's degree
○ UniversityL Master's degree
○ Doctoral degree
○ Other

What is your annual net income?

○ Information obtained with this question is needed for statistical purposes, we would appreciate very much your answer. Below EUR 15,000 per year.

- ○ EUR 15,000–30,000 per year
- ○ EUR 30,000–55,000 per year
- ○ EUR 55,000–75,000 per year
- ○ More than EUR 75,000 per year
- ○ Prefer not to say

Appendix D. Restoration Works

Table A2. Current restoration works in the Venice Lagoon as part of the REST-COAST project.

Saltmarsh	Surface (ha)	Sediment Volume Provisioning (m^3)	Responsible	Financing Resource	First Phase	Second Phase
Cornio 4	9.2	92,300	Provv. OO. PP.	MIT		
Gorna	7.5	52,600	Provv. OO. PP.	MIT		
Casonetto Zappa	6.5	59,000	Provv. OO. PP.	MIT		
Casone Zappa A	25.37	N/A	Provv. OO. PP.	MIT		
Lago della Pietra	5.78	37,000	Provv. OO. PP.	MIT		
Raina	5.37	N/A	CVN	Piano Europa	April 2023–June/2024	Not defined
Cornio 1	9.34	N/A	Provv. OO. PP.	MIT		
Cornio 2	21.92	N/A	Provv. OO. PP.	MIT		
Casone Zappa B	28.18	N/A	Provv. OO. PP.	MIT		
Cornio 3	12.39	30,000	Provv. OO. PP.	MIT		
Piovego	6.98	N/A	CVN	Piano Europa		
	138.43	270,900				

References

1. Viti, M.; Löwe, R.; Sørup, H.J.; Ladenburg, J.; Gebhardt, O.; Iversen, S.; McKnight, U.S.; Arnbjerg-Nielsen, K. Holistic valuation of Nature-Based Solutions accounting for human perceptions and nature benefits. *J. Environ. Manag.* **2023**, *334*, 117498. [CrossRef] [PubMed]
2. Booth, H.; Milner-Gulland, E.J.; McCormick, N.; Starkey, M. Operationalizing transformative change for business in the context of Nature Positive. *One Earth* **2024**, *7*, 1235–1249. [CrossRef]
3. Sewell, E.S.; Bouma, J.; van der Esch, S. *Investigating the Challenges and Opportunities for Scaling Up*; PBL Netherlands Environmental Assessment Agency: Hague, The Netherlands, 2016; Volume 2356, pp. 1–78.
4. Ermgassen, S.O.S.E.z.; Löfqvist, S. Financing ecosystem restoration. *Curr. Biol.* **2024**, *34*, R412–R417. [CrossRef] [PubMed]
5. Cohen-Shacham, E.; Walters, G.; Janzen, C.; Maginnis, S. *Nature-Based Solutions to Address Global Societal Challenges*; IUCN International Union for Conservation of Nature: Gland, Switzerland, 2016.
6. Somarakis, G.; Stagakis, S.; Chrysoulakis, N. (Eds.) (2019) Thinknature Nature-Based Solutions Handbook. Think Nature Project Funded by the EU Horizon 2020 Research and Innovation Programme. Available online: https://ec.europa.eu/research/participants/documents/downloadPublic?documentIds=080166e5c7061325&appId=PPGMS (accessed on 10 June 2024).
7. Raymond, C.M.; Frantzeskaki, N.; Kabisch, N.; Berry, P.; Breil, M.; Nita, M.R.; Geneletti, D.; Calfapietra, C. A framework for assessing and implementing the co-benefits of nature-based solutions in urban areas. *Environ. Sci. Policy* **2017**, *77*, 15–24. [CrossRef]
8. Calliari, E.; Staccione, A.; Mysiak, J. An assessment framework for climate-proof nature-based solutions. *Sci. Total Environ.* **2019**, *656*, 691–700. [CrossRef]
9. den Heijer, C.; Coppens, T. Paying for green: A scoping review of alternative financing models for nature-based solutions. *J. Environ. Manag.* **2023**, *337*, 117754. [CrossRef]
10. Vicarelli, M.; Sudmeier-Rieux, K.; Alsadadi, A.; Shrestha, A.; Schütze, S.; Kang, M.M.; Leue, M.; Wasielewski, D.; Mysiak, J. On the cost-effectiveness of Nature-based Solutions for reducing disaster risk. *Sci. Total Environ.* **2024**, *947*, 174524. [CrossRef]
11. Favero, F.; Hinkel, J. Key Innovations in Financing Nature-Based Solutions for Coastal Adaptation. *Climate* **2024**, *12*, 53. [CrossRef]
12. Li, C.Z.; Umair, M. Does green finance development goals affects renewable energy in China. *Renew. Energy* **2023**, *203*, 898–905. [CrossRef]
13. Bodin, B.; Garavaglia, V.; Pingault, N.; Ding, H.; Wilson, S.; Meybeck, A.; Gitz, V.; D'Andrea, S.; Besacier, C. A standard framework for assessing the costs and benefits of restoration: Introducing The Economics of Ecosystem Restoration. *Restor. Ecol.* **2022**, *30*, e13515. [CrossRef]
14. Brathwaite, A.; Pascal, N.; Clua, E. Private Capital to Improve Nature-Based Solutions for Coastal Protection: Time for a Boost. *Oceans* **2022**, *3*, 60–71. [CrossRef]
15. Iftekhar, M.S.; Polyakov, M.; Ansell, D.; Gibson, F.; Kay, G.M. How economics can further the success of ecological restoration. *Conserv. Biol.* **2017**, *31*, 261–268. [CrossRef]

16. Wild, T.; Baptista, M.; Wilker, J.; Kanai, J.M.; Giusti, M.; Henderson, H.; Rotbart, D.; Espinel, J.-D.A.; Hernández-Garcia, J.; Thomasz, O.; et al. Valuation of urban nature-based solutions in Latin American and European cities. *Urban For. Urban Green.* **2024**, *91*, 128162. [CrossRef]
17. McQuaid, S.; Rhodes, M.L.; Andersson, T.; Croci, E.; Feichtinger-Hofer, M.; Grosjean, M.; Lueck, A.E.; Kooijman, E.; Lucchitta, B.; Rizzi, D.; et al. From Nature-Based Solutions to Nature-Based Economy—Delivering the Green Deal for Europe. Draft White Paper for Consultation. Nature-Based Economy Working Group of EC Task Force III on Nature Based Solutions. 2021. Available online: https://networknature.eu/sites/default/files/images/NBE%20White%20Paper%20final%20.pdf (accessed on 10 June 2024).
18. McQuaid, S.; Kooijman, E.D.; Rhodes, M.L.; Cannon, S.M. Innovating with nature: Factors influencing the success of nature-based enterprises. *Sustainability* **2021**, *13*, 12488. [CrossRef]
19. Sánchez-Arcilla, A.; Cáceres, I.; Le Roux, X.; Hinkel, J.; Schuerch, M.; Nicholls, R.J.; Otero, d.M.; Staneva, J.; de Vries, M.; Pernice, U.; et al. Barriers and enablers for upscaling coastal restoration. *Nat.-Based Solut.* **2022**, *2*, 100032. [CrossRef]
20. Mayor, B.; Toxopeus, H.; McQuaid, S.; Croci, E.; Lucchitta, B.; Reddy, S.E.; Egusquiza, A.; Altamirano, M.A.; Trumbic, T.; Tuerk, A.; et al. State of the art and latest advances in exploring business models for nature-based solutions. *Sustainability* **2021**, *13*, 7413. [CrossRef]
21. Stork, S.; Morgenstern, R.; Pölling, B.; Feil, J.H. Holistic Business Model Conceptualisation—Capturing Sustainability Contributions Illustrated by Nature-Based Solutions. *Sustainability* **2023**, *15*, 14091. [CrossRef]
22. Chesbrough, H.; Lettl, C.; Ritter, T. Value Creation and Value Capture in Open Innovation. *J. Prod. Innov. Manag.* **2018**, *35*, 930–938. [CrossRef]
23. Saebi, T.; Foss, N.J. Business models for open innovation: Matching heterogeneous open innovation strategies with business model dimensions. *Eur. Manag. J.* **2015**, *33*, 201–213. [CrossRef]
24. Drupsteen, L.; Wakkee, I. Exploring Characteristics of Regenerative Business Models through a Delphi-Inspired Approach. *Sustainability* **2024**, *16*, 3062. [CrossRef]
25. Teece, D.J. Business models, business strategy and innovation. *Long Range Plann.* **2010**, *43*, 172–194. [CrossRef]
26. Osterwalder, A.; Pigneur, Y. *Business Model Generation: A Handbook for Visionaries, Game Changers, and Challengers*; Wiley: Hoboken, NJ, USA, 2010.
27. Baines, T.; Bigdeli, A.Z.; Bustinza, O.F.; Shi, V.G.; Baldwin, J.; Ridgway, K. Servitization: Revisiting the state-of-the-art and research priorities. *Int. J. Oper. Prod. Manag.* **2017**, *37*, 256–278. [CrossRef]
28. Foss, N.J.; Saebi, T. Fifteen Years of Research on Business Model Innovation: How Far Have We Come, and Where Should We Go? *J. Manag.* **2017**, *43*, 200–227. [CrossRef]
29. Tuli, K.R.; Kohli, A.K.; Bharadwaj, S.G. Rethinking customer solutions: From product bundles to relational processes. *J. Mark.* **2007**, *71*, 1–17. [CrossRef]
30. Visnjic, I.; Jovanovic, M.; Neely, A.; Engwall, M. What brings the value to outcome-based contract providers? Value drivers in outcome business models. *Int. J. Prod. Econ.* **2017**, *192*, 169–181. [CrossRef]
31. Stubbs, W.; Cocklin, C. Conceptualizing a 'sustainability business model'. *Organ. Environ.* **2008**, *21*, 103–127. [CrossRef]
32. Schaltegger, S.; Hansen, E.G.; Lüdeke-Freund, F. Business Models for Sustainability: Origins, Present Research, and Future Avenues. *Organ. Environ.* **2016**, *29*, 3–10. [CrossRef]
33. Pardalis, G.; Mahapatra, K.; Mainali, B. A business model canvas framework for sustainable one-stop-shops. *IOP Conf. Ser. Earth Environ. Sci.* **2022**, *1085*, 50DUMMY. [CrossRef]
34. Daou, A.; Mallat, C.; Chammas, G.; Cerantola, N.; Kayed, S.; Saliba, N.A. The Ecocanvas as a business model canvas for a circular economy. *J. Clean. Prod.* **2020**, *258*, 120938. [CrossRef]
35. Egusquiza, A.; Arana-Bollar, M.; Sopelana, A.; Almenar, J.B. Conceptual and operational integration of governance, financing, and business models for urban nature-based solutions. *Sustainability* **2021**, *13*, 11931. [CrossRef]
36. Zhu, L.; Ye, Q.; Yuan, J.; Hwang, B.G.; Cheng, Y. A scientometric analysis and overview of research on infrastructure externalities. *Buildings* **2021**, *11*, 630. [CrossRef]
37. Baas, J.; Schotten, M.; Plume, A.; Côté, G.; Karimi, R. Scopus as a curated, high-quality bibliometric data source for academic research in quantitative science studies. *Quant. Sci. Stud.* **2020**, *1*, 377–386. [CrossRef]
38. Rojas, R.; Bennison, G.; Gálvez, V.; Claro, E.; Castelblanco, G. Advancing collaborative water governance: Unravelling stakeholders' relationships and influences in contentious river basins. *Water* **2020**, *12*, 3316. [CrossRef]
39. Zingraff-Hamed, A.; Hüesker, F.; Lupp, G.; Begg, C.; Huang, J.; Oen, A.; Vojinovic, Z.; Kuhlicke, C.; Pauleit, S. Stakeholder mapping to co-create nature-based solutions: Who is on board? *Sustainability* **2020**, *12*, 8625. [CrossRef]
40. Sjödin, D.; Parida, V.; Jovanovic, M.; Visnjic, I. Value Creation and Value Capture Alignment in Business Model Innovation: A Process View on Outcome-Based Business Models. *J. Prod. Innov. Manag.* **2020**, *37*, 158–183. [CrossRef]
41. Janowski, L.; Madricardo, F.; Fogarin, S.; Kruss, A.; Molinaroli, E.; Kubowicz-Grajewska, A.; Tegowski, J. Spatial and temporal changes of tidal inlet using object-based image analysis of multibeam echosounder measurements: A case from the Lagoon of Venice, Italy. *Remote Sens.* **2020**, *12*, 2117. [CrossRef]
42. Tagliapietra, D.; Baldan, D.; Barausse, A.; Buosi, A.; Curiel, D.; Guarneri, I.; Pessa, G.; Rismondo, A.; Sfriso, A.; Smania, D.; et al. Protecting and Restoring the Salt Marshes and Seagrasses in the Lagoon of Venice. In *Management and Restoration of Mediterranean*

Coastal Lagoons in Europe. Included in the Project "LIFE Pletera (LIFE13 NAT/ES/001001); Càtedra d'Ecosistemes Litorals Mediterrànis i LIFE Pletera: Venice, Italy, 2018; pp. 39–65.
43. Bettinetti, A.; Pypaert, P.; Sweerts, J.P. Application of an integrated management approach to the restoration project of the lagoon of Venice. *J. Environ. Manag.* **1996**, *46*, 207–227. [CrossRef]
44. Carniello, L.; Defina, A.; D'Alpaos, L. Morphological evolution of the Venice lagoon: Evidence from the past and trend for the future. *J. Geophys. Res. Earth Surf.* **2009**, *114*, 4002. [CrossRef]
45. Ravera, O. The Lagoon of Venice: The result of both natural factors and human influence. *J. Limnol.* **2000**, *59*, 19–30. [CrossRef]
46. Rova, S.; Müller, F.; Meire, P.; Pranovi, F. Sustainability perspectives and spatial patterns of multiple ecosystem services in the Venice lagoon: Possible roles in the implementation of the EU Water Framework Directive. *Ecol. Indic.* **2019**, *98*, 556–567. [CrossRef]
47. Sun, C.; Fagherazzi, S.; Liu, Y. Classification mapping of salt marsh vegetation by flexible monthly NDVI time-series using Landsat imagery. *Estuar. Coast. Shelf Sci.* **2018**, *213*, 61–80. [CrossRef]
48. Tempesta, T.; Visintin, F.; Marangon, F.; Economics, A. Ecotourism demand in North-East Italy. In *Monitoring and Management of Visitor Flows in Recreational and Protected Areas*; Institute for Landscape Architecture and Landscape Management, Bodenkultur University: Vienna, Austria, 2002; pp. 373–379.
49. Anvarovich, K.A. Role of ecological tourism in the development of specially protected natural areas and the regional economy. *J. Econ. Tour. Serv.* **2024**, *3*, 104–107.
50. Mondino, E.; Beery, T. Ecotourism as a learning tool for sustainable development. The case of Monviso Transboundary Biosphere Reserve, Italy. *J. Ecotourism* **2018**, *18*, 107–121. [CrossRef]
51. Özkan, N. Evaluation of Bird Watching Tourism from Different Aspects. *Meriç Uluslararası Sos. Strat. Araştırmalar Derg.* **2023**, *7*, 202–221.
52. O'Leary, B.C.; Fonseca, C.; Cornet, C.C.; de Vries, M.B.; Degia, A.; Failler, P.; Furlan, E.; Garrabou, J.; Gil, A.; Hawkins, J.P.; et al. Nature-Based Solutions Embracing Nature-based Solutions to promote resilient marine and coastal ecosystems. *Nat.-Based Solut.* **2023**, *3*, 100044. [CrossRef]
53. Giordano, R.; Pluchinotta, I.; Pagano, A.; Scrieciu, A.; Nanu, F. Enhancing nature-based solutions acceptance through stakeholders' engagement in co-benefits identification and trade-offs analysis. *Sci. Total Environ.* **2020**, *713*, 136552. [CrossRef]
54. Brears, R.C. *Financing Nature-Based Solutions: Exploring Public, Private, and Blended Finance Models and Case Studies*; Springer International Publishing: Cham, Switzerland, 2022.
55. Böhnke-Henrichs, A.; Grechi, L.; Konosvska, I.; Mulder, S.; Duan, J.; de Groot, R.; Barausse, A.; Baldan, D.; Musner, T.; Palmeri, L. Cost benefit analysis report. Report on the cost of the low-impact soil bioengineering works versus current conservation measures, and the net socio-economic benefits of the integrated VIMINE approach. 2017; pp. 1–83.
56. Rova, S.; Stocco, A.; Pranovi, F. Ecosystem services' capacity and flow in the Venice Lagoon and the relationship with ecological status. *One Ecosyst.* **2022**, *7*, e79715. [CrossRef]
57. Stronza, A.L.; Hunt, C.A.; Fitzgerald, L.A. Ecotourism for conservation? In *Routledge Handbook of Ecotourism*; Routledge: London, UK, 2022; pp. 372–397. [CrossRef]
58. Lecchini, D.; Bertucci, F.; Schneider, D.; Berthe, C.; Gache, C.; Fogg, L.; Waqalevu, V.; Maueau, T.; Sturny, V.; Bambridge, T.; et al. Assessment of ecological status of the lagoon of Bora-Bora Island (French Polynesia). *Reg. Stud. Mar. Sci.* **2021**, *43*, 101687. [CrossRef]
59. Suana, I.W.; Ahyadi, H.; Hadiprayitno, G.; Amin, S.; Kalih, L.A.T.T.W.S.; Sudaryanto, F.X. Environment carrying capacity and willingness to pay for bird-watching ecotourism in Kerandangan Natural Park, Lombok, Indonesia. *Biodiversitas* **2020**, *21*, 2266–2274. [CrossRef]
60. Eggermont, H.; Balian, E.; Azevedo, J.M.N.; Beumer, V.; Brodin, T.; Claudet, J.; Fady, B.; Grube, M.; Keune, H.; Lamarque, P.; et al. Nature-based solutions: New influence for environmental management and research in Europe. *Gaia* **2015**, *24*, 243–248. [CrossRef]
61. Pereira, P.; Yin, C.; Hua, T. Nature-based solutions, ecosystem services, disservices, and impacts on well-being in urban environments. *Curr. Opin. Environ. Sci. Health* **2023**, *33*, 100465. [CrossRef]
62. Thompson, A.; Bunds, K.; Larson, L.; Cutts, B.; Hipp, J.A. Paying for nature-based solutions: A review of funding and financing mechanisms for ecosystem services and their impacts on social equity. *Sustain. Dev.* **2023**, *31*, 1991–2066. [CrossRef]
63. Regulation (EU) 2016/679 of the European Parliament and of the Council. Available online: https://eur-lex.europa.eu/eli/reg/2016/679/oj (accessed on 18 June 2024).

sustainability

MDPI

Article

A Biomechanical Study of Potential Plants for Erosion Control and Slope Stabilization of Highland in Thailand

Warakorn Mairaing [1], Apiniti Jotisankasa [2,*], Nisa Leksungnoen [3,*], Monir Hossain [2], Chatchai Ngernsaengsaruay [4], Prem Rangsiwanichpong [5], Jarunee Pilumwong [6], Sony Pramusandi [7], Surat Semmad [8] and Abu Noman Faruq Ahmmed [9]

1 Department of Civil Engineering, Faculty of Engineering at Kamphaeng Saen, Kasetsart University, Nakhon Pathom 73140, Thailand; mairaing@yahoo.com
2 Department of Civil Engineering, Kasetsart University, Bangkok 10900, Thailand; mdmonir.h@ku.th
3 Department of Forest Biology, Kasetsart University, Bangkok 10900, Thailand
4 Department of Botany, Kasetsart University, Bangkok 10900, Thailand; fsciccn@ku.ac.th
5 Department of Water Resources Engineering, Kasetsart University, Bangkok 10900, Thailand; prem.r@ku.ac.th
6 Highland Research and Development Institute (Public Organization), Chiang Mai 50200, Thailand; jp.foodbank@gmail.com
7 Department of Civil Engineering, State Polytechnic of Jakarta, Depok City 16425, Indonesia; sony.pramusandi@sipil.pnj.ac.id
8 Department of Construction Engineering, Rajamangala University of Technology Tawan-Ok, Bangkok 10330, Thailand; semmad.s@gmail.com
9 Department of Plant Pathology, Sher-e-Bangla Agricultural University, Dhaka 1207, Bangladesh; nomanfaruq@sau.edu.bd
* Correspondence: fengatj@ku.ac.th (A.J.); ffornsl@ku.ac.th (N.L.)

Abstract: Soil bioengineering provides a sustainable method for erosion control and soil slope stabilization using vegetation with multiple co-benefits. This study evaluated ten plant species in Thailand's highland regions for their soil bioengineering potential and additional benefits. Root architecture, tensile strength, and Young's modulus were measured to compare biomechanical traits. *G. sepium*, *F. griffithii*, *P. americana*, *B. asiatica*, and *C. arabica* exhibited H-type roots with wide lateral spread, while *M. denticulata* and *C. officinarum* had VH-type roots with deep taproots and wide lateral extent. *A. sutepensis* showed M-type roots with most root matrix in the top 0.3 m, where *C. cajan* and *C. sinensis* had R-type roots with deep, oblique growth. Most species showed a negative power relationship between the root strength and Young's modulus with the root diameter except *C. cajan* that showed a positive correlation. *P. americana*, *F. griffithii*, *C. officinarum*, and *C. arabica* showed relatively high values of 1 mm root tensile strength (exceeding 24 to 42 MPa), while *M. denticulata*, *G. sepium*, and *B. asiatica* exhibited intermediate root tensile strength (ranging from 8 to 19 MPa). *A. sutepensis*, *C. cajan*, and *C. sinensis* demonstrated the lowest root tensile strength, up to 7 MPa. It is advised to plan slope vegetation by selecting diverse plant species with varying root structures and benefits, addressing both engineering and socioeconomic needs of the sustainable nature-based solution.

Keywords: soil bioengineering; root biomechanics; root architecture

1. Introduction

Due to the current climate change and extreme weather conditions, land degradation and rainfall-triggered landslide are one of the most pressing environmental concerns particularly in mountainous highlands. These problems are accelerating in recent years due to land scarcity and the increasing pressure for food security that lead local farmers in highland to follow unsustainable agriculture practice such as slash and burn, associated with annual crops. In order to tackle this problem, a sustainable approach that promotes societal development while addressing economic and environmental needs is followed by Royal Project Foundation and Highland Research and Development Institute (HRDI) in Thailand [1]. Various long-term initiatives were developed to encourage local farmers in northern provinces to turn to more environmentally friendly agroforestry practice and reforestation. One of the philosophies proposed by the late King Bhumibol of Thailand regarding this issue is that of "three forests with four benefits". The three kinds of forests include species for (a) general usage such as timbers, (b) edible fruits, and (c) fuels, which can yield four benefits, including the three original purposes, plus the fourth one being soil and water conservation and reforestation. Such a sustainable approach of plant selection is mostly adopted in community capacity development projects in the northern part of Thailand.

The vegetation planted on slopes not only serves as soil and water conservation and erosion control, benefiting the environment, but also provides economic incentives to local farmers, making these practices sustainable. In this way, all three key aspects of sustainability can be realized, that is, the environmental, economic, and social development aspects. However, biomechanical studies on how these species reinforce the soil and their application in soil bioengineering are relatively scarce [2]. A holistic discussion of all aspects of sustainability, combined with biomechanical studies of plant roots, is also rare.

Soil bioengineering is the use of plants for erosion control and shallow slope stabilization, normally combined with engineering techniques such as erosion control blankets, with some forms of retaining structures, namely gabion walls and soil bags [3–6]. This technique has recently received considerable attention in infrastructure development due to its attributes of being relatively low cost, emitting low levels of carbon dioxide (CO_2), being aesthetically pleasing, and being environmentally friendly [3–17]. Plant roots can stabilize soil by providing mechanical reinforcement and creating a composite material with improved properties, such as enhanced resistance to tension [8,18,19]. The effectiveness of soil stabilization by plant roots depends on various factors including species, architectural shape, morphology, and interface properties of the root system, as well as their concentration, branching characteristics, and spatial distribution within the soil [20]. Essentially, it is the combination of root orientation, root concentration, and root tensile strength characteristics that collectively gives rise to the overall capacity of the root-reinforced soil [4].

Root strength and architecture are influenced by local soil and environmental conditions. Yen [21] classified root growth types into H-type, R-type, VH-type, V-type, and M-type. Styczen and Morgan [20] further elaborated on these root growth types, providing detailed explanations to enhance comprehension. According to Styczen and Morgan [20], H-type and VH-type plant roots are suitable for slope stabilization and wind resistance, while H-type and M-type roots are recommended for soil reinforcement [22]. Plant roots provide soil reinforcement by physically anchoring shallow soil to deeper layers, thereby preventing erosion and slope failures [23,24]. The size of roots has a significant impact on their effectiveness in soil reinforcement, with fibrous roots being particularly effective in creating a root–soil composite similar to fiber-reinforced soil [25,26]. Fibrous roots can exhibit greater cohesiveness compared with tap roots, even when considering the same root area ratio [27]. Smaller roots tend to have higher tensile strength than larger roots primarily due to their biological components such as cellulose [4,28]. Consequently, small roots can leverage their tensile strength to enhance soil–root friction, while larger roots intersect shear surfaces and mobilize soil–root friction force [24]. Root mechanical reinforcement is influenced by various factors, including root system morphology [29] and growth conditions [30], the number and relative amount of fine and coarse roots, the diameter [25,27,29,31], water content and suction [8,32–34], chemical composition [35], and tensile strength and stiffness, characterized by Young's modulus (the initial linear part of a tensile stress–strain curve) [19,25] and the loading rate of testing [36,37]. It is noteworthy that root moisture content and root suction were found to have profound effects on root tensile strength as well as the stability of the root-reinforced slope subjected to long-term drought followed by heavy rain [8].

The most commonly utilized soil bioengineering method in Thailand and other Southeast Asian countries is the use of the vetiver (*Chrysopogon zizanioides*) system, known for its cost-effectiveness and environmental friendliness [38]. In Thailand, the predominant species is *Chrysopogon zizanioides*, a lowland vetiver, while *Chrysopogon nemoralis*, an upland vetiver, is less favored due to its potential invasiveness and shorter root system [39]. Despite its advantages, vetiver grass exhibits shade intolerance and is occasionally viewed as a monospecies solution. The incorporation of local plant species in addition to the vetiver system not only fosters biodiversity but also reduces expenses associated with harvesting, handling, and transportation [40].

This study aims to investigate locally cultivated plant species commonly found in landslide-prone regions within four HRDI project areas in northern Thailand for their potential use in erosion control and soil bioengineering applications, while also considering their additional benefits. The primary goal is to analyze the root characteristics of these species and assess their biomechanical properties, specifically focusing on tensile strength and modulus. Additionally, a comparative evaluation was undertaken to gauge the suitability of each species for soil bioengineering applications based on their biomechanical attributes and other benefits, presenting a promising avenue for future research in this field.

2. Materials and Methods

2.1. Selection of Plant Species and Sampling

In selecting plant species for soil bioengineering applications in this study, careful consideration is given to various factors, including not only their effectiveness in resisting mechanical stresses but also the habitat of the species being both native and exotic species that are available in the study areas as well as their other uses. Furthermore, observations from local farmers and expertise from relevant authorities (HRDI) contributed to the selection process, ensuring other tangible benefits such as economic value, timber resources, and soil and water conservation. Various plant species (10 species in total; details of which are shown in Tables 1 and 2) were selected for root architecture study and tensile strength testing based on their availability in four project areas of HRDI in northern provinces of Thailand. In Tak province (Mae Song project area), *Cajanus cajan*, *Persea americana*, *Fraxinus griffithii*, and *Cinnamomum officinarum* were tested. In Chiang Rai province (Mae Salong project area), *Camellia sinensis*, *Aspidistra sutepensis*, and *Coffea arabica* were featured. The Sop Moei project area, Ma Hongsorn province, included *Macaranga denticulata* and *Buddleja asiatica*. The excavation and washing methods were employed to examine the root architecture of the selected species from the study areas and to collect root specimens for further biomechanical analysis. In spite of the risk of breaking smaller roots, even with careful handling [41,42], this technique was considered the most reliable option given the resources and tools available during this study.

Table 1. Selected plant species for this study and their family, tree height, age, perimeter, longest measurable root length, root type, and site coordinates.

Plant Species	Family	Site Cordinates	Age, Year	Height, m	Tree Perimeter, cm	Longest Measurable Root Length, m	Type of Root *
Persea americana	*Lauraceae*	17.507069°,98.216669°	10	1.35	31	3.37	H
Fraxinus griffithii	*Oleaceae*	17.503434°,98.216081°	2	2.2	5.87	2.3	H
Cajanus cajan	*Fabaceae*	17.519346°,98.098063°	0.75	3	19.64	1.3	R
Cinnamomum officinarum	*Lauraceae*	17.503434°,98.216081°	2	2.35	16	4.5	VH
Camellia sinensis	*Theaceae*	20.152026°,99.616434°	35	0.72	29.42	0.9	R
Aspidistra sutepensis	*Asparagaceae*	20.148751°,99.614794°	2	0.82	NA	0.3	M
Coffea arabica	*Rubiaceae*	20.148751°,99.614794°	2	2.26	5.76	1.07	H
Gliricidia sepium	*Fabaceae*	17.860048°,97.842293°	2	3.65	13.72	1.5	H
Macaranga denticulata	*Euphorbiaceae*	17.861017°,97.841098°	Unknown	1.8	5.59	2.1	VH
Buddleja asiatica	*Scrophulariaceae*	17.861017°,97.841098°	Unknown	2.55	5.2	1.4	H

* Root types according to Yen [21]. NA = Not applicable.

Table 2. Selected plant species for this study with additional information on vetiver, their distributions, origin, invasiveness, and references.

Scientific Names	Distributions	Native/Exotic Species	Invasive/Non-Invasive Species	References
Persea americana Mill.	Central Mexico to Costa Rica.	It is a non-invasive exotic species and is cultivated as a fruit tree, especially in the highland areas of northem Thailand.	Non-invasive species	[43]
Fraxims griffinh C. B. Clarke	Bangladesh to Nansei-shoto, Java to Lesser Sunda Islands (Bali).	It is a non-invasive exotic species and is cultivated for its wood in the highland habitats of northem Thailand	Non-invasive species	[43,44]
Cajanus cajan (L.) Huth	Indian Subcontinent.	It is a non-invasive exotic species and is used as animal food.	Non-invasive species	[43]
Cirnamonum officinarian (L.) J. Presl	Korea (Jeju-do), W. Central & S. Japan to E. & S. Taiwan.	It is a non-invasive exotic species.	Non-invasive species	[43,45]
Camellia sinensis (L.) Kuntze	E. Himalaya to S. China and N. Indo-China, Hainan.	*Camellia sinensis* (L.) Kuntze var. assamica (Royle ex Hook.) Steenis is a native variety. *Camellia sinensis* (L.) Kuntze var. sinensis (Royle ex Hook) Steenis is a non-invasive exotic variety. The native range of this variety is S. China.	Unknown	[43,46,47]
Aspidistra sutepensis K. Larsen	China (Yunnan) to N. Thailand.	It is anative species.	Unknown	[43,48]
Coffee arabica L.	E. South Sudan, SW. Ethiopia, N. Kenya (Mt. Marsibit).	It is a non-invasive exotic species and is cultivated for its seeds.	Non-invasive species	[43]
Gliricidia sepium (Jacq.) Kunth	Mexico to Colombia.	It is a non-invasive exotic species.	Non-invasive species	[43]
Macaranga denticulata (Blume) Müll. Arg.	S. China to Tropical Asia.	It is a native species.	Unknown	[43,49]
Buddleja asiatica Lour.	Central & S. China to Tropical Asia and Marianas.	It is a native species.	Invasive species	[43]
Chrysopogon zizanioides (L.) *Roberty Poaceae*	NE. India to Indo-China.	It is a native species.	Unknown	[43]

From Table 2, it is noted that the eight selected species—*Persea americana*, *Fraxinus griffithii*, *Cajanus cajan*, *Cinnamomum officinarum*, *Camellia sinensis*, *Coffea arabica*, and *Gliricidia sepium*—are all non-invasive exotic species. These plants have been primarily cultivated for various benefits, such as fruits, food, timber, and ornamental purposes, in addition to soil and water conservation. Over time, these exotic species have become naturalized to the environment in Thailand, integrating into the native ecosystem and demonstrating their ability to survive. This adaptation is likely reflected in their root tensile strength.

2.2. Measurements of Root Biomechanical Properties

After excavation and washing, the root specimens were placed in sealed plastic bags, transferred to the laboratory, and stored in a refrigerator at 8 °C. Subsequently, all specimens were tested at ambient temperature in the moisture condition they retained post-sampling. This condition is hereafter referred to as as-sampled moisture condition. However, for

Cinnamomum officinarum, additional series of tensile strength tests were carried out on root specimens that had been dried for 7 days in order to explore the effects of drying. In the tensile strength test, the root specimen was gripped using a drill-bit clamp connected to a load cell, and tension was applied using a loading frame at a constant speed of 4 mm/min [8]. To ensure secure gripping of fine roots, a customized gripping mechanism was employed, utilizing two chuck grips at the ends of the root. Epoxy resin or latex glue was applied to reinforce up to 5 mm at the root ends to prevent squeezing and pulling out during testing, and sometimes a rubber membrane and gauze cloth were also used between the root and the grip to avoid stress concentrations and improve the grip. After reinforcement of root ends (within 30 min), specimens were subjected to testing, with the root segment length between clamps set at 50 mm. Only test results of specimens with ruptures occurring near the middle part were considered reliable and included in the data analysis. In the case of any slippage, the tensile test was discarded. Figure 1 shows the complete assembly of the tensile testing machine used in this study. The root tensile strength (MPa) at maximum load and Young's modulus (MPa) were obtained from the stress–strain curve of the tested sample [30]. The tensile strength (T_r) at maximum load was calculated using Equation (1).

$$T_r = \frac{F}{\pi\left(\frac{d^2}{4}\right)} \quad (1)$$

where F is the maximum force (Newton) the root can sustain before breaking, and d is the root diameter (mm) (the arithmetic mean of three-point measurements made after the tensile test). Young's modulus (E_r) was calculated from the gradient of the stress–strain curve measured at the peak stress:

$$E_r = \frac{FL_0}{\pi\left(\frac{d^2}{4}\right)\Delta L} \quad (2)$$

where L_0 is the initial length (mm) of the root sample (i.e., 50 mm), and ΔL is the change in the root length (mm) at peak stress [27,30]. The root diameter is, therefore, a key parameter that can affect both the values of root tensile strength and Young's modulus (Equations (1) and (2)).

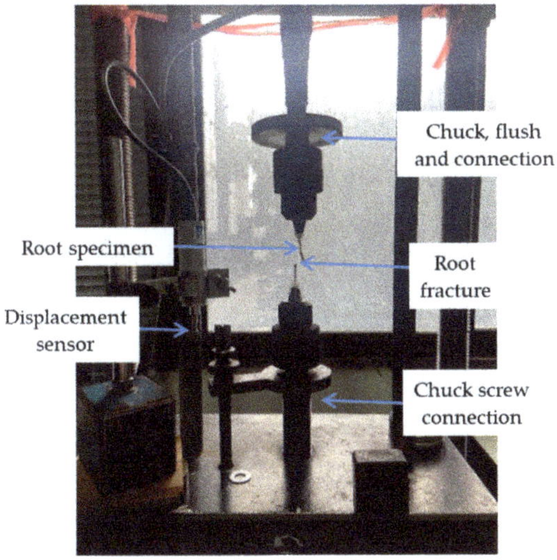

Figure 1. Root biomechanical testing equipment (root tensile strength test).

3. Results and Discussions

3.1. Root Morphology and Architecture

The morphology and architectural shape of plant roots (Figure 2) were characterized using patterns of root growth outlined by Yen [21] and Styczen & Morgan [20] in this study. According to Yen's classification, *Gliricidia sepium*, *Persea americana*, *Fraxinus griffithii*, *Coffea arabica*, and *Buddleja asiatica* exhibit H-type roots. These roots are characterized by a wide lateral extent, with 80% of the root matrix concentrated in the top 60 cm of soil, and moderate depth for maximum root length. On the other hand, *Macaranga denticulata* and *Cinnamomum officinarum* have VH-type roots, similar to H-type but with a main taproot penetrating deep into the ground while lateral roots extend horizontally. *Aspidistra sutepensis* possess M-type roots, where the main root profusely grows under the stump with a narrow lateral extent, and the majority of the root matrix accumulates in the top 30 cm of soil. Lastly, *Cajanus cajan* and *Camellia sinensis* are characterized by R-type roots, which have deep maximum root growth but grow obliquely, with only 20% of the root matrix found in the top 60 cm of soil and with main roots extending at right angles to the slope.

Understanding the root architecture of different tree species can greatly aid in the strategic arrangement of vegetation planting on slopes. Planting only a single tree species (monoculture) or a certain type of root morphology on slopes can pose risks to slope stability. Monocultures tend to have uniform root patterns, lacking the diverse root systems of mixed-species forests, thereby reducing their resistance to a variety of slope failure mechanisms. They are also more susceptible to diseases and pests, which can devastate the entire tree population and further compromise slope stability. In contrast, a diverse array of plant species with varied root architectures (H-, VH-, R-, and M-types) can enhance slope stability by addressing different types of mass movement, both surface and deep-seated. Additionally, such diversity improves ecological values, decreases vulnerability to diseases and pests, and provides economic and societal benefits. It should be noted that while the site conditions across the study areas are similar, i.e., hilly terrain, there may be variations in soil types and nutrient availability from one site to another. These factors can influence root architecture and the extent to which root types vary with site conditions. Further detailed studies on specific species are needed to elucidate these variations.

3.2. Root Tensile Strength

This study investigates the root tensile strength across 10 plant species, ranging in diameter from 0.22 mm to 3.43 mm (Table 3). Average tensile strengths for all root diameters are also shown to vary across the species, with values ranging from 4.06 MPa (*Aspidistra sutepensis*, the lowest value) to 36.82 MPa (*Fraxinus griffithii*, the highest value) in an as-sample root moisture condition. Notably, *Fraxinus griffithii* demonstrated the highest strength per individual root section, reaching 89.12 MPa for a 0.612 mm diameter root segment. The effect of root drying was clearly evident by comparing the root tensile strengths of *Cinnamomum officinarum* in as-sampled and in 7-day drying conditions. The average tensile strength increased from 19.11 MPa to 33.37 MPa as the roots dried for 7 days. This trend of increasing strength with decreasing moisture content is in line with the behavior of other woody species observed by Hales and Miniat [33] and Boldrin et al. [32], and herbaceous plants reported by Zhang et al. [34], while it contradicted that of the grass species reported by Mahannopkul and Jotisankasa [8]. It is noted that the range of root suction tested by Mahannopkul and Jotisankasa [8] was only up to 50 kPa, while the root suction during drying of *Cinnamomum officinarum* in this study was unknown. Table 3 presents the fitting equations based on the power law relationship, including the coefficients of determination (R^2) and p-values obtained from regression analysis and ANOVA in Microsoft Excel. For seven species—*Persea Americana*, *Fraxinus griffithii*, *Cajanus cajan*, *Cinnamomum officinarum* (7 day drying), *Camellia sinensis*, *Coffea Arabica*, and *Buddleja asiatica*—a significant relationship (p-value < 0.05) was found between the root diameter and tensile strength. In contrast, the relatively low R^2 values and higher p-values (>0.05) for *Cinnamomum officinarum* (as sampled), *Aspidistra sutepensis*, *Gliricidia sepium*, and *Macaranga denticulata* suggest that factors beyond the root diameter, such as root suction and moisture conditions, likely contribute to variations in tensile strength.

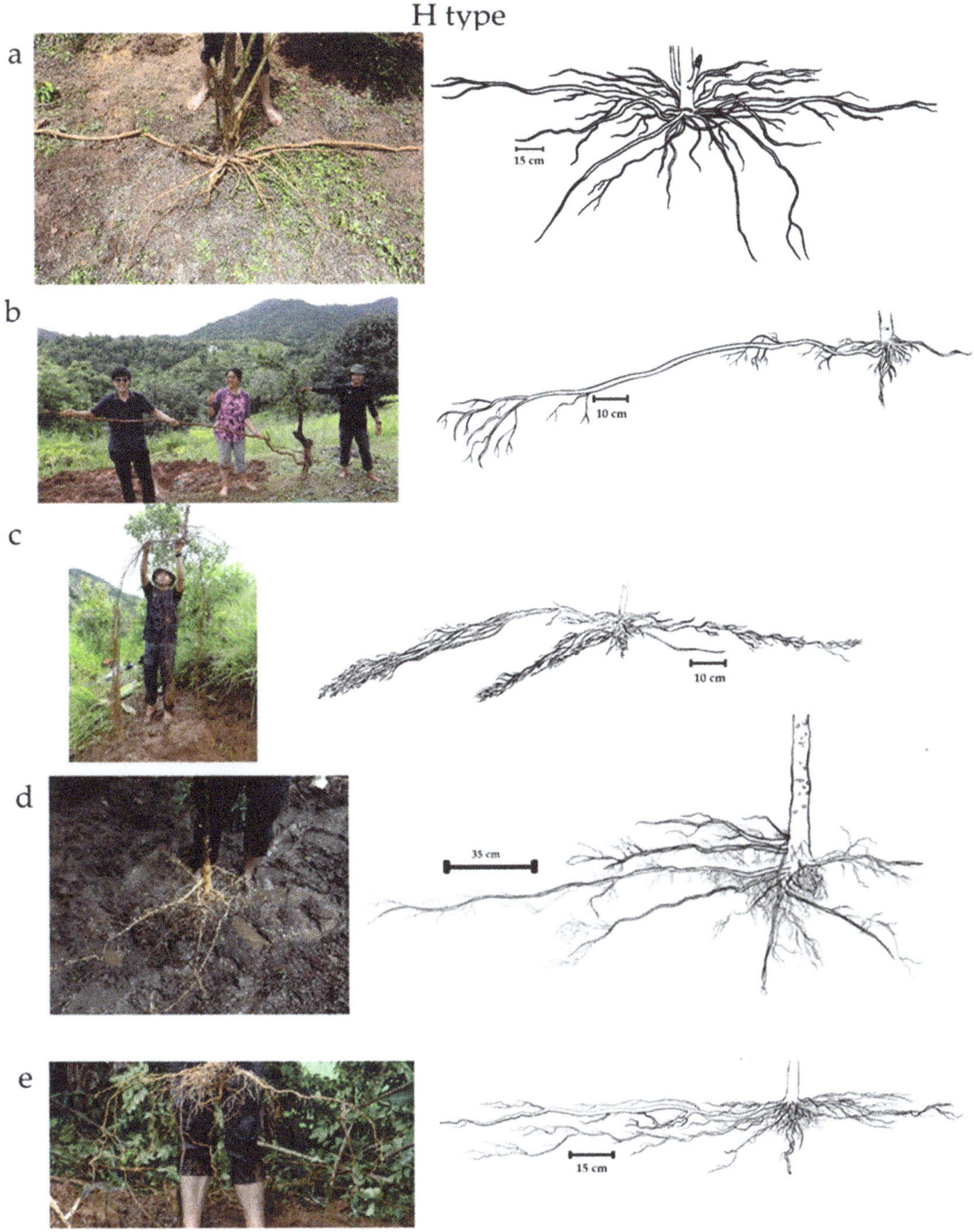

Figure 2. *Cont.*

Figure 2. *Cont.*

R type

i

j

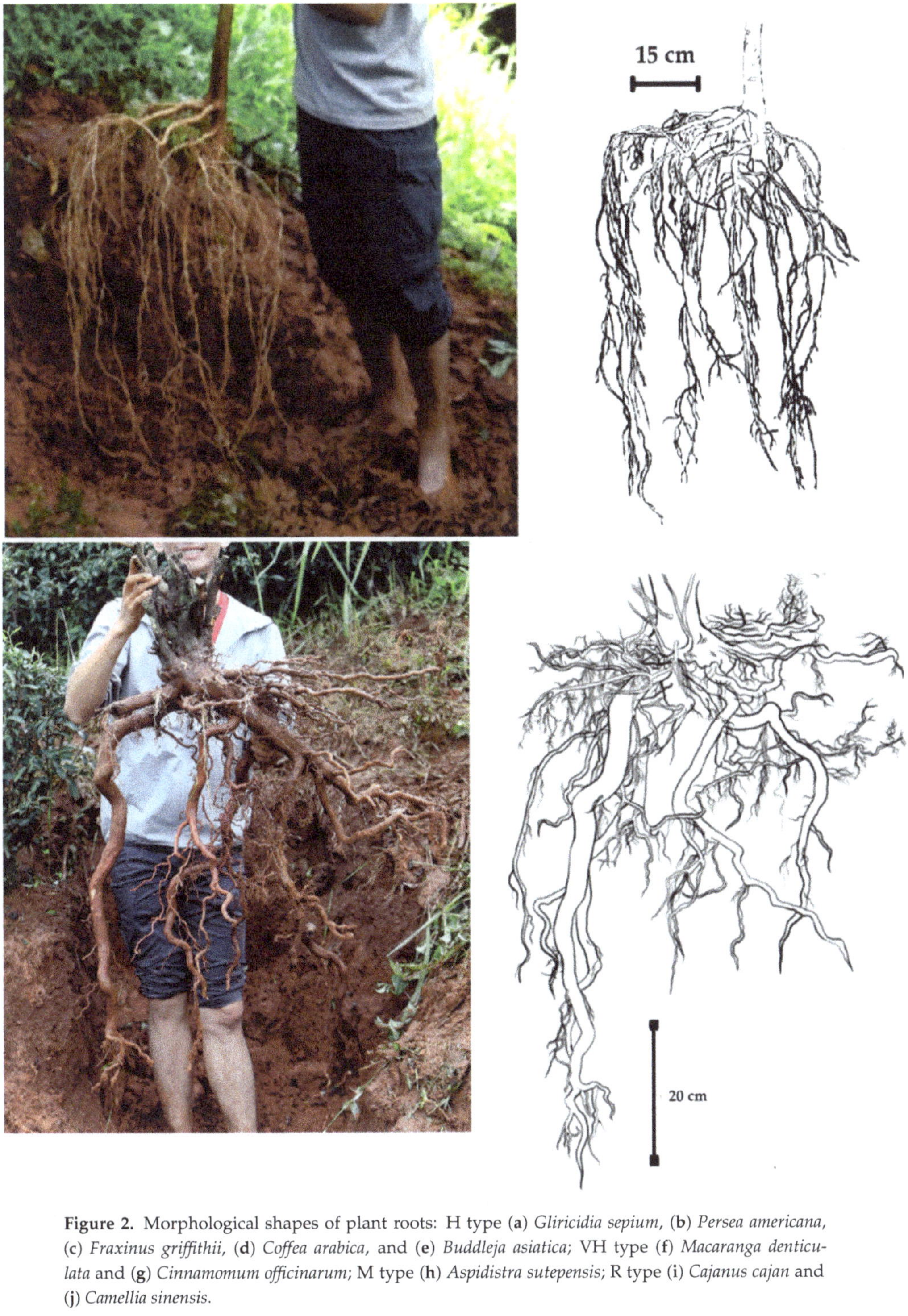

Figure 2. Morphological shapes of plant roots: H type (**a**) *Gliricidia sepium*, (**b**) *Persea americana*, (**c**) *Fraxinus griffithii*, (**d**) *Coffea arabica*, and (**e**) *Buddleja asiatica*; VH type (**f**) *Macaranga denticulata* and (**g**) *Cinnamomum officinarum*; M type (**h**) *Aspidistra sutepensis*; R type (**i**) *Cajanus cajan* and (**j**) *Camellia sinensis*.

Table 3. Summary of the data on root tensile strength and Young's modulus (Mean) per tested species. The best-fit equations, R^2, and p-values are given for the strength and Young's modulus–diameter relations.

Plant Species	Diameter Range, (mm)	No of Samples, n	Average Tensile Strength, (MPa)	Average Young's Modulus, (MPa)	Fitting Equation, R^2 and p-Values Tensile Strength	Young's Modulus
Persea americana	0.84–2.59	16	26.99	309.74	$T_r = 39.838d^{-1.109}$ ($R^2 = 0.277$, $p = 0.036$)	$M_r = 302.81d^{-0.538}$ ($R^2 = 0.044$, $p = 0.437$)
Fraxinus griffithii	0.58–2.85	10	36.82	615.63	$T_r = 34.442d^{-1.414}$ ($R^2 = 0.815$, $p = 0.0003$)	$M_r = 455.65d^{-2.71}$ ($R^2 = 0.768$, $p = 0.0009$)
Cajanus cajan	0.68–2.20	13	5.53	276.97	$T_r = 2.9691d^{1.5474}$ ($R^2 = 0.472$, $p = 0.0095$)	$M_r = 124.24d^{1.732}$ ($R^2 = 0.436$, $p = 0.014$)
*Cinnamomum officinarum**	0.86–1.79	8	33.37	476.03	$T_r = 42.002d^{-0.884}$ ($R^2 = 0.617$, $p = 0.021$)	$M_r = 482.87d^{-0.297}$ ($R^2 = 0.061$, $p = 0.743$)
Cinnamomum officinarum	0.48–3.43	10	19.11	269.28	$Tr = 17.89d^{-0.187}$ ($R^2 = 0.066$, $p = 0.472$)	$Mr = 252.63d^{0.0523}$ ($R^2 = 0.012$, $p = 0.760$)
Camellia sinensis	0.92–3.19	8	4.48	187.94	$T_r = 5.2934d^{-0.395}$ ($R^2 = 0.514$, $p = 0.045$)	$M_r = 260.97d^{-1.135}$ ($R^2 = 0.555$, $p = 0.034$)
Aspidistra sutepensis	1.03–2.78	5	4.06	62.90	$T_r = 6.3501d^{-0.882}$ ($R^2 = 0.569$, $p = 0.141$)	$M_r = 106.23d^{-1.064}$ ($R^2 = 0.618$, $p = 0.115$)
Coffea arabica	0.22–2.34	8	33.36	460.47	$T_r = 24.654d^{-0.483}$ ($R^2 = 0.506$, $p = 0.048$)	$M_r = 259.21d^{-0.865}$ ($R^2 = 0.646$, $p = 0.016$)
Gliricidia sepium	1.20–1.99	5	12.44	242.35	$T_r = 15.79d^{-0.566}$ ($R^2 = 0.331$, $p = 0.310$)	$M_r = 558.7d^{-3.087}$ ($R^2 = 0.1612$, $p = 0.503$)
Macaranga denticulata	1.24–2.80	8	10.97	242.07	$T_r = 16.995d^{-0.712}$ ($R^2 = 0.458$, $p = 0.065$)	$M_r = 343.67d^{-0.606}$ ($R^2 = 0.235$, $p = 0.224$)
Buddleja asiatica	1.10–1.69	9	7.40	106.82	$T_r = 13.107d^{-1.983}$ ($R^2 = 0.4491$, $p = 0.048$)	$M_r = 176.91d^{-2.084}$ ($R^2 = 0.213$, $p = 0.211$)

Note: *Cinnamomum officinarum** is tested after 7 days drying, T_r = root tensile strength, M_r = Young's modulus of root.

Interestingly, this study identified two distinct trends in the tensile strength–diameter relationship across species, which may be described by two different types of fitting equations (Figure 3; Table 3). Among the species, 9 show a negative power trend in the tensile strength–diameter relationship, indicating a decrease in tensile strength with an increasing root diameter, while the *Cajanus cajan* exhibit positive relationships between strength and diameter and show an increase in tensile strength with diameter. The positive correlation can be attributed to several factors, including changes in root structure, namely their relative cellulose and lignin percentages, with root diameter [50]. In the case that larger roots exhibit a higher proportion of the lignin/cellulose ratio, known to provide material strength, this results in positive correlations. For negative correlation, an increase in diameter may correspond to a lower lignin/cellulose ratio, typically associated with decreased strength as reported by Yang et al. [50]. However, there is still no root cross-sectional image studies to confirm this hypothesis. Further studies are required, in particular, to understand the basic cellular structure of dicot and monocot roots of these species.

Figure 4a presents the tested root tensile strength as a boxplot. The largest variation in strength with root diameter was found in *Fraxinus griffithii*, which was also reflected in the high exponent value (1.414) of its negative power law curve fitting as shown in Table 3. In other words, an increase in the root diameter can bring about a faster rate of reduction in tensile strength for the roots of *Fraxinus griffithii* as compared with other species. The roots of *Persea americana*, *Cinnamomum officinarum*, and *Coffea arabica* exhibited a more moderate variation, while the others showed a narrow variation in tensile strength. However, it should also be noted that the range of root diameters was different among the tested species. The narrow variation in tensile strength for some species can be due to the fact that no smaller root size was presented in the specimens tested, or these smaller roots were too soft to handle during the tests.

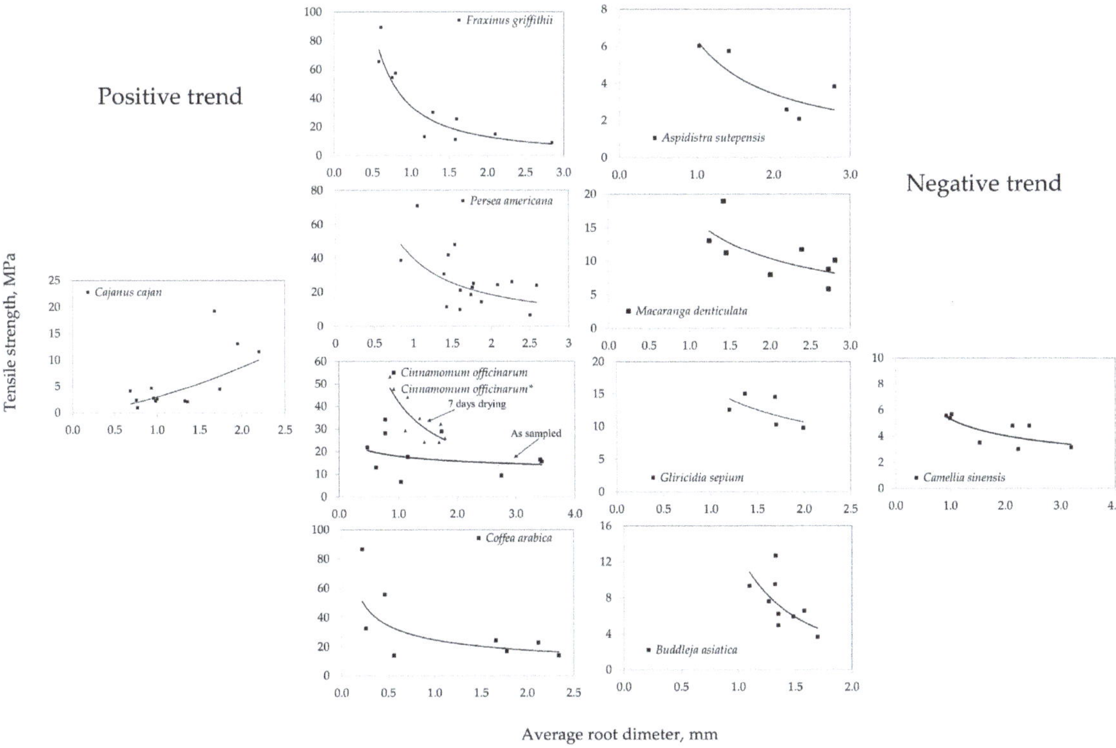

Figure 3. Root tensile strength versus diameter for the 10 species tested (also see Table 3) (*Cinnamomum officinarum** tested after 7 days drying).

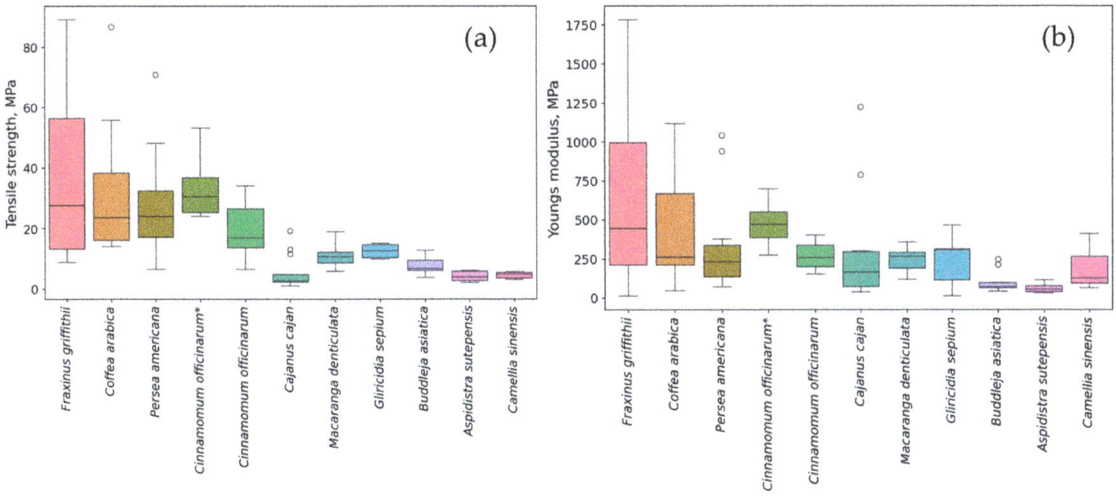

Figure 4. Boxplots of root tensile strength (**a**) and Young's modulus (**b**) tested species for different species. The bottom and top of the boxes represent the 25th and 75th percentiles, respectively, while the line within the box marks the median and the circle represents outlier (*Cinnamomum officinarum** refers to 7-day drying condition).

3.3. Young's Modulus of Roots

Figure 5 and Table 3 illustrate the relationship between Young's modulus of roots for each tested species. The average Young's modulus of various plant specimens significantly ranged from 62.90 MPa for *Aspidistra sutepensis* to 615.63 MPa for *Fraxinus griffithii*. For *Cinnamomum officinarum* under as-sampled condition, Young's modulus was measured at 269.28 MPa, a lower value than the drier roots, demonstrating the same trend as that of tensile strength as explained earlier. The tested species displayed a relationship between Young's modulus and the root diameter (Figure 5, Table 3), akin to trends observed in strength–diameter relationships. The modulus–diameter relationship, with p-values lower than 0.05 indicating significant relationships, was only found for *Fraxinus griffithii*, *Cajanus cajan*, *Camellia sinensis*, and *Coffea arabica*. Other species, including *Persea Americana*, *Cinnamomum officinarum*, *Aspidistra sutepensis*, *Gliricidia sepium*, *Macaranga denticulate*, and *Buddleja asiatica*, showing p-values higher than 0.05, demonstrated larger deviation of moduli from the fitting trend line, reflecting their intrinsic natural variability. Stiffer roots, characterized by higher Young's modulus values, are shown to effectively mobilize stress, potentially enhancing soil shear strength and stability. Figure 4b presents the tested root Young's modulus as a boxplot. A similar trend can be observed regarding the variation in the modulus as the tensile strength. The largest variation in the modulus with the root diameter was still found in *Fraxinus griffithii* followed by *Coffea arabica*, which showed a more moderate variation, while the other species exhibited a more constant modulus.

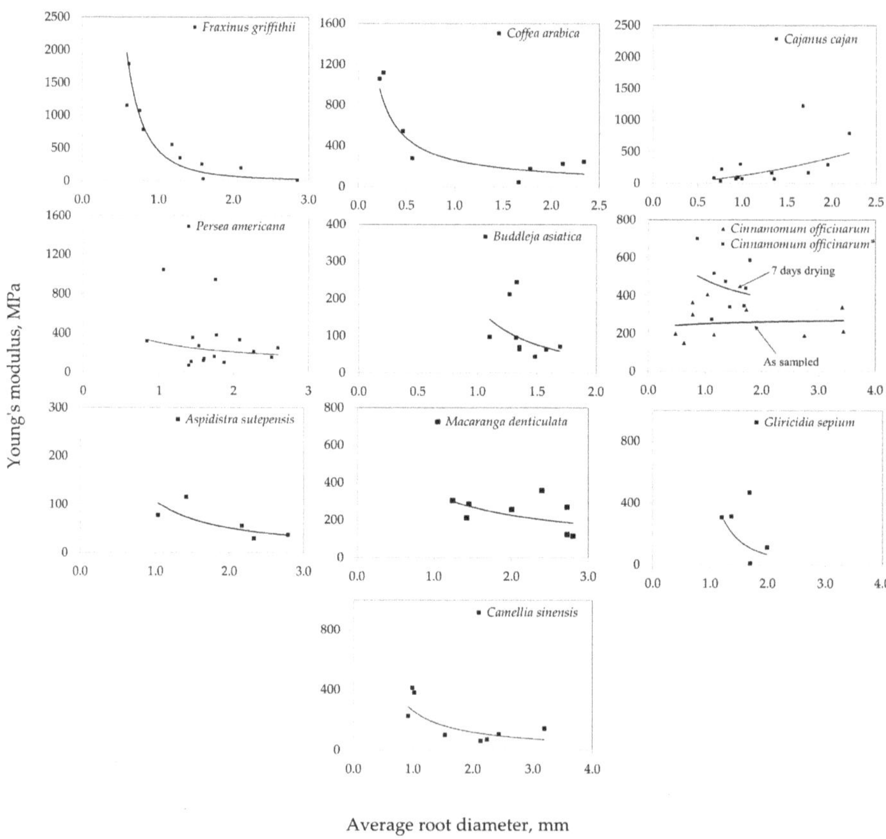

Figure 5. Young's modulus versus root diameter for the 10 species tested (*Cinnamomum officinarum** refers to 7-day drying condition).

Figure 6 focuses on comparing the root tensile strength among different plant species using a standardized root diameter of 1 mm. This 1 mm root tensile strength can be directly obtained from the fitting equations of trendlines as summarized in Table 3. The root diameter of 1mm is typically associated with fibrous root reinforcement, which contributes more significantly to the root reinforcement than thicker roots, as shown in previous studies [27,28]. Notably, exotic species such as *Cinnamomum officinarum*, *Persea americana*, *Fraxinus griffithii*, and *Coffea arabica* demonstrate relatively higher values of 1mm root tensile strength while not being native to Thailand. These exotic species, showing a similar range of tensile strength to *Chrysopogon zizanioides*, would be able to contribute to not only soil erosion control but also their usage for foods and economic purposes. Additionally, this study categorizes species into intermediate (*Macaranga denticulata*, *Gliricidia sepium*, and *Buddleja asiatica*) and low root tensile strength groups, with species like *Aspidistra sutepensis*, *Camellia sinensis*, and *Cajanus cajan* exhibiting the lowest root tensile strength. However, it is noted that even species with relatively lower root tensile strength can contribute to soil stability if their density (root biomass per soil volume) is sufficiently high [51,52].

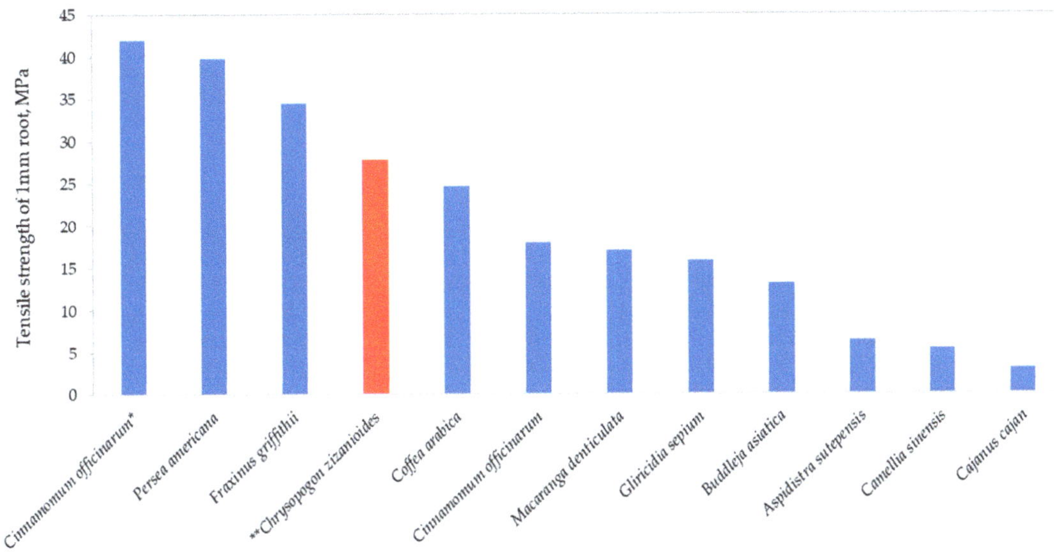

Figure 6. Root tensile strengths of 1 mm diameter root size of the studied species (blue bars), including data of ***Chrysopogon zizanioides* (red bar) from Mahannopkul and Jotisankasa [8] for comparison (*Cinnamomum officinarum** refers to 7-day drying condition; other cases referred to as as-sampled moisture condition).

Table 4 provides a concise overview of the soil reinforcing properties and diverse advantages of several plant species investigated in this study. *Persea americana* and *Fraxinus griffithii* stand out with higher tensile strength and are considered effective in soil and water conservation, with *Persea americana* also providing high economic and edible fruit benefits. *Cinnamomum officinarum*, also known for its commercial values and medicinal use, possesses moderate tensile strength and a dense root system, rendering it appropriate for reinforcing soil and conserving soil and water at a medium to high level. *Cajanus cajan* and *Aspidistra sutepensis* have low tensile strength but would still contribute to soil conservation, as their root density varies from medium to high. They also provide high edible fruit benefits. *Coffea arabica* is distinguished by relatively high tensile strength and root density, which shows its potential for soil reinforcement and conservation in addition to its well-known commercial use. Other species, like *Gliricidia sepium* and *Buddleja asiatica*, offer medium tensile strength and root density, with varying benefits for general use, economic use, and soil conservation.

Table 4. Summary of the strengthening properties of plant roots and their applications.

Plant Species	Soil Reingforcement		Benefits			
	Tensile Strength	Root Density	General Use	Economic Use	Edible Fruits	Soil and Water Conservation
Persea americana	High	Medium	Medium	High	High	Medim-High
Fraxinus griffithii	High	High	High	Unknown	Unknown	High
Cajanus cajan	Low	High	Medium	Unknown	High	Medium
Cinnamomum officinarum	Medium	High	High	Unknown	Medium	Medim-High
Camellia sinensis	Low	High	Low	High	High	Medium
Aspidistra sutepensis	Low	Medium	No	Unknown	High	Medium
Coffea arabica	High	High	Low	High	High	High
Gliricidia sepium	Medium	Medium	Medium	Unknown	Medium	Medium
Macaranga denticulata	Medium	Medium	Low	Unknown	Unknown	Medium
Buddleja asiatica	Medium	Medium	High	Unknown	Unknown	Medium

3.4. Plant Response to Biotic and Abiotic Stresses

Table 5 discusses the growth response of the tested species to biotic and abiotic stresses. *Gliricidia sepium* and *Cajanus cajan* are fast-growing legumes that enhance soil fertility through superior nitrogen fixation [53], even under drought conditions. *Gliricidia sepium* also offers high biomass and nutrient levels but does not significantly aid in erosion control or slope stabilization [54]. It tolerates frost and moderate salinity but struggles in waterlogged conditions like *Aspidistra sutepensis*, providing some protection against fungal, viral, and insect threats [55]. It also provides protection against certain fungal infections (rust), viral diseases (leaf spot), and insect infestations (stem borer) [56]. *Persea americana* faces growth challenges from water quality, drought, low soil fertility, and coffee berry disease [57,58]. *Fraxinus griffithii* is a saline-tolerant species capable of growing under drought stress, but it requires ample space due to its large canopy. However, *Coffea arabica* and *Camellia sinensis* thrive under shade but are highly susceptible to several pests and diseases mentioned in Table 5. As a medicinal plant, *Buddleja asiatica* leaves exhibit in vitro antifungal activities and possess antibacterial and cytotoxic properties, offering protection against pests such as rats [59] that help in slope stabilization. According to Joshi et al. [60], this shrub demonstrates antimicrobial activity against six bacteria (*Bacillus subtilis, Staphylococcus aureus, Pseudomonas aeruginosa, Escherichia coli, Proteus vulgaris*, and *Bacillus mycoides*) and two fungi (*Aspergillus niger* and *Candida albicans*). Additionally, it works as an antifungal agent against *Pseudomonas aeruginosa* and is effective against *Aspergillus niger* when used with Fluconazole [60]. Similarly, *Macaranga denticulata* and *Cinnamomum officinarum* possess antioxidant properties, though not as potent as *Buddleja asiatica*.

It is well recognized [61] that planting diversified species as vegetation islands and terraces can enhance the heterogeneity of the natural ecosystem recovery and help to accelerate successional processes, creating hospitable microclimates and soil microflora and fauna that can mimic natural processes. The 10 studied species in this research not only provide a diversified vegetation, which is valuable for ecological improvement, but also are beneficial socially and economically. Their implementation for slope stabilization projects is still underway, and the results will be presented in the future.

Table 5. Growth response to biotic and abiotic stresses.

Plant Species	Advantage	Disadvantage	References
Gliricidia sepium	Drought, frost and saline tolerant, enhancing soil fertility, resistant to rust and leaf spot diseases and stem borer insect.	Does not survive in waterlogged conditions	[54–56]
Persea americana	Improve soil physical properties, increase water holding capacity (WHC), mitigate nutrient loss.	Sensitive to water quality and few irrigations, drought, low soil fertility, susceptible to coffee berry disease.	[56,57,62]
Fraxinus griffithii	Drought and saline tolerant.	Requires ample space due to large canopy.	
Coffea arabica	Direct sunlight not necessary.	Susceptible to *Xylotrechus quadripes*, *Xylosandrus compactus*, and *Hemileia vastatrix*	[59]
Buddleja asiatica	Antifungal, antibacterial, and cytotoxic properties against rat, resistant against six bacteria and two fungi.	Sensitive to cold and dry/drought stress.	[59,60,63]
Macaranga denticulata	Antioxidant properties, survive in adverse soil and environment, have complex symbiotic mycorrhizal relationship that increase nutrient uptake.	Difficult management and hand weeding needed.	[64]
Cinnamomum officinarum	Antioxidant properties	No report found	
Aspidistra sutepensis	Grows under diverse environment	Does not survive in waterlogged soil.	[65]
Cajanus cajan	Drought, frost, saline tolerant, enhancing soil fertility, protest rust, leaf spot and stem borer.	Does not survive in waterlogged conditions	[65]
Camellia sinensis	Minimum runoff, high organic matter and water holding capacity, natural habitat.	Need a combination of cover crops, susceptible to *Xylotrechus quadripes*, *Xylosandrus compactus*, and *Hemileia vastatrix*	[59,66]

4. Conclusions

This study investigated ten plant species in Thailand's highland regions for their soil bioengineering potential and additional benefits. The root morphology along with the relationship between root tensile strength and Young's modulus with a diameter was extensively studied. The following conclusions can be drawn from this study:

- *G. sepium*, *F. griffithii*, *P. americana*, *B. asiatica*, and *C. arabica* possessed H-type roots and wide lateral spread. *M. denticulata* and *C. officinarum* demonstrated VH-type roots with deep taproots in addition to their lateral extent. *A. sutepensis* showed M-type roots with most root matrix in the top 0.3m, whereas *C. cajan* and *C. sinensis* had R-type roots with deep, oblique growth. This information is crucial for strategically arranging slope vegetation, enabling a more robust bioengineering system through diversified root reinforcement patterns and a variety of plant usages.
- The root tensile strength and Young's modulus–diameter relationships of the ten species highlighted two different trends. For nine of the species tested, *P. americana*, *F. griffithii*, *M. denticulata*, *G. sepium*, *A. sutepensis*, *C. arabica*, *C. sinensis*, *C. officinarum*, and *B. asiatica* demonstrated a decrease in tensile strength with the diameter, the commonly observed negative power law relationship with the exponent values ranging from −0.187 to −1.983. Conversely, *C. cajan* exhibited an increase in tensile strength

with the diameter, the positive power law with the exponent value of 1.5474 diverging from the typical trend.

- Plant species with relatively high values of 1-mm root tensile strength (exceeding 24 to 42 MPa) include *P. americana*, *F. griffithii*, *C. officinarum*, and *C. arabica*. Species exhibiting intermediate root tensile strength (ranging from 8 to 19 MPa) comprise *M. denticulata*, *G. sepium*, and *B. asiatica*. Conversely, *A. sutepensis*, *C. cajan*, and *C. sinensis* demonstrate the lowest root tensile strength, up to 7 MPa.
- It is recommended to design slope vegetation by strategically selecting species with diverse root morphologies and mechanical properties, as well as incorporating a variety of plant species that offer multiple benefits. This approach addresses both the engineering and socioeconomic requirements of the nature-based solution.

Author Contributions: Conceptualization, W.M. and A.J.; data curation, M.H., S.P. and S.S.; formal analysis, M.H. and A.J.; funding acquisition, A.J.; investigation, W.M., A.J., N.L., C.N., P.R., J.P., S.P. and S.S.; methodology, W.M., A.J. and N.L.; project administration, A.J. and J.P.; resources, A.J. and J.P.; supervision, A.J.; validation, A.J., N.L. and C.N.; writing—original draft, M.H. and A.J.; writing—review and editing, M.H., A.J., N.L., C.N. and A.N.F.A. All authors have read and agreed to the published version of the manuscript.

Funding: The research grant was provided by Highland Research and Development Institute (HRDI) and the Faculty of Engineering, Kasetsart University, Thailand.

Data Availability Statement: Data for the formal analysis are presented in the article. Raw data are available upon request from the corresponding author.

Acknowledgments: The authors express their gratitude for the grant provided by the Highland Research and Development Institute (HRDI) and the scholarship provided to the fourth author by Faculty of Engineering, Kasetsart University. They also acknowledge the students and staff at the Geotechnical Division of the Department of Civil Engineering at Kasetsart University and Green Ground Solutions, Co. Ltd., for their support.

Conflicts of Interest: The authors declare no conflicts of interest.

References

1. Gadhoke, P. Walking the Middle Path of Food Sovereignty, Food Security, Nutrition, and Health in Chiang Mai Province, Thailand. In *Transformations of Global Food Systems for Climate Change Resilience: Addressing Food Security, Nutrition, and Health*; CRC PRESS: Boca Raton, FL, USA, 2023; pp. 263–289. [CrossRef]
2. Jotisankasa, A.; Jamrueang, W.; Pramusandi, S.; Semmad, S.; Pilumwong, J. Field Observations of Soil Moisture, Suction and Movement of Cornfield in Tropical Highland with and without Vetiver System. In Proceedings of the 8th International Conference on Unsaturated Soils (UNSAT 2023), E3S Web of Conference, Milos, Greece, 2–5 May 2023; Volume 382, p. 24004. [CrossRef]
3. Coppin, N.J.; Barker, D.L.; Richards, I. *Use of Vegetation in Civil Engineering*; Butterworths: London, UK, 1990.
4. Gray, D.H.; Sotir, R.B. *Biotechnical and Soil Bioengineering Slope Stabilization: A Practical Guide for Erosion Control*; John Wiley & Sons: New York, NY, USA, 1996.
5. Stokes, A.; Norris, J.E.; Van Beek, L.P.H.; Bogaard, T.; Cammeraat, E.; Mickovski, S.B.; Jenner, A.; Di Iorio, A.; Fourcaud, T. How Vegetation Reinforces Soil on Slopes. *Slope Stab. Eros. Control Ecotechnol. Solut.* **2008**, *4*, 65–118.
6. Stokes, A.; Douglas, G.B.; Fourcaud, T.; Giadrossich, F.; Gillies, C.; Hubble, T.; Kim, J.; Loades, K.W.; Mao, Z.; McIvor, I.R.; et al. Ecological mitigation of hillslope instability: Ten key issues facing researchers and practitioners. *Plant Soil* **2014**, *377*, 1–23. [CrossRef]
7. Jotisankasa, A.; Sirirattanachat, T. Effects of Grass Roots on Soil-Water Retention Curve and Permeability Function. *Can. Geotech. J.* **2017**, *54*, 1612–1622. [CrossRef]
8. Mahannopkul, K.; Jotisankasa, A. Influence of Root Suction on Tensile Strength of *Chrysopogon zizanioides* Roots and Its Implication on Bioslope Stabilization. *J. Mt. Sci.* **2019**, *16*, 275–284. [CrossRef]
9. Inui, T.; Chau, C.; Soga, K.; Nicolson, D.; O'Riordan, N. Embodied Energy and Gas Emissions of Retaining Wall Structures. *J. Geotech. Geoenviron. Eng.* **2011**, *137*, 958–967. [CrossRef]
10. Liu, Y.J.; Wang, T.W.; Cai, C.F.; Li, Z.X.; Cheng, D.B. Effects of Vegetation on Runoff Generation, Sediment Yield and Soil Shear Strength on Road-Side Slopes under a Simulation Rainfall Test in the Three Gorges Reservoir Area, China. *Sci. Total Environ.* **2014**, *485*, 93–102. [CrossRef] [PubMed]
11. Liu, G.; Hu, F.; Zheng, F.; Zhang, Q. Effects and Mechanisms of Erosion Control Techniques on Stairstep Cut-Slopes. *Sci. Total Environ.* **2019**, *656*, 307–315. [CrossRef] [PubMed]

12. Kim, J.H.; Fourcaud, T.; Jourdan, C.; Maeght, J.L.; Mao, Z.; Metayer, J.; Meylan, L.; Pierret, A.; Rapidel, B.; Roupsard, O.; et al. Vegetation as a Driver of Temporal Variations in Slope Stability: The Impact of Hydrological Processes. *Geophys. Res. Lett.* **2017**, *44*, 4897–4907. [CrossRef]
13. Chen, Z.; Liu, X.; Ai, Y.; Chen, J.; Luo, X.; Chen, J.; Zhong, S. Effects and Mechanisms of Revegetation Modes on Cadmium and Lead Pollution in Artificial Soil on Railway Rock-Cut Slopes. *Sci. Total Environ.* **2018**, *644*, 1602–1611. [CrossRef]
14. Oshunsanya, S.O.; Li, Y.; Yu, H. Vetiver Grass Hedgerows Significantly Reduce Nitrogen and Phosphorus Losses from Fertilized Sloping Lands. *Sci. Total Environ.* **2019**, *661*, 86–94. [CrossRef]
15. Grima, N.; Edwards, D.; Edwards, F.; Petley, D.; Fisher, B. Landslides in the Andes: Forest can provide cost-effective landslide regulation services. *Sci. Total Environ.* **2020**, *745*, 141128. [CrossRef] [PubMed]
16. Wu, Z.; Leung, A.K.; Boldrin, D.; Ganesan, S.P. Variability in root biomechanics of *Chrysopogon zizanioides* for soil eco-engineering solutions. *Sci. Total Environ.* **2021**, *776*, 145943. [CrossRef]
17. Ng, C.W.W.; Leung, A.K.; Ni, J.J. *Plant-Soil Slope Interaction*; Taylor & Francis: New York, NY, USA, 2019.
18. Docker, B.B.; Hubble, T.C.T. Quantifying root-reinforcement of river bank soil by four Australian tree species. *Geomorphology* **2008**, *100*, 401–418. [CrossRef]
19. Boldrin, D.; Leung, A.K.; Bengough, A.G. Root biomechanical properties during establishment of woody perennials. *Ecol. Eng.* **2017**, *109*, 196–206. [CrossRef]
20. Styczen, M.E.; Morgan, R.P.C. Engineering properties of vegetation. In *Slope Stabilization and Erosion Control: A Bioengineering Approach*; Morgain, R.P.C., Rickson, R.J., Eds.; E & F N Spon: London, UK, 1995.
21. Yen, C. Tree Root Patterns and Erosion Control. 1987. Available online: https://www.cabidigitallibrary.org/doi/full/10.5555/19930667066 (accessed on 13 June 2024).
22. Fan, C.-C.; Chen, Y.-W. The effect of root architecture on the shearing resistance of root-permeated soils. *Ecol. Eng.* **2010**, *36*, 813–826. [CrossRef]
23. Capilleri, P.P.; Motta, E.; Raciti, E. Experimental Study on Native Plant Root Tensile Strength for Slope Stabilization. *Procedia Eng.* **2016**, *158*, 116–121. [CrossRef]
24. Persichillo, M.G.; Bordoni, M.; Meisina, C. The role of land use changes in the distribution of shallow landslides. *Sci. Total Environ.* **2017**, *574*, 924–937. [CrossRef]
25. Stokes, A.; Atger, C.; Bengough, A.G.; Fourcaud, T.; Sidle, R.C. Desirable plant root traits for protecting natural and engineered slopes against landslides. *Plant Soil* **2009**, *324*, 1–30. [CrossRef]
26. Norris, J.E.; Di Iorio, A.; Stokes, A.; Nicoll, B.C.; Achim, A. Species selection for soil reinforcement and protection. In *Slope Stability and Erosion Control: Ecotechnological Solutions*; Norris, J.E., Stokes, A., Mickovski, S.B., Cammeraat, E., van Beek, R., Nicoll, B.C., Achim, A., Eds.; Springer: Dordrecht, The Netherlands, 2008; pp. 167–210.
27. De Baets, S.; Poesen, J.; Reubens, B.; Wemans, K.; De Baerdemaeker, J.; Muys, B. Root tensile strength and root distribution of typical Mediterranean plant species and their contribution to soil shear strength. *Plant Soil* **2008**, *305*, 207–226. [CrossRef]
28. Genet, M.; Stokes, A.; Salin, F.; Mickovski, S.B.; Fourcaud, T.; Dumail, J.-F.; van Beek, R. The Influence of Cellulose Content on Tensile Strength in Tree Roots. *Plant Soil* **2005**, *278*, 1–9. [CrossRef]
29. Mao, Z.; Wang, Y.; McCormack, M.L.; Rowe, N.; Deng, X.; Yang, X.; Xia, S.; Nespoulous, J.; Sidle, R.C.; Guo, D.; et al. Mechanical traits of fine roots as a function of topology and anatomy. *Ann. Bot.* **2018**, *122*, 1103–1116. [CrossRef] [PubMed]
30. Loades, K.W.; Bengough, A.G.; Bransby, M.F.; Hallett, P.D. Biomechanics of nodal, seminal and lateral roots of barley: Effects of diameter, waterlogging and mechanical impedance. *Plant Soil* **2013**, *370*, 407–418. [CrossRef]
31. Ghestem, M.; Cao, K.; Ma, W.; Rowe, N.; Leclerc, R.; Gadenne, C.; Stokes, A. A framework for identifying plant species to be used as 'Ecological Engineers' for fixing soil on unstable slopes. *PLoS ONE* **2014**, *9*, e95876. [CrossRef] [PubMed]
32. Boldrin, D.; Leung, A.K.; Bengough, A.G. Effects of root de-hydration on bio-mechanical properties of woody roots of U. europaeus. *Plant Soil* **2018**, *431*, 347–369. [CrossRef]
33. Hales, T.C.; Miniat, C.F. Soil moisture causes dynamic adjustments to root reinforcement that reduce slope stability. *Earth Surf. Process. Landf.* **2016**, *42*, 803–813. [CrossRef]
34. Zhang, C.B.; Zhou, X.; Jiang, J.; Wei, Y.; Ma, J.; Hallett, P.D. Root moisture content influence on root tensile tests of herbaceous plants. *Catena* **2019**, *172*, 140–147. [CrossRef]
35. Loades, K.W.; Bengough, A.G.; Bransby, M.F.; Hallett, P.D. Effect of root age on the biomechanics of seminal and nodal roots of barley (*Hordeum vulgare* L.) in contrasting soil environments. *Plant Soil* **2015**, *395*, 253–261. [CrossRef]
36. Cofie, P.; Koolen, A.J. Test speed and other factors affecting the measurements of tree root properties used in soil reinforcement models. *Soil Tillage Res.* **2001**, *63*, 51–56. [CrossRef]
37. Zhang, C.; Chen, L.; Jiang, J.; Zhou, S. Effects of gauge length and strain rate on the tensile strength of tree roots. *Trees* **2012**, *26*, 1577–1584. [CrossRef]
38. Truong, P.; Van, T.T.; Pinners, E. Vetiver System Applications: Technical Reference Manual. *Vetiver Netw. Int.* **2008**, 89.

39. Cuong, D.C.; Truong, P.; Van Minh, V.; Minh, P.T.; Truong, P.N. Confusion between Chrysopogon nemoralis and Chrysopogon zizanioides at Bo Bo Mountain in Quang Nam Province, Vietnam. Available online: https://d1wqtxts1xzle7.cloudfront.net/88699934/2_20Doan_20Chi_20Cuong_20Paper-libre.pdf?1658101759=&response-content-disposition=inline;+filename=1_CONFUSION_BETWEEN_Chrysopogon_nemorali.pdf&Expires=1721788768&Signature=FKeOhqDSU6j4XPbuFlujtkUoam81QeLjFizuQgDIks9j1HmEJnpqANThkoaW4~-ykMIABiLi~rljJuRgc05uIhTOZbx3xvl0eOdvNFiCHqX4PDlXlIlk9JKaC0S-XKTzPRUu3I994MPab0qY2yUGkKDdQL5kR4fJWL1DhrwDfjvXfYKzAGQKyOYOMcE3DTucDhdXrZAT0P-coegf2JyVbOSd05XdZG2vJ--5Kv5NGMPZ3IsKM8kQq9Zpjf7JFAAgj8hWtp5JGQOi16rOfc32NaRDm2GYl9DcPMBySvx4Ybljc56Eh1XgC3pyyUcH~ZbTwXcRI8KuzBJJgVvmgRmviA__&Key-Pair-Id=APKAJLOHF5GGSLRBV4ZA (accessed on 13 June 2024).
40. Lewis, L. *Soil Bioengineering: An Alternative for Roadside Management*; US Department of Agriculture, Forest Service, Technology & Development Program: Washington, DC, USA, 2000.
41. Zhang, Z.; Fan, B.; Song, C.; Zhang, X.; Zhao, Q.; Ye, B. Advances in Root System Architecture: Functionality, Plasticity, and Research Methods. *J. Resour. Ecol.* **2023**, *14*, 15–24. [CrossRef]
42. Wang, J.B.; Zhang, X.J.; Wu, C. Advances in experimental methods for root system architecture and root development. *J. For. Res.* **2015**, *26*, 23–32. [CrossRef]
43. POWO. Plants of the World Online. Facilitated by the Royal Botanic Gardens, Kew. 2024. Available online: http://www.plantsoftheworldonline.org/ (accessed on 9 May 2024).
44. Chang, M.C.; Qiu, L.Q.; Green, P.S. Oleaceae. In *Flora of China*; Wu, C.Y., Raven, P.H., Hong, D.Y., Eds.; Science Press: Beijing, China; Missouri Botanical Garden Press: St. Louis, MO, USA, 1996; Volume 15, pp. 272–319.
45. Li, X.W.; Li, J.; Huang, P.H.; Wei, F.N.; Cui, H.B.; van der Werff, H. Lauraceae. In *Flora of China*; Wu, C.Y., Raven, P.H., Hong, D.Y., Eds.; Science Press: Beijing, China; Missouri Botanical Garden Press: St. Louis, MO, USA, 2008; Volume 7, pp. 102–254.
46. Keng, H. Theaceae. In *Flora of Thailand*; Smitinand, T., Larsen, K., Eds.; The Forest Herbarium, Royal Forest Department: Bangkok, Thailand, 1972; Volume 2, Part 2, pp. 142–158.
47. Bartholomew, B. Theaceae. In *Flora of China*; Wu, C.Y., Raven, P.H., Hong, D.Y., Eds.; Science Press: Beijing, China; Missouri Botanical Garden Press: St. Louis, MO, USA, 2007; Volume 12, pp. 366–478.
48. Ding, H.B.; Xiong, K.C.; Yang, B.; Yin, J.T.; Bouamanivong, S.; Tani, Y.H. New species and taxonomic notes of Aspidistra (Asparagaceae) for the flora of China and Laos. *Taiwania* **2021**, *66*, 439–449. [CrossRef]
49. van Welzen, P.C.; Chayamarit, K. Euphorbiaceae (Genera G–Z). In *Flora of Thailand*; Santisuk, T., Larsen, K., Eds.; The Forest Herbarium, National Park, Wildlife and Plant Conservation Department: Bangkok, Thailand, 2007; Volume 8, Part 2, pp. 305–592.
50. Yang, Y.; Chen, L.; Li, N.; Zhang, Q. Effect of root moisture content and diameter on root tensile properties. *PLoS ONE* **2016**, *11*, e0151791. [CrossRef]
51. Nilaweera, N.S. Effects of Tree Roots on Slope Stability—The Case of Khao Luang Mountain Area, Southern Thailand. Ph.D. Thesis, Asian Institute of Technology, Pathum Thani, Thailand, 1994; p. 452.
52. Nilaweera, N.S.; Nutalaya, P. Role of tree roots in slope stabilisation. *Bull. Eng. Geol. Environ.* **1999**, *57*, 337–342. [CrossRef]
53. Gargi, B.; Semwal, P.; Jameel Pasha, S.B.; Singh, P.; Painuli, S.; Thapliyal, A.; Cruz-Martins, N. Revisiting the Nutritional, Chemical and Biological Potential of *Cajanus cajan* (L.) Millsp. *Molecules* **2022**, *27*, 6877. [CrossRef]
54. Fagbola, O.; Osonubi, O.; Mulongoy, K.; Odunfa, S. Effects of drought stress and arbuscular mycorrhiza on the growth of *Gliricidia sepium* (Jacq). Walp, and *Leucaena leucocephala* (Lam.) de Wit. in simulated eroded soil conditions. *Mycorrhiza* **2001**, *11*, 215–223. [CrossRef]
55. Wartenberg, A.C.; Blaser, W.J.; Roshetko, J.M.; Van Noordwijk, M.; Six, J. Soil fertility and Theobroma cacao growth and productivity under commonly intercropped shade-tree species in Sulawesi, Indonesia. *Plant Soil* **2020**, *453*, 87–104. [CrossRef]
56. Elevitch, C.R.; Francis, J.K. *Gliricidia sepium* (gliricidia). *Species Profiles Pac. Isl. Agrofor.* **2006**, *2*, 1–8.
57. Juma, I. Production, local trade and diversity of avocado (*Persea americana* Mill.) in the southern highlands of Tanzania. *Acta Univ. Agric. Sueciae* **2020**, *2020*, 26.
58. Sina, B.; Demissie, H.; Rezene, Y. Assessment of the Constraints and Challenges in Avocado (*Persea americana* Mill.) Production and Marketing in Southern Ethiopia. *Int. J. Fruit Sci.* **2024**, *24*, 60–72. [CrossRef]
59. Liao, Y.H.; Houghton, P.J.; Hoult, J.R. Novel and known constituents from *Buddleja* species and their activity against leukocyte eicosanoid generation. *J. Nat. Prod.* **1999**, *62*, 1241–1245. [CrossRef] [PubMed]
60. Joshi, S.; Mishra, D.; Khetwal, K.S.; Bisht, G. Antibacterial and Antifungal Properties of Crude Extracts of *Buddleja asiatica* L. Aerial Parts. *J. Pharm. Res.* **2011**, *4*, 2282.
61. Schaefer, J.W.; Dhar, A.; Chanasyk, D.S.; Naeth, M.A. Reclamation of an eroded lakeshore slope using small vegetation islands and terraces. *Ecol. Eng.* **2024**, *206*, 107327. [CrossRef]
62. Atucha, A.; Merwin, I.A.; Brown, M.G.; Gardiazabal, F.; Mena, F.; Adriazola, C.; Lehmann, J. Soil erosion, runoff and nutrient losses in an avocado (*Persea americana* Mill) hillside orchard under different groundcover management systems. *Plant Soil* **2013**, *368*, 393–406. [CrossRef]
63. Kriticos, D.J.; Watt, M.S.; Potter, K.J.; Manning, L.K.; Alexander, N.S.; Tallent-Halsell, N. Managing invasive weeds under climate change: Considering the current and potential future distribution of Buddleja davidii. *Weed Res.* **2011**, *51*, 85–96. [CrossRef]
64. Changkija, S.; Thakuria, D.; Cynthia, A. Traditional Use of Macaranga Trees for Soil Fertility: By Naga shifting cultivators in northeast India. In *Farmer Innovations and Best Practices by Shifting Cultivators in Asia-Pacific*; CABI: New York, NY, USA, 2023; pp. 557–570.

65. Dwivedi, A.K.; Tripathi, B.D. Pollution tolerance and distribution pattern of plants in surrounding area of coal-fired industries. *J. Environ. Biol.* **2007**, *28*, 257–263.
66. Sahoo, D.C.; Madhu, M.G.; Bosu, S.S.; Khola, O.P. Farming methods impact on soil and water conservation efficiency under tea [*Camellia sinensis* (L.)] plantation in Nilgiris of South India. *Int. Soil Water Conserv. Res.* **2016**, *4*, 195–198. [CrossRef]

sustainability

MDPI

Article

Promising Nature-Based Solutions to Support Climate Adaptation of Arizona's Local Food Entrepreneurs and Optimize One Health

Yevheniia Varyvoda [1,*], Taylor Ann Foerster [2], Joona Mikkola [3] and Matthew M. Mars [4]

[1] Mel and Enid Zuckerman College of Public Health, The University of Arizona, Tucson, AZ 85724, USA
[2] College of Social & Behavioral Sciences, The University of Arizona, Tucson, AZ 85719, USA; tfoerster@arizona.edu
[3] Arid Lands Resource Sciences, The University of Arizona, Tucson, AZ 85719, USA; joonamikkola@arizona.edu
[4] College of Humanities, The University of Arizona, Tucson, AZ 85721, USA; mmars@arizona.edu
* Correspondence: varyvoda@arizona.edu

Abstract: This study explores the uptake and potential application of nature-based solutions (NbS) that are particularly promising for small-scale farmers, ranchers, and food entrepreneurs operating in arid and semi-arid regions. Studying the adoption of NbS by local food entrepreneurs (LFEs), including related strengths and limitations, remains an area of exploration due to their potential to optimize interventions that foster environmental sustainability at the intersection of people, animals, and natural ecosystems (i.e., One Health). A multi-method design was used, including literature review, questionnaires, and semi-structured key informant interviews to assess adaptation needs and NbS among a sample of LFEs located in Southern AZ, USA. The findings revealed that existing NbS have been introduced mostly through learning-by-doing practices that are bounded by economic and technological resource constraints. The paper describes a range of accessible approaches and practices that can be piloted and/or scaled up to enhance local food system resilience and contribute to the overlapping health of people, animals, and natural ecosystems. The priority adaptation pathways for NbS were identified to be funding and financing and the co-creation and sharing of knowledge through peer-to-peer and expert-to-peer approaches. The results suggested that AZ LFEs are likely to adopt NbS based on their capacity to address priority climate-driven issues, revenue generation potential, and seamless augmentation with existing food production and operational activities.

Keywords: nature-based solutions; local food systems; entrepreneurs; climate change; community innovation

1. Introduction

Studies have increasingly demonstrated that climate change has considerable impacts on food systems in arid and semi-arid regions, compromising the intersectional health of humans, domestic and wild animals, plants, and surrounding natural ecosystems, i.e., the One Health dynamic [1]. Arizona (AZ) is the sixth-largest state in the United States (US). AZ is located in the southwestern corner of the country where it is especially challenged by rapid population growth, shifting land use and land cover, increasingly scarce water resources, and consequently, long-term drought [2]. Indeed, much of AZ is characterized as arid to semiarid, with annual average precipitation ranging from less than 4 inches in the desert lowlands to around 40 inches in the mountain highlands. The deserts are some of the hottest and driest areas of the country, while the higher terrain of the Colorado Plateau has much colder winters and relatively mild summers [3]. AZ has warmed by about 2.5 °F (1.35 °C) over the last century. In the past two decades, AZ has experienced its highest recorded temperatures. Phoenix, AZ, had its all-time warmest summer on record in 2023, with an average summer temperature of 97.0 °F (36.11 °C), surpassing the previous record of 96.7 °F (35.94 °C) established just two years earlier in 2020 [4].

AZ is one of the states in the US that has been hardest hit by climate change to the detriment of human health and wellbeing, economic prosperity and social welfare, natural and modified ecosystems, food and water security, and cultural heritage. Symptoms of said climate change include rising air temperatures; changes to the timing, form, and amount of precipitation and associated flooding events; increases in extreme heat events; surface and groundwater reductions; heat stress and heat-related illnesses; and intensifying dust storms and wildfires [5,6]. Likewise, the economic and environmental sustainability and community impacts of AZ's local food systems are continually threatened by climate change. AZ's agri-food industry, valued at 23 billion USD, is currently facing significant climate-driven threats, with 90% of the state having been affected by "exceptional drought" over the past two decades [7].

Small-scale farmers, ranchers, and food purveyors who operate locally across AZ, whom together we refer to hereafter as local food entrepreneurs (LFEs), are confronted with drops in production and sales rates and operational inefficiencies that are perpetuated by socioeconomic constraints and climate uncertainties. As a result, LFEs are forced to downsize production (e.g., removing orchard trees and multi-year crops, selling off portions of herds and flocks), resulting in lost profits and declines in the availability and affordability of locally produced, highly nutritious foods [8].

Beyond the financial consequences, climate change also affects the physical heath and wellbeing of the LFEs. Specifically, climate change, and especially heat stress, can exacerbate a range of acute conditions that include heat stroke, dehydration, and exhaustion, as well as longer term cardiovascular and water- and zoonotic-borne diseases [9]. Further, the financial uncertainties that are being intensified by climate change make LFEs more prone to anxiety, depression, and other debilitating mental health challenges that can, in the extreme, lead to suicide [10].

The overall effects of climate change undermine the health and wellbeing of LFEs and the economic, environmental, and social sustainability of AZ's local food systems. Unfortunately, this wicked dynamic is repeated in countless other locales and regions, making AZ a highly relevant living laboratory in which to further describe the economic, environmental, and social impacts of climate change on local food systems, as well as to explore potential interventions. Climate adaptation interventions tailored to AZ's local and regional needs with health-enhancing benefits are urgently required to meet the climate-driven challenges and optimize the overlapping wellbeing and vibrancy, i.e., One Health, of people, animals, and ecosystems.

Here, we use a multi-method, single case study design to explore the potential effects of NbS on climate change adaptations carried out by LFEs and the One Health of the locales and regions they serve. The International Union of Conservation of Nature (ICUN) Global Standard defines NbS as "actions to protect, sustainable manage and restore natural and modified ecosystems in ways that address societal challenges effectively and adaptively, to provide both human well-being and biodiversity benefits" (IUCN Global Standard for NbS, 2020, p. 1) [11]. NbS for climate adaptation comprise a broad portfolio of interventions with particular attention being paid to fostering natural regeneration and restoration with native species [12]. Given the lack of consensus on which agricultural practices are considered NbS, there are no data available on how widely NbS are used in agriculture at any scale [13]. Accordingly, our current objective is to generate insights into the experiences and everyday realities of LFEs that stand to inform NbS development and implementation. We draw on the insights generated to propose nature-based climate adaptation strategies aimed at enhancing food system resilience.

2. Materials and Methods

We created a single case study that was designed to assess the likely viability of NbS in the Southern AZ local food system as viewed through the lenses of LFEs who are actively engaged in food production, storage and distribution, processing, retail, and marketing. For our purposes, Southern AZ was bounded to four neighboring counties: Cochise, Pima, Pinal, and Santa Cruz. Geographically, these counties are mostly rural and largely agriculture-based, though a large majority of the 1.65 million residents live in Pima County within the Tucson metroplex area [14]. Northern Pinal County stretches into the suburban outskirts of the Phoenix metroplex and is home to nearly 500,000 of the residents within the four-county study region. The rural–urban–suburban mix that characterizes this bounded area is consistent with the understanding that local food systems rely on short supply chains that typically connect rural production areas with urban and suburban consumption sites [15].

The research approach mobilized multiple disciplines (community development, communications, public health, sociology) and spanned a diverse array of food system activities (production, processing, distribution, consumption) and health, socioeconomic, and environmental outcomes. Data were generated through semi-structured interviews and questionnaires that together explored and examined the experiences and everyday realities that shape Southern AZ LFEs' perspectives on climate change, climate adaptation, One Health dynamics, and ultimately the relevancy and viability of NbS.

2.1. *Qualitative Method: Interview Protocol*

2.1.1. Sampling and Data Collection

Interviewees were purposefully selected using maximum variation and chain sampling strategies [16,17]. To foster maximum variation in the sample, we limited our selection to a single general criterion: active as an LFE in the Southern AZ local food system. The LFE role included producers, processors, and purveyors as previously indicated [18]. The sampling strategies generated a sample composed of 11 Southern AZ LFEs.

Each interview began by asking the participating LFE a series of demographic questions, including age, gender, level of education, relevant experience in agriculture or food production and distribution, and current activities as an LFE. From there, the LFE was asked questions pertaining to their current and anticipated climate related challenges, mitigation strategies used to cope with climate challenges, and familiarity with NbS. Thereafter, each interview was concluded with a series of questions aimed at identifying the need for resources, barriers to adoption of NbS, and perceived opportunities surrounding NbS. The interviews were conducted either in-person or over Zoom, audio recorded, and transcribed verbatim for analysis.

2.1.2. Qualitative Data Analysis and Trustworthiness

The interview data were analyzed inductively using a grounded, open-code approach to reveal initial codes and emergent trends [19]. Multiple rounds of ideographic and nomothetic analysis were then conducted, refining initial patterns and trends into salient themes [20]. Two forms of triangulation were used to enhance the trustworthiness of the findings [21,22]. First, researcher triangulation was conducted throughout the analytical process, which involved the members of the research team consistently discussing, debating, and reaching consensus on the insights as they emerged from the data. Second, data triangulation was used to compare, affirm, refine, and further contextual the findings developed from the interviews with those generated from the questionnaire described below. Further, the transferability of the qualitative elements of the findings are enhanced by presenting them in the Results and Discussion sections using rich, thick descriptions [23].

2.1.3. Quantitative Method: Questionnaire Design

The questionnaire was designed to gain insights into the constructs of the Theory of Planned Behavior (TPB). TPB posits that attitude, social norms, and perceived behavioral control predict behavioral intention [24]. Thus, the current research used these constructs to explore baseline data of the potential motivators of LFEs' intentions to adopt NbS into their practices.

We used an online questionnaire to capture participant demographic information and describe existing behaviors and attitudes pertaining to behavior, perceived social norms, and perceived behavioral control beliefs intentions, and perceptions of beneficial resources regarding the use of NbS in participants LFEs' practices.

Participants completed an online questionnaire consisting of 25 questions that spanned demographic-, psychological-, and resource-related items. Questions were designed in line with the TPB constructs [23] to gauge participant attitudes, perceived social norms, and perceived behavioral control, along with their intention to engage with NbS. Participants then ranked resources from most to least beneficial to the implementation of NbS in their own practice. A non-random convenience sample (n = 16) was generated by recruiting participants through researcher networks and online local food resource sites. Participants were required to be an adult working in the production, distribution, aggregation, processing, packaging, and/or retail selling of local food within the Southern AZ local food system (as previously defined). Data collection occurred between October 2023 and February 2024.

The questionnaire consisted of 26 questions and took participants on average around eight minutes to complete. The questionnaire was assessed for face validity through an expert review and all constructs were adapted from previous measures as indicated in the respective construct descriptions. Cronbach's alpha coefficients were used to determine construct reliability (Table 1). Unless otherwise noted, all variables were measured using a five-point Likert-type scale with scores ranging from one (strongly disagree) to five (strongly agree). If participants were not actively engaged with growing food items themselves, they were asked to provide their perspectives in relation to the products that they source.

Table 1. Cronbach's alpha reliability estimates by questionnaire constructs.

Construct	Cronbach's Alpha
Existing behavior	0.96
Attitudes	0.96
Perceived social norms	0.88
Perceived behavioral control	0.83
Behavioral intention	0.96

Demographics. At the beginning of the questionnaire, participants were asked about the following demographics: age, gender, highest level of education and discipline of degree if relevant, years of experience in the industry, and the industry that they were currently a part of (i.e., food production, aggregation and distribution, processing and packaging, retail, other).

Defining NbS. After measuring demographic information [25,26], participants were asked to read the following definition of NbS:

"Nature-based solutions are actions to protect, sustainably manage, or restore natural or modified ecosystems to address societal challenges, simultaneously providing benefits for people and the environment [27]. Solutions include protection or conservation of natural areas, soil health practices, water management, agroforestry, afforestation using native tree species, or other practices that contribute to climate change adaptation while improving the health and livelihoods of farmers, ranchers, and food entrepreneurs".

Following the definition, participants were asked to indicate that they had read the definition to proceed with the questionnaire.

Existing behavior. To measure participant's current engagement with NbS practices, two items were used. These items were rated using a five-point Likert-type scale to indicate frequency, with a score of one indicating that a participant never used NbS and a score of five indicating that participants always used NbS practices. The items included "In my current practice, I use NbS" and "I incorporate nature based solutions into my current activities as a local food entrepreneur".

Attitude. Participants' attitudes towards NbS were rated using three different five-point Likert-type scales adapted from Miller et al., (2019) [25]. Specifically, attitudes towards NbS were portrayed as a function of effectiveness of NbS, evaluations of NbS, and favorableness of using NbS. Each scale was proportionate to the other, with a score of one indicating NbS were "not at all" the quality of focus (i.e., effective, good, favorable) and a score of five indicating "Extremely" the quality of focus. The same item was used for each scale "For me, using NbS would be...".

Perceived social norms. Perceptions of social norms as they relate to NbS were measured using four items that were adapted from Miller et al. (2019) and Senger et al. (2017) [25,26]. The items captured both injunctive (two items) and descriptive (two items) norms to create the construct of social norms. Example items included "Most people who are important to me approve of the use of NbS" and "Most people who are important to me incorporate NbS in their practice".

Perceived behavioral control. The extent to which participants perceived their control over the adoption of NbS was measured using four items adapted from Miller et al., (2019) [25]. Example items included "If I wanted to, I could implement NbS" and "Using NbS in my practice is in my control".

Behavioral intention. Participant's intention to adopt NbS for themselves was measured using three items adapted from Senger et al., (2017) [26]. Example items included "I plan to use NbS in the future" and "I intend to adopt NbS in the next three years".

Resources. Participants were asked to rank four resources: (1) Funding, (2) Community (i.e., connections with other local food producers or related community resources), (3) Education, and (4) Additional labor (i.e., volunteers, employees) along with an "other" option with the opportunity to explain within the textbox provided under the option. Specifically, participants ranked the resources in order of most beneficial to least beneficial to them if they were to implement NbS in their own LFE activities. The resources ranked were informed by data from the interview portion of this research.

3. Results

In this section, we present the results of the questionnaire. We then use the Discussion section to contextualize and bring nuance to the said results through the integration of the insights generated from the analysis of the qualitative interview data.

3.1. Questionnaire Participant Demographics

The questionnaire sample ($n = 16$) consisted of more women (68.8%) than men (31.3%). The participant age range was from 25 to 69 years of age with an approximate mean age of 41 years old. The distribution of the sample by highest level of education obtained was as follows: high school education (6.3%), associate degrees (12.5%), bachelor's degree (37.5%), master's degrees (12.5%), a doctorate (6.3%), and other (e.g., some wrote college; 25%). Participants indicated that within their respective industries they had between one year and 26 years of experience. On average, participants had approximately seven years of experience. When asked about their current position as an LFE, 68.8% of participants selected food production, 31.3% food aggregation and distribution, 56.3% food processing and packaging, and 56.3% selected retail.

3.2. Descriptive Results

Given the low sample size ($n = 16$), descriptive statistics were used to provide baseline insights into the perceptions surrounding the adoption of NbS among the surveyed LFEs. All statistical analyses were performed using SPSS software (IBM Corp. Released 2023. IBM SPSS Statistics for Macintosh, Version 29.0.2.0 Armonk, NY, USA: IBM Corp). The results revealed that participants on average used (or sourced products from suppliers that use) NbS about half of the time in their current practice. Attitudes ($M = 4.02$) about using NbS were positive on average, with a score of four indicating very good, favorable, or effective evaluations of using NbS. Perceptions of social norms ($M = 3.59$) and behavioral control ($M = 3.52$) averaged slightly positive scores, with a score of three for both constructs indicating a neutral position (Neither agree not disagree) and a score of four indicating agreeing somewhat with the statements presented. Behavioral intention to adopt NbS was rated relatively high (*M* = 4.29), indicating that, on average, participants held positive intentions to engage with NbS in their own practice in some capacity. Table 2 provides the scores for each construct as well as the standard deviation.

Table 2. Descriptive statistics by construct.

Construct	M	SD
Existing behavior	3.16	1.31
Attitudes	4.02	1.16
Perceived social norms	3.59	0.79
Perceived behavioral control	3.52	0.89
Behavioral intention	4.29	0.95

Participants rated constructs on five-point Likert-type scales. The existing behavior items used a five-point frequency scale ranging from "never—one" to "always—five". The items used to measure attitude were rated on a five-point scale to provide an evaluative perception of NbS with a score of one indicating "not at all" the quality of focus (i.e., effective, good, favorable) and a score of five indicating that using NbS were rated as "Extremely" effective, good, or favorable. All other constructs were measured using a scale of agreement, such that higher scores indicate more agreement with the statements and thus higher levels of the construct.

3.3. Resources

Thirteen of the sixteen participants ranked the resources provided (i.e., funding, community, education, labor, and an other option) from most beneficial to least beneficial to them for implementing NbS in their LFE activities. Table 3 summarizes the rankings of the resources. Overall, Funding was ranked as the most beneficial resource more frequently than the other resources. Funding was followed by Community (i.e., connections with other local food producers or related community resources), which was most frequently rated as the second most beneficial resource by participants. One participant provided "equity share" as another resource. Resources were ranked from most beneficial (first) to least beneficial (fourth). The "other" option provided consistently ranked as fifth.

Table 3. Results of ranking resources for implementing NbS.

Resource	First	Second	Third	Fourth
Funding	9	4		
Community	3	6	3	1
Education	1	2	6	4
Additional labor		1	4	8

4. Discussion

4.1. Current Uptake of NbS

Over 40 NbS have been identified as evidence-based practices that are known to be effective in mitigating and adapting to high-priority climate-related issues identified by LFEs [22,28–61] (Supplementary Materials, Table S1). This list is not exhaustive and NbS such as cover crops and diversified crop rotations are already widely used practices in sustainable farming. US farmers are undertaking five broad categories of climate adaptation strategies based on NbS: water management, crop management, nutrient management, technological management, and financial management [62]. Our study has shown that many small-scale farmers, ranchers, and other LFE-types are familiar with and practicing NbS. However, they might not be fully aware of the benefits of One Health and limitations associated with NbS. For example, one participant stated, "I've heard the term, but I know a lot of people have different descriptions of it. That sounds like my standard farming practices". As such, the information presented can help LFEs to leverage specific NbS that are likely to deliver desirable climate adaptation outcomes. Table 4 summarizes coping strategies used by AZ local food entrepreneurs to address climate-driven threats.

Table 4. Coping strategies used by AZ local food entrepreneurs to address climate-driven threats.

Theme	Coping Strategies in Place to Address Climate-Driven Threats
Water management	Irrigation (Olla irrigation, drip irrigation, micro irrigation, irrigation scheduling, targeted irrigation), water harvesting and reuse system, dry farming, planting heritage crops, deep watering.
Crop management	Changing crop varieties, diversification in crops, wildlife habitat improvement, microclimate creation, planting crops in raising beds or pots, crop switching, crop rotation, creating the windbreaks, pollinators habitat creation, shifted planting dates.
Nutrient management	Cover crops, reduced or zero tillage, customized compost, organic fertilizers, rotational grazing, livestock integration on the field, biochar application, mulching, use of microorganisms for dry soil restoration.
Low-Tech management	Swamp cooling, evaporative coolers, optimization of energy efficiency, optimization of logistics, using the mechanical equipment that nature-positive (a roller crimper for no till planting), shading nets, moisturizing of the shading nets, hail nets, on-farm data collection, organza fruit netting bags.
Financial management	Developing a regenerative cuisine, collaboration with federal agencies and cooperative extension professionals.

There are numerous examples of NbS that are being successfully funded, permitted, and implemented across regions and sectors by many of the US' federal agencies and partners [17]. However, the full-scale successful adoption of NbS is influenced by financial and operational barriers that impede any stage of the implementation cycle of NbS. Overcoming said barriers is reliant on the level of awareness and perceptions of climate change and existing adaptation strategies by LFEs and the financial accessibility and longer term performance of the involved solutions.

In AZ, LFEs are critical stakeholders in and facilitators of NbS adoption. After analyzing the perceptions, preferences, and perspectives of Southern AZ LFEs on NbS, we identified that a lack of access to funding and financing, limitations in available labor and expertise, and insufficient awareness as principal barriers to NbS adoption. Such insights lead us to reinforce three main pathways are NbS adoption. Table 5 summarizes the main factors for NbS adoption among Southern AZ LFEs as reflected in the relevant literature and across our interview data.

Table 5. Enablers of and barriers to NbS adoption among AZ local food entrepreneurs.

Enablers	Barriers	References
Pathway 1: Funding and Financing of NbS		
Availability of national and local funding and financing drawn from public, private, and alternative sources to support the implementation and management of NbS. Strong consumer interest in locally and sustainably produced foods represents an important opportunity for greater financial support of LFEs in NbS adoption. The potential of NbS to create jobs and increase agriculture output and food security, while also contribute to climate adaptation and environmental and human health benefits.	The perceived high expenses associated with NbS implementation and maintenance. Farmers are unlikely to adopt climate-smart agricultural practices or actions unless they bring greater profitability compared to their current practices or if there are incentives provided. Limited awareness of available payment mechanisms for privately implemented NbS. The divide between landownership and farming activities can pose challenges for LFEs when it comes to long-term planning and investments on the farm. Additionally, the pressure for private land development and increasing land values can restrict access to forage and create obstacles for generational transfer and the adoption of new management practices. Farmers and ranchers are implementing various strategies such as reducing the size of their herds, purchasing hay, and investing in water conservation technologies to address the challenges of declining revenues and increasing debts. Existing market structures are not favorable towards nature-positive food production. There is currently no study on farmer's interest in alternative sources to finance climate adaptation practices i.e., crowdfunding.	[5,11,13,22, 62,63]
Pathway 2: Peer-to-peer (P2P) and Expert-to-peer (E2P) knowledge co-creation and exchange		
LFEs possess a keen awareness of weather patterns and climatic conditions that affect them on a daily basis. They continuously adapt to changes in both short-term and long-term weather and climate variability. The AZ Cooperative Extension organization plays a critical role in bringing research and innovative approaches to local-level stakeholders in all fifteen counties and from five Native nations. The exponential rise in informational resources and consulting services available about NbS. Understanding the value and benefits of NbS for climate adaptation and resilience. It is very likely that increased use of NbS practices, e.g., in relation to heat mitigation or weed and pest control, will reduce food entrepreneurs' exposure to chemicals and thus contribute to reducing the large number of associated negative acute and long-term health impacts. Over the years AZ's farmers and ranchers have successfully cultivated crops and raised herds of livestock in areas that are prone to drought and have increasing precipitation variability. It has equipped food producers in the region with a wealth of inter-generational knowledge, enabling them to adapt to changing climate conditions. As signs of climate change become apparent, ranchers and farmers are more engaged in having conversations about the issue, including its risks and possible solutions. Local food entrepreneurs' interest in NbS adoption.	Finding the specific information and tailored NbS food entrepreneurs need can be challenging due to the large amount of content available. Lack of a scientific understanding of NbS and their benefits and limitations. Limited knowledge/information on NbS tailored for small-scale food production in the Southwest. Within the food storage and distribution elements of the food systems, NbS are not considered as relevant or applicable. The length of time and labor intensity needed for NbS practices to deliver concrete benefits is uncertain. The riskiness of changing practices. Food producers prefer to implement reactive adaptation strategies that are informed by their previous experiences. Aging farming population. The effectiveness of implemented NbS is not well quantified. The population in Arizona continues to increase and residential areas continue expand, which leads to an additional demand for and reliance on natural resources.	[5,62,64–66]

4.2. NbS Adaptation Pathways

Pathway 1: Funding and Financing of NbS. NbS are recognized as cost-effective responses to climate change and environmental degradation that can simultaneously benefit One Health outcomes [67]. There are numerous examples of NbS that can support improved One Health outcomes (e.g., desert agroforestry, biochar application, rotational grazing, photo-selective shading made of natural materials). Such outcomes include reducing diseases, improving food security and nutrition, enhancing mental health, mitigating heat stress, improving soil health, and more. By protecting, conserving, restoring, and sustainably managing natural or modified ecosystems, NbS serve as essential tools in climate change adaptation, promoting biodiversity, ensuring clean air and water access, and strengthening food security [68].

Even when aware of the multiple benefits provided by NbS, LFEs remain reluctant to adopt said solutions due to having a limited understanding of the financial implications, specifically when it comes to who should bear the expenses and potential funding sources [69]. Further, small-scale farmers prioritize immediate financial returns over climate-smart agricultural practices because of the high risks and low returns associated with agricultural production [36]. Therefore, they require compensation for the environmental benefits they provide, assistance with acquiring affordable and less labor-intensive solutions, and access to the various financial resources needed to adopt climate-smart and regenerative farming techniques [70]. According to research conducted by Hovis et al. (2023) in eastern North Carolina, the primary factors influencing the willingness to accept payments for NbS are landowners who are younger, possess more wealth, and manage larger parcels of land [71].

NbS are typically small, stakeholder-driven initiatives tailored to local environmental and socio-economic conditions, as well as culture, values, and traditions [72]. Across the European Union, 72% of projects based on NbS cover less than 1 km^2 or 100 ha [73]. Returning to AZ, NbS are trending towards newer, smaller, and more urban operations; from 1997 to 2017, the average farm size in Arizona decreased by 57% while the number of farms increased by 127% [74]. Nearly one quarter (24%) of farmers are considered beginner farmers, meaning they have 10 years of operational experience or fewer [74,75].

Further consideration of funding and financing for NbS adoption in AZ agri-food sector is especially important given nearly 50% of farms have less than ten acres of land in their operation, with 88% of these same farms generating less than 25,000 USD in annual sales [76]. Indeed, AZ farmers and LFEs associate high costs with the implementation and maintenance of NbS as evidenced by the following participant quotes:

- "I don't think the problem is that farms don't have the right amount of information to solve problems. The problem is they don't have the resources to do it. And resources just in capital".
- "We also need to be able to get more money to the farmer, the farmer earns about seven cents out of every food dollar it spends and that's not enough for the farmer to be able to adapt to climate change of any type".

This perception is largely perpetuated by the lack of quantitative evidence for the benefits, costs, and overall effectiveness associated with NbS implementation and maintenance for climate adaptation. As an example, the integration of NbS with desert agroforestry and agricultural land and habitat restoration would require 1 to 1.5-acre feet (AF) per acre of water, as compared to an average of 4.5-AF-per-acre of water for conventionally grown AZ crops. Habitat restoration would use even less water if and when native desert vegetation is established [49]. Importantly, not all NbS have the same financial implications and are thus dependent on local and regional variations [77].

AZ's small-scale farmers and LFEs face a significant challenge in identifying and accessing funding and financing sources for the implementation of NbS. They lack a comprehensive understanding of the payment options available to them to incentivize NbS. Payment is defined as the funding, financing, and/or partnership mechanisms utilized in the implementation and management of NbS [78]. More specifically, funding mechanisms include grants and donations provided by federal, state, and philanthropic sources to support a one-time cost for specific NbS projects. Unlike loans, these funds do not need to be repaid by the recipient. While some programs may have match requirements for the recipients, they often waive or reduce those requirements for economically disadvantaged communities. This enables them to access the necessary funds without facing excessive financial burdens. Financing mechanisms such as loans and bonds can provide the supplementary project funds that are needed. However, they require repayment and interest for their use [79].

There are a variety of public, private, and alternative sources of funding and financing available to AZ's small-scale farmers and LFEs, though user awareness and the time required to seek and access said sources are remarkably low.

NBS projects generate a multitude of public benefits such as food and water security, disaster risk reduction, climate change mitigation and adaptation, reversing ecosystem degradation and biodiversity loss, human health, and socio-economic development. However, high sunk costs coupled with the lack of immediate revenue can pose a challenge to implementing NbS [80]. Furthermore, NbS often rely on public financing due to previously identified barriers like low revenue potential and susceptibility to market failures [67].

Public investments play a significant role in funding and financing NbS projects, they account for up to 82% (165 billion USD) of the funding allocated to NbS-related projects on a global scale. Over 71% (117 billion USD) of public finance for NbS is allocated to biodiversity and landscape protection and to sustainable agriculture, forestry, and fishing [81].

Federal Funding: To date, there are 140 federal funding programs offered by a diverse set of agencies that are currently supporting or have the potential to support NbS. These programs have primarily focused on providing extensive assistance for the planning, design, implementation, and construction stages of NbS, while operations, maintenance and monitoring are less supported [17]. Table 6 lists examples of funding opportunities offered by federal programs that could provide assistance to AZ's small-scale farmers, ranchers, and food entrepreneurs who are interested in the adoption of NbS.

There is a detailed guide to federal programs for sustainable agriculture, forestry, entrepreneurship, conservation, food systems, and community development that local small-scale farming and food entrepreneurs can use to leverage the existing resources for NbS implementation [82].

Financial support for NbS initiatives can be provided under various state-funded programs. For example, Arizona Department of Agriculture through the Resilient Food Systems Infrastructure program funding will help to build resilience in Arizona's local and regional food systems. The proposed program would assist small-scale farms and ranches, new and beginning farmers and ranchers, underserved producers, veteran producers, and underserved communities in the realization of infrastructure projects to make food production, aggregation, and processing more efficient. It will contribute to establishing more stabilized markets for producers and expand capacity for supplying culturally appropriate food through emergency food systems and programs [83]. These programs include the Graham & Greenlee County Micro-Loan Fund and the Arizona Natural Resources Conservation Service (NRCS), among others.

Private financing: According to R.C. Brears (2022, p. 34), private finance for NBS is the raising, provision, or management of private capital to conserve, restore, sustainably use, or avoid a negative footprint on biodiversity and ecosystem services. Private finance includes commercial banks and investment companies, developers and infrastructure operators, and private equity and infrastructure funds [84]. Approximately 18% (equivalent to 35 billion USD) of the total financial resources allocated to NbS come from private financing worldwide. In the United States, farmers contribute around 390–450 million USD through private investment to conservation agriculture [81].

Table 6. Examples of funding opportunities offered by federal programs [17] that could provide assistance to AZ's small-scale farmers, ranchers and food entrepreneurs in the adoption of NbS.

Agency	Funding Program	The Types of NbS That Are Eligible for Support	NbS project Lifecycles Supported by the Program *			
			1	2	3	4
USDA	Conservation Reserve Program (CRP) (https://www.fsa.usda.gov/programs-and-services/conservation-programs/conservation-reserve-program/index Accessed on 14 June 2023)	"Environmentally sensitive agricultural land conservation. Long-term, resource-conserving plant species such as approved grasses or trees (known as "covers") to control soil erosion, improve water quality, and develop wildlife habitat".	X	X	X	X
USDA	Emergency Conservation Program (https://www.fsa.usda.gov/programs-and-services/conservation-programs/emergency-conservation/index Accessed on 14 June 2023)	"Rehabilitation of farmland and conservation structures damaged by natural disasters and implementation of emergency water conservation measures in periods of severe drought".	X	X	–	–
USDA	The State Acres for Wildlife Enhancement (SAFE) Initiative (https://www.fsa.usda.gov/Assets/USDA-FSA-Public/usdafiles/FactSheets/archived-fact-sheets/state_acres_wildlife_enhancement_init_jul2015.pdf Accessed on 15 June 2023)	"Soil conservation, water quality protection, or wildlife habitat enhancement".	X	X	X	X
USDA	The Agricultural Conservation Easement Program (ACEP) (https://www.nrcs.usda.gov/programs-initiatives/acep-agricultural-conservation-easement-program Accessed on 15 June 2023)	"Protection of croplands and grasslands on working farms and ranches by limiting non-agricultural uses of the land through conservation easements".	X	X	–	X
USDA	Conservation Innovation Grant (CIG) (https://www.nrcs.usda.gov/programs-initiatives/cig-conservation-innovation-grants Accessed on 15 June 2023)	"The development of new tools, approaches, practices, and technologies to further natural resource conservation on private lands".	X	X	X	X
USDA	Conservation Stewardship Program (CSP) (https://www.nrcs.usda.gov/programs-initiatives/csp-conservation-stewardship-program Accessed on 15 June 2023)	"Development and implementation of practices and activities that expands on the benefits of cleaner water and air, healthier soil and better wildlife habitat, all while improving agricultural operations".	X	X	X	–
USDA	Environmental Quality Incentives Program (https://www.nrcs.usda.gov/programs-initiatives/eqip-environmental-quality-incentives Accessed on 26 June 2023)	"Technical and financial assistance to improve water and air quality; conserve ground and surface water, increase soil health; reduce soil erosion and sedimentation; improve or create wildlife habitat; mitigation against drought and increasing weather volatility".	X	X	X	X
USDA	Landscape Conservation Initiatives (https://www.nrcs.usda.gov/programs-initiatives/landscape-conservation-initiatives Accessed on 26 June 2023)	"Voluntary on-farm conservation initiatives to improve water and air quality, soil health, and preserve the wildlife habitats".	X	X	X	X

Table 6. *Cont.*

Agency	Funding Program	The Types of NbS That Are Eligible for Support	NbS project Lifecycles Supported by the Program *			
			1	2	3	4
USDA	National Water Quality Initiative (NWQI) (https://www.nrcs.usda.gov/programs-initiatives/national-water-quality-initiative Accessed on 26 June 2023)	"Voluntary on-farm conservation initiatives that promote soil health, reduce erosion and lessen nutrient runoff, such as filter strips, cover crops, reduced tillage and manure management".	X	X	X	X
USDA	Regional Conservation Partnership Program (RCPP) (https://www.nrcs.usda.gov/programs-initiatives/rcpp-regional-conservation-partnership-program Accessed on 26 June 2023)	"To support the adoption of climate-smart agriculture practices, which have direct climate mitigation benefits, advance a host of other environmental co-benefits, and offer farmers, ranchers and foresters new revenue streams".	X	X	–	–
USDA	Environmental Quality Incentives Program—WaterSMART Initiative (https://www.nrcs.usda.gov/programs-initiatives/eqip-watersmart/priority-areas Accessed on 26 June 2023)	"To increase water conservation and resilience to drought. Water conservation improvements may also improve soil health; reduce soil erosion, sediment, nutrient, and pathogen losses; protect crop health and productivity; and make using equipment, facilities, and agricultural operations more efficient".	X	X	X	X

* NbS project phases are 1—planning and design; 2—implementation or construction; 3—operations and maintenance; 4—monitoring (X—supported; – not supported).

The private-sector production, storage, distribution, and consumption components of the agri-food system have financial interests in achieving a more sustainable food production and consumption in Arizona [5]. These can include: (a) mid-size and large-scale farmers, ranchers, beef, dairy, and poultry producers operating with sustainability approaches; for example, Duncan Family Farms, Oatman Farms and Oatman Flats Ranch, Ramona Farms, the San Xavier Cooperative Farm; (b) agribusiness, especially innovators such as AZ Baking Company, Barrio Bread, the Food Conspiracy among others; (c) agricultural and conservation associations, such as Arizona Farm Bureau, Arizona Beef Council, Dairy associations, Southwest Black Ranchers, Agribusiness & Water Council of Arizona, Western Growers, Arizona Grain Research and Promotion Council, The Arizona Association of Conservation Districts, Ajo Center for Sustainable Agriculture, Baja Arizona Sustainable Agriculture, The Yuma Center of Excellence for Desert Agriculture, etc. Potential partners and investors can be engaged among the various actors that are engaged in food system activity.

However, it can be challenging for small-scale farmers and LFEs to navigate and secure federal and state public funding [71]. So, encouraging them to engage in NbS projects requires access to and the availability of alternative financing models. According to den Heijer C and Coppens, "Alternative financing is arrangements that draw on financial resources other than public budgets collected through general taxation. They could be sourced from institutional investors, businesses, and citizens" [67].

AZ's food producers and LFEs may consider various alternative financial models that they can apply when it comes to adopting NbS. Carbon markets. Carbon markets, albeit still immature, could become a significant source of funding for NbS. Different carbon market mechanisms vary from compliance programs, such as the European Union's emission-trading scheme (EU-ETS) and California's Cap-and-Trade program, to voluntary and international carbon market places such as the Voluntary Carbon Market (VCM) and Internationally Transferred Mitigation Outcomes (ITMOs) under the Paris agreement article 6.2.

There are four major carbon registries operating in the United States: ACR (formerly known as American Carbon Registry), Climate Action Reserve (CAR), Verra's Verified Carbon Standard (VCS), and Gold Standard. These four registries have issued more than 412 million credits corresponding to 412 million metrics tons of carbon dioxide (CO_2) equivalent to projects based in the United States for both voluntary and compliance markets [85,86].

Although some carbon market programs do not yet allow carbon offsets to be purchased from farmers, their role in funding sustainable food system transformation and NbS is likely to grow. The global carbon credit market demand was estimated at 181 million metric tons of CO_2 equivalents in 2023 and is expected to grow to 1.2 billion metric tons by 2030 and 5.4 gigatons by 2050 [87]. According to a study by McKinsey, the demand for carbon credits could grow by a factor of more than 15 by 2030 and by a factor of 100 by 2050 [88].

Commercialization of NbS initiatives: There are various biomass sources available, including crops grown specifically as biomass inputs, the leftovers from crop and wood harvesting, and manures, all of which can contribute to small-scale biochar production. Biochar technology can improve soil productivity and capture carbon. The integration of biochar technology with small-scale farm operations is especially promising given the associated esource constraints and available labor. Moreover, biochar technologies can increase water and fertilizer efficiencies, enhance the re-purposing of organic wastes, and improve crop productivity [89]. With the growing availability of lower cost commercial biochar equipment, such innovation stands to make small-scale farmers more environmentally and economically sustainable.

Philanthropy and charity funding: Partnerships with philanthropic organizations and private philanthropists who have an interest in NbS for adapting to climate change should be considered as a potential source of additional funding to supplement the resources provided by government agencies, private entities, and others [90]. There is a growing number of these foundations that focus on community-based sustainable agriculture, aiming to combat climate change, improve access to healthy and nutritious food, promote regenerative agriculture, and build local food systems resilience. Examples of such foundations include the Rockefeller Foundation, the Foundation for Food and Agriculture, the Gordon & Betty Moore Foundation, the Walton Family Foundation, the Howard G. Buffett Foundation, The Schmidt Family Foundation 11th Hour Project, IKEA Foundation, The David & Lucille Packard Foundation, the Bill and Melinda Gates Foundation, the Mars Wrigley Foundation, The Cedar Tree Foundation, etc.

Crowdfunding: Crowdfunding became widely popular thanks to the rise of social media and online platforms dedicated to directly connecting entrepreneurs with backers (i.e., crowd-funders, investors), such as crowdfunding websites [91]. There have been no studies conducted on the interest of LFEs in Arizona in using alternative finance options, such as crowdfunding, to support NbS. However, a study by Kragt et al. (2021) revealed that farmers with a financially secure farming business, previous exposure to crowdfunding, or a strong sense of responsibility to mitigate climate change are more interested in using crowdfunding [91].

Household/business investment: According to Burton and Otte (2022), the importance of being innovative and having a willingness to invest has been recognized as key drivers for successfully implementing climate mitigation and adaptation measures at the family farm level [92]. There is a progressive trend of declining climate mitigation/adaptation intentions throughout the life-cycle stages of the farm as a business (investment, debt, expansion, consolidation, contraction, exit) and due to generational aspects (early-stage developers, commercial developers, commercially disengaged and semi-retired withdrawers). For example, in the case of semi-retired withdrawers, promoting new investment in the farm may pose challenges due to the population of aging farmers and their low intentions to invest. The average age of food producers in Arizona is 59.4 years. With this insight in mind, we propose that, to effectively promote, design, and implement NbS while leveraging various funding sources, it is important to consider the life cycle of agri-food entrepreneurial activity.

Rebates, tax credits, or low-interest financing can incentivize NbS adoption. For instance, the Rainwater Harvesting rebate program, sponsored by Tucson Water, offers rebates of up to 2000 USD to individuals or small businesses who install a rainwater-harvesting system. Qualifying options include passive rainwater harvesting, which involves directing and retaining water in the landscape, as well as active rainwater harvesting, such as the use of storage tanks for future water usage [93].

Although there are various alternative financial models available for NbS adoption, none of them can completely substitute traditional public finance [33]. Yet, NbS' adoption in small-scale food systems have the potential to enhance individual farm's productivity, efficiency, and resilience while also delivering broader societal benefits like climate change mitigation, biodiversity, water conservation, and One Health support. It is essential to identify all potential private and public benefits that are linked to the implementation of NbS project in order to unlock funding sources.

Pathway 2: Peer-to-peer (P2P) and Expert-to-peer (E2P) knowledge co-creation and exchange: providing interactions in both the real world and online settings. Understanding the sources of information about NbS that LFEs and experts/advisors rely on and trust is valuable for scaling the adoption of NbS. This knowledge can help identify the most effective providers to disseminate reliable and trustworthy information about NbS. AZ small-scale farmers and LFEs have emphasized the value of peer-to-peer and expert-to-peer knowledge exchange. Examples of quotes from interview participants include the following:

- "It's being done around the world. It's being done right next door and you just don't know it".

- "So to have somebody to walk with the farmer, and say we're going to do this together".
- "To look more towards interpersonal relationships, whether that be a consultant, a neighbor, but that these relationships need to be more consistent, more streamlined and more accessible to everyone".

Peers are found to offer valuable guidance and assistance in the practical adoption of NbS due to their previous experiences with similar challenges e.g., loss of food production, reduced soil quality, variation in temperature and precipitation, and increased pests. This prevents costly trial-and-error learning and encourages the exploration of underutilized or neglected solutions for common problems.

Rust et al. (2022) found that, in a sample of farmers from Hungary and the UK, that participants placed the most trust in other farmers (i.e., peer-to-peer) when learning about new soil practices. In contrast, participants revealed that they were less trusting of institutions and research outside of the fields of agriculture [64].

Experts are found to offer valuable guidance and assistance in developing field-ready solutions, introducing innovative affordable climate smart approaches for water conservation, novel fertilizers, pathogen control, and real-time field monitoring.

According to Sutherland et al., (2017) small-scale farmers utilize three main types of networks to access different types of knowledge. One type is a centralized network, where farmers interact individually with an advisor who then interacts with other advisors. Distributed networks represent communities or networks of practice and involve peers exchanging personal knowledge to varying degrees. Decentralized networks involve knowledge from sources outside the peer group and are associated with acquiring potential knowledge, such as future or cutting-edge innovations. The structure of these networks reflects the farmers understanding of available sources of knowledge, and their credibility [94].

To ensure successful adoption and scaling-up of NbS within small-scale food systems, it is essential to engage all types of knowledge networks. A centralized network can be used to establish one-to-one interactions with formal advisors who can provide valuable information on funding and financing options, can assist in completing grant and loan applications, as well as help develop a compelling business case for NbS in small-scale agri-food systems. Additionally, distributed networks should be utilized to facilitate collaboration and the exchange of knowledge and best practices related to NbS implementation on-site. Lastly, decentralized networks play a key role in promoting advances in knowledge and practices regarding NbS.

Different levels and quality of informational support are needed for the planning, implementation, and stewardship of NbS. By leveraging these various types of knowledge networks, the adoption of NbS can be accelerated among AZ small-scale food producers and entrepreneurs.

The interventions implemented to ensure public health and safety during the COVID-19 pandemic (and now endemic) led to notable increases in the use of digital knowledge-exchange platforms. To stay up to date with this trend it is recommended to develop a peer-to-peer and expert-to-peer knowledge sharing, capacity building, and networking platform, which allows farmers and LFEs to connect and interact with each other, as well as with local subject matter experts to facilitate and accelerate the adoption of NbS, increasing Arizona's small-scale food systems resilience.

Although there is a growing availability of resources and information on the incorporation of NbS, there remains a lack of a user-friendly guide for practitioners, specifically LFEs [79]. This guide should contain a set of procedures to accompany each phase of NbS projects, including planning and design, implementation or construction, operations and maintenance, and monitoring. Having access to such a document would greatly support LFEs in developing and implementing evidence-based NbS and help to avoid maladaptation during the process.

To bridge the implementation gap, the eLearning training course "Advancing Arizona's agrifood systems climate adaptation through NbS was developed. It aims to improve awareness and understanding of the science, policy, practice, resourcing, and governance of NbS specific to Arizona, and support an adoption of NbS by small-scale food system actors to boost food and nutrition security and to strengthen the resilience of food systems. The course incentivizes continual learning about locally available and accessible NbS that work best to (i) enhance food security and the nutrition of individuals, households, and communities; (ii) benefit human, animal, and environmental health; (iii) promote climate resiliency; and (iv) generate economic viability. The course showcases a range of available, proven, accessible NbS that can be piloted and/or scaled up to enhance food security and the health and wellbeing of LFEs across Arizona.

4.3. Promoting the Adoption of NbS

The adoption of NbS relies on the synergy between stakeholders (e.g., local farmers, ranchers, LFEs), experts from many disciplines (e.g., public health, agriculture, food systems, environment), and affected communities, creating a value-added and participatory approach that can give rise to context-driven implementable solutions. Although the scientific community and policymakers recognize and promote the importance of NbS in climate adaptation, their practical implementation is struggling mainly due to a lack of awareness among the main stakeholders.

Currently, NbS have primarily been introduced through learning-by-doing practices and LFES actively seeking implementable solutions while working within economic and technological resource constraints. Limited finances, time, and labor resources, needs for infrastructural improvements, the availability and effectiveness of extension services and outreach, uncertainty regarding implementation processes and their effectiveness are the most frequently cited unmet needs of AZ's local food system stakeholders in response to climate change with NbS. Various factors, such as encouragement for NbS initiatives, recognition of diverse expertise, and the significance of incorporating different knowledge and practice-based perspectives, can impact the level to which LFEs will adopt NbS. AZ's LFEs are likely to adopt NbS based on their capacity to address priority climate-driven issues, revenue generation potential, and seamless augmentation with existing food production and operational activities. These key factors will play a critical role in the adoption process.

Based on a literature review, consultations with subject-matter experts and targeted outreach to stakeholders, the following gaps need to be bridged to promote the adoption of NbS among LFEs in Arizona.

- The need to analyze available private and alternative sources of funding and financing. The resource guide will increase the connectivity among potential investors, LFEs, and policymakers across the local food system and enable the expansion of investments to support and accelerate the adoption of NbS. Evidence-based and context-based quantification of the benefits, costs, and cost-effectiveness of NbS implementation and maintenance for climate adaptation are required. This information will help to mobilize financial resources and generate investible business cases to accelerate the implementation of NbS.
- The need to raise awareness among AZ's LFEs. It is important to emphasize that NbS can be integrated and effectively applied at various stages within the food system, not limited to just agricultural production. NbS hold the capacity to benefit various elements of the food system, including food production, storage and distribution, processing and packaging, retail and marketing. Farmers, processors, distributors, transporters, and retailers whose livelihoods, health, and wellbeing are threatened by climate change can gain advantages from NbS.
- The need to document on-farm/on-ranch coping strategies used by AZ LFEs and quantify the performance of NbS in addressing climate-driven threats. This includes assessing the benefits and challenges associated with NbS implementation.

- The need for support in the planning, implementation, and stewardship of NbS. In order to achieve substantive and lasting results, small-scale farmers, ranchers, and LFEs need to be supported and navigated by Cooperative Extension professionals. Close collaboration between LFEs and experts is needed to accelerate the adaptation of NbS and prevent wasted efforts on impractical or infeasible solutions. To achieve this, it is necessary to develop professional learning resources aimed at enhancing AZ's Cooperative Extension's capacity to transfer knowledge regarding NbS adoption at the local/state level.

Taking into consideration the existing uncertainties, this study provides AZ's local small-scale farmers, ranchers, and LFEs with the information and tools they need to immediately and proactively implement NbS with One Health benefits (e.g., use of microorganisms for dry soil restoration, construct nesting boxes to conserve raptors and reduce virus risk, apply multi-story cropping for desert adapted crops, etc.).

Our findings and recommendations are aligned with those observed in other regions across the world. They emphasize the critical role of finding and financing, knowledge and information, revenue generation, and the capacity for the seamless integration of NbS with existing food production and operational activities. This is essential for small-scale farmers, ranchers, and LFEs who are considering NbS implementation.

4.4. Study Limitations

A non-random sampling technique was used to conduct this research, limiting the generalizability of the results which remain limited to those who participated. Given that research on perceptions of NbS is currently scant, the current research provides valuable contributions to the understanding of this topic despite its limitations.

The sample size was relatively small, and this limits our ability to conduct advanced statistical relationship testing. Future research should inquire into perceptions about NbS with a larger sample size to identify key relationships between variables. For example, relationships between attitudes, social norms, perceived behavioral control, and intention, as described in the Theory of Planned Behavior [24]. Additionally, the current research looked at NbS as a whole, and disaggregating the concept into specific NbS would be helpful in understanding the adoption of NbS more comprehensively.

5. Conclusions

The study explores the implementation potential of NbS that are particularly promising for supporting climate adaptation among LFEs and carry the capacity to optimize One Health in Arizona. It reveals current climate-related challenges experienced by Southern AZ's LFEs who promote and support the local food systems through localized production, distribution, and consumption. The study also explores the coping mechanisms that have already been implemented, adaptation needs, and NbS that can be better harnessed to address food security and deteriorating health outcomes within arid and semi-arid regions such as AZ using available and accessible NbS. A total of over 40 promising NbS were identified and proposed that could advance Arizona's small-scale food system's resilience and optimize One Health in a context-specific and cost-effective manner. Small-scale farmers, ranchers, and LFEs have various options for adopting NbS, depending on their specific needs and available resources. To prevent maladaptation, it is necessary to generate additional evidence regarding NbS' value and limitations to address climate adaptation and sustainable balance and optimize the health of people, animals, and ecosystems. The information and tools herein hold the potential for replication and scaling by both practitioners and researchers across other arid and semi-arid regions, particularly in low- and middle-income communities and countries because they address food security and deteriorating health outcomes with high-value, cost-effective interventions arising from available and accessible ecosystem services.

Supplementary Materials: The following supporting information can be downloaded at: https://www.mdpi.com/article/10.3390/su16083176/s1, Table S1: Promising nature-based and climate-smart solutions to support climate adaptation of Arizona's LFEs and optimize One Health.

Author Contributions: Conceptualization, M.M.M., Y.V. and T.A.F.; methodology, M.M.M. and T.A.F.; formal analysis, T.A.F. and Y.V.; investigation, T.A.F. and Y.V.; resources, M.M.M. and Y.V.; data curation, T.A.F.; writing—original draft preparation, Y.V., T.A.F. and J.M.; writing—review and editing, M.M.M.; supervision, M.M.M.; project administration, Y.V.; funding acquisition, Y.V. and M.M.M. All authors have read and agreed to the published version of the manuscript.

Funding: One Health Research Initiative, The University of Arizona Health Sciences: 2023-One Health Initiative Pilot Grant.

Institutional Review Board Statement: The project was approved in June 2023 by the Institutional Review Board, The University of Arizona, USA.

Informed Consent Statement: All participants gave their informed consent for inclusion before they participated in this study.

Data Availability Statement: The original contributions presented in the study are included in the article/Supplementary Materials, further inquiries can be directed to the corresponding author.

Conflicts of Interest: The authors declare that they have no known competing financial interests or personal relationships that could have appeared to influence the work reported in this paper.

References

1. One Health High-Level Expert Panel; Adisasmito, W.B.; Almuhairi, S.; Behravesh, C.B.; Bilivogui, P.; Bukachi, S.A.; Casas, N.; Becerra, N.C.; Charron, D.F.; Chaudhary, A.; et al. One Health: A new definition for a sustainable and healthy future. *PLOS Pathog.* **2022**, *18*, e1010537. [CrossRef]
2. Garfin, G.; Jardine, A.; Feldman, D.L. *Overview. Assessment of Climate Change in the Southwest United States: A Report Prepared for the National Climate Assessment*; Garfin, G., Jardine, A., Merideth, R., Black, M., LeRoy, S., Eds.; Island Press/Center for Resource Economics: Washington, DC, USA, 2013; pp. 21–36.
3. Frankson, R.; Kunkel, K.E.; Stevens, L.E.; Easterling, D.R.; Brown, T.; Selover, N.; Saffell, E. *Arizona State Climate Summary 2022*; NOAA Technical Report NESDIS 150-AZ; NOAA/NESDIS: Silver Spring, MD, USA, 2022.
4. NOAA National Centers for Enviromental Information. State Climate Summaries 2022. Arizona. 2022. Available online: https://statesummaries.ncicsorg/chapter/az/ (accessed on 15 July 2023).
5. Condon, L.E.; Ruiz, J.; Antin, P.; Buizer, J.; Collinge, S.; Esquerra, N.L.; Ikner, L.A.; Megdal, S.B.; Patten, K.J. *The Presidential Advisory Commission on the Future of Agriculture and Food Production in a Drying Climate*; The University of Arizona: Tucson, AZ, USA, 2023.
6. White, D.D.; Elias, E.H.; Thomas, K.A.; Bradatan, C.E.; Brunson, M.W.; Chischilly, A.M.; Enquist, C.A.F.; Fisher, L.R.; Froehlich, H.E.; Koebele, E.A.; et al. Southwest. In *Fifth National Climate Assessment*; Crimmins, A.R., Avery, C.W., Easterling, D.R., Kunkel, K.E., Stewart, B.C., Maycock, T.K., Eds.; U.S. Global Change Research Program: Washington, DC, USA, 2023.
7. Kacira, M. Controlled Environment Agriculture. 2022. Available online: https://wrrc.arizona.edu/sites/wrrc.cals.arizona.edu/files/2022-09/Sustainability_MuratKacira.pdf (accessed on 25 November 2023).
8. AZFB. Assessing Western Drought Conditions 2022: Arizona Farm Bureau. 2022. Available online: https://www.azfb.org/Article/Assessing-Western-Drought-Conditions-2022 (accessed on 29 August 2023).
9. de Lima, C.Z.; Buzan, J.R.; Moore, F.C.; Baldos, U.L.C.; Huber, M.; Hertel, T.W. Heat stress on agricultural workers exacerbates crop impacts of climate change. *Environ. Res. Lett.* **2021**, *16*, 044020. [CrossRef]
10. Lookadoo, R.E.; Bell, J.E. Public Health Policy Actions to Address Health Issues Associated with Drought in a Changing Climate. *J. Law Med. Ethics* **2020**, *48*, 653–663. [CrossRef]
11. IUCN. *Guidance for Using the IUCN Global Standard for Nature-Based Solutions. A User-Friendly Framework for the Verification, Design and Scaling Up of Nature-Based Solutions*; IUCN: Gland, Switzerland, 2020.
12. Johnson, B.A.; Kumar, P.; Okano, N.; Dasgupta, R.; Shivakoti, B.R. Nature-based solutions for climate change adaptation: A systematic review of systematic reviews. *Nat. Based Solut.* **2022**, *2*, 100042. [CrossRef]
13. ILO; UNEP; IUCN. *Decent Work in Nature-Based Solutions 2022*; International Labour Organization: Geneva, Switzerland, 2022.
14. Census, U.S. QuickFacts. 2023. Available online: https://www.census.gov/quickfacts/fact/table/AZ/PST045223 (accessed on 23 February 2024).
15. Mars, M.M.; Schau, H.J. The jazziness of local food work: Organization level ingenuity and the entrepreneurial formation and evolution of local food systems. *Rural. Sociol.* **2019**, *84*, 257–283. [CrossRef]
16. Biernacki, P.; Waldorf, D. Snowball Sampling: Problems and Techniques of Chain Referral Sampling. *Sociol. Methods Res.* **1981**, *2*, 141–163. [CrossRef]

17. CEQ. *Nature-Based Solutions Resource Guide 2.0*; White House Council on Environmental Quality, White House Office of Science and Technology Policy: Washington, DC, USA, 2023.
18. Mars, M.M.; Schau, H.J. What is local food entrepreneurship? Variations in the commercially and socially oriented features of entrepreneurship in the Southeastern Arizona local food system. *Rural. Sociol.* **2018**, *83*, 568–597. [CrossRef]
19. Corbin, J.; Strauss, A. *Basics of Qualitative Research: Techniques and Procedures for Developing Grounded Theory*, 4th ed.; Sage Publications: Thousand Oaks, CA, USA, 2015.
20. Maxwell, J.A. *Qualitative Research Design: An Interactive Approach*, 3rd ed.; Sage Publications: Thousand Oaks, CA, USA, 2013.
21. Crouch, M.; McKenzie, H. The logic of small samples in interview-based qualitative research. *Soc. Sci. Inf.* **2016**, *45*, 483–499. [CrossRef]
22. Bainbridge, D.A. Buried clay pot irrigation: A little known but very efficient traditional method of irrigation. *Agric. Water Manag.* **2001**, *48*, 79–88. [CrossRef]
23. Lincoln, Y.S.; Guba, E.G. *Naturalistic Inquiry*; Sage Publications: Newbury Park, CA, USA, 1985.
24. Ajzen, I. The theory of planned behavior. *Organ. Behav. Hum. Decis. Process.* **1991**, *50*, 179–211. [CrossRef]
25. Miller, Z.D.; Freimund, W.; Metcalf, E.C.; Nickerson, N.; Powell, R.B. Merging elaboration and the Theory of Planned Behavior to understand bear spray behavior of day hikers in Yellowstone National Park. *Environ. Manag.* **2019**, *63*, 366–378. [CrossRef] [PubMed]
26. Senger, I.; Borges, J.A.R.; Machado, J.A.D. Using the Theory of Planned Behavior to understand the intention of small farmers in diversifying their agricultural production. *J. Rural. Stud.* **2017**, *49*, 32–40. [CrossRef]
27. White House Council on Environmental Quality; White House Office of Domestic Climate Policy; White House Office of Science and Technology Policy. *Opportunities for Accelerating Nature-Based Solutions: A Roadmap for Climate Progress, Thriving Nature, Equity, and Prosperity*; Report to the National Climate Task Force; The White House: Washington, DC, USA, 2022.
28. Acevedo, M.; Pixley, K.; Zinyengere, N.; Meng, S.; Tufan, H.; Cichy, K.; Bizikova, L.; Isaacs, K.; Ghezzi-Kopel, K.; Porciello, J. A scoping review of adoption of climate-resilient crops by small-scale producers in low- and middle-income countries. *Nat. Plants* **2020**, *6*, 1231–1241. [CrossRef] [PubMed]
29. Al-Dawood, A. Towards Heat Stress Management in Small Ruminants—A Review. *Ann. Anim. Sci.* **2017**, *17*, 59–88. [CrossRef]
30. Cibils, A.F.; Estell, R.E.; Spiegal, S.; Nyamuryekung'E, S.; McIntosh, M.M.; Duni, D.M.; Conegliano, O.A.H.; Almeida, F.A.R.; Estrada, O.R.; Blanco, L.J.; et al. Adapting to climate change on desert rangelands: A multi-site comparison of grazing behavior plasticity of heritage and improved beef cattle. *J. Arid. Environ.* **2023**, *209*, 10886. [CrossRef]
31. Defraeye, T.; Shoji, K.; Schudel, S.; Onwude, D.; Shrivastava, C. Passive evaporative coolers for postharvest storage of fruit and vegetables: Where to best deploy them and how well do they perform. *Front. Food Sci. Technol.* **2023**, *3*. [CrossRef]
32. Egan, P.A.; Chikoye, D.; Green, K.K.; Tamò, M.; Feit, B.; Kumar, P.; Bandyopadhyay, R.; Tepa-Yotto, G.; Ortega-Beltran, A.; Sæthre, M.-G.; et al. Harnessing nature-based solutions for smallholder plant health in a changing climate. *Environ. Sci. Agric. Food Sci.* **2021**.
33. Guinet, M.; Adeux, G.; Cordeau, S.; Courson, E.; Nandillon, R.; Zhang, Y.; Munier-Jolain, N. Fostering temporal crop diversification to reduce pesticide use. *Nat. Commun.* **2023**, *14*, 7416. [CrossRef] [PubMed]
34. Rosales, B.H.; Lutz, A. Assessing the Feasibility of Rooftop Rainwater Harvesting for Food Production in Northwestern Arizona on the Hualapai Indian Reservation. *Sustainability* **2023**, *15*, 2891. [CrossRef]
35. Huang, Y.; Tao, B.; Lal, R.; Lorenz, K.; Jacinthe, P.-A.; Shrestha, R.K.; Bai, X.; Singh, M.P.; Lindsey, L.E.; Ren, W. A global synthesis of biochar's sustainability in climate-smart agriculture-Evidence from field and laboratory experiments. *Renew. Sustain. Energy Rev.* **2023**, *172*, 113042. [CrossRef]
36. Idowu, O.; Grover, K. Principles of cover cropping for arid and semi-arid farming systems. In *Agricultural Experiment Station and Cooperative Extension Service, New Mexico State University Circular*; New Mexico State University: Las Cruces, NM, USA, 2014.
37. Jarimi, H.; Powell, R.; Riffat, S. Review of sustainable methods for atmospheric water harvesting. *Int. J. Low-Carbon Technol.* **2020**, *15*, 253–276. [CrossRef]
38. Lamonaca, E.; Bouzid, A.; Caroprese, M.; Ciliberti, M.G.; Cordovil, C.M.S.; Karatzia, M.-A.; Keskin, M.; Lazereg, M.; Lidga, C.; Panniello, U.; et al. A framework towards resilient Mediterranean eco-solutions for small-scale farming systems. *Agric. Food Secur.* **2022**, *11*, 65. [CrossRef]
39. Librán-Embid, F.; Olagoke, A.; Martin, E.A. Combining Milpa and Push-Pull Technology for sustainable food production in smallholder agriculture. A review. *Agron. Sustain. Dev.* **2023**, *43*, 45. [CrossRef]
40. Lord, J.; Thomas, A.; Treat, N.; Forkin, M.; Bain, R.; Dulac, P.; Behroozi, C.H.; Mamutov, T.; Fongheiser, J.; Kobilansky, N.; et al. Global potential for harvesting drinking water from air using solar energy. *Nature* **2021**, *598*, 611–617. [CrossRef] [PubMed]
41. Mahmood, A.; Hu, Y.; Tanny, J.; Asante, E.A. Effects of shading and insect-proof screens on crop microclimate and production: A review of recent advances. *Sci. Hortic.* **2018**, *241*, 241–251. [CrossRef]
42. Maraveas, C. The Sustainability of Plastic Nets in Agriculture. *Sustainability* **2020**, *12*, 3625. [CrossRef]
43. Mizik, T. How can precision farming work on a small scale? A systematic literature review. *Precis. Agric.* **2022**, *24*, 384–406. [CrossRef]
44. Mpanga, I. Cultivation of Mixed Summer Cover Crops (Buckwheat, Cowpea, and Teff Grass) In High Tunnels: The University of Arizona Cooperative Extension. 2022. Available online: https://extension.arizona.edu/sites/extension.arizona.edu/files/pubs/az1982-2022.pdf (accessed on 23 February 2024).

45. Mpanga, I.K.; Neumann, G.; Schuch, U.K.; Schalau, J. Sustainable Agriculture Practices as a Driver for Increased Harvested Cropland among Large-Scale Growers in Arizona: A Paradox for Small-Scale Growers. *Adv. Sustain. Syst.* **2020**, *4*, 1900143. [CrossRef]
46. Mpanga, I.; Walworth, J.; Braun, H. Zeolite application in crop production: Importance to nutrient, water, soil health, and environmental pollution management. Contract No.: az1851. 2020.
47. Mpanga, I.K.; Gaikpa, D.S.; Koomson, E.; Dapaah, H.K. Innovations in Water Management: Agriculture. In *The Palgrave Handbook of Global Sustainability*; Brinkmann, R., Ed.; Springer International Publishing: Cham, Switzerland, 2023; pp. 381–403.
48. Mpanga, I.K.; Schuch, U.K.; Schalau, J. Adaptation of resilient regenerative agricultural practices by small-scale growers towards sustainable food production in north-central Arizona. *Curr. Res. Environ. Sustain.* **2021**, *3*, 100067. [CrossRef]
49. Nabhan, G.P.; Richter, B.D.; Riordan, E.C.; Tornbom, C. *Future Scenarios Addressing Water Scarcity in the Lower Colorado River Basin*; Lincoln Institute of Land Policy: Cambridge, MA, USA, 2023.
50. Ntawuruhunga, D.; Ngowi, E.E.; Mangi, H.O.; Salanga, R.J.; Shikuku, K.M. Climate-smart agroforestry systems and practices: A systematic review of what works, what doesn't work, and why. *For. Policy Econ.* **2023**, *150*, 102937. [CrossRef]
51. Olle, M. Review: Bokashi technology as a promising technology for crop production in Europe. *J. Hortic. Sci. Biotechnol.* **2020**, *96*, 145–152. [CrossRef]
52. Danquah, F.O.; Twumasi, M.A.; Asare, I. Factors Influencing Zai Pit Technology Adaptation: The Case of Smallholder Farmers in the Upper East Region of Ghana. *Agric. Res. Technol. Open Access J.* **2019**, *21*, 556150. [CrossRef]
53. Rai, P.K.; Rai, A.; Sharma, N.K.; Singh, T.; Kumar, Y. Limitations of biofertilizers and their revitalization through nanotechnology. *J. Clean. Prod.* **2023**, *418*, 138194. [CrossRef]
54. Rao, A.P.; Patel, J.; Pradhan, A.K. Application of alternative sources of water in agricultural food production—Current trends and future prospects. *Curr. Opin. Food Sci.* **2022**, *47*, 100877. [CrossRef]
55. Sojka, M.; Saeid, A. Chapter 10—Bio-based products for agriculture. In *Smart Agrochemicals for Sustainable Agriculture*; Chojnacka, K., Saeid, A., Eds.; Academic Press: Cambridge, MA, USA, 2022; pp. 279–310.
56. Tariq, Z.; Iqbal, D.N.; Rizwan, M.; Ahmad, M.; Faheem, M.; Ahmed, M. Significance of biopolymer-based hydrogels and their applications in agriculture: A review in perspective of synthesis and their degree of swelling for water holding. *RSC Adv.* **2023**, *13*, 24731–24754. [CrossRef] [PubMed]
57. Thomas, G.; Rusman, Q.; Morrison, W.R.; Magalhães, D.M.; Dowell, J.A.; Ngumbi, E.; Osei-Owusu, J.; Kansman, J.; Gaffke, A.; Damodaram, K.J.P.; et al. Deciphering Plant-Insect-Microorganism Signals for Sustainable Crop Production. *Biomolecules* **2023**, *13*, 997. [CrossRef] [PubMed]
58. USDA. *Evaluation of Cool Season Cover Crops in the Southwest Region*; West National Technology Support Center: Portland, OR, USA, 2020.
59. USDA. Smart Nutrient Management (2022). Available online: https://www.farmers.gov/sites/default/files/2022-09/farmersgov-fact-sheet-smart-nutr-mgnt-9-2022.pdf (accessed on 23 February 2024).
60. WIPO. *Green Technology Book: Solutions for Climate Change Adaptation*; World Intellectual Property Organization: Geneva, Switzerland, 2022.
61. Zhao, G.; Zhu, X.; Zheng, G.; Meng, G.; Dong, Z.; Baek, J.H.; Jeon, C.O.; Yao, Y.; Xuan, Y.H.; Zhang, J.; et al. Development of biofertilizers for sustainable agriculture over four decades (1980–2022). *Geogr. Sustain.* **2024**, *5*, 19–28. [CrossRef]
62. Ishtiaque, A. US farmers' adaptations to climate change: A systematic review of adaptation-focused studies in the US agriculture context. *Environ. Res. Clim.* **2023**, *2*, 022001. [CrossRef]
63. Lynch, H.; Uchanski, M.; Patrick, M.; Wharton, C. Small Farm Sustainability in the Southwest: Challenges, Opportunities, and Best Practices for Local Farming in Arizona and New Mexico. *Food Stud. Interdiscip. J.* **2018**, *8*, 45–56. [CrossRef]
64. Rust, N.A.; Stankovics, P.; Jarvis, R.M.; Morris-Trainor, Z.; de Vries, J.R.; Ingram, J.; Mills, J.; Glikman, J.A.; Parkinson, J.; Toth, Z.; et al. Have farmers had enough of experts? *Environ. Manag.* **2022**, *69*, 31–44. [CrossRef] [PubMed]
65. Saliman, A.; Petersen-Rockney, M. Rancher Experiences and Perceptions of Climate Change in the Western United States. *Rangel. Ecol. Manag.* **2022**, *84*, 75–85. [CrossRef]
66. Brugger, J.; Crimmins, M. The art of adaptation: Living with climate change in the rural American Southwest. *Glob. Environ. Change* **2013**, *23*, 1830–1840. [CrossRef]
67. den Heijer, C.; Coppens, T. Paying for green: A scoping review of alternative financing models for nature-based solutions. *J. Environ. Manag.* **2023**, *337*, 117754. [CrossRef]
68. WHO; IUCN. *Highlights Brief on WHO-IUCN Report on Designing Nature—Based Solutions for Human Health*; WHO: Geneva, Switzerland; IUCN: Gland, Switzerland, 2023.
69. Brears, R.C. Introduction. In *Financing Nature-Based Solutions: Exploring Public, Private, and Blended Finance Models and Case Studies*; Brears, R.C., Ed.; Springer International Publishing: Cham, Switzerland, 2022; pp. 1–6.
70. Mizik, T. Climate-Smart Agriculture on Small-Scale Farms: A Systematic Literature Review. *Agronomy* **2021**, *11*, 1096. [CrossRef]
71. Hovis, M.; Cubbage, F.; Smith, G.; Zuniga-Teran, A.; Varady, R.; Shear, T.; Chizmar, S.; Lupek, M.; Baldwin, M.; Fox, A.; et al. Estimating landowners' willingness to accept payments for nature-based solutions in eastern North Carolina for flood hazard mitigation using the contingent valuation method. *J. Soil Water Conserv.* **2023**, *78*, 500–514. [CrossRef]

72. Salafsky, N.; Suresh, V.; Bierbaum, R.; Clarke, E.; Smith, M.S.; Whaley, C.; Margoluis, R. *Taking Nature-Based Solutions Programs to Scale*; Foundations of Success (FOS): Bethesda, MD, USA; The Science and Technical Advisory Panel (STAP) to the Global Environmental Facility (GEF): Washingthon, DC, USA; Gordon and Betty Moore Foundation (GBMF): Palo Alto, CA, USA, 2021.
73. EIB. *Investing in Nature-Based Solutions. State-of-Play and Way Forward for Public and Private Financial Measures in Europe*; European Investment Bank: Luxembourg, 2023.
74. NASS, U. 2017 Census of Agriculture, Census Data Query Tool (CDQT). 2019. Available online: https://www.nass.usda.gov/Publications/AgCensus/2017/ (accessed on 27 September 2023).
75. Mostafa, A.M.; La Ravia, T.; Harrington, K. *Urban, Small-Scale, and Beginner Farmer Needs Assessment*; The University of Arizona: Tucson, AZ, USA, 2022.
76. USDA. National Agricultural Statistics Service. 2018. Available online: https://www.nass.usda.gov/AgCensus/index.php (accessed on 7 October 2023).
77. Van Raalte, D.; Ranger, N. *Financing Nature-Based Solutions for Adaptation at Scale: Learning from Specialised Investment Managers and Nature Funds*; Global Center on Adaptation and Environmental Change Institute, University of Oxford: Oxford, UK, 2023.
78. Thompson, A.; Bunds, K.; Larson, L.; Cutts, B.; Hipp, J.A. Paying for nature-based solutions: A review of funding and financing mechanisms for ecosystem services and their impacts on social equity. *Sustain. Dev.* **2023**, *31*, 1991–2066. [CrossRef]
79. Pathak, A.; Glick, P.; Hansen, L.J.; Hilberg, L.E.; Ritter, J.; Stein, B.A. *Incorporating Nature-based Solutions in Community Climate Adaptation Planning. National Wildlife Federation*; National Wildlife Federation: Washington, DC, USA, 2022.
80. Mayor, B.; Toxopeus, H.; McQuaid, S.; Croci, E.; Lucchitta, B.; Reddy, S.E.; Egusquiza, A.; Altamirano, M.A.; Trumbic, T.; Tuerk, A.; et al. State of the Art and Latest Advances in Exploring Business Models for Nature-Based Solutions. *Sustainability* **2021**, *13*, 7413. [CrossRef]
81. UNEP. *State of Finance for Nature: The Big Nature Turnaround—Repurposing $7 Trillion to Combat Nature Loss*; UN Environment Programme: Nairobi, Kenia, 2023.
82. Richards, D.; Krone, M.; Hernandez, A. *Building Sustainable Farms, Ranches, and Communities: A Guide to Federal Programs for Sustainable Agriculture, Forestry, Entrepreneurship, Conservation, Food Systems, and Community Development*; Sustainable Agriculture Research & Education: College Park, MD, USA, 2020.
83. AZDA. Resilient Food Systems Infrastructure (RFSI) Program. 2023. Available online: https://agriculture.az.gov/about-us/divisions/agricultural-consultation-training/resilient-food-systems-infrastructure-rfsi (accessed on 7 October 2023).
84. Brears, R.C. Financing Nature-Based Solutions. In *Financing Nature-Based Solutions: Exploring Public, Private, and Blended Finance Models and Case Studies*; Brears, R.C., Ed.; Springer International Publishing: Cham, Switzerland, 2022; pp. 29–50.
85. USDA. *Report to Congress: A General Assessment of the Role of Agriculture and Forestry in U.S. Carbon Markets*; U.S. Department of Agriculture: Washington DC, USA, 2023.
86. So, I.S.; Haya, B.K.; Elias, M. Voluntary Registry Offsets Database v8. 2023. Available online: https://gspp.berkeley.edu/research-and-impact/centers/cepp/projects/berkeley-carbon-trading-project/offsets-database (accessed on 7 October 2023).
87. Bloomberg. Long-Term Carbon Offsets Outlook 2023. 2023. Available online: https://www.bloomberg.com/professional/blog/long-termcarbon-offsets-outlook-2023/ (accessed on 30 October 2023).
88. Blaufelder, C.L. A Blueprint for Scaling Voluntary Carbon Markets to Meet the Climate Challenge. 2021. Available online: https://www.mckinsey.com/capabilities/sustainability/our-insights/a-blueprint-for-scaling-voluntary-carbon-markets-to-meet-the-climate-challenge (accessed on 30 October 2023).
89. Amonette, J.E.; Yorgey, G.G.; Schiller, S. *Technical Assessment of Potential Climate Impact and Economic Viability of Biochar Technologies for Small-Scale Agriculture in the Pacific Northwest*; Pacific Northwest National Laboratory (PNNL): Richland, WA, USA, 2023.
90. Robinson, J.O. *Why Grassfed Is Best!: The Surprising Benefits of Grassfed Meats, Eggs, and Dairy Products*; Vashion Island Press: Vashon, WA, USA, 2000; p. 128.
91. Kragt, M.E.; Burton, R.; Zahl-Thanem, A.; Otte, P.P. Farmers' interest in crowdfunding to finance climate change mitigation practices. *J. Clean. Prod.* **2021**, *321*, 128967. [CrossRef]
92. Burton, R.J.F.; Otte, P.P. Promoting climate change mitigation in agriculture: Do we need to account for farm family life-cycle? *J. Rural. Stud.* **2022**, *96*, 270–281. [CrossRef]
93. Tucson, C.; Water, T. Rainwater Harvesting Rebate. 2023. Available online: https://www.tucsonaz.gov/files/sharedassets/public/v/1/water/documents/rebate-program/rwh-rebate-changes-july-2023-2.pdf (accessed on 12 December 2023).
94. Sutherland, L.-A.; Madureira, L.; Dirimanova, V.; Bogusz, M.; Kania, J.; Vinohradnik, K.; Creaney, R.; Duckett, D.; Koehnen, T.; Knierim, A. New knowledge networks of small-scale farmers in Europe's periphery. *Land Use Policy* **2017**, *63*, 428–439. [CrossRef]

land

MDPI

Article

Spatial Coupling and Resilience Differentiation Characteristics of Landscapes in Populated Karstic Areas in Response to Landslide Disaster Risk: An Empirical Study from a Typical Karst Province in China

Huanhuan Zhou [1], Sicheng Wang [1,*], Mingming Gao [1,2] and Guangli Zhang [1]

1 College of Architecture and Urban Planning, Guizhou University, Guiyang 550025, China; gs.hhzhou23@gzu.edu.cn (H.Z.); 11912024@zju.edu.cn (M.G.); gs.guanglizhang23@gzu.edu.cn (G.Z.)
2 College of Civil Engineering and Architecture, Zhejiang University, Hangzhou 310058, China
* Correspondence: scwang@gzu.edu.cn; Tel.: +86-16622519840

Abstract: Landslides pose a significant threat to the safety and stability of settlements in karst regions worldwide. The long-standing tight balance state of settlement funding and infrastructure makes it difficult to allocate disaster prevention resources effectively against landslide impacts. There is an urgent need to fully leverage the landscape resources of karst settlements and develop landslide risk prevention strategies that balance economic viability with local landscape adaptability. However, limited research has explored the differential resilience characteristics and patterns of landslide disaster risk and settlement landscapes from a spatial coupling perspective. This study, based on landslide disaster and disaster-adaptive landscape data from a typical karst province in China, employs the frequency ratio-random forest model and weighted variance method to construct landslide disaster risk (LDR) and disaster-adaptive landscape (DAL) base maps. The spatial characteristics of urban, urban–rural transition zones, and rural settlements were analyzed, and the resilience differentiation and driving factors of the LDR–DAL coupling relationship were assessed using bivariate spatial autocorrelation and geographical detector models. The key findings are as follows: (1) Urban and peri-urban settlements exhibit a high degree of spatial congruence in the differentiation of LDR and DAL, whereas rural settlements exhibit distinct divergence; (2) the Moran's I index for LDR and DAL is 0.0818, indicating that urban and peri-urban settlements predominantly cluster in H-L and L-L types, whereas rural settlements primarily exhibit H-H and L-H patterns; (3) slope, soil organic matter, and profile curvature are key determinants of LDR–DAL coupling, with respective influence strengths of 0.568, 0.555, and 0.384; (4) in karst settlement development, augmenting local vegetation in residual mountain areas and parks can help maintain forest ecosystem stability, effectively mitigating landslide risks and enhancing disaster-adaptive capacity by 6.77%. This study helps alleviate the contradiction between high LDR and weak disaster-adaptive resources in the karst region of Southwest China, providing strategic references for global karst settlements to enhance localized landscape adaptation to landslide disasters.

Keywords: landslide disaster risk; disaster-adaptive landscape; karst region; spatial differentiation; coupling coordination

 https://doi.org/10.3390/land14040847

1. Introduction

Karst regions cover approximately 15% of the Earth's land area and supply drinking water to nearly 25% of the global population, making them one of the planet's most vital ecosystems [1]. Between 2004 and 2016, complex topography and fragile geological conditions contributed to 4862 landslide events, resulting in 55,997 fatalities, with Asia experiencing the highest spatial concentration of these disasters [2]. Landslide risks in karst regions are further intensified by extreme weather events and human activities. Research indicates that each unit increase in monthly rainfall anomalies corresponds to a 99.7% rise in landslide damage [3]. Agricultural reclamation, urban expansion, and large-scale deforestation increased landslide susceptibility in the Three Gorges Dam reservoir area by 46% between 2015 and 2019 [4]. The karst region of Southwest China is the largest and most intensely developed ecological fragile area among the three major karst plateaus globally [5]. It also experiences the most widespread and severe karst landslides globally [6]. This region is home to approximately 222 million people, primarily engaged in low-productivity agriculture [7], characterized by low socio-economic development and severe human–environment conflicts. In recent years, China has consistently underscored the need to "fortify the national ecological security barrier." As one of China's four major ecologically fragile zones, Southwest China's karst region is pivotal to the nation's ecological civilization initiatives. Confronted with frequent landslides, limited disaster-adaptive resources, and mounting pressures from national ecological civilization demands, settlements in Southwest China's karst region must urgently pursue a self-sustaining strategy for landscape adaptation to landslides. However, policymakers' underestimation of landslide consequences coupled with inadequate risk monitoring and disaster prevention infrastructure due to socio-economic underdevelopment result in 80% of geological disasters occurring outside designated hazard zones annually [8]. A significant information gap exists between landslide disaster risk (LDR) and settlement landscapes in karst regions. Therefore, strengthening the synergy between LDR and disaster-adaptive landscape (DAL) and achieving their spatial integration are essential for enhancing the resilience of landscapes in populated karstic areas to landslide risks and advancing regional ecological civilization efforts.

Extensive research has been conducted on landslide prevention and control. From a risk management perspective, the primary focus is on systematically identifying, assessing, and controlling landslide risks to minimize their negative impact on exposed elements [9–11]. On the other hand, resilient cities emphasize their capacity for "reorganization", allowing for them to absorb, learn, adapt, and rapidly recover from landslide disruptions [12]. Both approaches prioritize proactive disaster adaptation, implementing preemptive measures to mitigate or even prevent disaster-related losses. Response strategies primarily involve landslide vulnerability assessments using various methods [13–15], along with terrain monitoring, slope displacement tracking, and remote sensing [16–19]. Research has also explored disaster early-warning systems triggered by rainfall and soil physical properties [20,21], engineering-based prevention and control methods [22–25], and emergency response strategies [26,27]. However, all these measures necessitate substantial capital investment, which remains misaligned with the economic underdevelopment of karst settlements. Conventional engineering-based disaster adaptation strategies often reinforce the antagonistic relationship between humans and the environment. Landslide disasters are inevitable; thus, coexistence remains the only viable approach [28]. Resilient landscapes, as integral spatial components of ecological systems, exhibit self-adaptive and self-restorative capacities. When subjected to external disturbances, such as pollution, geological disasters, floods, hurricanes, and droughts, resilient landscapes sustain the stability of ecosystems and landscape communities. Karst regions exhibit a greater diversity of landscape types. Leveraging landscapes as a tool for disaster adaptation presents novel strategies for disaster prevention and mitigation in karst settlements. However, the existing research on resilient landscapes predominantly examines resilience within urban public spaces. Through the establishment of indicator systems [29] and the development of design strategies [30], urban green systems' ability to withstand and recover from natural disasters has been enhanced. Few studies have investigated the spatially coupled resilience differentiation patterns of landslide risk and settlement landscapes.

In summary, to enhance the disaster-adaptive resilience of karst settlement landscapes and clarify the spatial coupling relationship between landslide disaster risks and settlement landscapes, this study takes Guizhou Province, located in the center of Southwest China's karst region, as an example. First, the frequency ratio-random forest model and the weighted variance method are applied to generate LDR and DAL base maps and examine their spatial characteristics across three settlement levels: urban, urban–rural transition zones, and rural. Next, bivariate spatial autocorrelation is employed to assess the spatial differentiation of LDR and DAL. Finally, the geographical detector model is utilized to identify the driving factors governing the LDR–DAL coupling. The findings bridge the information gap between LDR and DAL, providing strategic insights to reconcile the conflict between frequent landslide disasters and limited disaster prevention resources in Southwest China's karst regions. Furthermore, this study proposes a cost-effective strategy for global karst settlements to adapt local landscapes to landslide disasters.

2. Materials and Methods

2.1. Study Area

Guizhou Province lies at the heart of Southwest China's karst region, encompassing nine prefecture-level cities and spanning approximately 176,000 square kilometers (Figure 1). It is situated in the plateau slope zone, marking the transition from the Yunnan-Guizhou Plateau to the eastern low hills and mountains, characterized by a complex geological structure. It is among the regions in China with the most widespread karst topography. This geomorphological complexity not only shapes Guizhou's unique natural landscapes but also contributes to significant environmental challenges, as its intricate folds, fault structures, rugged terrain, exposed rock surfaces, and weak soil–rock cohesion make the region highly susceptible to landslides. In recent years, Guizhou's rapid economic development has intensified human-induced disruptions to the geological environment. The frequency of human-induced landslides has increased, posing a growing constraint on Guizhou's urban development.

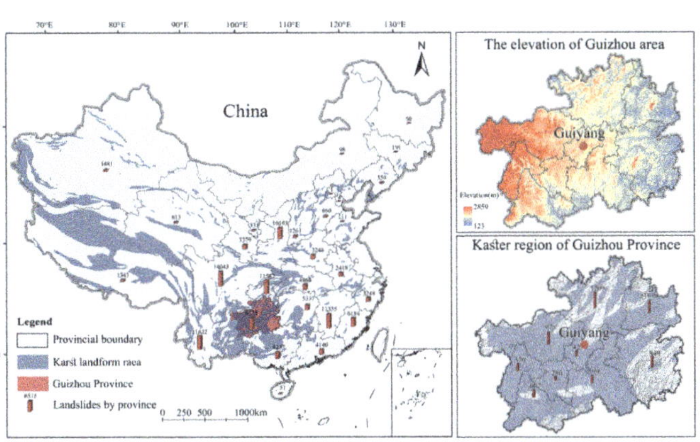

Figure 1. The research area.

2.2. Data Sources

Geological disaster data are obtained from the spatial distribution dataset of geological disaster points provided by the Institute of Geographic Sciences and Natural Resources Research, Chinese Academy of Sciences. The digital elevation model (DEM) data are obtained from the CDEMV3 30 m elevation dataset of the National Geospatial Data Cloud and are used to extract indicators, including slope, aspect, planar curvature, profile curvature, terrain roughness, terrain undulation, river network density, and the topographic wetness index (TWI). NDVI data are calculated from Landsat TM8 imagery. Precipitation and fault distribution data are obtained from the National Earth System Science Data Center. Soil erosion data are derived from the mainland China soil erosion dataset compiled by Yan Jining and colleagues on the Scientific Data Platform [31]. Geomorphological type is obtained from the Institute of Geographic Sciences and Natural Resources Research, Chinese Academy of Sciences. Stratigraphic lithology data are retrieved from the Geological Cloud of the China Geological Survey. Vegetation data are obtained from the vegetation map of the People's Republic of China [32]. Road network and point of interest (POI) vector data are extracted from the Gaode Open Platform via web crawling. Soil organic matter data are obtained from the National Tibetan Plateau Science Data Center [33].

2.3. Research Methods

As illustrated in Figure 2, this study is primarily divided into two parts: the construction of LDR and DRL base maps and the analysis of resilience differentiation characteristics and coupling driving factors of LDR and DRL. The LDR base map is generated using the frequency ratio-random forest method, whereas the DAL is derived from LDR results integrated with landscape elements through the weighted variance method. This study first examines the spatial characteristics of LDR and DAL across three settlement levels: urban, urban–rural transition zones, and rural. Next, bivariate spatial autocorrelation is applied to assess their resilience differentiation characteristics. Finally, the geographical detector model is utilized to identify the key driving factors influencing the LDR–DAL coupling.

2.3.1. Evaluation Index of LDR and DAL

Drawing on relevant literature on landslide risk assessment [2,34,35] and the characteristics of landslide development in Southwest China's karst landscapes [36], this study selects indicators from four key domains—topography and geomorphology, climate and hydrology, soil and vegetation, and human activities—to construct a landslide risk evaluation index system. The selected indicators are categorized as follows: Topography and geomorphology—elevation, slope, aspect, surface undulation, terrain roughness, planar curvature, profile curvature, lithology, distance to faults, and landform type; climate and hydrology—average monthly precipitation, river intensity index, and river network density; soil and vegetation—NDVI and soil erosion; human activities—distance to roads and POI density, comprising a total of 17 indicators.

Regarding landslide resistance, vegetation root systems enhance soil stability, mitigate surface water flow, and reduce erosion from rainwater scouring, thereby lowering the likelihood of landslides [37]. Additionally, various vegetation types exhibit differing soil retention capacities, influencing their effectiveness in landslide resistance [38,39]. Studies indicate that vegetation degradation is a major driver of frequent landslides in the mountainous regions of Southwest China [37]. With respect to landslide adaptation, road networks function as natural evacuation routes, buffers, and refuge zones during landslide events [40]. Different road classifications serve distinct disaster mitigation functions [41], and road quality directly influences a city's capacity to adapt to landslide hazards. Refuge sites are designated areas where disaster victims seek shelter from sudden natural disasters and accidents, equipped with rescue, resettlement, and medical support services [42]. In human settlements, green spaces, plazas, large public venues, and schools serve not only as recreational, aesthetic, and cultural hubs in normal times but also as critical temporary refuge sites during landslide disasters. Regarding landslide recovery, studies suggest that soil organic matter content not only influences the critical moisture threshold—directly affecting landslide occurrence—but also plays a vital role in soil retention and vegetation restoration post-landslide due to its nutrient composition [43,44]. In summary, this study incorporates various vegetation types and soil organic matter from natural landscapes, while roads, park green spaces, plazas, schools, large public venues, and hospitals are selected from artificial landscapes as DAL elements for constructing the DAL base map (Table 1).

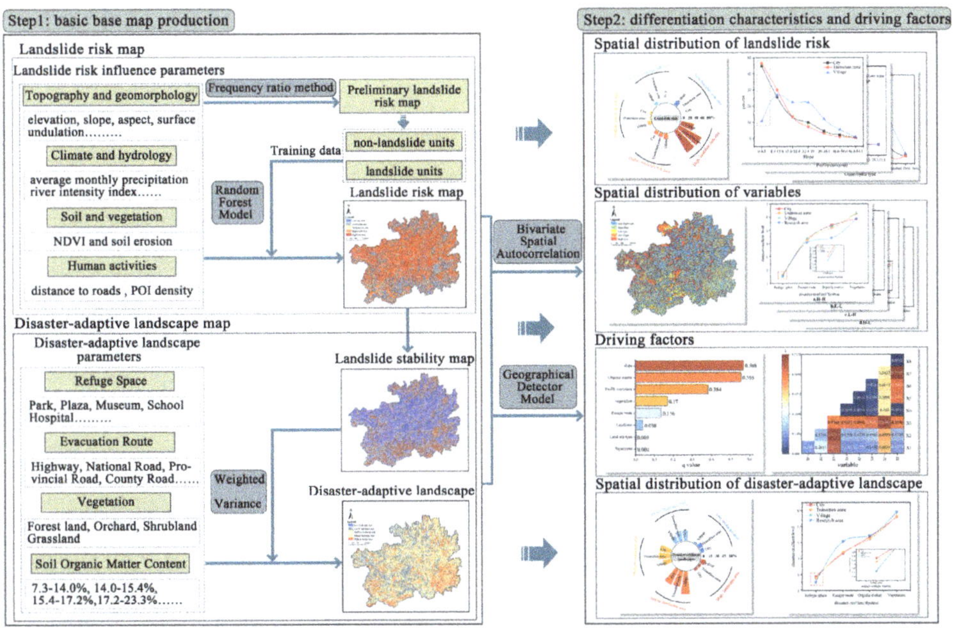

Figure 2. Research framework.

Table 1. Disaster-adaptive landscape indicators.

Refuge Space	Evacuation Route	Vegetation	Soil Organic Matter Content (%)
Park	Highway	Forest land	7.3–14.0
Plaza	National Road	Orchard	14.0–15.4
Museum	Provincial Road	Shrubland	15.4–17.2
School	County Road	Grassland	17.2–23.3
Hospital	Township Road	—	23.3–28.9
—	Other Roads	—	—

2.3.2. Frequency Ratio-Random Forest Model

Machine learning is a widely adopted approach for evaluating landslide risk. However, in machine learning-based landslide risk assessments, non-landslide units are typically selected randomly within the study area. This approach may cause non-landslide units to overlap with potential landslide zones, introducing bias into the model's evaluation results.

Thus, a more rigorous selection of non-landslide units is essential, and the frequency ratio-random forest (FR-RF) model effectively addresses this issue [45]. This method first employs the frequency ratio (*FR*) approach to generate a preliminary landslide risk map. Next, non-landslide units from low-risk areas, selected to match landslide disaster samples, are randomly chosen and integrated with known landslide samples to form the training and validation dataset for the random forest model. Finally, the trained random forest model produces a high-precision landslide risk map, substantially enhancing the model's predictive performance.

The frequency ratio (*FR*) calculation formula is as follows:

$$FR = \frac{F_j/F}{C_j/C} \tag{1}$$

where F_j represents the number of landslide disaster grids for a given factor within a classification interval, F denotes the total number of landslide disaster grids in the interval, C_j refers to the number of grids for a given factor within the classification interval, and C is the total number of grids in the study area.

Random forest is a sophisticated ensemble classification algorithm [46]. This algorithm utilizes the bootstrap resampling technique, randomly selecting n samples (constituting two-thirds of the total) from the sample set T to construct a training dataset. Each training sample is used to train a decision tree, ultimately forming an ensemble of n decision trees, collectively referred to as a "forest". During prediction, the final outcome is obtained by aggregating the results of the n decision trees, using majority voting for classification tasks and averaging for regression tasks. In summary, this study employs the frequency ratio-random forest model to construct the LDR base map.

2.3.3. Weighted Variance

The weighted variance method is a statistical technique for quantifying the contribution of different groups or categories to the overall variance in a dataset. This method assigns weights to each group's variance based on its sample size, reflecting its relative importance in the dataset.

The computation formulas are as follows:

$$W_i = \frac{\sum_{i=1}^{k} n_i \times \mathrm{var}(y_i)}{N} \tag{2}$$

$$\mathrm{var}(y_i) = \frac{\sum_{i=1}^{n} (y_i - \hat{y}_i)^2}{n-1} \tag{3}$$

where k denotes the number of classification intervals, n_i represents the sample size of the i-th classification interval, $\mathrm{var}(y_i)$ refers to the variance of the i-th classification interval, $\hat{y}_i$ is the mean of the dependent variable y in the i-th classification interval, and N is the total sample size. This study generates a landslide stability map by inverting the LDR base map and then constructs the DAL base map using the weighted variance method in combination with DAL elements. Here, W_i represents the capability of the settlement landscape to adapt to landslide disasters. A higher W_i value indicates a greater adaptive capacity to landslide disasters and corresponds to a higher DAL level.

2.3.4. Bivariate Spatial Autocorrelation

Spatial autocorrelation measures the degree of correlation between a geographic phenomenon or attribute within a spatial unit and the same phenomenon or attribute in adjacent units. It is typically assessed using global and local spatial autocorrelation metrics [47]. The global Moran's I index quantifies the spatial clustering tendency of a given attribute within the study area. Its calculation formula is as follows:

$$I = \frac{n}{\sum_{i=1}^{n}\sum_{j=1}^{n} W_{ij}} \times \frac{\sum_{i=1}^{n}\sum_{j=1}^{n} w_{ij}(x_i - \hat{x})(x_j - \hat{x})}{\sum_{i=1}^{n}(x_i - \hat{x})^2} \tag{4}$$

where x_i and x_j denote the landslide disaster or disaster-adaptive landscape values of the i-th and j-th spatial units, respectively; $\hat{x}$ represents the mean attribute value of all spatial units; and W_{ij} refers to the weight matrix between units i and j. The global Moran's I value generally falls within the range of -1 to 1. A value below 0 indicates a negative spatial correlation, meaning spatial units exhibit dissimilar attributes and a dispersed distribution. A value near 0 suggests a random distribution, while a value above 0 indicates a positive spatial correlation among spatial unit attributes. The statistical significance of the global Moran's I index is typically evaluated using the p-value.

The local Moran's I index identifies localized spatial correlation patterns in regional attribute values and is calculated as follows:

$$I_{kl}^{i} = Z_k^i \sum_{j=1}^{n} W_{ij} Z_l^j \tag{5}$$

where $Z_k^i = \frac{X_k^i - \hat{x}_k}{\gamma_k}$; $Z_l^i = \frac{X_l^i - \hat{x}_l}{\gamma_l}$; W_{ij} represents the weight matrix between spatial units i and j; X_k^i denotes the disaster-adaptive landscape value of unit i, while X_l^i denotes the landslide frequency value of unit i; $\hat{x}_k$ and X_l^i are the mean values of attributes k and l; and γ_k and γ_l are their respective variances.

This study investigates the spatial distribution patterns of LDR and DAL through bivariate spatial autocorrelation analysis. Based on the local Moran's I index, this study classifies areas into four spatial association types: H-H (High LDR, High DAL), L-L (Low LDR, Low DAL), H-L (High LDR, Low DAL), and L-H (Low LDR, High DAL).

2.3.5. Geographical Detector Model

The geographical detector is a quantitative analytical method designed to detect spatial differentiation and identify its underlying driving factors. With advantages such as the absence of linear assumptions, resistance to collinearity, and broad applicability, it has been widely utilized in investigating the factors and mechanisms driving spatial heterogeneity in geographic phenomena [48]. The method compares the within-group variance of an influencing factor across different partitions with the total variance of the entire study region. The calculation formula is as follows:

$$q = 1\frac{\sum_{h=1}^{L} N_h \sigma_h^2}{N\sigma^2} = 1\frac{SSW}{SST} \tag{6}$$

$$SSW = \sum_{h=1}^{L} N_h \sigma_h^2, SST = N\sigma^2 \tag{7}$$

where h (1, …, L) denotes the classification of the dependent variable Y or factor X; N_h and N represent the number of units in category h and the entire region, respectively; and σ_h^2 and σ^2 correspond to the variance of Y within category h and the entire region. SSW represents the sum of within-category variances, whereas SST denotes the total variance of the entire region.

This study applies the geographical detector model to examine the driving factors governing the coupling between LDR and DAL. The q value ranges from 0 to 1, with higher values indicating a stronger influence of the factor on the LDR–DAL coupling.

3. Results

3.1. LDR Map

3.1.1. Evaluation of the Frequency Ratio-Random Forest Model

Model performance is assessed using the area under the receiver operating characteristic (ROC) curve (AUC), accuracy, and recall. The AUC value typically falls between 0.5 and 1, with higher values indicating superior model performance. The ROC curve, generated using test samples, is presented in Figure 3, where the model attains an AUC of 0.949. Accuracy measures the proportion of correctly classified landslide and non-landslide units among all samples, with values ranging from 0 to 1. Recall quantifies the model's ability to correctly identify landslide units, with higher values signifying improved predictive performance. The frequency ratio-random forest model achieved an accuracy and recall rate of 0.91, indicating good predictive performance and its suitability for landslide risk prediction.

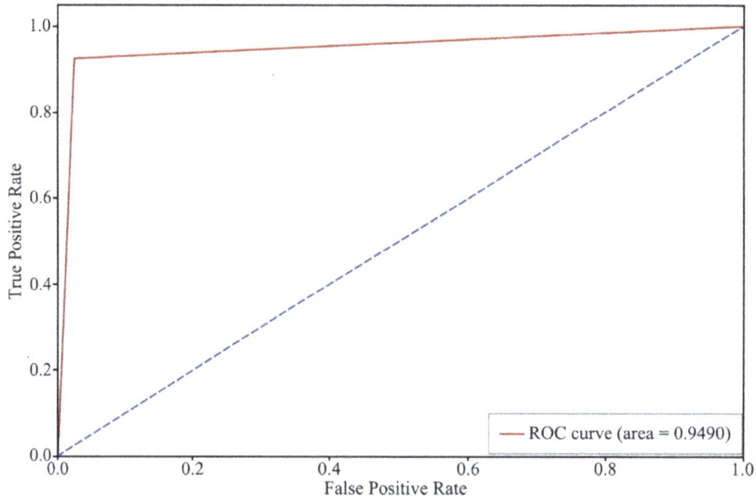

Figure 3. AUC curve of the random forest model.

3.1.2. Analysis of Risk Results

The landslide risk assessment results derived from the frequency ratio-random forest model are presented in Figure 4. The results reveal that high-risk landslide zones are predominantly concentrated in the central region of the study area, whereas low-risk zones are mainly distributed along the periphery, with high-risk zones covering 42.94% of the total area. At the settlement scale (Figure 5), high LDR zones are more concentrated in urban and urban–rural transition areas, comprising up to 80% of the total, whereas medium and low LDR zones constitute less than 10%. Rural areas demonstrate lower LDR values than urban and urban–rural transition zones, with medium and low LDR levels comprising up to 40%. Overall, low-risk zones in Guizhou Province are predominantly found in rural areas, while urban and urban–rural transition settlements represent critical regions for landslide disaster prevention and mitigation.

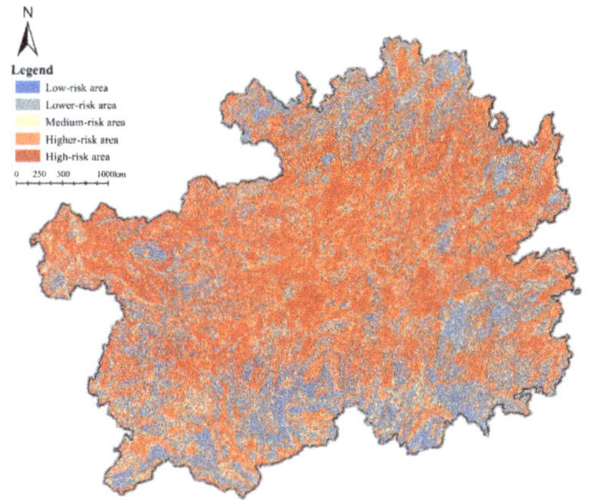

Figure 4. LDR zoning.

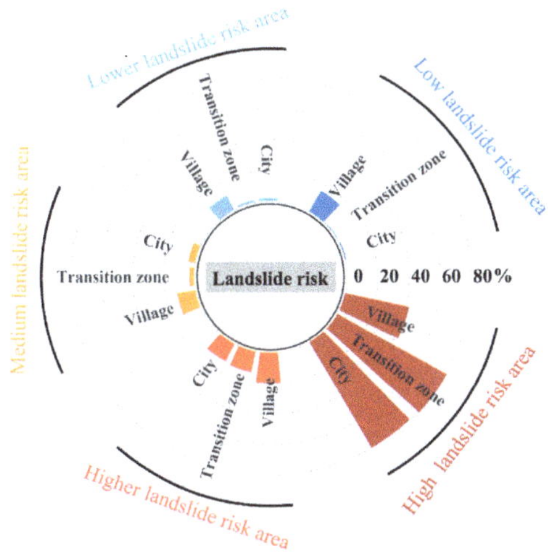

Figure 5. LDR settlement proportion.

Feature importance analysis from the random forest model (Figure 6) reveals that geomorphology, profile curvature, and slope are the three most influential factors in landslide occurrence, with respective importance values of 0.126, 0.118, and 0.113. A comparison of geomorphology, profile curvature, and slope distributions across different settlement types (Figure 7) shows that high-altitude plains and terraces, characterized by low profile curvature and gentle slopes, are more susceptible to landslides than other regions. This risk arises from limestone, the predominant rock type in karst regions, which is highly prone to dissolution by water, resulting in its gradual decomposition. Organic matter in the topsoil is transported by carbonate-rich water solutions into rivers, underground streams, and other hydrological systems. Consequently, steep high-altitude areas become barren as nutrients are transported to karst peak-cluster plains and terraces. Urban areas, however, are typically situated in relatively flat terrain. Compared to rural areas, urban settlements generally feature gentler slopes and thicker soil layers. During heavy rainfall events, increased soil moisture reduces soil strength, rendering urban areas more vulnerable to landslides. In summary, geomorphology, profile curvature, and slope should be prioritized in the planning and development of engineering activities in karst settlements to enhance landslide disaster prevention and mitigation.

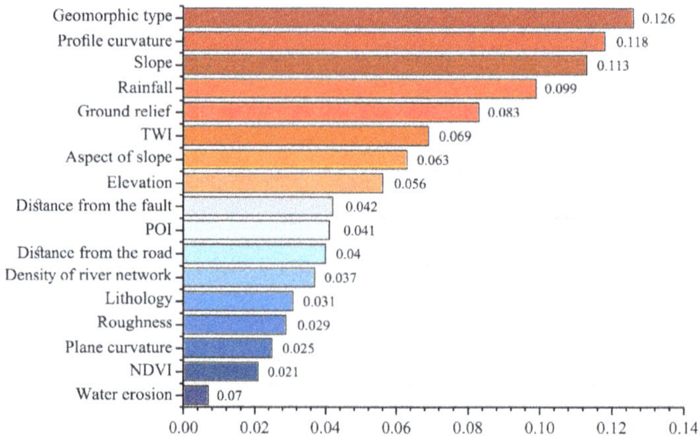

Figure 6. Ranking of importance of evaluation factors.

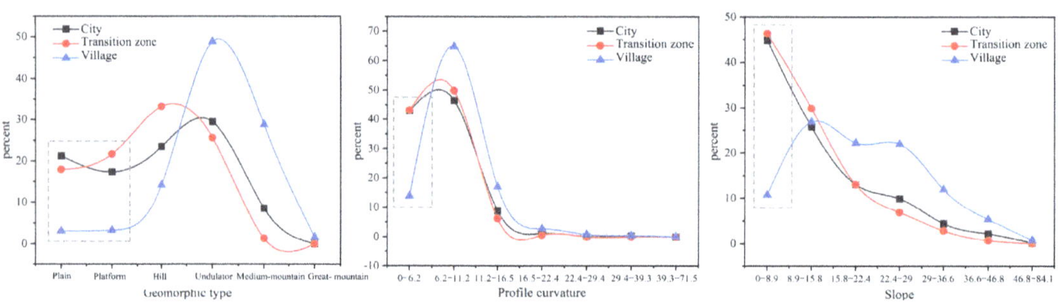

Figure 7. Proportion of settlement LDR influencing factors.

3.2. DAL Results

With landslide stability base map data as the dependent variables and DRL elements as independent variables, the weighted variance method was employed to assess the disaster-adaptive capacity of each landscape element. A spatial overlay analysis was subsequently performed to derive the disaster-adaptive landscape (DAL) levels across the study area (Figure 8). The DAL distribution in Guizhou Province presents a fragmented pattern, marked by large-scale dispersion and small-scale clustering in planar and linear formations. This spatial pattern corresponds to the rugged topography and the dispersed, limited flatlands characteristic of karst regions. High DAL areas and above constitute 64% of the total study area, reflecting a generally strong disaster-adaptive landscape capacity. At the settlement scale (Figure 9), high DAL areas constitute less than 5% in urban and urban–rural transition zones, whereas medium and low DAL areas make up as much as 60%. Conversely, over 80% of rural areas belong to the medium and high DAL categories. Compared to urban and urban–rural transition zones, rural areas contain a higher proportion of high DAL regions and fewer low DAL zones. These findings indicate that urban and urban–rural transition zones still require substantial enhancements in DAL development.

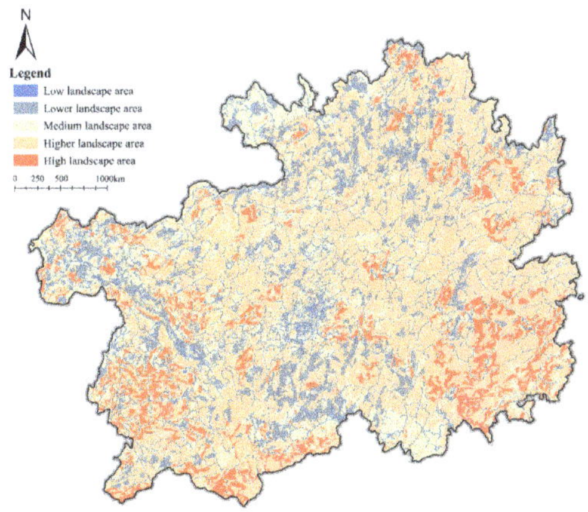

Figure 8. Spatial distribution of DAL.

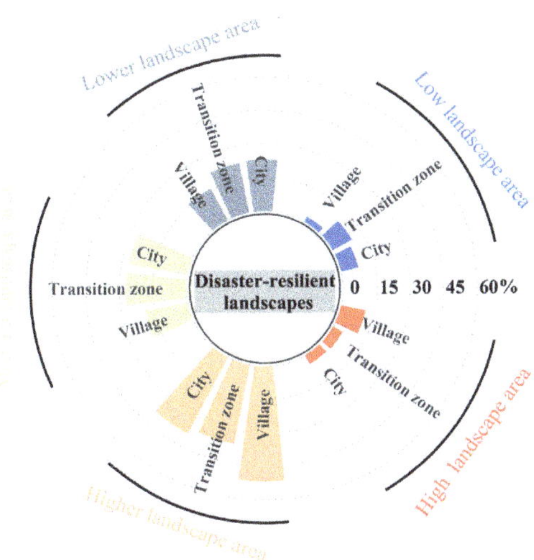

Figure 9. DAL settlement proportion.

From the perspective of DAL components (Figure 10), given that rural areas constitute a significant portion of Guizhou Province's land area, their DRL levels generally align with the overall average. In urban and urban–rural transition areas, only refuge spaces exceed the average, whereas evacuation routes, soil organic matter, and vegetation remain below average. This suggests that urban and urban–rural transition areas exhibit greater capacity for emergency evacuation following a landslide, whereas rural areas demonstrate stronger preventive capabilities. To support sustainable development, future urban planning should prioritize the development of disaster-adaptive landscape systems, such as evacuation routes and vegetation, while rural settlements should focus on expanding refuge space availability.

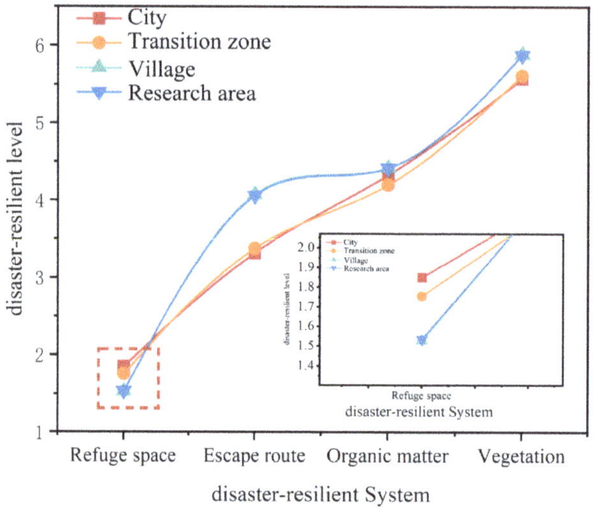

Figure 10. Settlement DRL distribution.

3.3. Spatial Autocorrelation Results

Using the GeoDa spatial analysis tool, a spatial weight matrix was developed to compute the global spatial autocorrelation index Moran's I for LDR and DAL. At a 99% confidence level, Moran's I was 0.0818, suggesting a positive spatial correlation between LDR and DAL. Among the four spatial association types (Figure 11), H-H is the predominant category in Guizhou Province, comprising 24.4%, followed by H-L at 18.7%. L-L and L-H types are less prevalent, constituting 9.3% and 7.5%, respectively. Overall, Guizhou Province presents a spatial coexistence of high LDR–high DAL and high LDR–low DAL patterns.

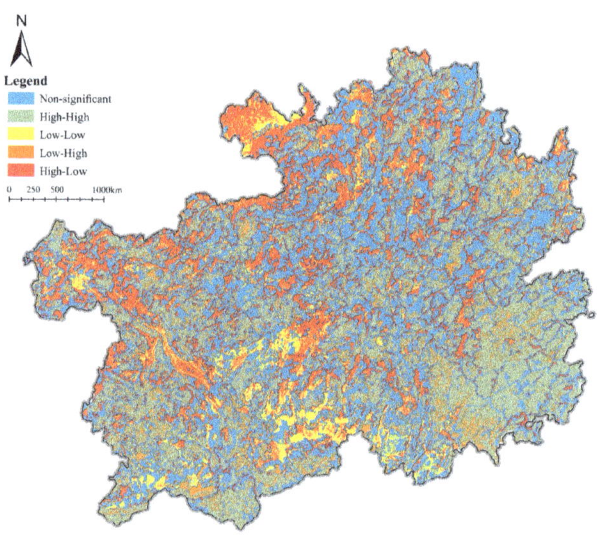

Figure 11. Bivariate spatial distribution.

At the settlement scale (Figure 12), urban and urban–rural transition areas are primarily classified as H-L, comprising 31%, followed by L-L at 17%, while H-H and L-H collectively constitute only 12%. The H-H and L-H types make up only 12%. This indicates that urban areas generally have low DAL levels, with a significant mismatch between LDR and DAL. As discussed in the LDR analysis, urban and urban–rural transition areas feature relatively gentle slopes, rendering them more vulnerable to landslide risks. Figure 13b,d illustrate that, in urban and urban–rural transition settlements with low DAL, all DAL elements—except refuge spaces—fall below the study area's average. Conversely, in areas with high DAL (Figure 13a,c), all DAL elements—except vegetation—surpass the average value. This implies that, in urban and urban–rural transition areas, refuge spaces have already reached saturation, indicating that further expansion will not substantially improve overall disaster-adaptive capacity. Instead, strengthening evacuation routes, soil organic matter, and vegetation components can enhance the overall disaster-adaptive capacity, particularly the soil organic matter system, which shows a significant gap from the average. Even in areas with high disaster-adaptive landscape levels, vegetation factors merely reach the average, highlighting the substantial potential for enhancing vegetation landscape systems in urban and urban–rural transition areas.

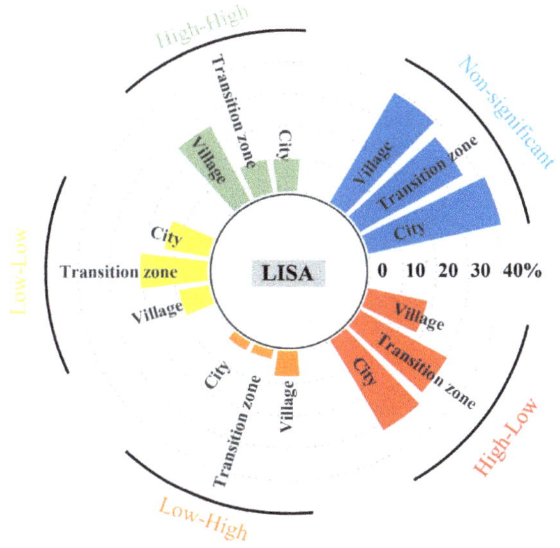

Figure 12. The proportion of different agglomeration types of settlements.

In rural areas, H-H is the most prevalent type, comprising 26%, followed by H-L at 18%, while L-L and L-H together constitute 16%. This suggests that LDR and DAL are relatively well integrated in rural areas, though certain regions still require further DAL development. A comparison in Figure 13 reveals substantial fluctuations in evacuation routes and soil organic matter components in rural areas. In high DAL regions, these two factors surpass the average, whereas in low DAL areas, they remain below the average. This implies that these two landscape components exert a strong influence on overall DAL levels in rural areas. Additionally, vegetation levels in low DAL areas remain below average, highlighting its role in shaping overall DAL levels in rural regions. Enhancing vegetation coverage in sparsely vegetated rural areas can significantly improve their disaster-adaptive capacity.

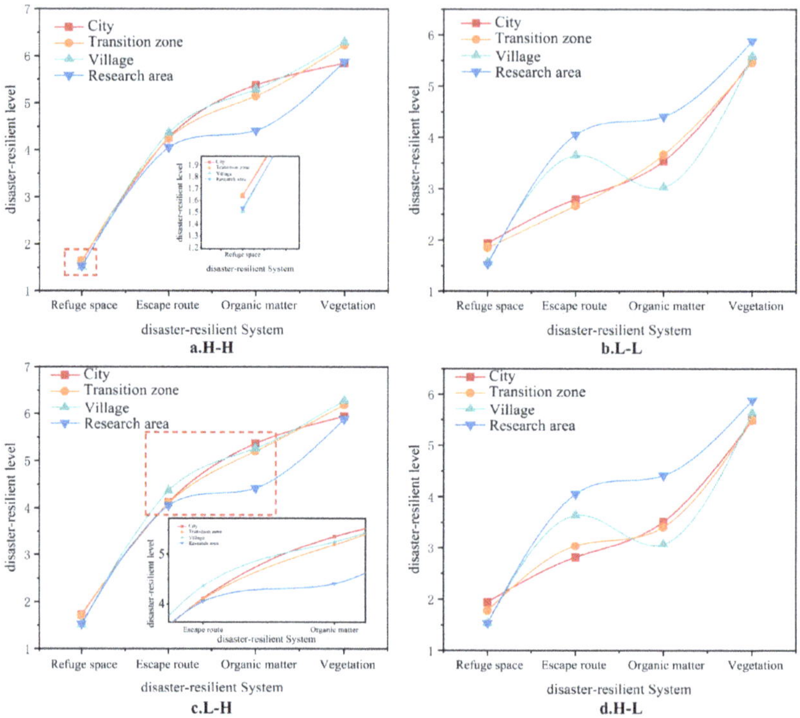

Figure 13. Distribution of settlement DAL across different spatial differentiation types: (**a**) H-H type; (**b**) L-L type; (**c**) L-H type; (**d**) H-L type.

3.4. Analysis of Driving Factors

To investigate the driving factors affecting the coupling of LDR and DAL in Guizhou Province, this study employs the coupling coordination degree of LDR and DAL as the dependent variable. Based on prior analysis, potential driving factors are categorized into three domains: disaster-adaptive landscape elements, human activity influences, and landslide disaster risk. Eight indicators were chosen as independent variables: X1 evacuation routes, X2 vegetation, X3 soil organic matter, X4 population density, X5 land use type, X6 geomorphology, X7 profile curvature, and X8 slope. The geographical detector model was utilized to conduct factor detection and interaction analysis, with the results presented below.

According to the single-factor detection results (Figure 14), landslide risk factors exhibit the strongest explanatory power in the coupling of LDR and DAL, followed by disaster-adaptive landscape factors. Among these, slope (X8) and soil organic matter (X3) exert the strongest driving forces on the coupling of LDR and DAL, with values of 0.568 and 0.555, respectively, identifying them as the primary influencing factors. These are followed by profile curvature (X7), vegetation (X2), and evacuation routes (X1), with corresponding values of 0.384, 0.17, and 0.136. Geomorphology (X6), land use type (X5), and population density (X4) exhibit the weakest influence.

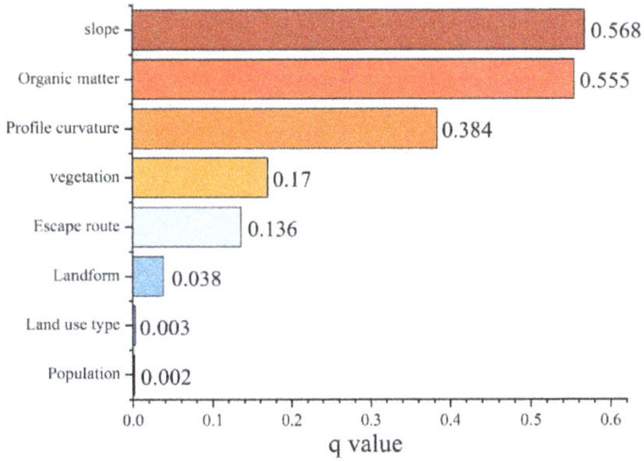

Figure 14. Factor detection results.

In karst regions, limestone readily decomposes. Due to elevation effects, organic matter from the topsoil is transported by erosion and accumulates in low-lying areas. Consequently, areas with gentle slopes and low-profile curvature tend to feature thick soil layers, high organic matter content, and lush vegetation, contributing to elevated DAL levels. Simultaneously, these areas are prone to rapid soil moisture increases during heavy rainfall, which, under gravitational influence, can trigger landslides, leading to high LDR values. In contrast, steeply sloped areas generally exhibit lower soil content, more exposed rock surfaces, and sparse vegetation, frequently leading to both low DRL and low LDR. Therefore, slope, soil organic matter, profile curvature, and vegetation serve as key drivers of DRL–LDR coupling.

The interaction detection results (Figure 15) reveal that the combination of different factors consistently strengthens the explanatory power of DRL–LDR spatial coupling. This effect encompasses 23 instances of bivariate enhancement and four instances of nonlinear enhancement. Slope (X8) and soil organic matter (X3) exhibit strong interactions with other factors. Among these, the interaction between slope (X8) and profile curvature (X7) demonstrates the highest explanatory power, with a value of 0.7105. The second and third strongest interactions occur between soil organic matter (X3) and profile curvature (X7) (0.6841) and soil organic matter (X3) and vegetation (X2) (0.6813). Notably, the interaction between soil organic matter (X3) and population density (X4) further strengthens the explanatory power for risk and spatial coupling.

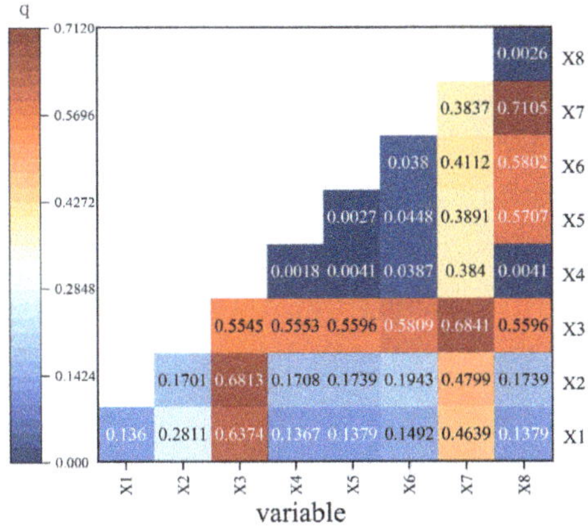

Figure 15. Interaction detection results.

Single-factor detection results indicate that slope (X8), soil organic matter (X3), profile curvature (X7), and vegetation (X2) are key drivers of LDR–DAL coupling. Consequently, their interactions further amplify this effect. The interactions between slope (X8) and land use type (X5) and slope (X8) and geomorphology (X6) significantly reinforce the explanatory power of the coupling. This phenomenon arises because urban and residential developments in karst regions are often situated in high-altitude plains or terraces with gentle slopes. These areas typically feature abundant vegetation, numerous refuge spaces, and well-developed evacuation routes, fostering strong LDR–DAL coupling. The interaction between soil organic matter (X3) and population density (X4) further reinforces the explanatory power of the coupling. This occurs because areas with high soil organic matter are typically found in karst peak-cluster depressions, which are preferred locations for karst settlements due to agricultural activities and urban development needs. Consequently, areas with high soil organic matter and population density generally exhibit strong LDR–DAL coupling.

4. Discussion

Significant differences exist in land use and ecological functions among urban, urban–rural transition, and rural areas. Therefore, this study classifies karst settlements into three zones—urban, urban–rural transition, and rural—to examine the spatial coupling between LDR and DAL. However, LDR and DAL analysis reveals that urban and urban–rural transition areas share similar spatial differentiation characteristics, whereas rural settlements exhibit significant differences. Consequently, this study primarily focuses on interpreting the LDR–DAL coupling relationship in urban and urban–rural transition areas versus rural settlements.

Local spatial autocorrelation analysis indicates that karst urban settlements generally exhibit low disaster-adaptive landscape levels, with a mismatch between LDR and DAL. High LDR–low DAL is the predominant clustering type. This mismatch is primarily due to the continuous encroachment of urban development on natural resources and engineering activities that reduce the stability of slope structures. Survey data from the Chinese Academy of Sciences' Huanjiang Karst Ecosystem Observation and Research Station indicate that, in landslide-prone karst cities, vegetation is predominantly grass or shrub-grass (85%), whereas landslides are less frequent in areas with tall trees (6%). Vegetation root systems reinforce soil shear strength, thereby enhancing slope stability. Therefore, integrating residual mountains and parks with suitable vegetation during development can improve soil structure and organic matter content, thereby enhancing soil shear strength and stability in karst settlements. During settlement development, minimizing disturbances to forest systems is essential. Comprehensive assessments should be conducted in areas with poor geomorphic conditions, followed by targeted protection and remediation measures, or development should prioritize geologically stable regions. Implementing these measures to optimize soil organic matter and vegetation coverage can effectively mitigate landslide disaster risks. As a result, the soil organic matter landscape level can be elevated from 3.4 to 4.5, and vegetation levels can be elevated from 5.6 to 6 (Figure 13b,d), leading to an overall 6.77% improvement in disaster-adaptive landscape capacity.

Although rural settlements exhibit better coupling between LDR and DRL compared to urban settlements, 18% still show a concentration of high LDR and low DRL. Rural settlements can mitigate economic losses caused by landslide disasters by reducing development activities in high LDR areas or by focusing on tree planting-based agricultural activities in these areas. This approach would enhance slope stability and reduce the likelihood of landslides.

This study elucidates the spatial coupling between landslide disaster risk (LDR) and disaster-adaptive landscapes (DAL) in karst regions, providing strategic insights for enhancing the localized landscape in karst settlements to adapt to landslide disasters. However, two limitations remain to be addressed in future research. First, this study exclusively examines landslide disaster risk, whereas cities contend with a broader spectrum of complex hazards. Future research should investigate urban landscape adaptability through a multi-hazard lens. Second, this study analyzed the spatial coupling between LDR and DAL at a macro settlement scale; however, socio-economic development and settlement construction exhibit significant disparities across prefecture-level cities in Guizhou Province. Refining the analysis to the county and township levels would facilitate the development of more targeted and effective strategies for strengthening settlement disaster resilience.

5. Conclusions

Effective utilization of localized landscapes in karst settlements can substantially mitigate landslide disaster risks. The coupling of landslide disaster risk (LDR) and disaster-adaptive landscapes (DAL) is crucial for strengthening both economic resilience and disaster adaptability in settlements. Using Guizhou Province, a representative karst region in China, as a case study, this study employs the frequency ratio-random forest model, bivariate spatial autocorrelation, and the geographical detector model to investigate the spatial coupling and resilience differentiation of LDR and DAL in karst settlements. The key findings are as follows:

(1) Urban and peri-urban settlements exhibit a high degree of spatial congruence in the differentiation of LDR and DAL, whereas rural settlements exhibit distinct divergence.

(2) The Moran's I index for LDR and DAL in Guizhou's karst settlements is 0.0818, indicating a weak positive spatial correlation. Urban and urban–rural transition areas predominantly exhibit H-L and L-L clustering, whereas rural settlements primarily demonstrate H-H and L-H patterns.

(3) Slope, soil organic matter, and profile curvature serve as the primary drivers of LDR–DAL coupling in karst settlements, with impact strengths of 0.568, 0.555, and 0.384, respectively. The synergistic effects of multiple factors further amplify the explanatory power of spatial coupling. The interaction between slope and profile curvature demonstrates the highest explanatory power (0.7105), particularly in areas with gentle slopes, high soil organic matter content, and low-profile curvature, where LDR and DAL exhibit strong coupling.

(4) In karst settlement development, increasing native vegetation in residual mountains and parks can enhance forest ecosystem stability, effectively mitigating landslide disaster risks and improving the disaster-adaptive landscape level by 6.77%.

Author Contributions: Conceptualization, H.Z. and S.W.; methodology, H.Z.; software, H.Z.; validation, H.Z., S.W. and G.Z.; formal analysis, H.Z.; investigation, H.Z.; resources, M.G.; data curation, H Z.; writing—original draft preparation, H.Z.; writing—review and editing, H.Z., S.W. and G.Z.; visualization, H.Z.; supervision, M.G.; project administration, M.G.; funding acquisition, S.W. All authors have read and agreed to the published version of the manuscript.

Funding: This research was funded by the "National Nature Science Foundation of China (52368004)", "National Nature Science Foundation of China (52368002)", and Guizhou Provincial Science and Technology Projects (Qian Ke He Ji Chu-ZK [2022] General 234).

Data Availability Statement: The datasets generated during and/or analyzed during the current study are available from the corresponding author upon reasonable request.

Acknowledgments: We are particularly grateful to the providers of information on the data researched for this paper. The editor's hard work on this paper and the reviewers' valuable comments on this paper are also gratefully acknowledged.

Conflicts of Interest: The authors declare no conflicts of interest.

Abbreviations

The following abbreviations are used in this manuscript:

LDR —Landslide disaster risk
DAL —Disaster-adaptive landscape

References

1. Kalhor, K.; Ghasemizadeh, R.; Rajic, L.; Alshawabkeh, A. Assessment of groundwater quality and remediation in karst aquifers: A review. *Groundw. Sustain. Dev.* **2019**, *8*, 104–121. [CrossRef]
2. Froude, M.J.; Petley, D.N. Global fatal landslide occurrence from 2004 to 2016. *Nat. Hazards Earth Syst. Sci.* **2018**, *18*, 2161–2181. [CrossRef]
3. Fang, Z.; Morales, A.B.; Wang, Y.; Lombardo, L. Climate change has increased rainfall-induced landslide damages in central China. *Int. J. Disaster Risk Reduct.* **2025**, *119*, 105320. [CrossRef]
4. Li, Y.; Wang, X.; Mao, H. Influence of human activity on landslide susceptibility development in the Three Gorges area. *Nat. Hazards* **2020**, *104*, 2115–2151. [CrossRef]
5. Weici, S. Rocky Land Desertification and Its Controlling Measurements in the Karst Mountainous Region, Southwest of China. *J. Soil Water Conserv.* **2002**, *2*, 29–32+79. [CrossRef]
6. Huang, M.; Qi, S.; Shang, G. Karst landslides hazard during 1940-2002 in the mountainous region of Guizhou Province, Southwest China. *Nat. Hazards* **2012**, *60*, 781–784. [CrossRef]
7. He, X.J.; Wang, L.; Ke, B.; Yue, Y.M.; Wang, K.L.; Cao, J.H.; Xiong, K.N. Progress on ecological conservation and restoration for China Karst. *Acta Ecol. Sin.* **2019**, *39*, 6577–6585.
8. Deng, L.Z.; Yuan, H.Y.; Zhang, M.Z.; Chen, J.G. Research progress on landslide deformation monitoring and early warning technology. *J. Tsinghua Univ. (Sci. Technol.)* **2023**, *63*, 849–864. [CrossRef]
9. Xu, J.; Sun, D.; Zhang, H.; Wen, H.; Wu, J.; Hang, Y. Multi-scale ecological risk assessment and risk management oflandslide disasters: A case study of Wanzhou-Wushan section in theThree Gorges Reservoir area. *Acta Ecol. Sin.* **2023**, *43*, 4594–4603.
10. Wu, J.; Luo, Y.; Wang, X.; Zhao, Y. Uncertainty analysis and risk management of ecological risk of urban landslide disaster: A case of Shenzhen City. *Acta Ecol. Sin.* **2020**, *40*, 3612–3621.
11. Ma, Y.; Li, C. Construction of Geological Hazard Risk Management Framework andDisaster Prevention and Mitigation System in Mountain Tourism AreasA Case Study of Leshan City, Sichuan Province. *Sci. Technol. Eng.* **2023**, *23*, 9719–9727.
12. Zhou, L. Resilient City:Risk Management and Its Indicators. *J. Beijing Adm. Inst.* **2016**, *2*, 13–20. [CrossRef]
13. Wang, H.; Zhang, L.; Yin, K.; Luo, H.; Li, J. Landslide identification using machine learning. *Geosci. Front.* **2021**, *12*, 351–364. [CrossRef]
14. Kavzoglu, T.; Sahin, E.K.; Colkesen, I. Landslide susceptibility mapping using GIS-based multi-criteria decision analysis, support vector machines, and logistic regression. *Landslides* **2014**, *11*, 425–439. [CrossRef]
15. Hong, H.; Pourghasemi, H.R.; Pourtaghi, Z.S. Landslide susceptibility assessment in Lianhua County (China): A comparison between a random forest data mining technique and bivariate and multivariate statistical models. *Geomorphology* **2016**, *259*, 105–118. [CrossRef]
16. Ye, B.F.; Qiu, H.J.; Tang, B.Z.; Liu, Y.; Liu, Z.J.; Jiang, X.Y.; Yang, D.D.; Ullah, M.; Zhu, Y.R.; Kamp, U. Creep deformation monitoring of landslides in a reservoir area. *J. Hydrol.* **2024**, *632*, 130905. [CrossRef]
17. Uhlemann, S.; Smith, A.; Chambers, J.; Dixon, N.; Dijkstra, T.; Haslam, E.; Meldrum, P.; Merritt, A.; Gunn, D.; Mackay, J. Assessment of ground-based monitoring techniques applied to landslide investigations. *Geomorphology* **2016**, *253*, 438–451. [CrossRef]
18. Tofani, V.; Raspini, F.; Catani, F.; Casagli, N. Persistent Scatterer Interferometry (PSI) Technique for Landslide Characterization and Monitoring. *Remote Sens.* **2013**, *5*, 1045–1065. [CrossRef]
19. Martelloni, G.; Segoni, S.; Fanti, R.; Catani, F. Rainfall thresholds for the forecasting of landslide occurrence at regional scale. *Landslides* **2012**, *9*, 485–495. [CrossRef]
20. Thiebes, B.; Bell, R.; Glade, T.; Jäger, S.; Mayer, J.; Anderson, M.; Holcombe, L. Integration of a limit-equilibrium model into a landslide early warning system. *Landslides* **2014**, *11*, 859–875. [CrossRef]

21. Intrieri, E.; Gigli, G.; Mugnai, F.; Fanti, R.; Casagli, N. Design and implementation of a landslide early warning system. *Eng. Geol.* **2012**, *147*, 124–136. [CrossRef]
22. Yang, Y.L.; Xie, F.H.; Wu, Q.H.; Wan, S.M. Stability Analysis and Treatment of a Landslide Engineering Along a Highway. In Proceedings of the 2nd International Conference on Civil Engineering, Architecture and Building Materials (CEABM 2012), Yantai, China, 25–27 May 2012; pp. 1076–1081.
23. Xiao, J. The Typical Landslide Accelerating Deformation Period Emergency Treatment Engineering Technology And Its Effect Analysis. In Proceedings of the 2nd International Conference on Civil Engineering and Transportation (ICCET 2012), Guilin, China, 27–28 October 2012; pp. 2692–2699.
24. Wang, C.T.; Wang, H.; Qin, W.M.; Wei, S.; Tian, H.N.; Fang, K. Behaviour of pile-anchor reinforced landslides under varying water level, rainfall, and thrust load: Insight from physical modelling. *Eng. Geol.* **2023**, *325*, 107293. [CrossRef]
25. He, H.; Xu, Y.; Huang, Z.Z.; Li, Z. A Study on Micro Pile Reinforcement Technology for the Accumulative landslide of Expansive Soil. In Proceedings of the 4th International Conference on Civil Engineering, Architecture and Building Materials (CEABM), Haikou, China, 24–25 May 2014; pp. 544–551.
26. Wang, M.Y.; Wang, W.K.; Dai, C.Y.; Ma, C.L.; Luo, Y.; Xu, M. Risk analysis and evaluation of emergency rescue in landslide disaster. *Nat. Hazards* **2024**, *120*, 14809–14835. [CrossRef]
27. Huang, Y.J.; Yi, S.; Li, Z.Y.; Shao, S.G.; Qin, X.C. Design of highway landslide warning and emergency response systems based on UAV. In Proceedings of the 17th China Conference on Remote Sensing—Remote Sensing of the Environment, Hangzhou, China, 27–31 August 2010.
28. Chae, B.G.; Park, H.J.; Catani, F.; Simoni, A.; Berti, M. Landslide prediction, monitoring and early warning: A concise review of state-of-the-art. *Geosci. J.* **2017**, *21*, 1033–1070. [CrossRef]
29. Ciftcioglu, G.C. Evaluating resilience for the management of social-ecological production landscapes and seascapes in Lefke Region of North Cyprus through adaptive comanagement. *Sustain. Sci.* **2019**, *14*, 1117–1130. [CrossRef]
30. Ahern, J. Urban landscape sustainability and resilience: The promise and challenges of integrating ecology with urban planning and design. *Landsc. Ecol.* **2013**, *28*, 1203–1212. [CrossRef]
31. Yan, J.N.; Wang, S.; Feng, J.X.; He, H.X.; Wang, L.Z.; Sun, Z.Y.; Zheng, C.M. The 30 m annual soil water erosion dataset in Chinese mainland from 1990 to 2022. *Sci. Data Bank* **2024**. [CrossRef]
32. Zhang, X.S. *Vegetation Map of the People's Republic of China (1:1,000,000)*; Geology Press: Beijing, China, 2007. [CrossRef]
33. Shangguan, W.; Dai, Y.J.; Liu, B.Y.; Zhu, A.X.; Duan, Q.Y.; Wu, L.Z.; Ji, D.Y.; Ye, A.Z.; Yuan, H.; Zhang, Q.; et al. A China dataset of soil properties for land surface modelling (version 2, CSDLv2). *Earth Syst. Sci. Data* **2013**, *17*, 517–543. [CrossRef]
34. Reichenbach, P.; Rossi, M.; Malamud, B.D.; Mihir, M.; Guzzetti, F. A review of statistically-based landslide susceptibility models. *Earth-Sci. Rev.* **2018**, *180*, 60–91. [CrossRef]
35. Fidan, S.; Tanyas, H.; Akbas, A.; Lombardo, L.; Petley, D.N.; Görüm, T. Understanding fatal landslides at global scales: A summary of topographic, climatic, and anthropogenic perspectives. *Nat. Hazards* **2024**, *120*, 6437–6455. [CrossRef]
36. Yue, X.-L.; Wu, S.-H.; Huang, M.; Gao, J.-B.; Yin, Y.-H.; Feng, A.-Q.; Gu, X.-P. Spatial association between landslides and environmental factors over Guizhou Karst Plateau, China. *J. Mt. Sci.* **2018**, *15*, 1987–2000. [CrossRef]
37. Du, Y.Y.; Peng, J.; Zhao, S.Q.; Hu, Z.C.; Wang, Y.L. Ecological risk assessment of landslide disasters in mountainous areasof Southwest China:A case study in Dali Bai Autonomous Prefecture. *Acta Geogr. Sin.* **2016**, *71*, 1544–1561.
38. Liu, L.; Bian, Z.; Ding, S. Effects of landscape spatial heterogeneity on the generation andprovision of ecosystem services. *Acta Ecol. Sin.* **2018**, *18*, 6412–6421.
39. Chen, L.; Xie, G.; Pei, X.; Zhang, C.; Fan, N.; Zhang, C.; Li, S. Ecosystem's soil conservation function and its spatial distribution inLancang River Basin, Southwest China. *Chin. J. Appl. Ecol.* **2012**, *23*, 2249–2256. [CrossRef]
40. Li, Y.Y.; Zhao, W.M. Research on Moutainous Urban Space Adapting to Disasters: Problemshouahts and Theoretical Framework. *Urban Dev. Stud.* **2017**, *24*, 54–62.
41. Jiao, H.; Zou, T.; Li, Y. Simulation of Urban Disaster Prevention Spatial Unit Risk ZoningPrevention and Control Algorithm. *Comput. Simul.* **2021**, *38*, 465–469.
42. Zhou, Y.; Liu, J.; Liang, J. Comprehensive Suitability Evaluation for Emergency Shelters Based onImproved Gray Relational Analysis Method in Fuzhou. *Geogr. Geo-Inf. Sci.* **2018**, *34*, 63–70+61.
43. Satgada, C.P.; Egusa, T.; Yamakawa, Y.; Imaizumi, F. Forest harvesting affects soil organic carbon and total nitrogen transports by facilitating landslides. *Catena* **2023**, *233*, 107517. [CrossRef]
44. Blonska, E.; Lasota, J.; Piaszczyk, W.; Wiechec, M.; Klamerus-Iwan, A. The effect of landslide on soil organic carbon stock and biochemical properties of soil. *J. Soils Sediments* **2018**, *18*, 2727–2737. [CrossRef]
45. Zhai, W.H.; Wang, X.D.; Wu, M.T.; Wu, X.L.; Li, Q.Q. Geological disaster susceptibility evaluation based on coupling of frequency ratio model and random forest model. *J. Nat. Disasters* **2022**, *32*, 74–82. [CrossRef]
46. Belgiu, M.; Dragut, L. Random forest in remote sensing: A review of applications and future directions. *ISPRS J. Photogramm. Remote Sens.* **2016**, *114*, 24–31. [CrossRef]

47. Wu, Y.; Wu, Y.M.; Gao, B.P.; Li, C.; Zheng, K.J.; Li, C. Spatial Relationship Between Human Activity Intensities and Ecosystem Services Value in Chengdu-Chongqing Urban Agglomeration. *Res. Soil Water Conserv.* **2023**, *30*, 173–182. [CrossRef]
48. Wang, J.F.; Xu, C.D. Geodetector: Principle and prospective. *Acta Geogr. Sin.* **2017**, *72*, 116–134.

forests

MDPI

Article

Assessing the Sustainability of Timber Production Under Policy-Driven Logging: A Spatial Analysis from Southwestern Japan

Yusuke Yamada [1,*], Hidesato Kanomata [2], Katsuto Shimizu [3], Wataru Murakami [4] and Yuichi Yamaura [3]

1 Department of Forest Management, Forestry and Forest Products Research Institute, 1 Matsunosato, Tsukuba 305-8687, Japan
2 Department of Forest Policy and Economics, Forestry and Forest Products Research Institute, 1 Matsunosato, Tsukuba 305-8687, Japan; kanomata@ffpri.affrc.go.jp
3 Shikoku Research Center, Forestry and Forest Products Research Institute, 2-915 Asakuranishi, Kochi 780-8077, Japan; katsutoshimizu@ffpri.affrc.go.jp (K.S.); yamaura@ffpri.affrc.go.jp (Y.Y.)
4 Department of Disaster Prevention, Meteorology and Hydrology, Forestry and Forest Products Research Institute, 1 Matsunosato, Tsukuba 305-8687, Japan; wmura@ffpri.affrc.go.jp
* Correspondence: yamayu@ffpri.affrc.go.jp; Tel.: +81-29-829-8313

Abstract: Promoting nature-positive forestry requires sustainable timber production that aligns with ecosystem service (ES) conservation. However, Japan's recently implemented top-down timber production policy may undermine sustainability in local forest landscapes. We assessed the spatial sustainability of plantation forestry by comparing actual logged areas (2000–2019) with allowable logging areas. Logged areas were identified using satellite imagery analysis, while allowable logging areas were estimated by excluding forests at high risk of landslides or with unclear ownership and dividing the remaining area by the standard logged age. While total logged area remained below the experience-based sustainable threshold, logging in profitable forests exceeded allowable levels in recent years. Forests with higher profitability experienced concentrated logging after 2015, indicating the strong influence of the national policy. This spatial imbalance threatens long-term sustainability by depleting productive forest patches while ignoring underutilized unprofitable forests. Our findings demonstrate the risks of uniform, production-oriented policies and highlight the need for adaptive, locally responsive forest governance. By integrating ecological and social constraints into spatial analysis, this study proposes a new sustainability measurement in line with nature-based solutions. Future forest policy must incorporate local knowledge and participatory decision-making to sustain forest ESs and timber supply under changing social and environmental conditions.

Keywords: timber production; top-down policy; spatial analysis; ecosystem services; harvesting patterns; nature-based solutions

1. Introduction

In recent years, there has been a notable increase in the focus on nature-based solutions (NbS), which are seen as a means of addressing environmental issues while simultaneously promoting biodiversity and human well-being [1,2]. The necessity of establishing clear pathways for the successful implementation and scaling of NbS is required broadly [3]. Since forests are major providers of ecosystem services (ESs) [4,5], appropriate forest management plays a crucial role in balancing biodiversity conservation and social benefits, which are essential for realizing NbS [6]. Among forest ESs, timber production is particularly significant, not only as a key provisioning service but also for its broader ecological and economic implications [7]. The regulation of timber harvesting, therefore, has a profound impact on both the availability of timber and the maintenance of forest ecosystems, influencing the long-term sustainability of forest NbS [8–10].

The Japanese government has been promoting timber production in the abundant coniferous forests that were intensively planted in the 1960s. To revitalize the forestry industry, it has introduced high-performance forestry machinery, encouraged the establishment of an efficient supply chain for the harvested wood products, and invested in human resource development [11]. This policy is aimed at establishing a cyclical forestry system that involves planting, thinning, and clear-cutting, particularly in low-slope areas and in areas near roads where high-performance machinery can be used efficiently. As a result, the total logged area and national timber supply have increased, and the rate of lumber self-sufficiency has reached the highest level in decades [12].

In response to the rapid increase in timber production, concerns have been raised about the sustainability of timber supplies. Central government policies, which often overlook local conditions, may lead to excessive logging and consequent resource shortages in local forests [13,14]. Although these concerns spurred initiatives to promote reforestation after logging, forests require a long maturation period before they can be harvested. With the government's recommendation of a harvesting age of 40 years [11], any imbalance between the pace of logging and the rate of forest maturation could disrupt the stability of timber supply. Such disruptions threaten the viability of forestry enterprises and lumber companies, ultimately weakening the local forestry industry [15]. Therefore, if central policies harm local sustainability, scientific validation and timely warnings are essential.

When assessing the sustainability of timber production, we must consider the effect on other ESs [16,17]. Although logging may have positive effects on forest ecosystems in certain contexts [18], it often causes harms by altering forest structures. For instance, logging can reduce the retention strength of the root systems that help prevent shallow landslides on steep slopes [19]. Such economic activities that lack social responsibility and fail to contribute to a nature-positive society should be discouraged [20]. To ensure the continuous provision of various forest ESs, trade-offs among these services must be carefully managed [21]. Consequently, sustainability assessments should be conducted under the assumption that areas critical for sustaining other ESs are excluded from timber production.

A critical question is which forest types have been logged? Previous studies have shown that forests with higher profitability are more likely to be logged [22,23]. Selective logging of these highly profitable areas may result in a landscape dominated by remaining mature forests with lower profitability. Consequently, the economic benefits derived from timber production could be diminished, ultimately undermining the long-term sustainability of the forestry sector. Yamada and Fukumoto [24] analyzed logging trends in Japan using satellite imagery and revealed that logging occurs predominantly in forests with gentle slopes and in areas adjacent to roads. Furthermore, the spatial bias in logging varies significantly among regions [24]. Thus, incorporating local spatial logging patterns into sustainability assessment is imperative, as it enables a more comprehensive evaluation of the economic viability of timber production. This study focuses on the municipal scale where spatial logging tendencies are generally homogeneous [25].

Evaluating the sustainability of timber production requires a simultaneous consideration of both the spatial distribution of logging—an indicator of economic efficiency—and the risk of losing critical ESs. Owing to the rapid increase in timber production brought about by the policy, there is an urgent need to properly assess the sustainability of the ES supply. Here, the objective of this study is to reveal the impacts of the timber production promotion policy by the central government on local ES sustainability. We examined the local sustainability of timber production by comparing actual logging areas with the estimated annual allowable logging area that accounts for forestry profitability and ES preservation. We estimated this allowable area—the maximum area that can be logged annually without impairing the provision of ecosystem services—and then compared it with actual logged areas. Forest land was categorized into groups based on profitability, and the sustainability of each group was verified. Through these analyses, we quantified the impact of top-down policies aimed at increasing timber production and the associated risk of depleting local forest resources.

First, we identified the locations of logged forests through satellite image analysis and created a logging sites dataset for each 5-year period from 2000 to 2019 to visualize the impact of increased timber production. Next, we developed a profitability map of the study area and classified forest lands by their economic viability. For each classified group, we calculated the area of forests with a high risk of shallow landslides and other reasons that hinder harvest, considering these unsuitable for timber production. The annual allowable logging area was then estimated by dividing the timber production area by a standard harvest period, which was based on the time required for forests to mature or reach the typical logged age in the study area. Finally, the actual logged area was compared with the estimated allowable logging area. If the actual logged area exceeds the allowable area, local sustainability is likely to be in jeopardy.

2. Materials and Methods

2.1. Study Area

This study focuses on the municipality level because spatial logging patterns have been shown to differ by region [25]. The study area comprises plantation forests owned by small-scaled forest owners in Saiki municipality, Oita Prefecture, in southwestern Japan (Figure 1). Covering a total area of 90,314 ha, of which 78,757 ha is forested, this region is predominantly mountainous, where forested areas have a mean slope of 23.9°. With the exception of urban and agricultural areas, nearly the entire study area is mountainous. The region is dominated by plantations of Japanese cedar (*Cryptomeria japonica*), Japanese cypress (*Chamaecyparis obtusa*), and red pine (*Pinus densiflora*), along with patches of deciduous broad-leaved forests, such as those dominated by sawtooth oak (*Quercus acutissima*). Within this region, 82% of the forested area is privately owned; notably, 60% of forest owners have <1 ha each. Forestry is a major industry in the region. Government policy and subsidies have led to a significant increase in high-performance forestry machinery, such as harvesters and forwarders, resulting in a rise in clear-cut timber production in recent years. There are no strong governmental regulations on logging in the study area, particularly for privately owned forest.

The locations of the study area were identified from a forest GIS map maintained by the local government. A digital elevation model was generated from aerial laser scanning data provided by Oita prefecture. The data were reprojected and resampled at a resolution of 30 m × 30 m, matching the resolution of the Landsat satellite imagery used to extract the logging sites.

2.2. Logging Locations

Forests logged between 2000 and 2019 were identified by using country-wide harvest mapping data published by Shimizu and Saito [26], which detects forest changes by analyzing time-series trends in spectral responses with a random forest model applied to annual Landsat image composites in Google Earth Engine platform [27,28]. On the basis of this logging location dataset, the study area was classified into areas logged in 2000–2004, 2005–2009, 2010–2014, 2015–2019 or never.

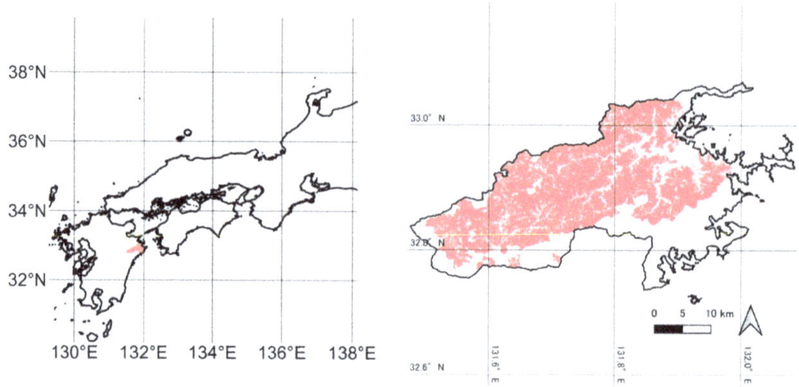

(a) (b)

Figure 1. Location of study area: (**a**) Saiki municipality; (**b**) enlarged view of the study area with privately owned plantation forests.

2.3. Profitability Analysis

The profitability of each forest cell was estimated from net income, incorporating the hypothetical benefits and costs associated with forestry operations from replantation to harvest. We assumed a sequential operational scheme, as outlined in Table A1, which was based on interviews with local forestry companies. These operations were defined solely to evaluate the relative profitability of each forest parcel; consequently, the logging age used in this analysis does not correspond to the standard logging age defined later.

Tree height was estimated as a function of forest age [29] as:

$$\text{Tree height} = 27.47 \times (1 - \exp(-0.03 \times \text{Age})), \tag{1}$$

To estimate both diameter at breast height (DBH) and tree volume, we used a stand density control diagram—a set of calculation formulas that derive tree dimensions from tree height and stand density [30]. The parameters used in Equation (1) were derived from statistical analyses of survey data on even-aged plantation forests in the Kyushu region, which includes our study area. Thus, the equation is appropriate for assessing forest growth under local conditions; however, its applicability may be limited in areas with different forest types or management practices [29,30]. To simplify the calculation, we ignored the spatial distribution of the forest biomass productivity.

The cost of each operation was estimated from tree dimensions and spatial variables using published equations and parameters [31]. The costs, which cover labor, machinery, and materials, tend to be higher in forests located further from roads (Table A1). Furthermore, the study area was classified into six groups based on annual net income: (1) <0 JPY, (2) 0–10,000 JPY, (3) 10,000–20,000 JPY, (4) 20,000–30,000 JPY, and (5) >30,000 JPY.

2.4. Landslide Risk Analysis

Forest root systems enhance the frictional forces with the soil, thereby helping to prevent shallow landslides [19,32]. This service is especially critical in areas with steep slopes or complex terrain, where landslides are more likely to occur [33]. We evaluated the landslide risk of each forest cell against an assessment chart published by the Japan Forestry Agency [34]. This chart, which evaluates hazard risk on the basis of geology, topography, soil depth, and forest age, was created with expert knowledge and Hayashi's Quantification Method-II. Each forest cell was scored according to its spatial features

(Table A2) and classified as a high-risk area when the score was >125. We excluded high-risk areas from our sustainability calculations because logging should be avoided in such areas [32,35].

2.5. Other Areas to Be Excluded

Even within plantation forests, some areas are no longer managed for timber production owing to changes in social factors or the owners' intentions [36,37]. Such areas should also be excluded from the sustainability calculation. However, it is nearly impossible to identify the locations of abandoned plantation forests, because no comprehensive data exist. Therefore, we estimated the area where clear ownership is not established. Unclear ownerships often result from omissions during the registration process at the time of inheritance, for example, and represent an important social issue in Japan [38]. The Ministry of Land, Infrastructure, Transport and Tourism of Japan has reported the area ratio of forests with unclear ownership as 28.2% [39], so we used this value in our calculations.

2.6. Sustainability Analysis

We analyzed the sustainability of timber production in the study area excluding areas with landslide risk and unidentified ownership as:

$$R = (1 - L) \times (1 - O)/A \quad (2)$$

where R is the annual allowable logging area ratio, L is the ratio of forest area with a high risk of shallow landslide, O is the ratio of forest area with unclear ownership, and A is the standard logging age. R was calculated by excluding the ratios of forests with high landslide risk (L) and unclear ownership (O) from the total forest area, representing the proportion of land suitable for sustainable timber production. We assessed the sustainability of timber production by comparing the actual logged area with the estimated annual allowable logging area (Equation (2)). Cyclical forestry was assumed as the baseline concept to define sustainable harvesting. The annual allowable logging area was thus calculated as the area of forest available for timber production—excluding zones with landslide risk or with unclear ownership—divided by the standard logging age.

We used two types of standard logging age. The first was defined by the local forest owners' association as 50 years, based on their experience and general tree growth rates. The allowable logging area estimated under this value is referred to as the Annual Allowable Logging Area based on Experience (AAE).

The second type was determined by analyzing changes in the forest age–class distribution between 2015 and 2020. The decreasing area of forests over >40 years old during this period could be considered as logged area. The weighted average of the logged age was taken as the typical logged age, and the corresponding allowable area was referred to as Annual Allowable Logging Area based on Logged Age Distribution (AAL).

3. Results

3.1. Logged Locations

We identified the spatial distribution of logged forests in the study area (Figure A1), revealing a gradual increase in logged area over time. The logged area increased from approximately 1070 ha in 2000–2004, to 1550 ha in 2005–2009, 1710 ha in 2010–2014, and peaked at 2470 ha in 2015–2019. Overall, this represents a 130% increase in logged area over the two-decade span. The most pronounced rise occurred between 2010 and 2014 and 2015 and 2019, with an increase of about 44% during this final period. This trend suggests an acceleration in logging activities in recent years, likely influenced by national policies aimed at promoting timber production and increased harvesting in economically viable forest areas.

3.2. Profitability Analysis

Forest net income in the study area was estimated by using a forest growth model and an operational cost model (Figure A2). Net income was negative in 52.1% of the study area (Figure 2). No significant differences were found in the area distributions among the other profitability groups: 9.7% in the 0–10,000 JPY group, 12.6% in the 10,000–20,000 JPY group, 11.5% in the 20,000–30,000 JPY group, and 14.1% in the >30,000 JPY group.

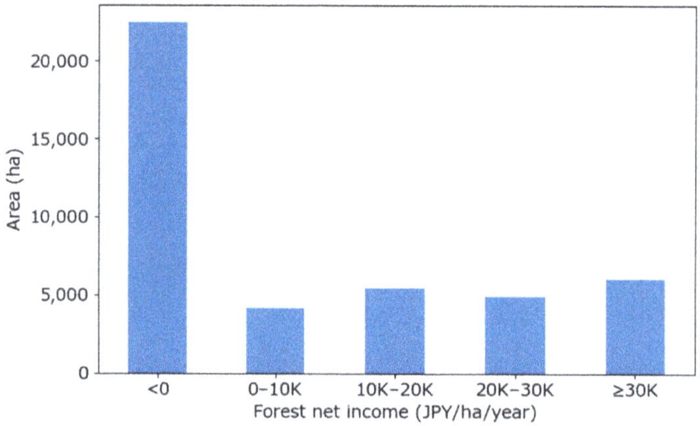

Figure 2. Area distribution of each profitability group.

3.3. Areas Excluded from Sustainability Analysis

Areas with a high risk of shallow landslide, areas lacking subsequent reforestation following logging, and areas with unidentified ownership were excluded from the sustainability analysis. In each profitability group, forests deemed high-risk due to landslide susceptibility accounted for 15%–20% (Table 1). As 28.2% of the forest area is classified as having unclear ownership, the forest area available for sustainable timber production was estimated to comprise between 58% and 61% of the total (Table 1).

Table 1. Forest area ratio available for sustainable timber production.

	Forest Net Income (JPY)				
	<0	0–10,000	10,000–20,000	20,000–30,000	30,000≤
High-risk of landslide	14.8%	17.4%	15.8%	16.3%	19.6%
Unclear ownership			28.2%		
Timber producing area	61.2%	59.3%	60.5%	60.1%	57.7%
AAE (5 years)	6.1%	5.9%	6.0%	6.0%	5.8%
AAL (5 years)	5.0%	4.8%	4.9%	4.9%	4.7%

3.4. Sustainability Analysis

Across all forests, the logged-area ratio rose steadily from about 2.4 % in 2000–2004 to 5.8 % in 2015–2019, thereby exceeding the AAL but remaining just below the AAE (6.0%) (Figure 3a). In the lowest-profitability class (<0 JPY), logging stayed below both thresholds until 2015–2019, when it surpassed the AAL but remained below the AAE (Figure 3b). Similarly, the 0–10,000, 10,000–20,000, and 20,000–30,000 JPY groups showed gradual increases and exceeded the AAL in 2015–2019, and the 0–10,000 JPY and 10,000–20,000 JPY groups also exceeded the AAE in 2015–2019 (Figure 3c–e). The >30,000 JPY class had the most pronounced rise: ratios exceeded both the AAL and AAE in 2015–2019. These patterns indicate that, although the study area as a whole has not yet transgressed the experience–derived sustainable limit, actual harvest rates in more profitable forest stands are already outpacing both empirically and experientially defined allowable rates.

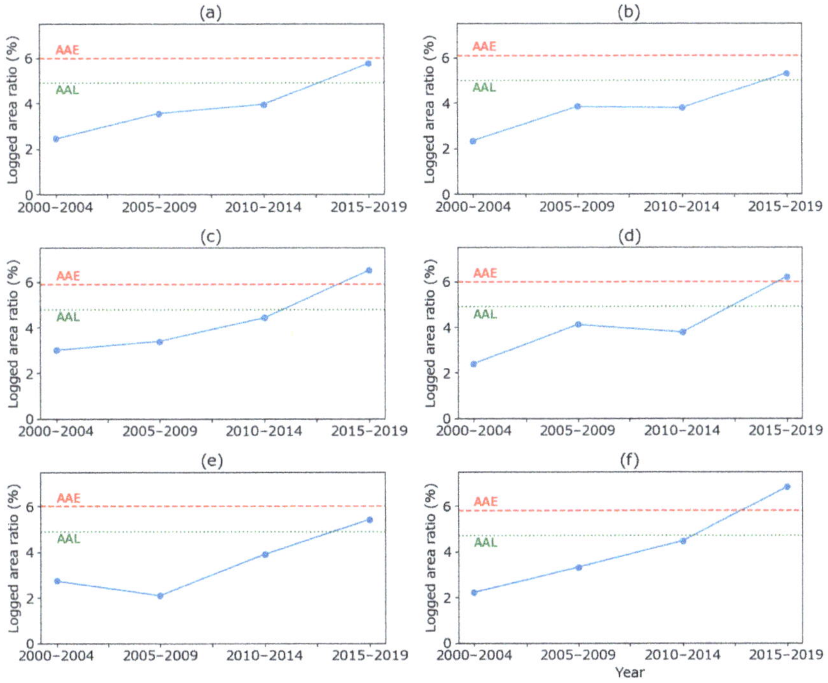

Figure 3. Comparison among logged area trends and annual allowable logging area in each profit group: (**a**) whole study area, (**b**) <0 JPY, (**c**) 0–10,000 JPY, (**d**) 10,000–20,000 JPY, (**e**) 20,000–30,000 JPY, and (**f**) ≥30,000 JPY. Blue lines, observed logged-area ratio (%); red lines, AAE; green lines, AAL (see Table 1).

4. Discussion

For the promotion of a nature-positive society, sustainable timber production in planted forests is fundamental [40]. However, national policies aimed at increasing timber production without sufficient consideration of local conditions may undermine the sustainability of local ESs [41,42]. Therefore, accurately assessing the impacts of such top-down policies on local timber production and ES provision is essential. We compared actual harvested areas with the annual allowable logging area, which was based on logging age while accounting for other ES supplies. Logged areas have significantly increased in recent years, and in some highly profitable forest regions the sustainability of timber production and associated ESs may have been compromised.

To better understand the underlying factors contributing to this potential decline in sustainability, we examined the relationship between forest profitability and logging activities. There was no clear relationship between profitability and logged area before 2010 (Figure 3). However, since 2010, economically advantageous forest areas have been more frequently logged. This aligns with the central government's timber production promotion policy, which emphasizes logging in profitable forest regions. Previous studies have suggested that logging tends to occur in areas with lower operational costs and better access to roads [43,44]. Such trends have become more apparent in the study area in recent years. These findings are consistent with those reported by Yamada and Fukumoto [24], who examined the broad-scale impacts of national policies in the same region. Overall, the results of this study clearly reveal the influence of central government policies on local logging patterns. In other words, this study empirically verifies the impact of top-down policy implementation on local forest management and ESs.

The sustainability assessment focused on plantation forests, excluding areas with a high risk of shallow landslides. Timber production can often be in a trade-off relationship with other ESs and may even exacerbate disaster risks [45,46]. Although high-risk forest areas have been logged in practice, we consider their exclusion in this analysis appropriate, as the sustainable provision of ESs is a fundamental goal of forest management. In fact, even in the study area, logging would not have exceeded the allowable limits if landslide risk had been disregarded (Table 1); however, such an approach would not guarantee the sustainable supply of ESs. Previous studies have also shown that spatially restricting logging to safer forest areas can improve ES outcomes [21,47]. With increasing recognition of the importance of NbS, it is essential to evaluate sustainability not merely in terms of resource quantities [48], but also in terms of whether timber production avoids undermining the balance among various ESs.

Similarly, social conditions are also critical for evaluating sustainability [25]. We excluded forests with unknown ownership under the assumption that they are unlikely to be actively managed. In addition, some forest lands are not used for timber production for social reasons such as landowners' intentions or conflicts with the provision of other ESs [49]. Therefore, incorporating the lives and behaviors of local residents is also essential for the effective implementation of NbS. Taken together, ecological and social dimensions must be jointly considered when we evaluate the sustainability of NbS, calling for indicators that can capture the complexity of human–nature interactions in forest landscapes [50]. The sustainability assessed here took into account both ESs and local social conditions and provides an example of a potential indicator for NbS in sustaining forest ES supplies.

The assessment highlighted deterioration in the sustainability of timber production in specific parts of the study area. Although the total timber production across the entire region did not exceed the allowable area based on experience, increased harvesting led to threshold violations in some economically profitable forests. This contrast is due to relatively low logging activity in unprofitable forests, which account for more than half of the study area. Focusing solely on the overall figures while overlooking spatial details may result in misjudgments regarding local sustainability. The concentration of logging in profitable areas could lead to resource depletion in those zones. Consequently, local forestry industries may face disruptions due to the unavailability of profitable timber [51]. Even in the presence of sufficient forest resources, timber production may stagnate under such conditions. While previous studies have evaluated the sustainability of timber production in terms of profitability [52,53] or current harvesting volumes [54], no studies to date have assessed sustainability by linking spatial logging patterns and profitability at the local scale, as conducted here.

These findings suggest that relying solely on top-down, production-oriented policies may overlook critical local dynamics. While timber production in the study area has increased as intended by the central government's policy, local logging patterns have inadvertently begun to compromise sustainability. The spatial imbalance in logging activities highlights the limitations of static, uniform policy frameworks. Adaptive management—including prediction, assessment, and improvement of policy effects [3,55]—offers an effective way to avoid such outcomes. Our results also show that top-down policies can have unintended consequences on local forest management. Therefore, future forest policy should incorporate more decentralized and participatory elements, empowering local actors to co-design and adjust management strategies [56,57].

5. Conclusions

This study assessed the sustainability of timber production in plantation forests by comparing actual logged areas with empirically and statistically defined allowable limits, while accounting for ecological risks and local social conditions. Our findings reveal a substantial increase in logging activities over the past two decades, with particularly intensive harvesting in economically profitable areas. Although overall timber production in the study area has not yet exceeded the allowable threshold based on experiential knowledge, logging in specific profitable zones has already surpassed sustainable limits, raising concerns over long-term forest resource depletion.

This study highlights the influence of top-down national policies on local forest management. The spatial imbalance in logging patterns, shaped by policy-driven incentives, underscores the limitations of uniform strategies that overlook local ecological and social complexities. By integrating environmental constraints such as landslide risk and social factors like unknown forest ownership, our sustainability assessment demonstrates the need for more nuanced, multi-dimensional indicators in evaluating NbS.

Whereas this study focused on spatial patterns of timber harvesting at the municipal scale, future research should evaluate sustainability within broader forest networks and in connection with climate change scenarios. A more comprehensive understanding of long-term sustainability will require integration of landscape-scale dynamics and climate resilience considerations [58]. In addition, further quantification of socio-political factors—such as the influence of subsidies and the environmental awareness of local communities—is essential for a more nuanced analysis of policy impacts [59]. Ultimately, the development of more multidimensional analytical approaches will be critical for proposing concrete management guidelines that support the realization of NbS in practice.

Author Contributions: Conceptualization, Y.Y. (Yusuke Yamada); methodology, Y.Y. (Yusuke Yamada), H.K., K.S., W.M.; formal analysis, Y.Y. (Yusuke Yamada), K.S.; writing—original draft preparation, Y.Y. (Yusuke Yamada); writing—review and editing, Y.Y. (Yuichi Yamaura), K.S.; visualization, Y.Y. (Yusuke Yamada); supervision, Y.Y. (Yuichi Yamaura); funding acquisition, Y.Y. (Yusuke Yamada). All authors have read and agreed to the published version of the manuscript.

Funding: This work was supported by JSPS KAKENHI Grant Number 23K11552.

Data Availability Statement: Logging location data are published in Zenodo at https://doi.org/10.5281/zenodo.4654619. The other data used in this study were obtained from local and prefectural government agencies and are not publicly available. However, they may be made available upon reasonable request and with permission from the respective data providers. Those interested in acquiring the data are encouraged to contact the corresponding author.

Acknowledgments: The authors thank the staff of Oita prefecture and Saiki municipality for providing information about the local forestry situations. During the preparation of this manuscript, the authors used ChatGPT (GPT-4-turbo) for the purposes of English editing. The authors have reviewed and edited the output and take full responsibility for the content of this publication.

Conflicts of Interest: The authors declare no conflicts of interest.

Abbreviations

The following abbreviations are used in this manuscript:

ES	Ecosystem service
NbS	Nature based solutions
AAE	Annual Allowable Logging Area based on Experience
AAL	Annual Allowable Logging Area based on Logged Age Distribution

Appendix A

Table A1. A sequential operational scheme to calculate forest net income.

Number of initial plantation trees	2500 trees/ha
Thinning ages	20, 30, 40 years
Thinning ratios	0.4, 0.4, 0.3
Logging age	50 years
Log price	12,500 JPY/m^3
Seedling price	75 JPY/tree
Labor cost	27,900 JPY/person·day
Machinery cost	50,000 JPY/day
Cutting time (C)	$N\ (107.3\ e^{0.0266D} + 90.2\ e^{0.0366D})$ s
Yarding time (Y)	$\frac{V}{8}\ (\frac{2}{3}R + 430)$ s
Working days	$\frac{C+Y}{3600*5*4}$ day
Transporting cost	2500 V JPY

N, numbers of trees; D, mean DBH (cm); V, tree volume (m^3); R, distance from roads (m).

Table A2. Score table for evaluating shallow landslide risk.

Terrain Element	Category	Geological Classification		
		Sedimentary	Volcanic	Hypabyssal
Slope	0%–30%	0	0	0
	31%–50%	12	31	23
	51%–70%	30	61	49
	71%–90%	42	71	66
	>91%	52	61	57
Profile curvature	<−0.01	22	31	28
	−0.01–+0.01	12	18	17
	>+0.01	4	16	0
Plan curvature	<150°	22	31	28
	150°–210°	12	12	15
	>211°	0	0	0
Soil depth	<0.5 m	0	0	0
	0.5–1.0 m	3	4	4
	1.0–2.0 m	7	10	9
	>2.0 m	15	20	19
Forest age	<5 years	25	35	32
	5–15 years	37	51	47
	15–35 years	28	37	34
	35–55 years	25	35	32
	>55 years	22	31	28

Table modified from MAFF [34].

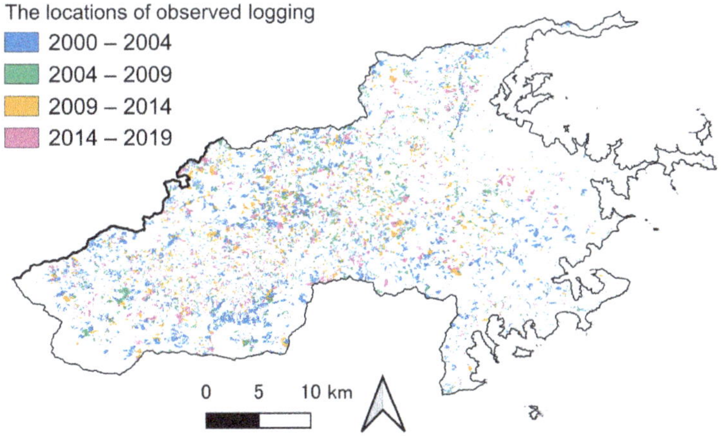

Figure A1. Locations of observed logging from 2000 to 2019.

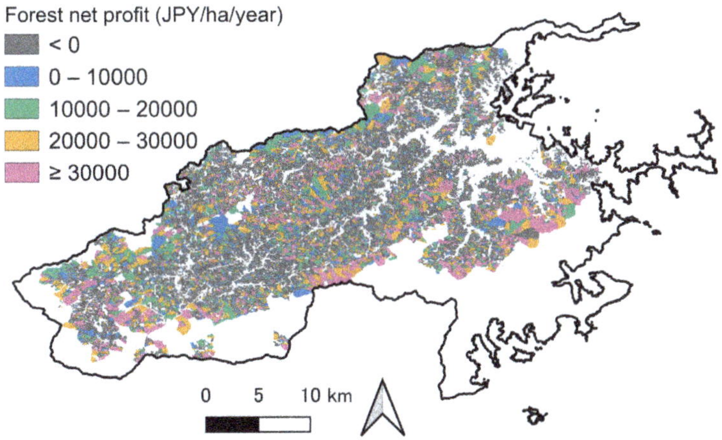

Figure A2. Evaluation map of forest net profit.

References

1. Kalantari, Z.; Ferreira, C.S.S.; Pan, H.; Pereira, P. Nature-Based Solutions to Global Environmental Challenges. *Sci. Total Environ.* **2023**, *880*, 163227. [CrossRef]
2. Seddon, N.; Daniels, E.; Davis, R.; Chausson, A.; Harris, R.; Hou-Jones, X.; Huq, S.; Kapos, V.; Mace, G.M.; Rizvi, A.R.; et al. Global Recognition of the Importance of Nature-Based Solutions to the Impacts of Climate Change. *Glob. Sustain.* **2020**, *3*, e15. [CrossRef]
3. Cohen-Shacham, E.; Andrade, A.; Dalton, J.; Dudley, N.; Jones, M.; Kumar, C.; Maginnis, S.; Maynard, S.; Nelson, C.R.; Renaud, F.G.; et al. Core Principles for Successfully Implementing and Upscaling Nature-Based Solutions. *Environ. Sci. Policy* **2019**, *98*, 20–29. [CrossRef]
4. Brockerhoff, E.G.; Barbaro, L.; Castagneyrol, B.; Forrester, D.I.; Gardiner, B.; González-Olabarria, J.R.; Lyver, P.O.B.; Meurisse, N.; Oxbrough, A.; Taki, H.; et al. Forest Biodiversity, Ecosystem Functioning and the Provision of Ecosystem Services. *Biodivers. Conserv.* **2017**, *26*, 3005–3035. [CrossRef]
5. Harrison, P.A.; Vandewalle, M.; Sykes, M.T.; Berry, P.M.; Bugter, R.; de Bello, F.; Feld, C.K.; Grandin, U.; Harrington, R.; Haslett, J.R.; et al. Identifying and Prioritising Services in European Terrestrial and Freshwater Ecosystems. *Biodivers. Conserv.* **2010**, *19*, 2791–2821. [CrossRef]
6. IPBES. *IPBES Global Assessment Report on Biodiversity and Ecosystem Services of the Intergovernmental Science-Policy Platform on Biodiversity and Ecosystem Services*; IPBES: Bon, Germany, 2019.

7. Santika, T.; Wilson, K.A.; Budiharta, S.; Kusworo, A.; Meijaard, E.; Law, E.A.; Friedman, R.; Hutabarat, J.A.; Indrawan, T.P.; St. John, F.A.V.; et al. Heterogeneous Impacts of Community Forestry on Forest Conservation and Poverty Alleviation: Evidence from Indonesia. *People Nat.* **2019**, *1*, 204–219. [CrossRef]
8. Knoke, T.; Messerer, K.; Paul, C. The Role of Economic Diversification in Forest Ecosystem Management. *Curr. For. Rep.* **2017**, *3*, 93–106. [CrossRef]
9. Pohjanmies, T.; Triviño, M.; Le Tortorec, E.; Mazziotta, A.; Snäll, T.; Mönkkönen, M. Impacts of Forestry on Boreal Forests: An Ecosystem Services Perspective. *Ambio* **2017**, *46*, 743–755. [CrossRef]
10. Gauthier, S.; Bernier, P.; Kuuluvainen, T.; Shvidenko, A.Z.; Schepaschenko, D.G. Boreal Forest Health and Global Change. *Science* **2015**, *349*, 819–822. [CrossRef]
11. Japan Forestry Agency. *The Basic Plan on Forest and Forestry*; Japan Forestry Agency: Tokyo, Japan, 2021.
12. Japan Forestry Agency. *Annual Report on Forest and Forestry in Japan (Fiscal Year 2023)*; Japan Forestry Agency: Tokyo, Japan, 2023.
13. Börner, J.; Schulz, D.; Wunder, S.; Pfaff, A. The Effectiveness of Forest Conservation Policies and Programs. *Annu. Rev. Resour. Econ.* **2020**, *12*, 45–64. [CrossRef]
14. Macpherson, E.; Mortiaux, R.; Jorgensen, E. Scale-Sensitive Marine Law and Policy Design: Towards Ecosystem-Based Management across Spatial and Temporal Scales. *Wiley Interdiscip. Rev. Energy. Environ.* **2024**, *13*, e537. [CrossRef]
15. Jensen, T.; Lussier, J.M. Economic Impacts of Partial Harvesting: Mitigating Mid-Term Timber Supply Shortages as a Result of Pest Outbreaks. *For. Chron.* **2021**, *97*, 271–276. [CrossRef]
16. Olschewski, R.; Klein, A.M.; Tscharntke, T. Economic Trade-Offs between Carbon Sequestration, Timber Production, and Crop Pollination in Tropical Forested Landscapes. *Ecol. Complex.* **2010**, *7*, 314–319. [CrossRef]
17. Carpentier, S.; Filotas, E.; Handa, I.T.; Messier, C. Trade-Offs between Timber Production, Carbon Stocking and Habitat Quality When Managing Woodlots for Multiple Ecosystem Services. *Environ. Conserv.* **2017**, *44*, 14–23. [CrossRef]
18. Jucker, T.; Bouriaud, O.; Avacaritei, D.; Coomes, D.A. Stabilizing Effects of Diversity on Aboveground Wood Production in Forest Ecosystems: Linking Patterns and Processes. *Ecol. Lett.* **2014**, *17*, 1560–1569. [CrossRef]
19. Noviandi, R.; Gomi, T.; Pratama, M.; Ritonga, R.P.; Fathani, T.F.; Maylda Pratama, G. Understanding the Role of Vegetation Root Systems in the Initiation of Rainfall-Induced Shallow Landslides: Scaling Perspectives. *J. For. Res.* **2025**, *30*, 165–178. [CrossRef]
20. WWF. *Global Roadmap for a Nature-Positive Economy an Economic and Financial Reform Agenda to Meet Narure and Climate Goals*; WWF: Gland, Switzerland, 2024.
21. Yamaura, Y.; Yamada, Y.; Matsuura, T.; Tamai, K.; Taki, H.; Sato, T.; Hashimoto, S.; Murakami, W.; Toda, K.; Saito, H.; et al. Modeling Impacts of Broad-Scale Plantation Forestry on Ecosystem Services in the Past 60 Years and for the Future. *Ecosyst. Serv.* **2021**, *49*, 101271. [CrossRef]
22. Polyakov, M.; Wear, D.N.; Huggett, R.N. Harvest Choice and Timber Supply Models for Forest Forecasting. *For. Sci.* **2010**, *56*, 344–355. [CrossRef]
23. Prestemon, J.P.; Wear, D.N. Linking Harvest Choices to Timber Supply. *For. Sci.* **2000**, *46*, 377–389. [CrossRef]
24. Yamada, Y.; Fukumoto, K. Spatial Relationships between Annual Timber Productions and Logging Locations in the Prefectures of Kyushu Island. *J. Jpn. For. Soc.* **2023**, *105*, 259–263. [CrossRef]
25. Yamada, Y.; Yamaura, Y.; Shimizu, K.; Murakami, W.; Nanko, K.; Takayama, N. Conflicts among Ecosystem Services May Depend on Environmental Awareness: A Multi-Municipality Analysis. *Forestry* **2024**, *97*, 424–435. [CrossRef]
26. Dataset: Country-Wide Mapping of Harvest Areas and Post-Harvest Forest Recovery Using Landsat Time Series Data in Japan. Available online: https://zenodo.org/records/11634214 (accessed on 22 February 2023).
27. Shimizu, K.; Saito, H. Country-Wide Mapping of Harvest Areas and Post-Harvest Forest Recovery Using Landsat Time Series Data in Japan. *Int. J. Appl. Earth Obs. Geoinf.* **2021**, *104*, 102555. [CrossRef]
28. Gorelick, N.; Hancher, M.; Dixon, M.; Ilyushchenko, S.; Thau, D.; Moore, R. Google Earth Engine: Planetary-Scale Geospatial Analysis for Everyone. *Remote Sens. Environ.* **2017**, *202*, 18–27. [CrossRef]
29. Japan Forestry Agency. *Procedures for Harvest Forecasts Using Forest Stand Density Control Diagram*; Japan Forestry Agency: Tokyo, Japan, 2024.
30. Nishizono, T.; Inoue, A.; Hosoda, K. Relationship between Relative Yield Index and Relative Spacing Index: Theoretical Derivation of the Relationship and Its Characteristics. *J. Jpn. For. Plann.* **2013**, *47*, 16–28. [CrossRef]
31. Oka, M. Studies on the Analysis and Evaluation on Mechanized Yarding. Ph.D. Thesis, University of Tokyo, Tokyo, Japan, 2006.
32. Saito, H.; Murakami, W.; Daimaru, H.; Oguchi, T. Effect of Forest Clear-Cutting on Landslide Occurrences: Analysis of Rainfall Thresholds at Mt. Ichifusa, Japan. *Geomorphology* **2017**, *276*, 1–7. [CrossRef]
33. Shuin, Y.; Turumi, K.; Matsue, K.; Aruga, K.; Tasaka, T. Evaluation of Soil Reinforcement by Tree Root System Using a Distributed Shallow Landslides Model. *J. Jpn. Soc. Reveget. Technol.* **2009**, *35*, 9–14. [CrossRef]
34. MAFF. *Investigation Method of the Disaster Risk District in the Mountain*; MAFF: Tokyo, Japan, 2016.
35. Shuin, Y.; Hotta, N.; Matsue, K.; Aruga, K.; Tasaka, T. Simulating Effects of Spatial Distribution of Different Stand Age of Hinoki Cypress on the Location of Shallow Landslides. *J. Jpn. Soc. Reveget. Technol.* **2011**, *37*, 102–107. [CrossRef]

36. Heinonen, T.; Pukkala, T.; Asikainen, A. Variation in Forest Landowners' Management Preferences Reduces Timber Supply from Finnish Forests. *Ann. For. Sci.* **2020**, *77*, 31. [CrossRef]
37. Onda, N.; Ochi, S.; Tsuzuki, N. Examination of Social Factors Affecting Private Forest Owners' Future Intentions for Forest Management in Miyazaki Prefecture: A Comparison of Regional Characteristics by Forest Ownership Size. *Forests* **2023**, *14*, 309. [CrossRef]
38. Yoshihara, S. Realities and Challenges of Land Issues in the Era of Depopulation. *Jpn. J. Real. Estate Sci.* **2017**, *31*, 79–83. [CrossRef]
39. The Ministry of Land Infrastructure Transport and Tourism of Japan. *Special Measures Law for Facilitating the Use of Land with Unknown Owners*; The Ministry of Land Infrastructure Transport and Tourism of Japan: Tokyo, Japan, 2019.
40. Demarais, S.; Verschuyl, J.P.; Roloff, G.J.; Miller, D.A.; Wigley, T.B. Tamm Review: Terrestrial Vertebrate Biodiversity and Intensive Forest Management in the U.S. *For. Ecol. Manag.* **2017**, *385*, 308–330. [CrossRef]
41. Benra, F.; Nahuelhual, L.; Gaglio, M.; Gissi, E.; Aguayo, M.; Jullian, C.; Bonn, A. Ecosystem Services Tradeoffs Arising from Non-Native Tree Plantation Expansion in Southern Chile. *Landsc. Urban. Plan.* **2019**, *190*, 103589. [CrossRef]
42. Blattert, C.; Eyvindson, K.; Hartikainen, M.; Burgas, D.; Potterf, M.; Lukkarinen, J.; Snäll, T.; Toraño-Caicoya, A.; Mönkkönen, M. Sectoral Policies Cause Incoherence in Forest Management and Ecosystem Service Provisioning. *For. Policy. Econ.* **2022**, *136*, 102689. [CrossRef]
43. Levers, C.; Verkerk, P.J.; Müller, D.; Verburg, P.H.; Butsic, V.; Leitão, P.J.; Lindner, M.; Kuemmerle, T. Drivers of Forest Harvesting Intensity Patterns in Europe. *For. Ecol. Manag.* **2014**, *315*, 160–172. [CrossRef]
44. Poje, A.; Pezdevšek Malovrh, Š.; Krč, J. Factors Affecting Harvesting Intensity in Small-Scale Private Forests in Slovenia. *Small-scale For.* **2016**, *15*, 73–91. [CrossRef]
45. Bennett, E.M.; Peterson, G.D.; Gordon, L.J. Understanding Relationships among Multiple Ecosystem Services. *Ecol. Lett.* **2009**, *12*, 1394–1404. [CrossRef] [PubMed]
46. Temperli, C.; Blattert, C.; Stadelmann, G.; Brändli, U.B.; Thürig, E. Trade-Offs between Ecosystem Service Provision and the Predisposition to Disturbances: A NFI-Based Scenario Analysis. *For. Ecosyst.* **2020**, *7*, 27. [CrossRef]
47. Frizzle, C.; Fournier, R.A.; Trudel, M.; Luther, J.E. Towards Sustainable Forestry: Using a Spatial Bayesian Belief Network to Quantify Trade-Offs among Forest-Related Ecosystem Services. *J. Environ. Manag.* **2022**, *301*, 113817. [CrossRef]
48. Leiter, M.; Neumann, M.; Egusa, T.; Harashina, K.; Hasenauer, H. Assessing the Resource Potential of Mountainous Forests: A Comparison between Austria and Japan. *Forests* **2022**, *13*, 891. [CrossRef]
49. Ficko, A. Private Forest Owners' Social Economic Profiles Weakly Influence Forest Management Conceptualizations. *Forests* **2019**, *10*, 956. [CrossRef]
50. Palomo, I.; Locatelli, B.; Otero, I.; Colloff, M.; Crouzat, E.; Cuni-Sanchez, A.; Gómez-Baggethun, E.; González-García, A.; Grêt-Regamey, A.; Jiménez-Aceituno, A.; et al. Assessing Nature-Based Solutions for Transformative Change. *One Earth* **2021**, *4*, 730–741. [CrossRef]
51. Bowe, S.A.; Ballweg, J.M.; Blinn, C.R.; Nolle, D.A.; Smail, R.A. The Logging Sector in Minnesota and Wisconsin: Status, Issues, and Challenges in 2021. *J. For.* **2025**, *123*, 41–62. [CrossRef]
52. Zhang, J.; Alavalapati, J.R.R.; Shrestha, R.K.; Hodges, A.W. Economic Impacts of Closing National Forests for Commercial Timber Production in Florida and Liberty County. *J. For. Econ.* **2005**, *10*, 207–223. [CrossRef]
53. Alounsavath Master, P.; Kim, S. Bin Economic Feasibility of a Sustainable Production Forest Management System in Xaibouathong Forest Management Area, Khammouan Province, Lao PDR. *Forest Sci. Technol.* **2021**, *17*, 119–124. [CrossRef]
54. Macpherson, A.J.; Carter, D.R.; Schulze, M.D.; Vidal, E.; Lentini, M.W. The Sustainability of Timber Production from Eastern Amazonian Forests. *Land Use Policy* **2012**, *29*, 339–350. [CrossRef]
55. Deitch, M.J.; Gancel, H.N.; Croteau, A.C.; Caffrey, J.M.; Scheffel, W.; Underwood, B.; Muller, J.W.; Boudreau, D.; Cantrell, C.G.; Posner, M.J.; et al. Adaptive Management as a Foundational Framework for Developing Collaborative Estuary Management Programs. *J. Environ. Manag.* **2021**, *295*, 113107. [CrossRef]
56. Reed, M.S. Stakeholder Participation for Environmental Management: A Literature Review. *Biol. Conserv.* **2008**, *141*, 2417–2431. [CrossRef]
57. Rico, M.; González, A. Social Participation into Regional Forest Planning Attending to Multifunctional Objectives. *For. Policy Econ.* **2015**, *59*, 27–34. [CrossRef]
58. Maxwell, C.J.; Scheller, R.M.; Wilson, K.N.; Manley, P.N. Assessing the Effectiveness of Landscape-Scale Forest Adaptation Actions to Improve Resilience under Projected Climate Change. *Front. For. Glob. Change* **2022**, *5*, 740869. [CrossRef]
59. Miranda, K.M.; Parry, I.W.H.; Gupta, S. Public Expenditure Policy and the Environment: A Review and Synthesis. *IMF Work. Pap.* **1993**, *93*, 42. [CrossRef]

sustainability

MDPI

Article

Prioritisation of Native Tree Species for Biodiversity Conservation, Carbon Capture, and Livelihoods Improvement in Shade-Grown Coffee Regions of Chiapas, Mexico

María Guadalupe Chávez Hernández [1], César Mateo Flores-Ortiz [1,*], Robert Hunter Manson [2], María Toledo-Garibaldi [2], Maraeva Gianella [3,4] and Tiziana Ulian [3,4,*]

1 Laboratorio de Fisiología Vegetal, Unidad de Biotecnología y Prototipos (UBIPRO), FES Iztacala, Universidad Nacional Autónoma de México (UNAM), Tlalnepantla 54090, Mexico; biomariachavez@gmail.com
2 Red de Ecología Funcional, Instituto de Ecología, A.C., Carretera Antigua a Coatepec No. 351, El Haya, Veracruz 91073, Mexico
3 Department of Life Sciences and Systems Biology, University of Turin, 10124 Turin, Italy
4 Royal Botanic Gardens, Kew, Wakehurst, Ardingly RH17 6TN, UK
* Correspondence: cmflores@unam.mx (C.M.F.-O.); tiziana.ulian@unito.it (T.U.)

Abstract

Coffee production, particularly in shade-grown farms, plays a crucial role in the livelihoods of Mexican farmers. Shade-grown coffee systems are also recognised for supporting biodiversity and enhancing carbon capture. Nevertheless, the geographical heterogeneity of Mexico makes the selection of tree species in these agroforestry systems challenging. This study develops region-specific priority lists to conserve biodiversity, improve carbon capture, and support the livelihoods of producers across nine coffee-growing regions within the state of Chiapas. We identified the tree species distributed in each region using an extensive dataset from the Global Biodiversity Information Facility and a novel approach that enhanced spatial resolution of the prioritisation process, despite biases in collection efforts. A set of 23 criteria, including conservation status, carbon content, and documented uses by local communities, was compiled from databases and literature reviews and used to calculate a priority score for each species. Based on these scores, a list of 20 recommended species was generated for each region. However, additional participatory validation is needed to translate these lists into practice. A similarity analysis revealed that geographically proximate regions shared similar species composition. Overall, this study provides a transparent framework for regionally tailored shade-tree selection to inform conservation and restoration planning in coffee agroforestry landscapes.

Keywords: agroforestry; Arabic coffee; arboreal diversity; small-scale coffee production; useful plants

1. Introduction

Mexico is one of the leading coffee producers in Latin America, with 15 coffee-growing states and approximately 564,445 hectares of productive land [1]. In addition to its production volume, Mexico is recognised for its high-quality coffee, which significantly contributes to national income through both domestic and export markets [2,3]. However, this sector has faced persistent challenges over recent decades, including producers abandoning their land or switching to other crops, coffee leaf rust [4], fluctuating coffee prices exacerbated by the COVID-19 pandemic, and the increasing impacts of climate change [4–7]. These pressures align with national and international policy priorities that promote climate-resilient agricultural systems and nature-based solutions in productive landscapes.

Most coffee produced in Mexico is *Coffea arabica* L. (85%) and is managed as rustic systems or traditional and commercial polycultures, characterised by varying proportions of native and introduced shade trees and different levels of canopy cover [8]. These agroforestry systems maintain a diverse and relatively dense tree cover, with <10% of the coffee farms in Mexico managed as sun coffee [9]. These shade-grown systems are predominantly managed by smallholder farmers (90% <2.5 ha), often from indigenous communities, whose practices support both agricultural resilience and socio-cultural continuity [10]. Because most shade coffee is produced in smallholder and indigenous territories, interventions promoting native shade trees can also contribute to rural development, equity, and culturally appropriate resource management policies.

Unlike coffee sun coffee or shade monocultures, coffee farms with diversified shade offer substantial environmental benefits. They support native biodiversity, provide critical ecosystem services such as carbon capture, and enable income diversification through the sustainable use of native tree species [9,11–13]. Additionally, diversified coffee agroforestry systems enhance soil health and boost ecosystem services and resilience, as the specific composition of shade trees determines soil-function co-benefits—such as enhanced nutrient cycling through deep-root capture and soil aggregation via consistent organic inputs—which are essential for sustaining the biodiversity and carbon storage potential of coffee agroforestry systems [14].

Mexico's principal coffee-growing regions overlap with areas of high biological and biocultural richness, highlighting the importance of promoting reforestation and biodiversity conservation strategies within these agroecosystems [11,15]. These multiple benefits directly support international commitments on biodiversity and climate, including the Global Biodiversity Framework and national climate strategies that emphasise sustainable land use and ecosystem service protection.

Tree diversification contributes to biodiversity conservation, ecological restoration, and the enhancement of carbon capture and can provide local communities with valuable resources such as edible or medicinal plants [16–19]. Diversifying the shade trees in these environments not only increases tree diversity but also promotes interactions with other groups of organisms, such as birds and mycorrhizal fungi, as has been reported for the Soconusco region in Chiapas [20]. Such ecological outcomes are increasingly recognised in restoration-oriented policies that prioritise native species, ecosystem functionality, and biodiversity monitoring in productive landscapes.

In this context, enhancing native tree species in coffee farms has gained attention as a promising approach for strengthening agroforestry systems, due to tree diversity's link with resilience to climate change in forest ecosystems [21]. The increasing recognition of shade-grown coffee as a biodiversity-friendly product has encouraged the development of certifications and market-based incentives aimed at improving environmental and social outcomes. For instance, the recent "Café de Sombra Natural" (CSN) certification by Mexico's Ministry of Environment and Natural Resources [22] includes a catalogue of native tree species to guide reforestation and biodiversity conservation on certified farms. However, operationalising such schemes requires transparent, regionally appropriate, species selection tools that balance conservation priorities, carbon outcomes, and farmer utility. A growing number of projects have been carried out to select suitable tree species for reforestation in coffee farms, proposing methodologies to prioritise native taxa with high carbon capture capacity and documented local uses [23–25]. Shade-tree diversification and participation in payment programmes for environmental services, such as

 https://doi.org/10.3390/su18073511

carbon or biodiversity credits, have also been proposed as viable strategies to enhance producers' income and prevent the intensification of coffee production or the conversion of shade coffee into other crop monocultures [13]. Priority lists grounded in biodiversity data and evidence of local-use can improve the credibility and adoption of these policy instruments.

Chiapas is Mexico's leading coffee-producing state, accounting for 41% of the country's total coffee production [3]. Coffee farms in this state have an average shade density of 47 to 53% [26]. Chiapas is also one of Mexico's two hotspots of tree biodiversity, making it a priority region for targeted conservation and reforestation actions [27].

Given their geographical extent, high arboreal diversity, and overlap with montane cloud forests, shade-grown coffee areas in Chiapas represent excellent opportunities for mitigating climate change and protecting native biodiversity [28]. However, achieving these outcomes requires regional species selection aligned with national and international commitments/goals, such as biodiversity conservation, climate change mitigation, and sustainable rural development. Additionally, practical considerations, such as the extent of canopy cover, must be considered to avoid compromising coffee yields due to excessive shade [13].

Selecting regionally suitable tree species for shade coffee farms is a key step in successful diversification while maintaining coffee productivity and quality at small-scale coffee farms [29,30]. To achieve this, developing a regional prioritisation method that incorporates species records is a critical initial step, both increasing methodological resolution and enhancing the efficiency of the selection process. The objective of this study was to provide an initial regional prioritisation for the nine coffee-growing areas in Chiapas, focusing on optimising tree cover with three goals: (1) preserving native and endangered biodiversity, (2) enhancing carbon capture, and (3) improving the livelihoods of producers through selecting useful taxa. By providing region-specific, evidence-based species lists, this study offers decision support for certification standards, restoration programmes, and incentive schemes targeting sustainable coffee landscapes.

2. Materials and Methods

2.1. Study Area

Species prioritisations were developed using the same algorithm for each of the nine coffee regions in Chiapas: Altos, Centro, Fraylesca, Fronteriza, Istmo-Costa, Norte, Selva, Sierra, and Soconusco (Figure 1, Table 1). These regions were proposed by Hernández-Martínez et al. [26] and reflect a diverse range of agro-climatic, production, and socioeconomic conditions in each coffee-producing region (Table 1).

Table 1. Climatic information and the number of records and species obtained from tree databases to develop the prioritisations for the nine coffee regions selected in Chiapas, Mexico. Sources: Hernández-Martínez et al. [26], GBIF [31], SEMARNAT [22].

Region	Mean Annual Rainfall (mm)	Mean Annual Temperature (°C)	Average Altitude (m)	Number of Records	Density of Records per Hectare	Native Tree Species Considered in Prioritisations	Native Tree Species Reported in Coffee Farms
Altos	1894	20	1332	4253	0.24	416	149
Centro	1405	23	978	7523	0.61	860	238
Fraylesca	1903	22	1252	7195	0.30	588	194
Fronteriza	2632	22	920	3914	0.36	537	176
Istmo-Costa	2423	24	795	430	0.02	182	84

https://doi.org/10.3390/su18073511

Table 1. *Cont.*

Region	Mean Annual Rainfall (mm)	Mean Annual Temperature (°C)	Average Altitude (m)	Number of Records	Density of Records per Hectare	Native Tree Species Considered in Prioritisations	Native Tree Species Reported in Coffee Farms
Norte	2113	22	1047	3759	0.17	463	171
Selva	2312	22	988	7643	0.15	728	209
Sierra	1762	21	1285	3868	0.09	584	195
Soconusco	3754	24	776	6177	0.18	670	220
Total	-	-	-	44,762		1362	286

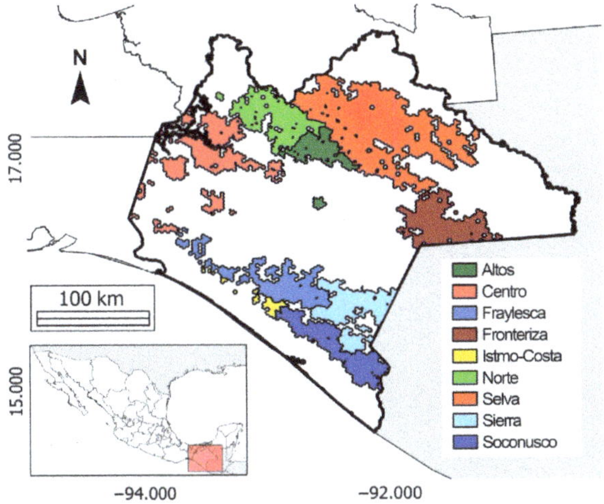

Figure 1. Map of nine coffee-producing regions studied in the state of Chiapas (thicker line), Mexico, identified by Hernández-Martínez et al. [26].

2.2. Data Collection

The list of native tree species (*n* = 2934 taxa for Mexico) used in this study is an updated version of the checklist published by Telléz et al. [27], adding taxa described in recent years and new records from bibliographic reviews. Georeferenced records of individuals of each tree species were retrieved from the Global Biodiversity Information Facility (GBIF) [31], and the dataset from Téllez et al. [27]. All records were cleaned using the R (version 2.6-8) package "CoordinateCleaner" [32], which deletes duplicates and records in the sea (scale of the default reference = 110), using country centroids (buffer = 1 km), near country capitals (buffer = 10 km), and within museums or botanic gardens (buffer = 100 m). Subsequently, cleaned records (*n* = 956,410) were filtered to the polygons of the coffee regions and used to identify the native species distributed in each one of the nine coffee regions using QGIS 3.36.3 [33] (Figure 2). An approximation of the regional abundance of the tree native species was obtained from the number of records of the species within each region and considered a relevant criterion for the species selection [34], together with previous reports of their use as shade trees in coffee farms (Table 2, Supplementary Material Table S1). Previous analyses were conducted considering only the presence/absence of species in coffee-growing regions resulting in algorithms that lacked sufficient resolution due to incidental records of species with high utility values. Species reported in coffee farms were obtained from an exhaustive bibliographic review [35–51] and are included in the Supplementary Material Table S1. Additional data were incorporated for all taxa on the list to provide the criteria for prioritisation and to align with the three main goals of the

https://doi.org/10.3390/su18073511

project: Biodiversity conservation, Carbon capture, and Improving livelihoods. No specific adjustments were made for missing values, since scores were calculated using only the available information for each species.

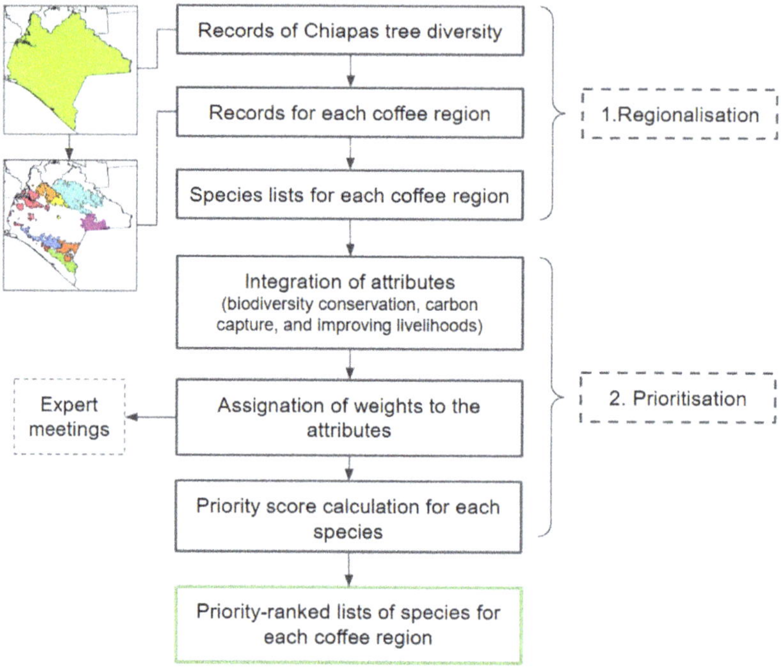

Figure 2. Flow chart of the proposed methodology to generate specific lists of priority tree species in each coffee region of Chiapas.

2.2.1. Biodiversity Conservation Goal

A total of six attributes related to biodiversity conservation were considered (Table 2). These included: the distribution and endemism in Mexico and Chiapas obtained from Villaseñor [55]; species conservation status from the IUCN Red List of Threatened Species [56], and the Mexican regional assessments were recovered from the "Norma Oficial Mexicana NOM-059-SEMARNAT-2010" [57]; the status of protection in situ and ex situ was downloaded using the BGCI tool "Global Tree Portal" [58].

2.2.2. Carbon Capture Goal

Seven attributes were used to describe variation in carbon capture potential (Table 2). Carbon content (the fraction of carbon in dry biomass) of each tree species was obtained from the GLOWCAD database [52]. Values reported as a percentage (0 to 100) were divided by 100 to standardise values prior to prioritisation. Three growth rate categories (fast, medium and slow) were proposed following the family information from Condit et al. [59], and the report of use as fuel was derived from the World Checklist of Useful Plant Species [60]. Mean biomass was calculated for species with data reported in the National Forest and Soil Inventory from 2015 to 2020 [53]. Wood density data were obtained from the public dataset published by Ricker et al. [54]. The biomass and wood density data were filtered to include only those measurements that met the definition of a tree proposed by Téllez et al. [27]. Means per species were then calculated, and the data were subsequently standardised using the method min-max normalisation prior to prioritisation. High carbon capture capacity and recommendations for reforestation information were incorporated

 https://doi.org/10.3390/su18073511

into the dataset as binary variables (0–1) based on an intensive literature review building on the revision made by Pompa-García and Sigala-Rodríguez [61], including more than 75 papers.

Table 2. Attributes considered in the prioritisations and associated normalised values obtained for all tree species considered in the study from 0 to 1. CR: Critically Endangered. EN: Endangered. VU: Vulnerable. NT: Near Threatened. See the text for more information on each attribute.

Goal	Attribute	Attribute Values			
Common	1. Reported in shade-grown coffee farms (SC)	Reported (1)		Not reported (0)	
	2. Regional abundance (RA)	Normalised abundance per species per region (0–1)			
Biodiversity conservation (BC)	3. Endemism to Mexico (EM)	Endemic (1)		Not endemic (0)	
	4. Endemism to Chiapas (EC)	Endemic (1)		Not endemic (0)	
	5. IUCN category (RL)	CR (1)	EN (0.75)	VU (0.5)	NT (0.25)
	6. NOM-059 category (NC)	P (1)	A (0.66)		Pr (0.33)
	7. Protected in situ (Pi)	Not protected (1)		Protected (0)	
	8. Protected ex situ (Pe)	Not protected (1)		Protected (0)	
Carbon capture (CC)	9. Carbon content (CCo)	Value given by Doraisami et al.'s [52] dataset (0–1)			
	10. Growth rate (GR)	Fast (1)	Medium (0.5)		Slow (0)
	11. Use as fuel (UF)	Not used (1)		Used (0)	
	12. Biomass per species (B)	Standardised value given by CONAFOR's [53] dataset (0–1)			
	13. Mean wood density (WD)	Standardised value given by Ricker's et al. [54] dataset (0–1)			
	14. High carbon capture (HC)	Yes (1)		No (0)	
	15. Recommended for reforestation (RR)	Yes (1)		No (0)	
Improving livelihoods (IL)	16. Use as animal food (AF)	Reported (1)		Not reported (0)	
	17. Use as food (F)	Reported (1)		Not reported (0)	
	18. Use as material (M)	Reported (1)		Not reported (0)	
	19. Use as medicine (Me)	Reported (1)		Not reported (0)	
	20. Use as poison (P)	Reported (1)		Not reported (0)	
	21. Social uses (SU)	Reported (1)		Not reported (0)	
	22. Nitrogen fixing (NF)	Reported (1)		Not reported (0)	
	23. Presence in cloud forest (CF)	Present (1)		Absent (0)	

2.2.3. Improving Livelihoods Goal

Eight additional attributes were used to describe information relative to improving livelihoods. Information on particular uses was obtained from the World Checklist of Useful Plant Species [60], considering six of the ten categories proposed: animal food, food, materials, medicines, poisons, and social uses. Taxa belonging to the Fabaceae family were considered potential nitrogen-fixers. Lastly, species' presence in cloud forests was added from an extensive bibliographical review and field observations made by the authors.

2.3. Prioritisation and Weighting of the Criteria

Nine independent prioritisations were generated using the 23 attributes described in Table 2, one for each coffee region in Chiapas. Three online meetings were held to determine

https://doi.org/10.3390/su18073511

the weighting of the criteria according to the study's goals (October and November 2024). All co-authors (researchers with extensive experience in coffee agroforestry, forest ecology, ecosystem services, and sustainable land management) participated in the assignment of attribute weights. These meetings followed the Delphi methodology modified from Garson [62], as well as the protocol established in Flores-Ortiz et al. [25]. After at least three iterative rounds of discussion, new versions of species lists, incorporating participant feedback, were presented. The process was repeated until the divergence between expert opinions was reduced and a consensus was reached. Since this study focused on coffee farms, the first criterion, "Reported in shade-grown coffee farms", was relevant to the project's three goals. The second criterion, "Regional abundance", was also relevant to all the goals. Table 3 and Formulas (1)–(4) summarise the final weighting used on the prioritisations. To avoid bias each goal was assigned 33% of the total influence on the final priority score.

Table 3. Consensus weighting of attributes defined in expert meetings and subsequently used in prioritisations of tree species for enriching shade cover in nine coffee regions in Chiapas, Mexico.

Attribute/Goal	Biodiversity Conservation	Carbon Capture	Improving Livelihoods
Reported in shade coffee farms (SC)	0.1	0.1	0.15
Regional abundance (RA)	0.1	0.1	0.1
Endemism to Mexico (EM)	0.1	0	0
Endemism to Chiapas (EC)	0.2	0	0
IUCN category (RL)	0.2	0	0
NOM-059 category (NC)	0.1	0	0
Protected in situ (Pi)	0.1	0	0
Protected ex situ (Pe)	0.1	0	0
Carbon content (CCo)	0	0.1	0
Growth rate (GR)	0	0.175	0
Use as fuel (UF)	0	0.1	0
Biomass per species (B)	0	0.1	0
Mean wood density (WD)	0	0.05	0
High carbon capture (HC)	0	0.175	0
Recommended for reforestation (RR)	0	0.1	0
Use as animal food (AF)	0	0	0.1
Use as food (F)	0	0	0.15
Use as material (M)	0	0	0.05
Use as medicine (Me)	0	0	0.1
Use as poison (P)	0	0	0.05
Social uses (SU)	0	0	0.1
Nitrogen fixing (NF)	0	0	0.15
Presence in cloud forest (CF)	0	0	0.05
TOTAL	1	1	1

Formula to calculate the biodiversity conservation score:

$$BCs = (SC*0.1) + (RA*0.1) + (EM*0.1) + (EC*0.2) + (RL*0.2) + (NC*0.1) + (Pi*0.1) + (Pe*0.1) \quad (1)$$

Formula to calculate the carbon capture score:

$$CCs = (SC*0.1) + (RA*0.1) + (CCo*0.1) + (GR*0.175) + (UF*0.1) + (B*0.1) + (WD*0.05) + (HC*0.175) + (RR*0.1) \quad (2)$$

Formula to calculate the improving livelihoods score:

$$ILs = (SC*0.15) + (RA*0.1) + (AF*0.1) + (F*0.15) + (M*0.05) + (Me*0.1) + (P*0.05) + (SU*0.1) + (NF*0.15) + (CF*0.05) \quad (3)$$

https://doi.org/10.3390/su18073511

Formula to calculate the final priority score:

$$PS = (BCs + CCs + ILs)/3 \quad (4)$$

2.4. Species Similarity Analysis

A similarity analysis was carried out to visualise the similarity of the species composition in each of the nine regions. This analysis was performed using the Bray–Curtis Dissimilarity coefficient to compare both the presence/absence of the species and their abundance in each region [63]. The analysis was carried out in R (version 2.6-8) using the package "vegan" [64]. Hierarchical clustering was performed using a UPGMA (Unweighted Pair Group Method with Arithmetic Mean) analysis to obtain the final dendrogram from the same software.

3. Results

3.1. Status of the Chiapas Arboreal Diversity and Its Presence in the Coffee Regions

A total of 1653 native tree species were reported for Chiapas, of which 1362 were distributed within the coffee-growing regions, representing 82% of the state's total native arboreal flora. The 1362 species distributed in coffee regions belong to 119 families, with Fabaceae being the most diverse with 215, Rubiaceae with 110, and Lauraceae with 85 taxa. At the genus level, 457 genera were reported. *Quercus* L. stood out with 37 species, followed by *Eugenia* P.Micheli ex L. with 36 and *Lonchocarpus* Kunth with 30.

From these taxa, 286 were previously reported in the literature as companion trees of coffee farms, for example, *Cedrela odorata* L., *Ceiba pentandra* (L.) Gaertn., or *Enterolobium cyclocarpum* (Jacq.) Griseb. Approximately 94% of the total number of tree species have been evaluated by the IUCN, and around 23% are either considered to be at risk (Critically Endangered, Endangered, or Vulnerable) or in the category of Near Threatened. For example, the micro endemic species *Amyris chiapensis* Lundell, *Arachnothryx sousae* Borhidi, or *Calliandra ricoana* H.M.Hern. & R.Duno. A total of 648 of the 1362 species had reported values of biomass, while 364 had reported values of wood density. The uses that stood out in the greatest number of species were food, with 32% of the taxa, and medicinal use, including 30% of the species distributed in the coffee regions. Some of those species are widely known and utilised, such as *Bixa orellana* L., *Persea americana* Mill., and *Psidium guajava* L.

3.2. Regional Prioritisation

Nine different prioritisations are presented, one for each one of Chiapas's coffee regions. The 20 taxa with the highest priority are presented in Table 4.

A total of 39 species were included among the top 20 species comprising the nine prioritisation lists (taxonomic information is included in Appendix A, Table A1). Fabaceae had the highest number of species, with 12 taxa (30%), followed by Malvaceae with 4 (10%). Inga (4 spp.) stood out as the most represented genus in the prioritisations. Key species were identified for their presence in more than one region. For example, five species were considered in the nine coffee regions: *Bursera simaruba* (L.) Sarg., *Guazuma ulmifolia* Lam., *Inga vera* Willd., *Manilkara zapota* (L.) P.Royen, and *Trema micranthum* (L.) Blume. The specific contribution of each goal to the overall priority score is reported visually for each region in the graphs presented in Appendix B (Figures A1–A9). The priority species lists for each one of the three goals can be consulted in Appendix C (Table A2), Appendix D (Table A3) and Appendix E (Table A4).

https://doi.org/10.3390/su18073511

Table 4. Top 20 ranked native tree species from the prioritisation conducted to enrich shade cover in each of the nine coffee-growing regions of Chiapas, Mexico. Taxonomic information of the 39 native tree species included in the regional prioritisations is reported in Appendix A.

Rank	Altos	Centro	Fraylesca	Fronteriza	Istmo-Costa	Norte	Selva	Sierra	Soconusco
1	*Leucaena leucocephala*	*Gliricidia sepium*	*Gliricidia sepium*	*Inga inicuil*	*Gliricidia sepium*	*Gliricidia sepium*	*Gliricidia sepium*	*Gliricidia sepium*	*Gliricidia sepium*
2	*Litsea glaucescens*	*Bursera simaruba*	*Leucaena leucocephala*	*Swietenia macrophylla*	*Bursera simaruba*	*Leucaena leucocephala*	*Leucaena leucocephala*	*Leucaena leucocephala*	*Leucaena leucocephala*
3	*Inga vera*	*Leucaena leucocephala*	*Inga inicuil*	*Liquidambar styraciflua*	*Cecropia obtusifolia*	*Liquidambar styraciflua*	*Bursera simaruba*	*Bursera simaruba*	*Inga inicuil*
4	*Guazuma ulmifolia*	*Guazuma ulmifolia*	*Bursera simaruba*	*Bursera simaruba*	*Manilkara zapota*	*Bursera simaruba*	*Swietenia macrophylla*	*Inga inicuil*	*Cecropia obtusifolia*
5	*Psidium guajava*	*Swietenia macrophylla*	*Cecropia obtusifolia*	*Cecropia obtusifolia*	*Liquidambar styraciflua*	*Cedrela odorata*	*Inga inicuil*	*Cedrela odorata*	*Bursera simaruba*
6	*Bursera simaruba*	*Cedrela odorata*	*Guazuma ulmifolia*	*Brosimum alicastrum*	*Inga vera*	*Dendropanax arboreus*	*Cedrela odorata*	*Cecropia obtusifolia*	*Guazuma ulmifolia*
7	*Cecropia obtusifolia*	*Cecropia obtusifolia*	*Cedrela odorata*	*Cedrela odorata*	*Guazuma ulmifolia*	*Guazuma ulmifolia*	*Guazuma ulmifolia*	*Guazuma ulmifolia*	*Cedrela odorata*
8	*Manilkara zapota*	*Cordia alliodora*	*Hymenaea courbaril*	*Litsea glaucescens*	*Cordia alliodora*	*Hymenaea courbaril*	*Dendropanax arboreus*	*Cordia alliodora*	*Trema micranthum*
9	*Trema micranthum*	*Trema micranthum*	*Cordia alliodora*	*Inga vera*	*Hymenaea courbaril*	*Cordia alliodora*	*Cecropia obtusifolia*	*Vachellia pennatula*	*Cordia alliodora*
10	*Persea americana*	*Inga vera*	*Liquidambar styraciflua*	*Manilkara zapota*	*Brosimum alicastrum*	*Manilkara zapota*	*Brosimum alicastrum*	*Hymenaea courbaril*	*Hymenaea courbaril*
11	*Liquidambar styraciflua*	*Manilkara zapota*	*Trema micranthum*	*Cordia alliodora*	*Malpighia glabra*	*Litsea glaucescens*	*Cordia alliodora*	*Dendropanax arboreus*	*Manilkara zapota*
12	*Prunus serotina*	*Hymenaea courbaril*	*Inga vera*	*Dendropanax arboreus*	*Trema micranthum*	*Inga vera*	*Manilkara zapota*	*Litsea glaucescens*	*Inga vera*
13	*Inga punctata*	*Brosimum alicastrum*	*Dendropanax arboreus*	*Guazuma ulmifolia*	*Inga punctata*	*Trema micranthum*	*Inga vera*	*Manilkara zapota*	*Brosimum alicastrum*
14	*Trophis racemosa*	*Vachellia pennatula*	*Litsea glaucescens*	*Trophis racemosa*	*Psidium guajava*	*Ceiba pentandra*	*Trema micranthum*	*Inga vera*	*Carpinus caroliniana*
15	*Vachellia pennatula*	*Dendropanax arboreus*	*Inga punctata*	*Theobroma cacao*	*Spondias mombin*	*Prunus serotina*	*Ceiba pentandra*	*Carpinus caroliniana*	*Dendropanax arboreus*
16	*Carica papaya*	*Hamelia patens*	*Manilkara zapota*	*Ceiba pentandra*	*Hamelia patens*	*Theobroma cacao*	*Inga punctata*	*Ceiba pentandra*	*Litsea glaucescens*
17	*Acaciella angustissima*	*Litsea glaucescens*	*Byrsonima crassifolia*	*Trema micranthum*	*Plumeria rubra*	*Vachellia pennatula*	*Trophis racemosa*	*Trema micranthum*	*Theobroma cacao*
18	*Dendropanax arboreus*	*Byrsonima crassifolia*	*Ceiba pentandra*	*Malpighia glabra*	*Inga flexuosa*	*Ochroma pyramidale*	*Litsea glaucescens*	*Brosimum alicastrum*	*Ceiba pentandra*
19	*Calliandra houstoniana*	*Ceiba pentandra*	*Brosimum alicastrum*	*Psidium guajava*	*Carpinus caroliniana*	*Psidium guajava*	*Malpighia glabra*	*Liquidambar styraciflua*	*Enterolobium cyclocarpum*
20	*Plumeria rubra*	*Malpighia glabra*	*Psidium guajava*	*Vachellia pennatula*	*Palicourea padifolia*	*Persea americana*	*Prunus serotina*	*Theobroma cacao*	*Schizolobium parahyba*

https://doi.org/10.3390/su18073511

3.3. Species Similarity Analysis

This analysis assessed the species composition across the nine coffee regions in Chiapas, considering both the presence of taxa and their abundance, measured by the number of records for each species in the area. The dendrogram (Figure 3) shows that areas such as Centro and Selva have similar species compositions. Istmo-Costa stands out as the most dissimilar region, suggesting it has a notably distinct tree flora. For example, the presence of species such as *Spondias mombin* L., *Inga flexuosa* Schltdl., and *Palicourea padifolia* (Willd. ex Schult.) C. M. Taylor & Lorence. On the other hand, Sierra, Fraylesca, and Soconusco form a cohesive group, indicating similarity between them.

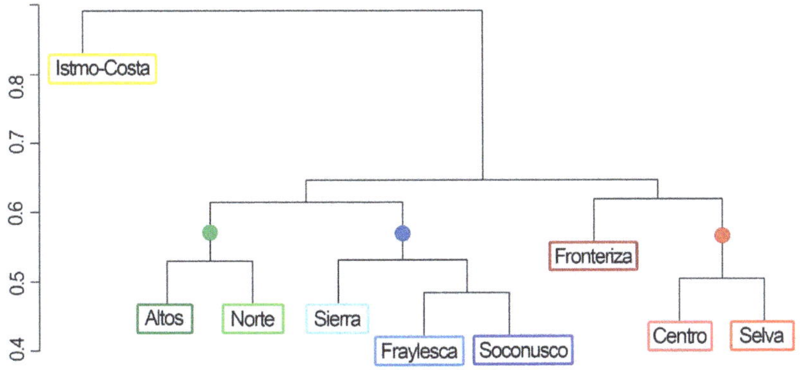

Figure 3. Similarity dendrogram of native tree species (n = 1362) for the nine coffee regions in Chiapas, Mexico.

4. Discussion

4.1. Status of the Chiapas Arboreal Diversity and Its Presence in the Coffee Regions

Chiapas stands out as one of the most biodiverse states in Mexico, itself a megadiverse country, with around 55% of the national tree diversity [27,55]. The percentage of these taxa overlapping with the coffee regions is also significant, accounting for 82% of the total tree diversity of the state (n = 1362 taxa). This could be related to their reported overlap with cloud forests, as well as the arboreal variability across the elevation gradient of the regions [28]. The high concentration of tree diversity within coffee landscapes highlights their strategic importance for national and subnational conservation planning.

While there have been previous studies reporting the tree diversity in Chiapas coffee regions, these appear to be an underestimate. For example, Reyes-Reyes et al. [65] reported 23 species in coffee farms of the Soconusco region in Chiapas. In comparison, Soto-Pinto et al. [35,36] reported 61 to 77 woody species in farms of Northern Chiapas, including both native and introduced species. However, the extensive bibliographic review carried out for this work revealed the presence of 286 species of native trees reported in the coffee farms, contributing to the knowledge of diversity in coffee farms and highlighting the relevance of these farms in the maintenance of the arboreal diversity at the landscape scale. These findings reinforce the role of shade coffee systems as complementary conservation areas within broader forested mosaics.

4.2. Prioritisation Goals

Since species selection plays a crucial role in conservation and reforestation projects, various methodologies have been proposed, each based on different criteria tailored to the specific study objectives [66,67]. This prioritisation method using a balanced approach was designed to fulfil the three goals of the project simultaneously of biodiversity conserva-

https://doi.org/10.3390/su18073511

tion, carbon capture, and improving livelihoods. The selected species have high priority scores across all three objectives (Appendix B), making them excellent candidates for diversification projects in coffee farms. This integrated approach is consistent with current policy frameworks that promote multifunctional landscapes delivering environmental and socio-economic benefits.

In terms of biodiversity conservation, promoting the propagation and use of species that are under a threat category could be useful to preserve the population of the endangered tree species in situ, even though they might be underestimated compared to species that have historically been used on coffee farms [25]. Incorporating threatened taxa into productive systems may therefore complement formal protected-area strategies.

In Mexico, agroforestry has been shown to enhance carbon capture, not only by promoting the growth of new trees but also by preventing the burning of these areas [38,68,69]. In Chiapas, this same result has been reported for silvopastoral systems [70]. The selected species' rapid growth, significant carbon storage capacity, and the fact that they are not used as fuel contribute to their potential for capturing atmospheric carbon and storing it in the soil [52]. Farm diversification also contributes to storing carbon in the soil, beyond its benefits on aboveground carbon capture. The selection of specific shade-tree species is critical because their unique traits drive essential soil-function pathways: deep-root systems facilitate nutrient cycling via 'nutrient pumping' from subsoil layers, while diverse litter inputs and root turnover promote soil aggregation and porosity [14]. These characteristics support the integration of shade coffee systems into climate mitigation and associated carbon capture mechanisms.

Additionally, the reported uses for the livelihood improvement goal could help diversify the income of coffee producers and be utilised within farms to support the growth of coffee plants [60]. Nevertheless, an essential part of the selection process should be the interest of local communities in the tree species, whether this is because their growth does not interfere with the coffee plants' development or because they could be useful for purposes such as food or medicine [71,72]. Participatory validation therefore remains essential to refine the results of this study and ensure long-term adoption by growers.

4.3. Regional Prioritisation

This is the first work that presents individual species prioritisations for all the coffee regions in Chiapas. Furthermore, the analysis included a considerable number of records for each region (Table 1), adding a precision component that previous studies had lacked by considering the total biodiversity of a state or country [25]. Although the number of records reflects the sampling intensity in each region and has biases, historical data remains our best alternative for understanding the distribution patterns and abundance of species [34]. In data-limited contexts, such approaches provide a pragmatic basis for spatially explicit decision-making.

Selections of suitable trees for shade-grown coffee farms have been done for the Ixhuatán and Tapalapa farms, reporting 25 species with a higher potential to be used as shade trees in the farms [73]. Some of those taxa, such as *Cordia alliodora* (Ruiz & Pav.) Oken, *Persea americana* and various *Inga* species were identified and are also reported in the lists proposed in this paper, supporting the results of our analysis. In contrast to the methodology employed in this paper, Yépez and collaborators conducted interviews with coffee producers to gather data on the attributes used in their prioritisation. While this approach provides valuable insights into the needs and preferences of farmers, it requires intensive fieldwork and is more challenging to scale across larger regions. Our approach complements participatory methods by providing an initial evidence-based

 https://doi.org/10.3390/su18073511

shortlist for further local validation, such as the approach carried out in some coffee regions in Veracruz [25].

Fabaceae stood out as the family with the highest number of relevant species in our study, with 11 taxa. This group has been previously documented for its consistent presence in shade-grown coffee farms, as well as the significance of its species to coffee producers [23,39]. Due to its nitrogen fixation ability, Fabaceae species enrich the soil with this essential nutrient for coffee plants and positively influence the biodiversity associated with the organic matter [74,75]. Moraceae and Malvaceae have also been reported for their species' presence in coffee farms in Mexico, and even in African countries such as Ethiopia [75,76].

Inga Mill., on the other hand, was the genus with more relevant species in the prioritisations (4 spp.): *Inga flexuosa* Schltdl., *I. inicuil* Schltdl. & Cham. ex G. Don, *I. punctata* Willd., and *I. vera*. In Veracruz, Mexico, previous prioritisations have suggested *Inga inicuil*, I. vera, and *I. punctata* as optimal species for reforestation [25]. *I. inicuil* in particular has also been studied to understand its germination under different temperatures and the effect of climate change on its distribution [77]. In other countries, such as Ethiopia and Costa Rica, it has been shown that farms with *Inga densiflora* Benth. promote carbon stock, improve water uptake without competing with coffee plants, and provide a better microclimate for the farms without compromising the farmers' yield [78–80].

4.4. Species Similarity Analysis

In addition to its high tree diversity, Chiapas' arboreal composition differs across regions and climate zones, making it more complicated to establish general species selection methodologies [81–83]. In this work, a practical classification of the coffee areas in the state was used to prioritise specific tree species for each geographical region. This procedure assumed that coffee farms in the same region are more similar to each other than those in different areas. In a country like Mexico, where environmental conditions tend to be more influenced by altitude than by geographical distance or political boundaries, this approach is highly recommended.

Our similarity analysis of the arboreous flora, based on the checklist of native tree species and the number of records for each taxon in each area, supports the assumption that floristic composition is more similar among geographically related regions, such as Altos and Norte in northern Chiapas, and Sierra, Fraylesca, and Soconusco in the south of the state (Figure 3). The marked dissimilarity of the Istmo-Costa region compared to the other eight areas may be attributed to its lower mean altitude (75 m a.s.l.) and land use patterns, as it hosts only 655 reported coffee farms, in contrast to the over 50,000 farms documented in the Altos and Norte regions (Table 1) [26]. Reported differences in the environmental, altitudinal, and edaphic variables in Chiapas and their influence on the distribution of tree species support the need to generate specific lists for each of the coffee areas and thus suggest trees that are in line with the characteristics of the study area [83]. The similarity of the species in the top 20 was not assessed because the criteria used in the prioritisation method aim to influence the relevance of the species, and this would no longer reflect the biogeographic relationships of the coffee regions. This distinction is important when interpreting prioritisation outputs for management purposes.

4.5. Relevance of Regional Prioritisations and Future Perspectives

Preserving biodiversity, including arboreal diversity, is one of the most urgent priorities of our time [27]. Since deforestation in Chiapas has been increasing in recent decades, the change in land use or the intensification of management in coffee farms are serious concerns [9,84]. The decrease in shaded coffee farms has direct implications for the

https://doi.org/10.3390/su18073511

loss of biodiversity, highlighting the urgency of finding alternative methods to ensure its permanence [85,86].

Planting a new tree represents a significant commitment for producers, particularly when farm sizes are limited, and the farm already provides substantial shade. In Mexico, most rustic coffee farms are smaller than 2 hectares, which is an important consideration in the species selection process [87]. Filtering out native species that are not reported in the area makes the process more precise and targeted while also aiding in the restoration of the original composition of these disturbed ecosystems.

This methodology can be applied to all coffee regions in Mexico or any other region if a list of tree species is available for the area of study. The values assigned to each criterion could also be modified to fulfil the goals of specific projects. For example, the benefit for growers could be maximised by giving the improving livelihoods criterion a higher weighting. Similar preferences could be given to conservation or carbon capture/storage, and the shift of the prioritised list should be explored in future studies. Considering priority shifts, the ability to shifts weights assigned to a set of criteria, should be useful for decision makers in distinct contexts, such as social programs and environmental legislation. Such flexibility enhances the relevance of the framework for diverse policy and funding contexts.

4.6. Limitations of This Study

Additional field work and feedback from coffee producers are required to improve the prioritisation lists. Multicriteria analyses with input from local producers have proven successful for diversification efforts in coffee farms in the coffee-producing state of Veracruz [25]. Further traits and information, such as nursery availability, establishment difficulty, and shade architecture compatibility, could be considered in future prioritisation efforts. However, this information could be scattered in the literature or simply not available for most taxa, which may reduce suitability for prioritisation efforts in regions with large numbers of species.

Since the main contribution of this study promoting a regional approach, additional fieldwork and flora documentation are recommended to reduce sampling biases and to ensure that the recommended trees are native and have an original distribution in the region, particularly in states as diverse and heterogeneous as Chiapas.

5. Conclusions

This study demonstrates that region-specific, evidence-based prioritisation is an effective tool for guiding shade-tree selection in coffee agroforestry systems. By integrating biodiversity data, expert knowledge, and multi-criteria evaluation, our approach provides tailored species lists that balance biodiversity conservation, carbon storage, and livelihood improvement across the diverse regions of Chiapas. Study results highlight the role of shade-grown coffee landscapes as multifunctional systems that support national and international environmental and development commitments and that can strengthen certification, restoration, and incentive schemes.

Regionally adapted species recommendations improve the effectiveness and adoption of agroforestry interventions by aligning ecological suitability with farmer needs, particularly in environmentally heterogeneous areas. The flexible structure of the framework allows decision-makers to adjust priorities according to policy objectives and local contexts. To maximise impact, future programmes should combine these tools with participatory validation, extension services, and long-term monitoring. Overall, the study provides a scalable decision-support approach to promote climate-resilient, biodiversity-friendly, and socially inclusive coffee landscapes in Mexico and beyond.

https://doi.org/10.3390/su18073511

Supplementary Materials: The following supporting information can be downloaded at: https://www.mdpi.com/article/10.3390/su18073511/s1, Table S1: List of species reported in shade-grown coffee farms.

Author Contributions: Conceptualization, M.G.C.H., C.M.F.-O. and T.U.; methodology, M.G.C.H.; software, M.G.C.H. and M.G.; validation, R.H.M. and M.T.-G.; formal analysis, M.G.C.H.; investigation, M.G.C.H.; resources, T.U. and M.G.; data curation, M.G.C.H. and M.G.; writing—original draft preparation, M.G.C.H.; writing—review and editing, C.M.F.-O., R.H.M., M.T.-G., M.G. and T.U.; visualization, M.G.C.H.; supervision, C.M.F.-O.; project administration, M.G. and T.U.; funding acquisition, T.U. All authors have read and agreed to the published version of the manuscript.

Funding: This work was funded by the project "Enhancing carbon sequestration and improving livelihoods in shade-grown coffee farms in the State of Veracruz, México", led by the Royal Botanical Garden of Kew, in collaboration with the Faculty of Higher Studies Iztacala of the National Autonomous University of México, in alliance with Pronatura Veracruz A.C., the Institute of Ecology A.C., and the Institute of Social Research (UNAM). It was funded by UK PACT Mexico (FCDO Project Number 30149), the Aldama Foundation, and the Emberson Foundation, and is supported by the British Embassy in Mexico.

Institutional Review Board Statement: Not applicable.

Informed Consent Statement: Not applicable.

Data Availability Statement: The raw data supporting the conclusions of this article will be made available by the authors on request.

Acknowledgments: To Andres De la Rosa Portilla and the Cafecol A.C. team for providing the coffee region shapefiles.

Conflicts of Interest: The authors declare no conflicts of interest.

Appendix A

Table A1. Taxonomic information of the 39 native tree species included in the nine regional prioritisations.

Family	Species Name	Taxonomic Author
Altingiaceae	*Liquidambar styraciflua*	L.
Anacardiaceae	*Spondias mombin*	L.
Apocynaceae	*Plumeria rubra*	L.
Araliaceae	*Dendropanax arboreus*	(L.) Decne. & Planch.
Betulaceae	*Carpinus caroliniana*	Walter
Boraginaceae	*Cordia alliodora*	(Ruiz & Pav.) Oken
Burseraceae	*Bursera simaruba*	(L.) Sarg.
Cannabaceae	*Trema micranthum*	(L.) Blume
Caricaceae	*Carica papaya*	L.
Fabaceae	*Acaciella angustissima*	(Mill.) Britton & Rose
Fabaceae	*Calliandra houstoniana*	(Mill.) Standl.
Fabaceae	*Enterolobium cyclocarpum*	(Jacq.) Griseb.
Fabaceae	*Gliricidia sepium*	(Jacq.) Kunth
Fabaceae	*Hymenaea courbaril*	L.
Fabaceae	*Inga flexuosa*	Schltdl.
Fabaceae	*Inga inicuil*	Schltdl. & Cham. ex G.Don
Fabaceae	*Inga punctata*	Willd.
Fabaceae	*Inga vera*	Willd.
Fabaceae	*Leucaena leucocephala*	(Lam.) de Wit

https://doi.org/10.3390/su18073511

Table A1. *Cont.*

Family	Species Name	Taxonomic Author
Fabaceae	*Schizolobium parahyba*	(Vell.) S.F.Blake
Fabaceae	*Vachellia pennatula*	(Schltdl. & Cham.) Seigler & Ebinger
Lauraceae	*Litsea glaucescens*	Kunth
Lauraceae	*Persea americana*	Mill.
Malpighiaceae	*Byrsonima crassifolia*	(L.) Kunth
Malpighiaceae	*Malpighia glabra*	L.
Malvaceae	*Ceiba pentandra*	(L.) Gaertn.
Malvaceae	*Guazuma ulmifolia*	Lam.
Malvaceae	*Ochroma pyramidale*	(Cav. ex Lam.) Urb.
Malvaceae	*Theobroma cacao*	L.
Meliaceae	*Cedrela odorata*	L.
Meliaceae	*Swietenia macrophylla*	King
Moraceae	*Brosimum alicastrum*	Sw.
Moraceae	*Trophis racemosa*	(L.) Urb.
Myrtaceae	*Psidium guajava*	L.
Rosaceae	*Prunus serotina*	Ehrh.
Rubiaceae	*Hamelia patens*	Jacq.
Rubiaceae	*Palicourea padifolia*	(Willd. ex Schult.) C.M.Taylor & Lorence
Sapotaceae	*Manilkara zapota*	(L.) P.Royen
Urticaceae	*Cecropia obtusifolia*	Bertol.

Appendix B

Graphs displaying the 20 prioritised species for each coffee region, along with the contribution of each goal's value to the overall priority score.

Figure A1. Contribution of each goal's value to the overall priority score in the Altos region.

https://doi.org/10.3390/su18073511

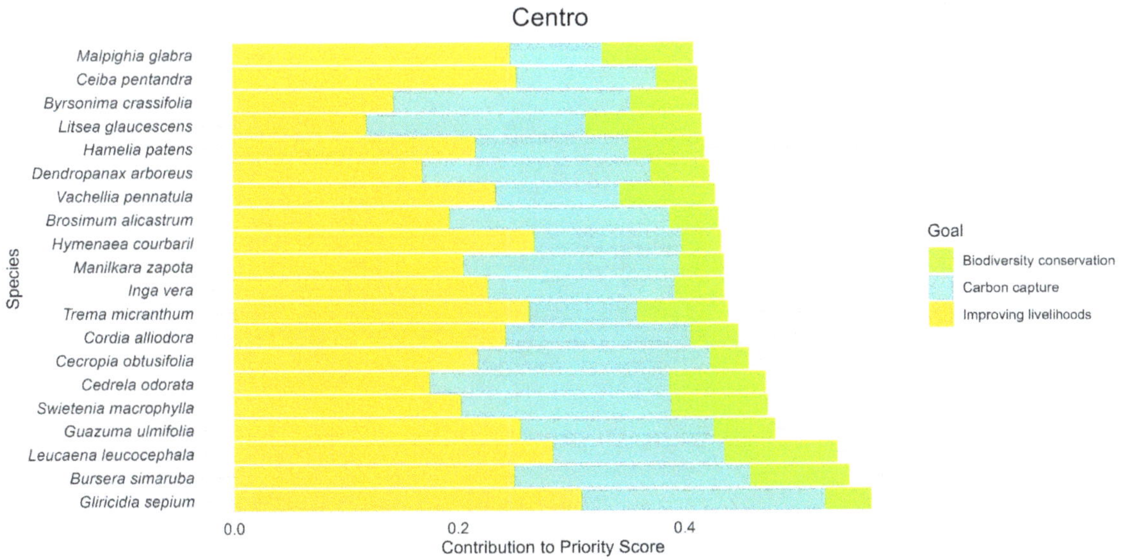

Figure A2. Contribution of each goal's value to the overall priority score in the Centro region.

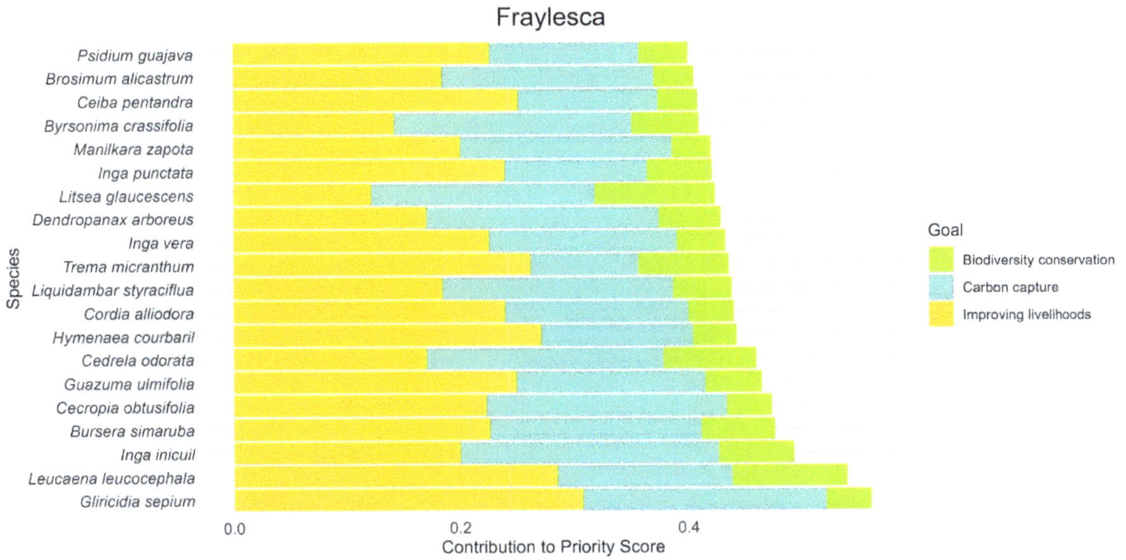

Figure A3. Contribution of each goal's value to the overall priority score in the Fraylesca region.

https://doi.org/10.3390/su18073511

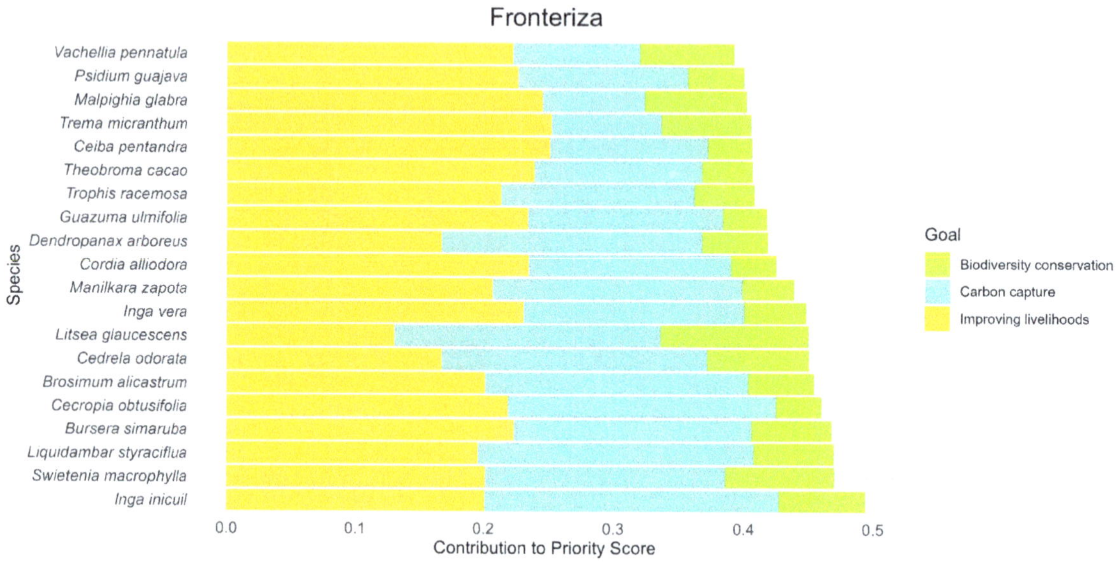

Figure A4. Contribution of each goal's value to the overall priority score in the Fronteriza region.

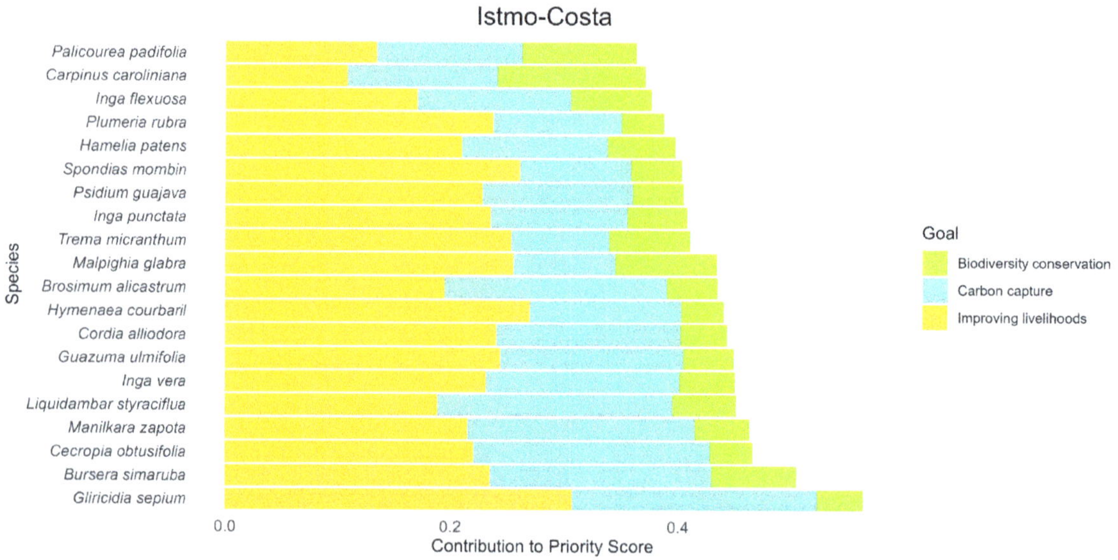

Figure A5. Contribution of each goal's value to the overall priority score in the Istmo-Costa region.

https://doi.org/10.3390/su18073511

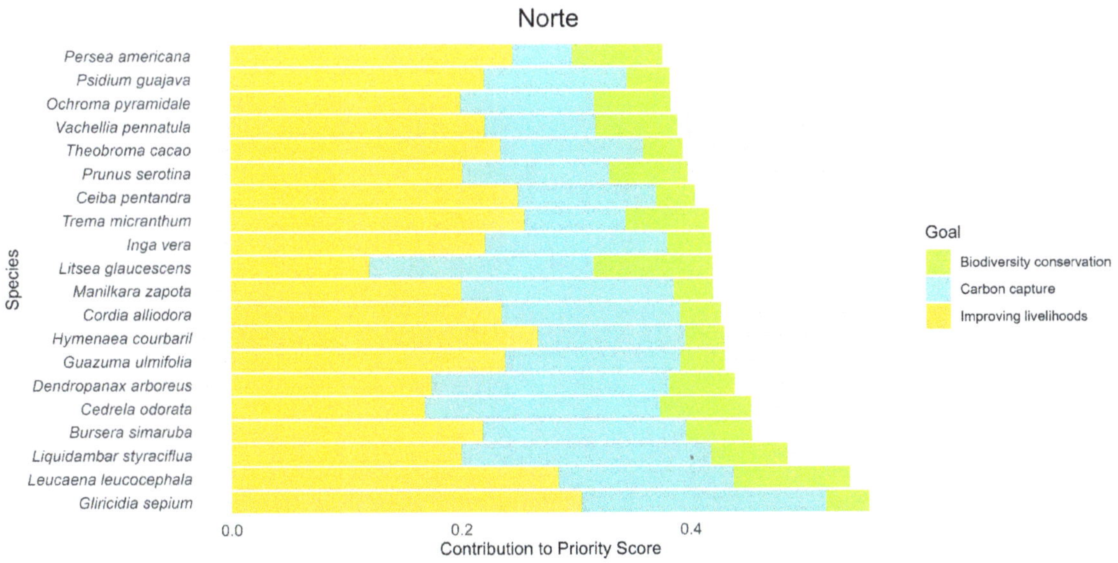

Figure A6. Contribution of each goal's value to the overall priority score in the Norte region.

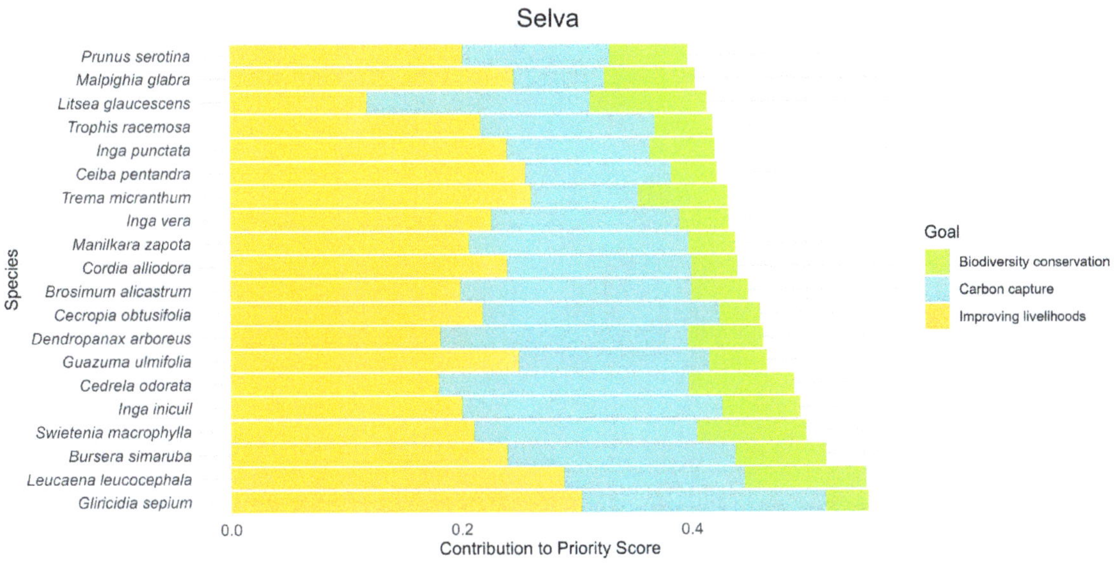

Figure A7. Contribution of each goal's value to the overall priority score in the Selva region.

https://doi.org/10.3390/su18073511

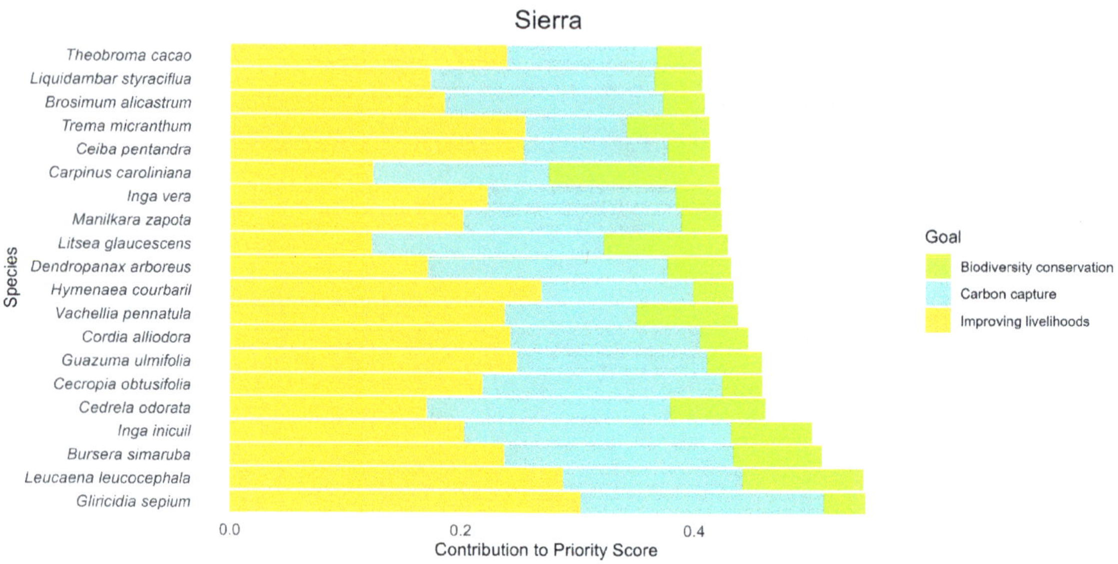

Figure A8. Contribution of each goal's value to the overall priority score in the Sierra region.

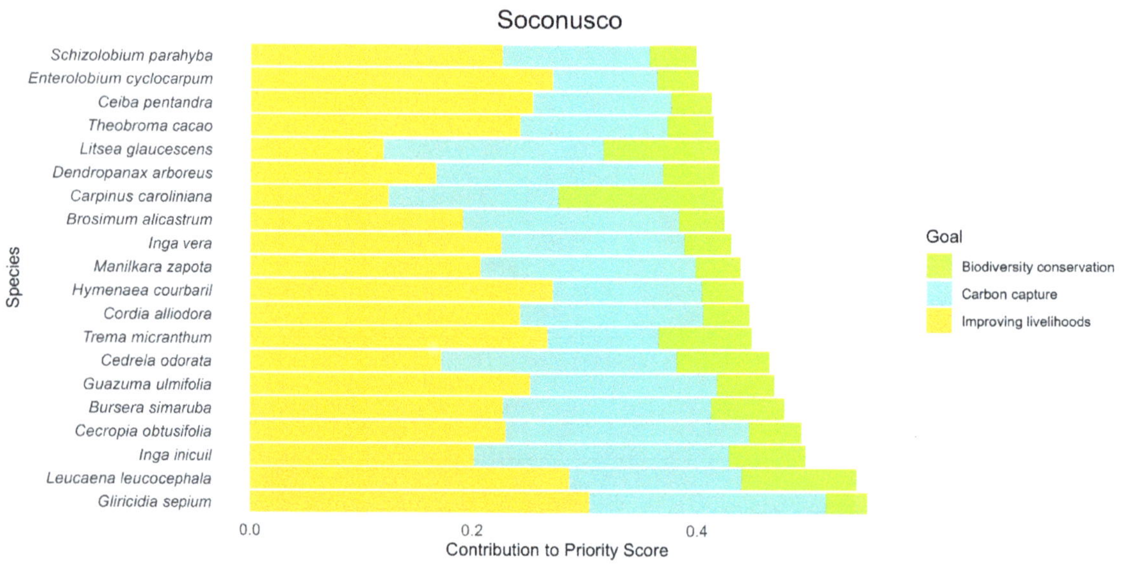

Figure A9. Contribution of each goal's value to the overall priority score in the Soconusco region.

https://doi.org/10.3390/su18073511

Appendix C

Table A2. Top 20 prioritised species for each coffee region in Chiapas according to the biodiversity conservation score.

Ranking	Altos	Centro	Fraylesca	Fronteriza	Istmo-Costa	Norte	Selva	Sierra	Soconusco
1	*Clethra chiapensis*	*Lonchocarpus foveolatus*	*Ateleia tenorioi*	*Magnolia mayae*	*Aiouea chiapensis*	*Lonchocarpus foveolatus*	*Calliandra ricoana*	*Ateleia tenorioi*	*Arachnothryx sousae*
2	*Picramnia deflexa*	*Amyris chiapensis*	*Maytenus stipitata*	*Damburneya leucocome*	*Ctenardisia purpusii*	*Hampea breedlovei*	*Hampea breedlovei*	*Aiouea chiapensis*	*Sommera parva*
3	*Aiouea chiapensis*	*Magnolia perezfarrerae*	*Aiouea chiapensis*	*Hampea montebellensis*	*Pseudomiltemia davidsonii*	*Orthion montanum*	*Magnolia mayae*	*Ctenardisia purpusii*	*Eugenia ovandensis*
4	*Chamaedorea cataractarum*	*Mortoniodendron ocotense*	*Chamaedorea cataractarum*	*Dalbergia stevensonii*	*Carpinus caroliniana*	*Erythrina pudica*	*Sommera parva*	*Daphnopsis flavida*	*Aiouea chiapensis*
5	*Juniperus gamboana*	*Orthion montanum*	*Ardisia breedlovei*	*Ardisia breedlovei*	*Saurauia madrensis*	*Magnolia sharpii*	*Lonchocarpus comitensis*	*Licaria glaberrima*	*Ctenardisia purpusii*
6	*Licaria glaberrima*	*Picramnia deflexa*	*Ctenardisia purpusii*	*Licaria glaberrima*	*Chamaedorea tepejilote*	*Ocotea matudae*	*Damburneya leucocome*	*Pseudomiltemia davidsonii*	*Daphnopsis flavida*
7	*Magnolia sharpii*	*Robinsonella pilosissima*	*Daphnopsis flavida*	*Palicourea breedlovei*	*Prunus tetradenia*	*Wimmeria acuminata*	*Orthion montanum*	*Wimmeria acuminata*	*Licaria glaberrima*
8	*Ocotea matudae*	*Hampea montebellensis*	*Eugenia breedlovei*	*Magnolia montebelloensis*	*Persea schiedeana*	*Yucca lacandonica*	*Juglans pyriformis*	*Yucca lacandonica*	*Ocotea matudae*
9	*Wimmeria acuminata*	*Gliricidia robusta*	*Licaria glaberrima*	*Wimmeria montana*	*Gonzalagunia tacanensis*	*Lonchocarpus santarosanus*	*Hampea montebellensis*	*Cordia colimensis*	*Ateleia glabrata*
10	*Ocotea psychotrioides*	*Lonchocarpus sumiderensis*	*Pseudomiltemia davidsonii*	*Carpinus caroliniana*	*Dendropanax populifolius*	*Hampea nutricia*	*Dalbergia stevensonii*	*Wimmeria montana*	*Wimmeria montana*
11	*Lonchocarpus santarosanus*	*Aiouea chiapensis*	*Yucca lacandonica*	*Magnolia schiedeana*	*Trema micranthum*	*Lonchocarpus latimarginatus*	*Chamaedorea cataractarum*	*Carpinus caroliniana*	*Carpinus caroliniana*
12	*Ocotea congregata*	*Bunchosia breedlovei*	*Carpinus caroliniana*	*Saurauia madrensis*	*Malpighia glabra*	*Euonymus chiapensis*	*Ardisia breedlovei*	*Magnolia schiedeana*	*Beaucarnea goldmanii*

https://doi.org/10.3390/su18073511

Table A2. *Cont.*

Ranking	Altos	Centro	Fraylesca	Fronteriza	Istmo-Costa	Norte	Selva	Sierra	Soconusco
13	*Rogiera breedlovei*	*Stenanona migueliana*	*Saurauia madrensis*	*Mortoniodendron guatemalense*	*Inga flexuosa*	*Mortoniodendron uxpanapense*	*Magnolia sharpii*	*Beaucarnea goldmanii*	*Dalbergia tucurensis*
14	*Wimmeria montana*	*Chamaedorea cataractarum*	*Maytenus matudae*	*Litsea glaucescens*	*Persea americana*	*Palicourea thornei*	*Wimmeria acuminata*	*Saurauia madrensis*	*Saurauia madrensis*
15	*Carpinus caroliniana*	*Juniperus gamboana*	*Hesperocyparis lusitanica*	*Saurauia villosa*	*Vachellia cornigera*	*Wimmeria montana*	*Yucca lacandonica*	*Prunus tartarea*	*Coccoloba chiapensis*
16	*Dalbergia tucurensis*	*Erythrina pudica*	*Leucaena leucocephala*	*Myrica cerifera*	*Aiouea montana*	*Magnolia faustinomi-randae*	*Ocotea psychotrioides*	*Hesperocyparis lusitanica*	*Mortoniodendron guatemalense*
17	*Leucaena leucocephala*	*Eugenia breedlovei*	*Litsea glaucescens*	*Vatairea lundellii*	*Myriocarpa longipes*	*Carpinus caroliniana*	*Lonchocarpus santarosanus*	*Leucaena leucocephala*	*Prunus erythroxylon*
18	*Litsea glaucescens*	*Licaria glaberrima*	*Saurauia villosa*	*Chamaedorea tepejilote*	*Coccoloba barbadensis*	*Dalbergia tucurensis*	*Damburneya inconspicua*	*Litsea glaucescens*	*Hesperocyparis lusitanica*
19	*Chrysophyllum cainito*	*Magnolia sharpii*	*Chamaedorea tepejilote*	*Inga xalapensis*	*Sideroxylon portoricense*	*Saurauia madrensis*	*Arachnothryx lineolata*	*Saurauia villosa*	*Leucaena leucocephala*
20	*Myrica cerifera*	*Parathesis cintalapana*	*Yucca gigantea*	*Pleuranthodendron lindenii*	*Palicourea padifolia*	*Aiouea breedlovei*	*Ocotea congregata*	*Myrica cerifera*	*Litsea glaucescens*

Appendix D

Table A3. Top 20 prioritised species for each coffee region in Chiapas according to the carbon capture score.

Ranking	Altos	Centro	Fraylesca	Fronteriza	Istmo-Costa	Norte	Selva	Sierra	Soconusco
1	*Cecropia obtusifolia*	*Gliricidia sepium*	*Inga inicuil*	*Inga inicuil*	*Gliricidia sepium*	*Gliricidia sepium*	*Inga inicuil*	*Inga inicuil*	*Inga inicuil*
2	*Litsea glaucescens*	*Cecropia obtusifolia*	*Gliricidia sepium*	*Cecropia obtusifolia*	*Cecropia obtusifolia*	*Cedrela odorata*	*Gliricidia sepium*	*Gliricidia sepium*	*Gliricidia sepium*
3	*Manilkara zapota*	*Cedrela odorata*	*Cecropia obtusifolia*	*Cedrela odorata*	*Manilkara zapota*	*Litsea glaucescens*	*Cecropia obtusifolia*	*Cecropia obtusifolia*	*Cecropia obtusifolia*

https://doi.org/10.3390/su18073511

Table A3. *Cont.*

Ranking	Altos	Centro	Fraylesca	Fronteriza	Istmo-Costa	Norte	Selva	Sierra	Soconusco
4	*Pinus pseudostrobus*	*Litsea glaucescens*	*Cedrela odorata*	*Litsea glaucescens*	*Brosimum alicastrum*	*Manilkara zapota*	*Cedrela odorata*	*Cedrela odorata*	*Cedrela odorata*
5	*Liquidambar styraciflua*	*Manilkara zapota*	*Litsea glaucescens*	*Manilkara zapota*	*Liquidambar styraciflua*	*Pinus pseudostrobus*	*Litsea glaucescens*	*Litsea glaucescens*	*Litsea glaucescens*
6	*Dendropanax arboreus*	*Brosimum alicastrum*	*Manilkara zapota*	*Brosimum alicastrum*	*Bursera simaruba*	*Liquidambar styraciflua*	*Manilkara zapota*	*Manilkara zapota*	*Manilkara zapota*
7	*Byrsonima crassifolia*	*Pinus pseudostrobus*	*Brosimum alicastrum*	*Pinus pseudostrobus*	*Inga vera*	*Dendropanax arboreus*	*Brosimum alicastrum*	*Brosimum alicastrum*	*Brosimum alicastrum*
8	*Quercus xalapensis*	*Liquidambar styraciflua*	*Pinus pseudostrobus*	*Liquidambar styraciflua*	*Chrysophyllum mexicanum*	*Byrsonima crassifolia*	*Pinus pseudostrobus*	*Pinus pseudostrobus*	*Dendropanax arboreus*
9	*Bursera simaruba*	*Dendropanax arboreus*	*Liquidambar styraciflua*	*Dendropanax arboreus*	*Cordia alliodora*	*Bursera simaruba*	*Liquidambar styraciflua*	*Liquidambar styraciflua*	*Byrsonima crassifolia*
10	*Inga vera*	*Byrsonima crassifolia*	*Dendropanax arboreus*	*Byrsonima crassifolia*	*Guazuma ulmifolia*	*Simira salvadorensis*	*Dendropanax arboreus*	*Dendropanax arboreus*	*Bursera simaruba*
11	*Chrysophyllum mexicanum*	*Swietenia macrophylla*	*Byrsonima crassifolia*	*Swietenia macrophylla*	*Aiouea montana*	*Inga vera*	*Byrsonima crassifolia*	*Byrsonima crassifolia*	*Simira salvadorensis*
12	*Leucaena leucocephala*	*Bursera simaruba*	*Quercus xalapensis*	*Bursera simaruba*	*Ficus yoponensis*	*Chrysophyllum mexicanum*	*Swietenia macrophylla*	*Quercus xalapensis*	*Inga vera*
13	*Guazuma ulmifolia*	*Simira salvadorensis*	*Bursera simaruba*	*Simira salvadorensis*	*Ficus obtusifolia*	*Cordia alliodora*	*Bursera simaruba*	*Bursera simaruba*	*Chrysophyllum mexicanum*
14	*Ficus cotinifolia*	*Inga vera*	*Inga vera*	*Inga vera*	*Inga flexuosa*	*Leucaena leucocephala*	*Simira salvadorensis*	*Inga vera*	*Cordia alliodora*
15	*Alnus acuminata*	*Chrysophyllum mexicanum*	*Chrysophyllum mexicanum*	*Chrysophyllum mexicanum*	*Hymenaea courbaril*	*Guazuma ulmifolia*	*Inga vera*	*Chrysophyllum mexicanum*	*Leucaena leucocephala*
16	*Trophis racemosa*	*Cordia alliodora*	*Cordia alliodora*	*Cordia alliodora*	*Carpinus caroliniana*	*Ficus cotinifolia*	*Chrysophyllum mexicanum*	*Cordia alliodora*	*Guazuma ulmifolia*

https://doi.org/10.3390/su18073511

Table A3. *Cont.*

Ranking	Altos	Centro	Fraylesca	Fronteriza	Istmo-Costa	Norte	Selva	Sierra	Soconusco
17	*Pinus oocarpa*	*Leucaena leucocephala*	*Leucaena leucocephala*	*Guazuma ulmifolia*	*Psidium oligospermum*	*Heliocarpus appendicula-tus*	*Cordia alliodora*	*Leucaena leucocephala*	*Ficus cotinifolia*
18	*Inga flexuosa*	*Guazuma ulmifolia*	*Guazuma ulmifolia*	*Alnus acuminata*	*Platanus mexicana*	*Aiouea montana*	*Swartzia cubensis*	*Guazuma ulmifolia*	*Alnus acuminata*
19	*Quercus rugosa*	*Ficus cotinifolia*	*Ficus cotinifolia*	*Heliocarpus appendicula-tus*	*Psidium guajava*	*Pseudolmedia glabrata*	*Leucaena leucocephala*	*Ficus cotinifolia*	*Heliocarpus appendicula-tus*
20	*Prunus serotina*	*Alnus acuminata*	*Heliocarpus appendiculatus*	*Pseudolmedia glabrata*	*Sideroxylon portoricense*	*Trophis racemosa*	*Guazuma ulmifolia*	*Alnus acuminata*	*Aiouea montana*

Appendix E

Table A4. Top 20 prioritised species for each coffee region in Chiapas according to the improving livelihoods score.

Ranking	Altos	Centro	Fraylesca	Fronteriza	Istmo-Costa	Norte	Selva	Sierra	Soconusco
1	*Leucaena leucocephala*	*Gliricidia sepium*	*Gliricidia sepium*	*Ceiba pentandra*	*Gliricidia sepium*	*Gliricidia sepium*	*Gliricidia sepium*	*Gliricidia sepium*	*Gliricidia sepium*
2	*Spondias mombin*	*Leucaena leucocephala*	*Leucaena leucocephala*	*Spondias mombin*	*Hymenaea courbaril*	*Leucaena leucocephala*	*Leucaena leucocephala*	*Leucaena leucocephala*	*Leucaena leucocephala*
3	*Trema micranthum*	*Hymenaea courbaril*	*Hymenaea courbaril*	*Trema micranthum*	*Spondias mombin*	*Hymenaea courbaril*	*Samanea saman*	*Hymenaea courbaril*	*Hymenaea courbaril*
4	*Guazuma ulmifolia*	*Enterolobium cyclocarpum*	*Enterolobium cyclocarpum*	*Erythrina berteroana*	*Trema micranthum*	*Ceiba pentandra*	*Enterolobium cyclocarpum*	*Enterolobium cyclocarpum*	*Enterolobium cyclocarpum*
5	*Plumeria rubra*	*Cassia grandis*	*Ceiba pentandra*	*Cordia alliodora*	*Cordia alliodora*	*Spondias mombin*	*Ceiba pentandra*	*Ceiba pentandra*	*Cassia grandis*
6	*Carica papaya*	*Ceiba pentandra*	*Spondias mombin*	*Guazuma ulmifolia*	*Guazuma ulmifolia*	*Trema micranthum*	*Myroxylon balsamum*	*Myroxylon balsamum*	*Ceiba pentandra*
7	*Bixa orellana*	*Myroxylon balsamum*	*Trema micranthum*	*Theobroma cacao*	*Plumeria rubra*	*Erythrina berteroana*	*Spondias mombin*	*Spondias mombin*	*Myroxylon balsamum*

https://doi.org/10.3390/su18073511

Table A4. *Cont.*

Ranking	Altos	Centro	Fraylesca	Fronteriza	Istmo-Costa	Norte	Selva	Sierra	Soconusco
8	*Malpighia glabra*	*Spondias mombin*	*Cordia alliodora*	*Plumeria rubra*	*Malpighia glabra*	*Cordia alliodora*	*Trema micranthum*	*Trema micranthum*	*Spondias mombin*
9	*Sapindus saponaria*	*Trema micranthum*	*Guazuma ulmifolia*	*Carica papaya*	*Calliandra houstoniana*	*Guazuma ulmifolia*	*Erythrina berteroana*	*Erythrina berteroana*	*Trema micranthum*
10	*Calliandra houstoniana*	*Erythrina berteroana*	*Theobroma cacao*	*Bixa orellana*	*Persea americana*	*Theobroma cacao*	*Cordia alliodora*	*Cordia alliodora*	*Erythrina berteroana*
11	*Persea americana*	*Cordia alliodora*	*Plumeria rubra*	*Malpighia glabra*	*Cecropia obtusifolia*	*Carica papaya*	*Guazuma ulmifolia*	*Guazuma ulmifolia*	*Cordia alliodora*
12	*Cecropia obtusifolia*	*Guazuma ulmifolia*	*Carica papaya*	*Sapindus saponaria*	*Bursera simaruba*	*Bixa orellana*	*Theobroma cacao*	*Theobroma cacao*	*Guazuma ulmifolia*
13	*Bursera simaruba*	*Theobroma cacao*	*Bixa orellana*	*Calliandra houstoniana*	*Inga vera*	*Malpighia glabra*	*Plumeria rubra*	*Plumeria rubra*	*Theobroma cacao*
14	*Inga vera*	*Plumeria rubra*	*Malpighia glabra*	*Cecropia obtusifolia*	*Psidium guajava*	*Sapindus saponaria*	*Carica papaya*	*Carica papaya*	*Plumeria rubra*
15	*Psidium guajava*	*Carica papaya*	*Sapindus saponaria*	*Bursera simaruba*	*Inga punctata*	*Calliandra houstoniana*	*Bixa orellana*	*Bixa orellana*	*Carica papaya*
16	*Tecoma stans*	*Bixa orellana*	*Calliandra houstoniana*	*Inga vera*	*Cestrum nocturnum*	*Annona squamosa*	*Malpighia glabra*	*Malpighia glabra*	*Bixa orellana*
17	*Inga punctata*	*Malpighia glabra*	*Annona squamosa*	*Schizolobium parahyba*	*Manilkara zapota*	*Persea americana*	*Sapindus saponaria*	*Sapindus saponaria*	*Malpighia glabra*
18	*Vachellia pennatula*	*Sapindus saponaria*	*Persea americana*	*Psidium guajava*	*Euphorbia pulcherrima*	*Bursera simaruba*	*Calliandra houstoniana*	*Calliandra houstoniana*	*Sapindus saponaria*
19	*Cestrum nocturnum*	*Calliandra houstoniana*	*Cecropia obtusifolia*	*Tecoma stans*	*Brosimum alicastrum*	*Inga vera*	*Annona squamosa*	*Persea americana*	*Calliandra houstoniana*
20	*Manilkara zapota*	*Annona squamosa*	*Bursera simaruba*	*Inga punctata*	*Hamelia patens*	*Psidium guajava*	*Persea americana*	*Cecropia obtusifolia*	*Annona squamosa*

https://doi.org/10.3390/su18073511

References

1. INEGI. Censo Agropecuario 2022. Ciudad de México: Instituto Nacional de Estadística y Geografía (INEGI). Available online: https://www.inegi.org.mx/programas/ca/2022/ (accessed on 27 January 2026).
2. Servín-Juárez, R.; Trejo-Pech, C.J.O.; Pérez-Vásquez, A.Y.; Reyes-Duarte, Á. Specialty Coffee Shops in Mexico: Factors Influencing the Likelihood of Purchasing High-Quality Coffee. *Sustainability* **2021**, *13*, 3804. [CrossRef]
3. SADER Cultivo de Café en México. Secretaría de Agricultura y Desarrollo Rural. 2022. Available online: https://www.gob.mx/agricultura/articulos/cultivo-de-cafe-en-mexico (accessed on 15 January 2026).
4. Torres Castillo, N.E.; Melchor-Martínez, E.M.; Ochoa Sierra, J.S.; Ramirez-Mendoza, R.A.; Parra-Saldívar, R.; Iqbal, H.M.N. Impact of Climate Change and Early Development of Coffee Rust—An Overview of Control Strategies to Preserve Organic Cultivars in Mexico. *Sci. Total Environ.* **2020**, *738*, 140225. [CrossRef]
5. Gay, C.; Estrada, F.; Conde, C.; Eakin, H.; Villers, L. Potential Impacts of Climate Change on Agriculture: A Case of Study of Coffee Production in Veracruz, Mexico. *Clim. Change* **2006**, *79*, 259–288. [CrossRef]
6. Renard, M.-C. The Mexican Coffee Crisis. *Lat. Am. Perspect.* **2010**, *37*, 21–33. [CrossRef]
7. Mestries-Benquet, F. Crisis cafetalera y migración internacional en Veracruz. *Migr. Int.* **2003**, *2*, 121–148.
8. USDA. *Report Coffee Annual Mexico*; United States Department of Agriculture (Foreign Agricultural Service): Washington, DC, USA, 2024.
9. Ellis, E.A.; Baerenklau, K.A.; Marcos-Martínez, R.; Chávez, E. Land Use/Land Cover Change Dynamics and Drivers in a Low-Grade Marginal Coffee Growing Region of Veracruz, Mexico. *Agrofor. Syst.* **2010**, *80*, 61–84. [CrossRef]
10. Toledo, V.; Moguel, P. Coffee and Sustainability: The Multiple Values of Traditional Shaded Coffee. *J. Sustain. Agric.* **2012**, *36*, 353–377. [CrossRef]
11. Moguel, P.; Toledo, V.M. Biodiversity Conservation in Traditional Coffee Systems of Mexico. *Conserv. Biol.* **1999**, *13*, 11–21. [CrossRef]
12. Bandeira, F.P.; Martorell, C.; Meave, J.A.; Caballero, J. The Role of Rustic Coffee Plantations in the Conservation of Wild Tree Diversity in the Chinantec Region of Mexico. *Biodivers. Conserv.* **2005**, *14*, 1225–1240. [CrossRef]
13. Hernandez-Aguilera, J.N.; Conrad, J.M.; Gómez, M.I.; Rodewald, A.D. The Economics and Ecology of Shade-Grown Coffee: A Model to Incentivize Shade and Bird Conservation. *Ecol. Econ.* **2019**, *159*, 110–121. [CrossRef]
14. Manono, B.O.; Mwami, B. Agroforestry and Soil Health: A Review of Impacts and Potential for Sustainable Agriculture. *Earth* **2026**, *7*, 31. [CrossRef]
15. Boege, E. *El Patrimonio Biocultural de Los Pueblos Indígenas de México. Hacia la Conservación In Situ de la Biodiversidad y Agrodiversidad en Los Territorios Indígenas*; Instituto Nacional de Antropología e Historia: Mexico City, Mexico, 2010; ISBN 978-968-03-0385-4.
16. Gerlicz, A.; Méndez, V.; Conner, D.; Baker, D.; Christel, D. Use and Perceptions of Alternative Economic Activities among Smallholder Coffee Farmers in Huehuetenango and El Quiché Departments in Guatemala. *Agroecol. Sustain. Food Syst.* **2018**, *43*, 310–328. [CrossRef]
17. Anderzén, J.; Luna, A.G.; Luna-González, D.V.; Merrill, S.C.; Caswell, M.; Méndez, V.E.; Jonapá, R.H.; Mier y Terán Giménez Cacho, M. Effects of On-Farm Diversification Strategies on Smallholder Coffee Farmer Food Security and Income Sufficiency in Chiapas, Mexico. *J. Rural Stud.* **2020**, *77*, 33–46. [CrossRef]
18. Toledo-Aceves, T.; Toledo-Garibaldi, M. Tree Species Diversity Increases Carbon Stocks in Tropical Montane Cloud Forests across Successional Stages. *For. Ecol. Manag.* **2025**, *578*, 122480. [CrossRef]
19. Ruiz-García, P.; Monterroso-Rivas, A.I.; Conde-Álvarez, A.C. Nature-Based Solutions in Coffee Agroecosystems in Mexico: Evaluation for Adoption. *Front. Sustain. Food Syst.* **2025**, *9*, 1604520. [CrossRef]
20. Zarza, E.; López-Pastrana, A.; Damon, A.; Guillén-Navarro, K.; García-Fajardo, L.V. Fungal Diversity in Shade-Coffee Plantations in Soconusco, Mexico. *PeerJ* **2022**, *10*, e13610. [CrossRef]
21. Poncet, V.; van Asten, P.; Millet, C.P.; Vaast, P.; Allinne, C. Which Diversification Trajectories Make Coffee Farming More Sustainable? *Curr. Opin. Environ. Sustain.* **2024**, *68*, 101432. [CrossRef]
22. SEMARNAT. *Catálogo de Árboles Multipropósito Presentes en Los Cafetales Bajo Sombra*; Primera Edición, Agosto 2024; Secretaría de Medio Ambiente y Recursos Naturales: Mexico City, Mexico, 2024; p. 214.
23. Valencia, V.; West, P.; Sterling, E.; García-Barrios, L.; Naeem, S. The Use of Farmers' Knowledge in Coffee Agroforestry Management: Implications for the Conservation of Tree Biodiversity. *Ecosphere* **2015**, *6*, art122. [CrossRef]
24. Smith Dumont, E.; Gassner, A.; Agaba, G.; Nansamba, R.; Sinclair, F. The Utility of Farmer Ranking of Tree Attributes for Selecting Companion Trees in Coffee Production Systems. *Agrofor. Syst.* **2019**, *93*, 1469–1483. [CrossRef]
25. Flores-Ortiz, C.M.; Davila, P.; Rodríguez-Arevalo, I.; Manson, R.H.; Toledo-Garibaldi, M.; Cabrera-Santos, D.; Salguero, M.A.; Vázquez, F.G.; Cobos-Silva, J.; Gianella, M.; et al. Prioritisation of Native Trees for Enhancing Carbon Sequestration in Shade-Grown Coffee Plantations in the State of Veracruz (México): Linking Conservation and Ecological Traits to Community Needs. *Agrofor. Syst.* **2025**, *99*, 55. [CrossRef]

 https://doi.org/10.3390/su18073511

26. Hernández-Martínez, G.; Vera-Espíndola, R.; De la Rosa-Portilla, A. *The Coffee of Mexico. The Richness of Its Regions*; Centro Agroecológico del Café: Xalapa, Mexico, 2018; p. 84.
27. Tellez, O.; Mattana, E.; Diazgranados, M.; Kühn, N.; Castillo-Lorenzo, E.; Lira, R.; Montes-Leyva, L.; Rodriguez, I.; Ortiz, C.M.F.; Way, M.; et al. Native Trees of Mexico: Diversity, Distribution, Uses and Conservation. *PeerJ* **2020**, *8*, e9898. [CrossRef]
28. Verdugo Morales, E.; Yáñez, I.; Alanís-Rodríguez, E.; González-Rodríguez, H.; Garza-Ocañas, F.; Larramendi, L. Riqueza de Especies, Estructura Horizontal y Diversidad Arbórea en Un Gradiente Altitudinal en Cafetales de Chiapas. *Rev. Mex. Cienc. For.* **2025**, *16*, 148–173. [CrossRef]
29. Alemu, M.M. Effect of Tree Shade on Coffee Crop Production. *J. Sustain. Dev.* **2015**, *8*, 66. [CrossRef]
30. Boreux, V.; Vaast, P.; Madappa, L.P.; Cheppudira, K.G.; Garcia, C.; Ghazoul, J. Agroforestry Coffee Production Increased by Native Shade Trees, Irrigation, and Liming. *Agron. Sustain. Dev.* **2016**, *36*, 42. [CrossRef]
31. GBIF Occurrence Download. Available online: https://doi.org/10.15468/Dl.2z7ark (accessed on 6 March 2025).
32. Zizka, A.; Silvestro, D.; Andermann, T.; Azevedo, J.A.; Ritter, C.; Edler, D.; Farooq, H.; Herdean, A.; Ariza, M.; Scharn, R.; et al. CoordinateCleaner: Standardized Cleaning of Occurrence Records from Biological Collection Databases. *Methods Ecol. Evol.* **2019**, *10*, 744–751. [CrossRef]
33. QGIS. QGIS Geographic Information System. Open Source Geospatial Foundation Project. 2024. Available online: https://qgis.org/download/ (accessed on 10 September 2024).
34. Gotelli, N.J.; Booher, D.B.; Urban, M.C.; Ulrich, W.; Suarez, A.V.; Skelly, D.K.; Russell, D.J.; Rowe, R.J.; Rothendler, M.; Rios, N.; et al. Estimating Species Relative Abundances from Museum Records. *Methods Ecol. Evol.* **2023**, *14*, 431–443. [CrossRef]
35. Soto-Pinto, L.; Perfecto, I.; Castillo-Hernandez, J.; Caballero-Nieto, J. Shade Effect on Coffee Production at the Northern Tzeltal Zone of the State of Chiapas, Mexico. *Agric. Ecosyst. Environ.* **2000**, *80*, 61–69. [CrossRef]
36. Soto-Pinto, L.; Romero-Alvarado, Y.; Caballero-Nieto, J.; Segura Warnholtz, G. Woody Plant Diversity and Structure of Shade-Grown-Coffee Plantations in Northern Chiapas, Mexico. *Rev. Biol. Trop.* **2001**, *49*, 977–987.
37. López-Gómez, A.M.; Williams-Linera, G.; Manson, R.H. Tree Species Diversity and Vegetation Structure in Shade Coffee Farms in Veracruz, Mexico. *Agric. Ecosyst. Environ.* **2008**, *124*, 160–172. [CrossRef]
38. Soto-Pinto, L.; Anzueto, M.; Mendoza, J.; Ferrer, G.J.; de Jong, B. Carbon Sequestration through Agroforestry in Indigenous Communities of Chiapas, Mexico. *Agrofor. Syst.* **2010**, *78*, 39–51. [CrossRef]
39. Soto-Pinto, L.; Villalvazo-López, V.; Jimenez-Ferrer, G.; Marcial, N.; Gómez, G.; Sinclair, F. The Role of Local Knowledge in Determining Shade Composition of Multistrata Coffee Systems in Chiapas, Mexico. *Biodivers. Conserv.* **2007**, *16*, 419–436. [CrossRef]
40. Alvarez-Alvarez, E.; Almazán-Núñez, R.; González-García, F.; Brito-Millán, M.; Méndez Bahena, A.; García-Ibáñez, S. Shade Coffee Plantations Maintain Woody Plant Diversity and Structure in a Cloud Forest Landscape of Southern Mexico. *J. For. Res.* **2021**, *32*, 637–648. [CrossRef]
41. Aguirre, S.; Piraneque, N.; Abaunza, C. Especies Con Potencial Para Sistemas Agroforestales en El Departamento Del Magdalena, Colombia. *Inf. Tecnológica* **2021**, *32*, 13–28. [CrossRef]
42. García Mayoral, L.E.; Valdez Hernández, J.I.; Luna Cavazos, M.; López Morgado, R. Estructura y diversidad arbórea en sistemas agroforestales de café en la Sierra de Atoyac, Veracruz. *Madera Bosques* **2015**, *21*, 69–82. [CrossRef]
43. Garza-Lau, R.; Maldonado-Torres, R.; Álvarez-Sánchez, M.E.; Torres-Rivera, J.A.; Garza-Lau, R.; Maldonado-Torres, R.; Álvarez-Sánchez, M.E.; Torres-Rivera, J.A. Characterization of Tree Species Associated with Coffee Cultivation. *Rev. Mex. Cienc. Agrícolas* **2020**, *11*, 25–32. [CrossRef]
44. Goodall, K.E.; Bacon, C.M.; Mendez, V.E. Shade Tree Diversity, Carbon Sequestration, and Epiphyte Presence in Coffee Agroecosystems: A Decade of Smallholder Management in San Ramón, Nicaragua. *Agric. Ecosyst. Environ.* **2015**, *199*, 200–206. [CrossRef]
45. Manson, R.; Hernández-Ortiz, V.; Gallina, S.; Mehltreter, K. *Agroecosistemas Cafetaleros de Veracruz: Biodiversidad, Manejo y Conservación*; Universidad Veracruzana: Veracruz, Mexico, 2008; ISBN 978-970-709-112-2.
46. Méndez, V.E.; Gliessman, S.R.; Gilbert, G.S. Tree Biodiversity in Farmer Cooperatives of a Shade Coffee Landscape in Western El Salvador. *Agric. Ecosyst. Environ.* **2007**, *119*, 145–159. [CrossRef]
47. Ordoñez-Jurado, H.R.; Martínez, D.L.; Cerón Martínez, M. Woody Species Associated with Coffee Production Systems in Southern Colombia. *Rev. Cienc. Agrícolas* **2021**, *38*, 4–16. [CrossRef]
48. Roncal-García, S.; Soto-Pinto, L.; Castellanos-Albores, J.; Ramírez-Marcial, N.; de Jong, B. Sistemas agroforestales y almacenamiento de carbono en comunidades indígenas de Chiapas, México. *Interciencia* **2008**, *33*, 200–206.
49. Suárez-Islas, A.; Williams-Linera, G.; Trejo, C.; Valdez-Hernández, J.; Cetina, V.; Vibrans, H. Local Knowledge Helps Select Species for Forest Restoration in A Tropical Dry Forest of Central Veracruz, Mexico. *Agrofor. Syst.* **2012**, *85*, 1–21. [CrossRef]
50. Villavicencio-Enríquez, L. Caracterización agroforestal en sistemas de café tradicional y rústico, en San Miguel, Veracruz, México. *Rev. Chapingo Ser. Cienc. For. Y Ambiente* **2013**, *19*, 67–80. [CrossRef]

 https://doi.org/10.3390/su18073511

51. Sánchez-Hernández, S.; Mendoza-Briseño, M.A.; García-Hernández, R.V. Diversificación de la sombra tradicional de cafetales en Veracruz mediante especies maderables. *Rev. Mex. Cienc. For.* **2017**, *8*, 7–18.
52. Doraisami, M.; Kish, R.; Paroshy, N.J.; Domke, G.M.; Thomas, S.C.; Martin, A.R. A Global Database of Woody Tissue Carbon Concentrations. *Sci. Data* **2022**, *9*, 284. [CrossRef]
53. CONAFOR. Inventario Nacional Forestal y de Suelos 2015–2020. Available online: https://snmf.cnf.gob.mx/Principaleindicadoresforestalesciclo-2015-2020/ (accessed on 10 February 2025).
54. Ricker, M.; Castillo-Santiago, M.Á.; Gutiérrez-García, G.; Martínez-Salas, E.M.; Mondragón, E. Dataset about Mexico's Forest Diversity: Site Locations of Tree Species, Wood Densities, and Geographic Database of Forest-Vegetation Provinces. *Data Brief.* **2024**, *53*, 110186. [CrossRef]
55. Villaseñor, J.L. Checklist of the Native Vascular Plants of México. *Rev. Mex. Biodivers.* **2016**, *87*, 559–902. [CrossRef]
56. IUCN. The IUCN Red List of Threatened Species. IUCN Red List of Threatened Species. Available online: https://www.iucnredlist.org/en (accessed on 12 March 2025).
57. Semarnat Norma Oficial Mexicana Nom-059-Semarnat-2010. Procuraduría Federal de Protección al Ambiente. 2010. Available online: https://www.gob.mx/profepa/documentos/norma-oficial-mexicana-nom-059-semarnat-2010 (accessed on 12 March 2025).
58. BGCI. Global Tree Portal. Botanic Gardens Conservation International. Available online: https://www.bgci.org/Resources/Bgci-Databases/Globaltree-Portal/Country-Search/?c=MX (accessed on 24 February 2025).
59. Condit, R.; Hubbell, S.P.; Foster, R.B. Identifying Fast-Growing Native Trees from the Neotropics Using Data from a Large, Permanent Census Plot. *For. Ecol. Manag.* **1993**, *62*, 123–143. [CrossRef]
60. Diazgranados, M.; Allkin, B.; Black, N.; Cámara-Leret, R.; Canteiro, C.; Carretero, J.; Eastwood, R.; Hargreaves, S.; Hudson, A.; Milliken, W.; et al. World Checklist of Useful Plant Species. 2020. Available online: https://kew.iro.bl.uk/concern/datasets/7243d727-e28d-419d-a8f7-9ebef5b9e03e (accessed on 12 March 2025).
61. Pompa-Garcia, M.; Rodríguez, J.Á.S. Variation of Carbon Uptake from Forest Species in Mexico: A Review. *Madera Bosques* **2017**, *23*, 225–235. [CrossRef]
62. Garson, G.D. *The Delphi Method in Quantitative Research*; Statistical Associates Blue Book Series; Statistical Associates Publishers: Asheboro, NC, USA, 2013.
63. Legendre, P.; De Cáceres, M. Beta Diversity as the Variance of Community Data: Dissimilarity Coefficients and Partitioning. *Ecol. Lett.* **2013**, *16*, 951–963. [CrossRef]
64. Oksanen, J. vegan: Community Ecology Package, R Package Version 2.6-8. 2024. Available online: https://cran.r-project.org/package=vegan (accessed on 6 May 2025).
65. Reyes, S.R.; Olvera, M.A.P.; Palma, G.I.; Rodríguez, J.A.C.; Vibrans, H.; Sánchez, D.F. Diversidad y uso tradicional de árboles de sombra en cafetales agroecológicos. *Rev. Geogr. Agrícola* **2020**, *64*, 259–273. [CrossRef]
66. Ávila-Bello, C.H.; Hernández-Romero, Á.H.; Vázquez-Luna, D.; Lara-Rodríguez, D.A.; Martínez-Jerónimo, A.; Meneses-García, B.N.; Sánchez-Sandoval, X.M. Design of Complex Agroecosystems: Traditional and Formal Knowledge to Conserve Agrobiodiversity in the Santa Marta Mountains, Veracruz, México. *Environ. Dev. Sustain.* **2024**, *26*, 7129–7161. [CrossRef]
67. Chávez-Hernández, M.G.; Gómez-Barreiro, P.; White, J.D.M.; Way, M. Sonoran Desert Ex Situ Conservation Gap Analysis: Charting the Path towards Conservation. *bioRxiv* **2024**. [CrossRef]
68. de Jong, B.; Montoya, G.; Nelson, K.C.; Soto-Pinto, L.; Taylor, J.; Tipper, R. Community Forest Management and Carbon Sequestration: A Feasibility Study from Chiapas, Mexico. *J. Interciencias* **1995**, *20*, 409–416.
69. Masera, O.R.; Bellon, M.R.; Segura, G. Forest Management Options for Sequestering Carbon in Mexico. *Biomass Bioenergy* **1995**, *8*, 357–367. [CrossRef]
70. Aryal, D.R.; Gómez-González, R.R.; Hernández-Nuriasmú, R.; Morales-Ruiz, D.E. Carbon Stocks and Tree Diversity in Scattered Tree Silvopastoral Systems in Chiapas, Mexico. *Agrofor. Syst.* **2019**, *93*, 213–227. [CrossRef]
71. Souza, H.N.; Cardoso, I.M.; Fernandes, J.M.; Garcia, F.C.P.; Bonfim, V.R.; Santos, A.C.; Carvalho, A.F.; Mendonça, E.S. Selection of Native Trees for Intercropping with Coffee in the Atlantic Rainforest Biome. *Agrofor. Syst.* **2010**, *80*, 1–16. [CrossRef]
72. Lamond, G.; Sandbrook, L.; Gassner, A.; Sinclair, F.L. Local Knowledge of Tree Attributes Underpins Species Selection on Coffee Farms. *Exp. Agric.* **2019**, *55*, 35–49. [CrossRef]
73. Yépez, C.; Muschler, R.; Benjamín, T.; Musálem, M. Selección de especies para sombra en cafetales diversificados de Chiapas, México. *Agrofor. Am.* **2003**, *9*, 55–61.
74. Carvalho, T.; Reis, P.; Bernardi, L.; Marafeli, P.; Martinez, P. Edaphic Mites and Their Response to the Incorporation of Organic Matter from Various Species of Fabaceae into the Soil beneath Coffee Trees. *Acarina* **2019**, *26*, 183–195. [CrossRef]
75. Getachew, M.; Verheyen, K.; Tolassa, K.; Tack, A.J.M.; Hylander, K.; Ayalew, B.; Boeckx, P.; Landuyt, D.; De Frenne, P. Effects of Shade Tree Species on Soil Biogeochemistry and Coffee Bean Quality in Plantation Coffee. *Agric. Ecosyst. Environ.* **2023**, *347*, 108354. [CrossRef]

 https://doi.org/10.3390/su18073511

76. Beyene, G.; Mulatu, K.; Mekonnen, D.F. *Comparative Study of Woody Species Composition, Diversity, Structure and Regeneration Status in Traditionally Managed Coffee Agro-Forestry System Adjacent to Natural Forest in Gimbo District, Southwest Ethiopia*; Research Square Platform LLC: Durham, NC, USA, 2022.
77. Sampayo-Maldonado, S.; Cabrera-Santos, D.; Dávila-Aranda, P.; Rodríguez-Arévalo, N.I.; Orozco-Segovia, A.; Gianella, M.; Bell, E.; Way, M.; Manson, R.H.; Quintas, G.S.; et al. Using the Optimal Seed Germination Temperature Approach to Determine the Potential Distribution of Inga Jinicuil in Mexico under Climate Change Scenarios. *Sci. Rep.* **2025**, *15*, 3951. [CrossRef]
78. Siles, P.; Harmand, J.-M.; Vaast, P. Effects of Inga Densiflora on the Microclimate of Coffee (*Coffea arabica* L.) and Overall Biomass under Optimal Growing Conditions in Costa Rica. *Agrofor. Syst.* **2010**, *78*, 269–286. [CrossRef]
79. Cannavo, P.; Sansoulet, J.; Harmand, J.-M.; Siles, P.; Dreyer, E.; Vaast, P. Agroforestry Associating Coffee and *Inga densiflora* Results in Complementarity for Water Uptake and Decreases Deep Drainage in Costa Rica. *Agric. Ecosyst. Environ.* **2011**, *140*, 1–13. [CrossRef]
80. Hergoualc'h, K.; Blanchart, E.; Skiba, U.; Hénault, C.; Harmand, J.-M. Changes in Carbon Stock and Greenhouse Gas Balance in a Coffee (*Coffea arabica*) Monoculture versus an Agroforestry System with *Inga densiflora*, in Costa Rica. *Agric. Ecosyst. Environ.* **2012**, *148*, 102–110. [CrossRef]
81. Breedlove, D.E. The Phytogeography and Vegetation of Chiapas (Mexico). In *Vegetation and Vegetational History of Northern Latin America*; Elsevier: Amsterdam, The Netherlands, 1973; pp. 149–165.
82. Wootton, A.; Enríquez, P.L.; Navarrete-Gutiérrez, D. Regional Patterns of Vegetation, Temperature, and Rainfall Trends in the Coastal Mountain Range of Chiapas, Mexico. *Atmósfera* **2023**, *36*, 91–122. [CrossRef]
83. González-Espinosa, M.; Benayas, J.; Marcial, N.; Huston, M.; Golicher, D. Tree Diversity in the Northern Neotropics: Regional Patterns in Highly Diverse Chiapas, Mexico. *Ecography* **2004**, *27*, 741–756. [CrossRef]
84. Richter, M. The Ecological Crisis in Chiapas: A Case Study from Central America. *Mt. Res. Dev.* **2000**, *20*, 332–339. [CrossRef]
85. Cayuela, L.; Golicher, D.J.; Rey-Benayas, J.M. The Extent, Distribution, and Fragmentation of Vanishing Montane Cloud Forest in the Highlands of Chiapas, Mexico. *Biotropica* **2006**, *38*, 544–554. [CrossRef]
86. Haggar, J.; Medina, B.; Aguilar, R.M.; Munoz, C. Land Use Change on Coffee Farms in Southern Guatemala and Its Environmental Consequences. *Environ. Manag.* **2013**, *51*, 811–823. [CrossRef]
87. Harvey, C.A.; Pritts, A.A.; Zwetsloot, M.J.; Jansen, K.; Pulleman, M.M.; Armbrecht, I.; Avelino, J.; Barrera, J.F.; Bunn, C.; García, J.H.; et al. Transformation of Coffee-Growing Landscapes across Latin America. A Review. *Agron. Sustain. Dev.* **2021**, *41*, 62. [CrossRef]

 https://doi.org/10.3390/su18073511

land

MDPI

Article

Evaluation of Habitat Quality in Karst Mountainous Areas of Guanling County Based on InVEST and MGWR Models

Shuanglong Du [1,2], Zhongfa Zhou [1,2,3,*], Denghong Huang [1,2], Fei Dong [1,2], Xiandan Du [1,2], Yining Luo [1,2], Qingqing Dai [1,2] and Yue Yang [1,2]

[1] School of Karst Science, Guizhou Normal University, Guiyang 550001, China; 222100170570@gznu.edu.cn (S.D.); hdh@gznu.edu.cn (D.H.); 232100170578@gznu.edu.cn (F.D.); dxd@gznu.edu.cn (X.D.); luoyining@gznu.edu.cn (Y.L.); 232100170576@gznu.edu.cn (Q.D.); 222100170578@gznu.edu.cn (Y.Y.)
[2] State Engineering Technology Institute for Karst Desertification Control, Guiyang 550001, China
[3] School of Geography & Environmental Science, Guizhou Normal University, Guiyang 550001, China
* Correspondence: fa6897@gznu.edu.cn

Abstract: As a core karst region in Southwest China, Guanling County plays a crucial role in regional ecological governance. This study integrates the InVEST model, landscape pattern index analysis, and the MGWR spatial model to systematically explore the dynamic mechanisms of habitat quality in Guanling's karst mountains. Key findings include: (1) Landscape pattern alterations exhibit significant impacts on habitat quality, characterized by strong spatial heterogeneity; (2) Expansion of forest and grassland effectively buffers the negative effects of construction land expansion, forming an ecological compensation mechanism through enhanced landscape connectivity; (3) Between 2000 and 2020, the proportion of high-importance habitat quality zones increased from 54.79% to 56.16%, with moderate-importance zones stabilizing at approximately 7.80% and general-importance zones growing to 2.46%. The results provide a multi-scale analytical framework for habitat protection and land use optimization in fragile karst ecosystems.

Keywords: guanling county; karst mountainous area; habitat quality; landscape pattern index; MGWR model; InVEST model

1. Introduction

Karst landforms have garnered global attention owing to their unique geological structures and fragile ecosystems [1,2]. In southern China, karst landforms not only constitute important natural landscapes [3] but also serve as statistically significant ecological protection barriers for the upper reaches of the Yangtze River and the Pearl River [4–6]. The environmental vulnerability of karst areas has become increasingly evident due to intensified global climate change and human activities [7–10]. The terrain in karst mountainous areas is highly undulating, with pronounced spatial heterogeneity. With the rapid progression of urbanization and industrialization, the regional landscape configuration has experienced substantial transformations [11,12]. These alterations in landscape patterns have given rise to a series of environmental challenges [13,14], with soil erosion and desertification being the most pressing issues in this area [15,16], but also threaten biodiversity maintenance. To address the increasingly severe environmental problems, the Chinese government has implemented a series of ecological restoration plans [17–19]. At the same time, landscape pattern changes also threaten biodiversity maintenance [20,21], seriously endangering regional quality of ecological habitats. Therefore, understanding the impact of landscape pattern changes on quality of ecological habitats is crucial for formulating reasonable natural resource management and protection measures. Guanling County, a representative karst region, is dominated by extensive karst landforms. Fluctuations in its landscape patterns have exerted diverse impacts on habitat quality, threatening the sustainability of socio-economic and ecological functions. Thus, investigating habitat quality and deciphering the response mechanisms of landscape patterns in Guanling County can facilitate a more comprehensive assessment of ecological conditions across the broader southwest karst region. This research offers critical reference value for habitat quality studies in similar karst environments.

In recent years, studies on the impact of landscape pattern changes on quality of ecological habitats have received extensive attention [22]. For instance, Zhu et al. [21] explored the effects of urbanization and landscape patterns on quality of ecological habitats in Hangzhou using OLS and GWR models. They found that rapid urbanization had a statistically significant negative impact on quality of ecological habitats, with the degree and direction of landscape pattern influence varying spatiotemporally. Li et al. [23] studied the spatiotemporal evolution of landscape patterns and their impacts on quality of ecological habitats in Nanchang City, discovering that the rapid expansion of low-habitat-suitability land utilization types (e.g., construction land) increased landscape fragmentation and reduced stability. Gu et al. [24] analyzed quality of ecological habitats and landscape pattern evolution characteristics in Shaanxi Province, finding a clear spatial correlation between them, with statistically significant spatial clustering effects that weakened as urbanization intensified. In related research, models such as OLS [21], GWR [25], and MGWR [26] have been used to analyze landscape changes, each with unique characteristics and application contexts. The OLS model is simple and efficient but lacks precision in areas with strong spatial heterogeneity [27]. The GWR model addresses OLS limitations [28] through localized parameter estimation, though it may still struggle to capture spatial variability for all variables [21]. The MGWR model further enhances GWR functionality [26], allowing explanatory variables to exhibit diverse spatial heterogeneities across scales. This improvement in model flexibility and accuracy [29] makes MGWR particularly suitable for uncovering spatial instabilities in landscape change drivers, offering novel insights for research in karst mountainous regions. The MGWR model effectively captures dynamic processes under spatial heterogeneity [29]. For example, Tang et al. [30] used MGWR to comprehensively study the spatial heterogeneity of climate and landscape change impacts on soil conservation services at county, township, and watershed scales. Rong et al. [31] applied MGWR to explore driving factors of soil and water conservation services in the Beijing-Tianjin-Hebei region, discussing multi-scale mechanisms. Zhou et al. [32] combined Geographical Detector (GD), Random Forest (RF), and MGWR to reveal PM2.5 concentration changes in the Niger River Basin. Hu et al. [29] integrated GTWR and MGWR to explore spatiotemporal driving mechanisms of landscape patterns on quality of ecological habitats in Nanjing.

The InVEST (Integrated Valuation of Ecosystem Services and Tradeoffs) model, as an ecosystem service assessment tool, has been widely applied in ecological research at various scales [33,34]. Its quality of ecological habitats module quantifies the impact of human activities on biodiversity [35], assessing quality of ecological habitats by considering threat sources and their impacts on specific ecosystems [36]. For instance, Wu et al. [37] used InVEST to study spatiotemporal quality of ecological habitats changes in the Guangdong-Hong Kong-Macao Greater Bay Area. Wang et al. [38] explored regional impacts, and Wei et al. [39] analyzed changes in the Aibih Lake Basin. However, for complex terrains such as karst mountainous areas, a single InVEST model may insufficiently represent reality. Integrating InVEST with MGWR facilitates more refined habitat quality assessments in highly heterogeneous landscapes, providing new perspectives for studies on fragmented ecosystems.

This study is intended to establish a quality of ecological habitats assessment system for karst mountainous areas using InVEST and MGWR spatial examination. It analyzes quality of ecological habitats changes under different landscape patterns and explores response mechanisms to landscape pattern changes, providing a comprehensive framework for karst quality of ecological habitats research. Through case studies, we aim to offer scientific support for ecological protection and policy formulation in these regions.

2. Materials and Methods

2.1. Study Area

Guanling Buyi and Miao Autonomous County (hereafter 'Guanling County') is situated in west-central Guizhou Province, China (Figure 1). Its geographical coordinates range from 105°21′ E to 105°56′ E and 25°29′ N to 25°55′ N. The total area of this county is approximately 1468 square kilometers. The terrain is highly undulating, with elevations averaging 800–1300 m. The region has a subtropical humid monsoon climate. The annual average temperature is about 14 °C, and the annual precipitation is around 1100 mm. The frost-free period is relatively long, which is conducive to the growth of various plants. This unique geographical location and climate conditions, combined with abundant natural resources, have enabled Guanling County to have a relatively high biodiversity and ecosystem service functions. The unique geological conditions of karst areas hinder restoration following soil erosion or land degradation [40,41]. In addition, affected by both natural and human activities, the phenomenon of rocky desertification in this region is very serious, causing varying degrees of interference to the integrity and stability of the regional ecosystem. In recent years, with the implementation of land management and ecological protection measures, the area of rocky desertification has statistically significantly decreased, dropping from 47.41% to below 10%. Therefore, this area is not only an ideal place for studying the ecosystem service functions of karst landforms, but also an important region for exploring the interaction relationship between human activities and the natural environment [42].

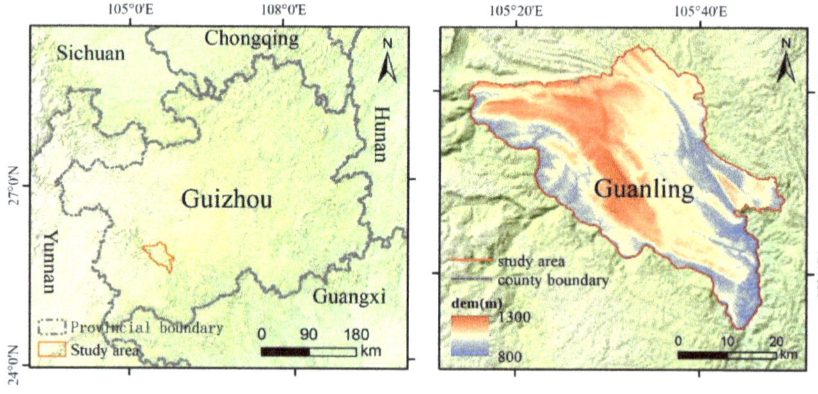

Figure 1. Location of the study area.

2.2. Data Sources

Land use data for 2000, 2010, and 2020 were obtained from the Resource and Environmental Science Data Center of the Chinese Academy of Sciences (http://www.resdc.cn (accessed on 1 December 2024)), with a spatial resolution of 30 m [43]. The data were preprocessed and reclassified using ArcGIS 10.8. The DEM data were sourced from the Geospatial Data Cloud Platform (http://www.gscloud.cn (accessed on 1 December 2024)), with a resolution of 10 m. Based on the ecological context of southern China's karst mountainous areas, we identified four threat sources: cultivated land, urban land, rural settlements, and transportation land. Following the InVEST Model Manual [44] and prior studies [6,45,46], we determined the maximum impact distances, weights, spatial attenuation types, habitat suitability, and sensitivity of each land utilization type to threat sources. Tables 1 and 2.

Table 1. Threat factor parameters.

Threat Sources	Maximum Impact Distance (km)	Weight	Spatial Attenuation Type
arable land	1	0.7	Linear
urban land	5	1	Exponential
rural settlements	3	0.6	Exponential
transport land	3	0.5	Exponential

Table 2. Sensitivity of different landscape types to habitat threat factors.

Land-Use Types (Level I)	Land-Use Types (Level II)	Urban Land	Cropland	Urban Land	Rural Settlements	Transport Land
Cropland	paddy	0.4	0.3	0.7	0.6	0.5
	dry land	0.2	0.3	0.7	0.6	0.5
Woodland	arbor forest	1	0.6	0.8	0.3	0.7
	shrubwood	0.8	0.6	0.6	0.3	0.5
Grassland	grassland	0.7	0.5	0.6	0.5	0.3
Waterbody	river	0.8	0.6	0.6	0.5	0.3
	lake and reservoir	0.9	0.7	0.7	0.6	0.4
Built-up land	urban land	0	0	0	0	0
	rural settlements	0	0	0	0	0
	other built-up land	0	0	0	0	0

2.3. Methods

2.3.1. InVEST Model

The InVEST quality of ecological habitats module evaluates quality of ecological habitats on the basis of land utilization data, incorporating the relative sensitivity of habitat types to threats, threat factor data, maximum impact distances, and weights [47]. The quality of ecological habitats index ranges from 0 to 1. The larger the value, the better the quality of ecological habitats [48]. The calculation formula is:

$$Q_{xj} = H_j \left[1 - \left(\frac{D_{xj}^Z}{D_{xj}^Z + k^Z} \right) \right] \tag{1}$$

$$D_{xj} = \sum_{r=1}^{R} \sum_{y=1}^{Y_r} (\omega_r / \sum_{r=1}^{R} \omega_r) r_y i_{rxxy} \beta_x S_{jr} \tag{2}$$

$$i_{rxy} = 1 - \left(d_{xy} / d_{rmax} \right) \tag{3}$$

where Q_{xj} is the quality of ecological habitats index; H_j is the habitat suitability of the jth place class; D_{xj} is the habitat degradation index of the xth raster in the jth place class; k is the half-saturation constant; Z is the model transformation coefficient; R is the total number of threat sources; r is the rth threat source; Yr is the total number of rasters for the rth threat source; y is the raster cell where the rth threat source is located; and ω_r is the weight of the

rth threat source; i_{rxy} is the maximum coercive distance of the threat source of category r; β_x is the accessibility of the threat source to the xth grid; S_{jr} is the sensitivity of the jth land class to the threat source of category r; d_{xy} denotes the distance between grids x and y, and d_{rmax} is the maximum influence range of the threat factor r.

2.3.2. Spatial Autocorrelation

Spatial autocorrelation can quantitatively describe the degree of similarity of attribute values and their spatial correlation patterns of spatially proximate regional units [49], which is employed to explore the spatial distribution characteristics of a certain geographic phenomenon and the degree of aggregation among variables between neighbouring regions [50], and is divided into global autocorrelation and local autocorrelation. In order to measure and test the degree of spatial dependence and spatial variation of quality of ecological habitats in the study area, this research effort conducted spatial autocorrelation analyses on the distribution pattern of quality of ecological habitats indices of karst mountainous areas in 2000, 2010, and 2020 to explore the spatial and temporal heterogeneity of the quality of ecological habitats in karst mountainous areas, and the indices were calculated to include the global spatial autocorrelation Moran's I index, as well as local spatial autocorrelation. Moran's I index, calculated as [51]:

$$I = \frac{N\sum_{i=1}^{N}\sum_{j=1}^{N}\omega_{ij}(x_i-\overline{x})(x_j-\overline{x})}{\left(\sum_{i=1}^{N}\sum_{j=1}^{N}\omega_{ij}\right)\sum_{i=1}^{N}(x_j-\overline{x})^2} \quad (4)$$

where I is the global Moran's I in the range $[-1, 1]$; N refers to the total number of spatially observed objects in the study area; x_i and x_j represent the observations at the ith and jth spatial locations, respectively; $\ddot{x}$ is the mean value of all the observations; and ω_{ij} is then used as a spatial weight matrix to describe the spatial proximity of the ith and jth monitoring objects.

Local spatial autocorrelation analysis was performed using LISA clusters. LISA clustering can indicate correlation between spatial variables at a certain level of significance [52], and four spatial distribution patterns were generated by local Moran's I, including high-high, high-low, low-high, and low-low aggregation [53].

2.3.3. The MGWR Model

Multi-scale geographically weighted regression (MGWR), an extension of traditional GWR [54], models spatial variability by allowing variable-specific bandwidths. Its ability to capture the spatial heterogeneity of data is improved by its ability to have different coefficients in different geographical units [55]. The formula is as follows [56]:

$$\gamma_i = \sum_{i=1}^{n}\beta_{bwj}(u_i, v_i)X_{ij} + \varepsilon_i \quad (5)$$

where: γ_i denotes quality of ecological habitats; x_{ij} denotes natural or socio-economic factors; ε_i denotes the random error term; (u_i, v_i) denotes the spatial coordinates of the sample points; β_{bwj} is the regression coefficient of fitting variable j using a specific bandwidth; the larger the absolute value of the regression coefficient, the stronger the effect on the quality of ecological habitats [57].

Reference to existing studies [58–60], six landscape pattern indices, PD (patch density), LPI (maximum type patch index), LSI (landscape shape index), SHDI (Shannon diversity index), SHEI (Shannon homogeneity index), and AI (agglomeration index), were selected for data processing and examination

2.3.4. Research Framework

The InVEST model was employed to quantify the spatial distribution of quality of ecological habitats. The quality of ecological habitats index, serving as the dependent variable of the MGWR model, was calculated using InVEST for the years 2000, 2010, and 2020 to generate raster data. Taking a 500 m × 500 m grid as the unit, the average quality of ecological habitats of each grid was extracted as the response variable for MGWR, further revealing the scale dependency of landscape patterns on quality of ecological habitats. The research framework is illustrated in Figure 2:

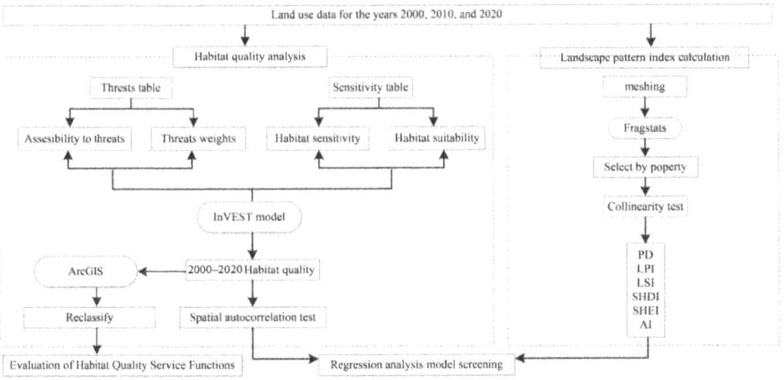

Figure 2. Technology roadmap.

3. Results

3.1. Analysis of Spatial and Temporal Evolution of Habitat Quality

The quality of ecological habitats index effectively reflects the suitability of the regional ecosystem. Using the InVEST model, the quality of ecological habitats of the study area was analyzed and classified into four intervals: low (0–0.25), relatively low (0.25–0.5), relatively high (0.5–0.75), and high (0.75–1).

From 2000 to 2020, the proportions of quality of ecological habitats levels in Guanling County, ranked in descending order, were high-level quality of ecological habitats, relatively high-level quality of ecological habitats, relatively low-level quality of ecological habitats, and low-level quality of ecological habitats (Table 3) Among them, the distribution range of high-level and relatively high-level quality of ecological habitats in Guanling County was larger than that of low-level and relatively low-level quality of ecological habitats. The high-level quality of ecological habitats increased by 0.19% over the 20-year period, and the relatively high-level quality of ecological habitats increased by 1.38%. The continuous expansion of high-level and relatively high-level quality of ecological habitats areas contributed to an overall improvement in regional quality of ecological habitats. The relatively low-level quality of ecological habitats decreased by 2.54%, indicating that the low-level quality of ecological habitats improved year by year. However, low-level quality of ecological habitats increased by 0.97%, indicating an expanding trend, though its impact on overall quality of ecological habitats remained relatively minor.

Table 3. Percentage of habitat quality classes, 2000, 2010 and 2020.

Habitat Quality Level	2000/%	2010/%	2020/%
Low	1.39	1.79	2.36
Relatively low	36.13	35.31	33.59
Relatively high	54.78	55.12	56.16
High	7.70	7.79	7.89

In terms of space (Figures 3 and 4), low-level quality of ecological habitats was concentrated in the construction area, low relatively low-level quality of ecological habitats was concentrated in the farmland area, relatively high-level quality of ecological habitats was concentrated in the grassland area, high-level quality of ecological habitats was widely distributed in the forest and water area, and the quality of ecological habitats in the natural ecological areas with less human activities was relatively high. The areas with frequent human activities in the social and economic development regions were the concentrated areas of low quality of ecological habitats. This indicates that the spatial distribution of different levels of quality of ecological habitats in Guanling County has a strong correlation with land utilization, and the quality of ecological land is positively correlated with the quality of habitat, while the result of construction land is the opposite.

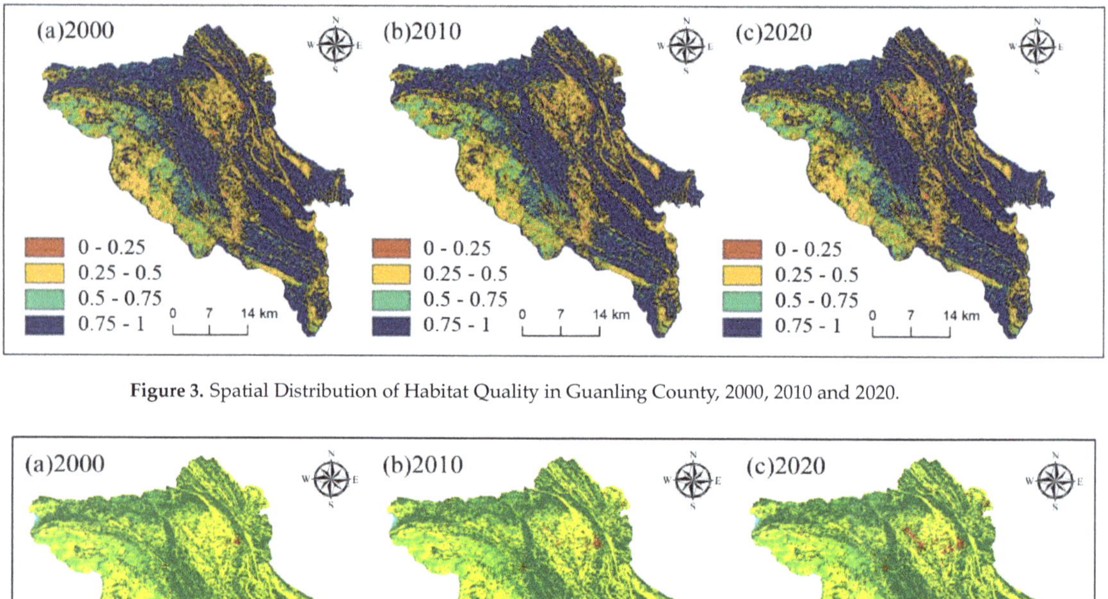

Figure 3. Spatial Distribution of Habitat Quality in Guanling County, 2000, 2010 and 2020.

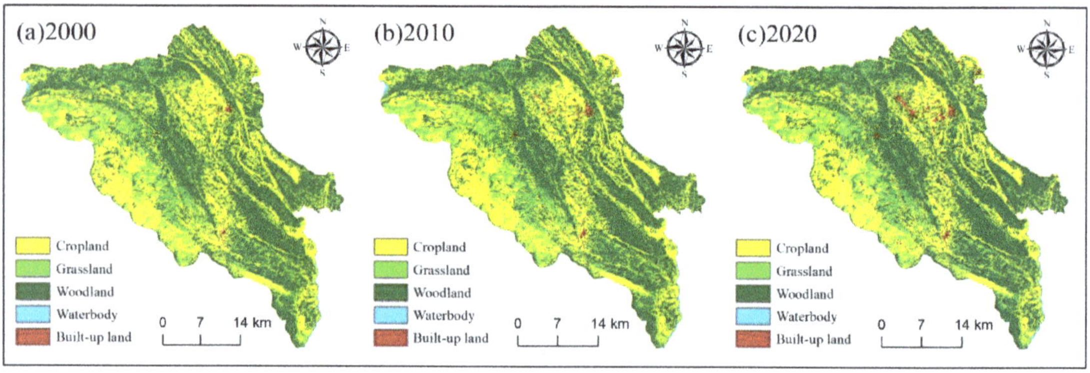

Figure 4. Spatial distribution of land use types in Guanling County, 2000–2020.

3.2. Spatial Autocorrelation Test for Habitat Quality

Prior to applying the MGWR model, spatial autocorrelation of regional quality of ecological habitats was assessed. Using ArcGIS, the study area was divided into 500 m × 500 m grids, and the quality of ecological habitats of each grid was extracted using the Multi-Value Extract to Point tool. Subsequently, spatial autocorrelation examination was performed to derive descriptive statistical parameters for the spatial effects of quality of ecological habitats. As shown in Table 4, the Moran's I values from 2000 to 2020 all passed the significance test ($p < 0.01$), with the numerical range being between 0.70 and 0.71, and the Z values were all greater than 25. This suggests that quality of ecological habitats in Guanling County exhibits statistically significant spatial autocorrelation. Further examination revealed a decline in Moran's I values from 2000 to 2020, indicating a weakening trend in the spatial aggregation of quality of ecological habitats.

Table 4. Spatial Autocorrelation Parameters of Habitat Quality in Guanling County, 2000, 2010, and 2020.

Year	Moran's I	Z Value	*p* Value
2000	0.71	78.25	0
2010	0.71	77.89	0
2020	0.70	77.14	0

In order to better study the internal spatial clustering characteristics of quality of ecological habitats in Guanling County, it is necessary to conduct local spatial autocorrelation examination of the study area. By identifying hotspots (high-high clustering), coldspots (low-low clustering), and spatial outliers (high-low clustering and low-high clustering) [60], the spatial distribution patterns of geographical phenomena in Guanling County can be studied (Figure 5). Among them, hotspots are mainly distributed in forestland and grassland areas, and the quality of ecological habitats shows a high-high clustering state; coldspots are mainly concentrated in cultivated land and construction areas, and the quality of ecological habitats shows a low-low clustering state. Overall, from 2000 to 2020, the quality of ecological habitats in Guanling County has a high degree of spatial clustering in forestland and grassland areas.

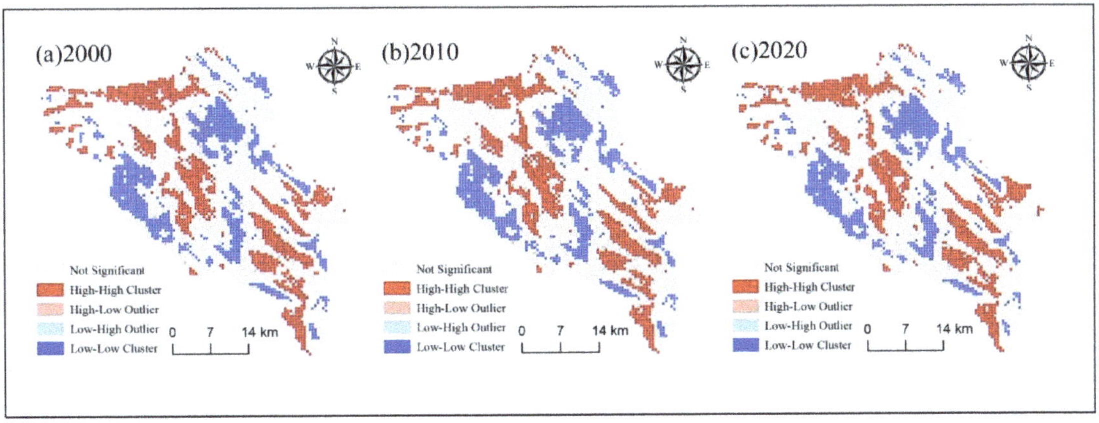

Figure 5. Spatial Aggregation Characteristics of Habitat Quality in Guanling County, 2000, 2010, and 2020.

The spatial heterogeneity of quality of ecological habitats in Guanling Karst mountainous area requires full consideration of regional land policies. In high-high aggregation areas, more attention should be paid to natural restoration to enhance the resilience of the ecosystem and consolidate the achievements of ecological protection. For instance, adhering to the policies of returning farmland to forest and protecting natural forests, strictly controlling the frequency of human activities to maintain the stability of regional ecological environment. On this basis, appropriate development of ecological tourism and other projects can be carried out as special channels for economic development. In low-low aggregation areas, strict control of environmental pollution and strengthening land utilization management are needed; an ecological compensation mechanism should be established to counterbalance the weakening of habitats caused by the expansion of construction land. For example, adjusting the industrial structure, developing low-carbon industries, building green industrial parks, and developing clean energy can promote the coordinated development of regional human and land resources. In areas with unstable quality of ecological habitats, ecological connections between different regions need to be strengthened. For instance, constructing ecological corridors to enhance the connectivity of habitats to improve the regional habitat function; strengthening land management, etc., targeted ecological restoration should be carried out to gradually improve the regional quality of ecological habitats.

3.3. Landscape Pattern Index Evolution

The landscape pattern indices exhibit distinct spatial distribution patterns in Guanling County (Figure 6). Taking 2020 as an example, the study found that the high values of PD, LSI, SHDI and SHEI were concentrated in the western, southern and northeastern parts of the study area. The land utilization types in these areas were mainly construction land, cultivated land, grassland and water bodies. The low values were mainly distributed in the areas with concentrated forest land. The high and low value distribution patterns of LPI and AI were exactly opposite to those of the above indices. As shown in Table 5, landscape-level calculations reveal statistically significant downward trends in PD, LPI, and LSI values over the past two decades, while the SHDI, SHEI and AI values have shown an upward trend.

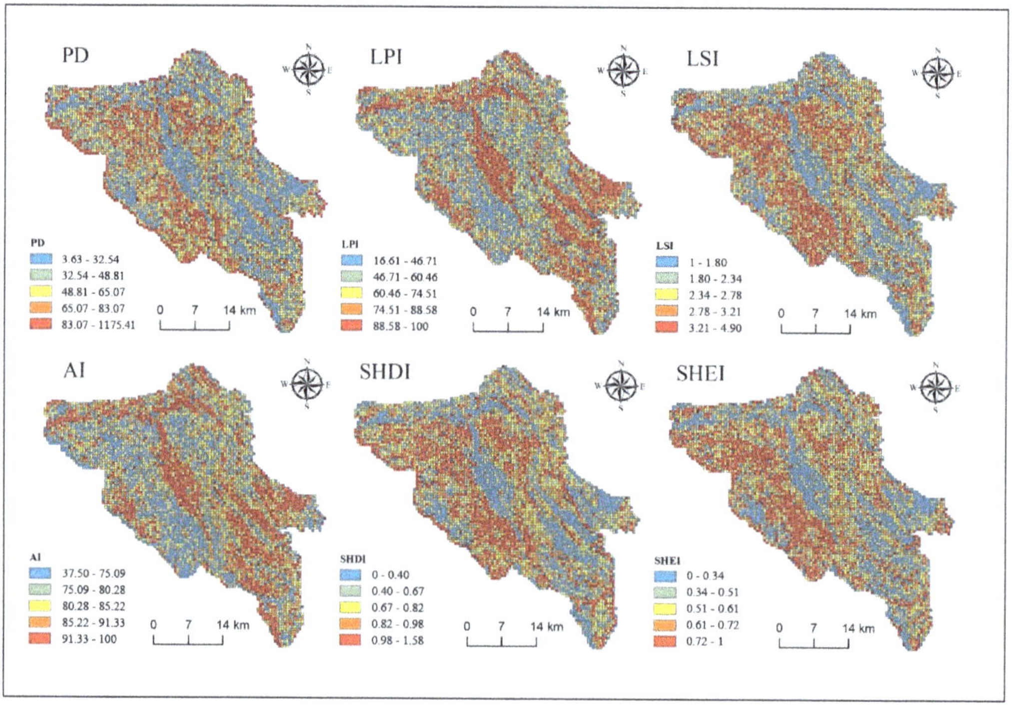

Figure 6. Spatial Distribution of Landscape Pattern Index in Guanling County.

Table 5. Guanling County Landscape Pattern Index, 2000, 2010, and 2020.

Year	PD	LPI	LSI	SHDI	SHEI	AI
2000	38.53	23.72	137.78	1.06	0.54	79.25
2010	38.00	22.61	135.80	1.08	0.55	79.56
2020	35.16	21.37	128.54	1.10	0.57	80.67

A decrease in PD value indicates that the number of patches within the same area has decreased, which usually reflects a decline in the overall heterogeneity or fragmentation degree of the landscape. For instance, large-scale monoculture plantations make the originally diverse landscape more uniform, reducing the number of patches and thereby increasing the average area of each patch. The decrease in the PD value indicates that Guanling County has been affected by natural processes such as forest restoration, or human activities such as agricultural integration. As a result, multiple small patches have merged into larger ones, leading to a reduction in the total number of patches. A decrease in the LPI value indicates that the proportion of the largest patches has decreased relative to the total area of the landscape. A decrease in LSI value means that the shapes of landscape patches become simpler or more regular, which usually reflects the smoothing of patch edges or the tendency of patch shapes to be rectangular or circular, indicating that the shape of the landscape in Guanling County has been affected to some extent by agricultural activities, nature conservation or other human interventions. Among them, large-scale land leveling and farmland integration agricultural activities reduce patch edges, making patch shapes more regular; nature conservation projects promote the restoration of forests or other natural landscapes, reducing fragmentation. An increase in the Shannon diversity index (SHDI) reflects greater diversity in patch types across Guanling County. This is due to the introduction of new land utilization types or ecological restoration measures leading to an increase in the number of different types of patches, making the landscape more diverse. From 2010 to 2020, the emergence of new residential areas around the county towns led to an increase in the diversity of patch types. The fragmentation expansion of construction land within the county was also an important reason for the rise in SHDI. An increase in Shannon evenness index (SHEI) indicates a more uniform distribution of different patch types in the landscape, with no single type of patch occupying an overwhelming proportion, reflecting a more balanced distribution of patch types in Guanling County. An increase in Aggregation Index (AI) means that similar types of patches are more inclined to cluster together, forming larger continuous areas rather than being scattered throughout the landscape.

Collectively, these trends indicate an evolutionary process toward greater diversity and structural complexity in the landscape of Guanling's karst mountainous area. This evolution is conducive to enhancing the service functions of the ecosystem, such as providing more habitats, strengthening the stability of the ecosystem and its ability to resist disturbances. However, such changes also need to be analyzed in the light of specific circumstances, such as whether they have had adverse effects on specific species and whether they are in line with the local ecological planning goals. If these changes are achieved through sustainable means and taking into account the characteristics and needs of the local ecosystem, they are usually positive. Conversely, if these changes lead to the disruption of certain key ecological processes, such as urban expansion, economic development, ecological protection, land policies and other activities related to them, they may bring negative effects.

3.4. Analysis of the Relationship Between Habitat Quality and Landscape Pattern

The MGWR model demonstrated high explanatory power (>89%), confirming the suitability of the selected influencing factors for modeling. Table 6 presents the descriptive statistics of the regression coefficients of the six landscape indices. The average and median values of the regression coefficients of PD, SHDI, and SHEI are close, and their impacts on quality of ecological habitats tend to be homogeneous spatially. The average and median values of the regression coefficients of LPI, LSI, and AI differ statistically significantly, and their impacts on quality of ecological habitats have strong heterogeneity spatially. The variation range of the regression coefficients of each landscape index is large, indicating that the differences in the direction and intensity of their spatial impacts are statistically significant, and their impacts on quality of ecological habitats have statistically significant spatial heterogeneity (Figure 7). The six landscape indices have positive or negative impacts on the quality of ecological habitats of different areas in the study area. Among them, PD, LPI, LSI, SHEI, and AI have positive impacts on quality of ecological habitats in the study area, accounting for 59.87%, 51.59%, 54.42%, 81.07%, and 59.14% of the total area, respectively. Among them, the positive impacts of PD and LPI are highly consistent with the distribution areas of forests, gradually weakening outward from the center, and the heterogeneity environment has a positive impact on the quality of ecological habitats of forests. While LSI and AI have different impacts on quality of ecological habitats in different land utilization types. The positive impact of SHEI is mainly concentrated in the cultivated land and grassland areas, indicating that the more balanced the landscape patches are, the more positive the impact on regional quality of ecological habitats is. SHDI has a negative impact on quality of ecological habitats as the main factor, accounting for 79.00% of the total area. Overall, landscape diversity has the greatest negative impact on the quality of ecological habitats of forest and grassland, gradually weakening outward from the center to the areas with forest and grassland, and has positive impacts on construction land and cultivated land, gradually weakening outward from the center. It indicates that the quality of ecological habitats in these areas is relatively higher, but decreases with the reduction of landscape diversity. The main reason for this phenomenon lies in the strong connectivity of ecological land such as forests and grasslands in Guanling Karst mountainous area. The fragmented and diverse landscape patterns to some extent will disrupt the connectivity of the ecosystem, and isolated ecological land is relatively fragile, with a greatly reduced ecological restoration capacity. Eventually, its ecological service functions are reduced. While construction land and cultivated land such as economic construction land are distributed relatively scattered under the influence of the spatial geography of the Karst mountainous area, isolated construction land and cultivated land are bound to be affected in economic development. However, they have great significance in stabilizing the environmental quality of regional species habitats and maintaining biodiversity.

Table 6. Statistical description of regression coefficients of MGWR model.

Landscape Index	Mean	Min	Median	Max
PD	0.06	−1.44	0.07	1.64
LPI	−0.02	−1.22	0.02	1.41
LSI	−0.02	−10.03	0.10	9.15
SHDI	−0.39	−2.23	−0.34	2.20
SHEI	0.23	−0.98	0.19	1.78
AI	0.05	−10.60	0.21	9.70

3.5. Analysis of the Importance of Habitat Quality Functions

To systematically and accurately assess the significance of quality of ecological habitats service functions in Guanling County, and to provide scientific basis for the coordinated development of land utilization management, social and economic construction, and ecological environment protection in typical karst areas. Based on the quality of ecological habitats index (0–1) output by InVEST, using ArcGIS 10.8 software, the research results on carbon storage and quality of ecological habitats were reclassified. Using the natural break point method (Jenks), they were reclassified into 4 grades, corresponding to the gradient of the importance of ecological service functions", the importance of the ecosystem service functions of the quality of ecological habitats in Guanling County was divided into four grades: generally important, relatively important, moderately important, and highly important (as shown in Table 7).

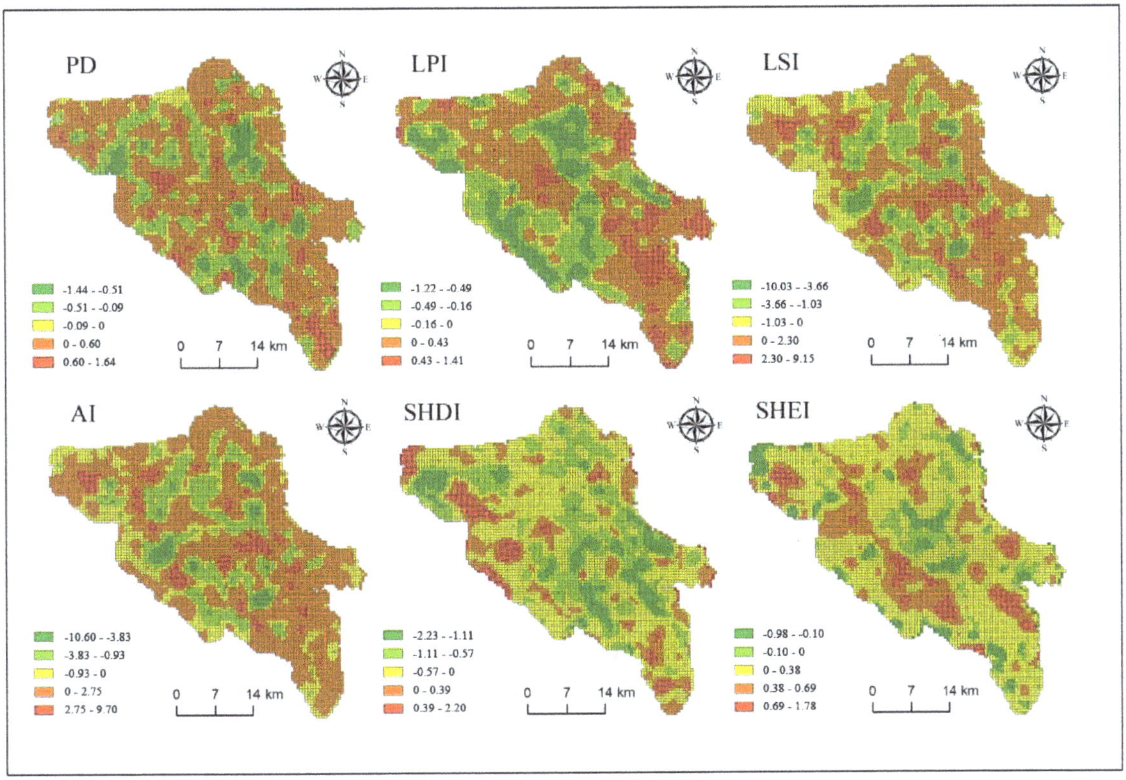

Figure 7. Spatial Distribution of Model Coefficients of MGWR Model in 2020.

Table 7. Importance Grades of Carbon Storage and Habitat Quality Services in Guanling County.

	General Important	More Important	Moderate Important	Highly Important
Habitat quality	≤0.25	0.25–0.50	0.5–0.75	≥0.75

As shown in Figure 8, the spatial distribution of the importance grades of quality of ecological habitats in Guanling County has a very high correlation with land utilization. Among them, the general importance area of quality of ecological habitats is mainly distributed in the construction area, the more important area is mainly concentrated in the grassland area, the moderate importance area is concentrated in the farmland area, and the high importance area is mainly distributed in the forestland and water area.

An examination was conducted on the importance grades of quality of ecological habitats in Guanling County from 2000 to 2020. The general importance area, moderate importance area and high importance area of quality of ecological habitats showed an increasing trend, and the average area increased by 13.81 km^2, 3.25 km^2 and 20.17 km^2 respectively in 20 years, with proportions of 0.94%, 0.22% and 1.37% respectively (Table 8). However, the more important area of quality of ecological habitats showed a decreasing trend, and it decreased by 37.23 km^2 in 20 years, with a proportion of 2.54%. Due to the influence of regional land utilization changes and the fragmentation of grassland landscape, the area of the more important area of quality of ecological habitats decreased statistically significantly, but the reduction of the more important area compensated for the moderate and high importance areas of quality of ecological habitats to some extent, promoting the overall improvement of the quality of ecological habitats service function in the karst mountainous area of Guanling County.

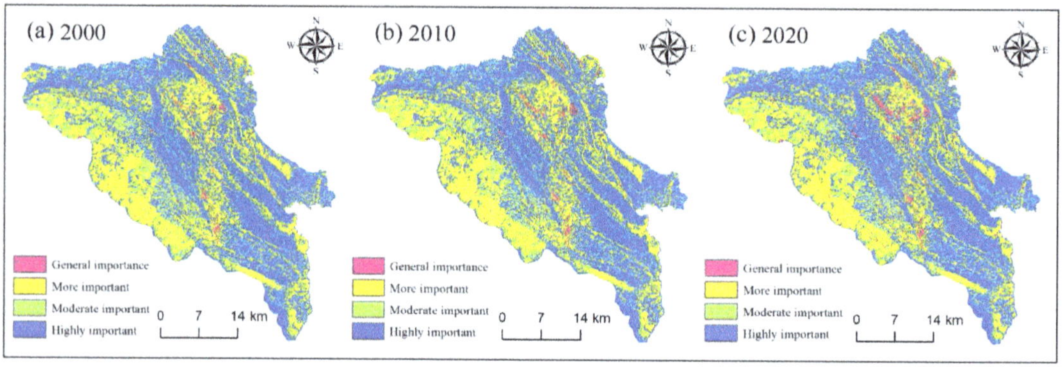

Figure 8. Spatial Distribution of the Importance Grades of Habitat Quality in Guanling County from 2000 to 2020.

Table 8. Importance Area and Proportion of Habitat Quality in Guilin County.

Year	General Important	More Important	Moderate Important	Highly Important
2000/km^2	22.24	530.33	111.18	804.24
%	1.51%	36.13%	7.57%	54.79%
2010/km^2	28.22	518.34	112.24	809.19
%	1.92%	35.31%	7.65%	55.12%
2020/km^2	36.05	493.11	114.43	824.41
%	2.46%	33.59%	7.80%	56.16%

4. Discussion

4.1. Impact of Land Use Change on Habitat Quality

The study used the spatial examination tools of ArcGIS 10.8 software to statistically analyze the average quality of ecological habitats across different land utilization/cover types in Guanling County. As shown in Table 9 and Figure 9.

Table 9. Average Habitat Quality Index of Different Land Use/cover Change Types.

Land Use	2000	2010	2020
Cropland	0.13	0.13	0.13
Grassland	0.20	0.20	0.20
Woodland	0.41	0.41	0.41
Waterbody	0.03	0.03	0.03
Built-up land	0.00	0.00	0.00

From the perspective of land utilization, with the land utilization changes in Guanling County over the past 20 years, the regional quality of ecological habitats has also changed [61], showing an overall upward trend. This is related to the expansion of land suitable for various habitats such as forest land, grassland and water bodies [62,63]. Since construction land has the lowest habitat suitability among various land utilizations, the threat sources of quality of ecological habitats mainly come from the expansion of construction land [64,65]. However, the expansion of ecological land such as forest land, grassland and water bodies can, to a certain extent, offset the negative impacts of construction lnd [62]. Over the past 20 years, the influence of urbanization has led to the continuous expansion of construction land occupying the cultivated land with higher habitat suitability, which is the main factor causing the expansion of low-level quality of ecological habitats in the region [66]. With the acceleration of urbanization construction and the implementation of the protection policy for cultivated land, construction land is bound to expand towards forest land and grassland with higher habitat suitability, and the range of low-level quality of ecological habitats will further expand to the range of high-level quality of ecological habitats, resulting in a rapid decline trend of the overall quality of ecological habitats in Guanling County [67]. As Liu et al. [68] have confirmed, ecological policies such as returning farmland to forest and grassland can effectively expand the area of forest land and grassland, thereby improving the regional quality of ecological habitats. In future land utilization planning and management, it is necessary to strengthen the implementation of ecological policies such as returning farmland to forest and grassland, increase the coverage of forest land and grassland, and thereby improve the quality of ecological habitats of the karst mountainous area in Guanling. In the process of urbanization, it is necessary to appropriately allocate corresponding ecological land, strengthen the management measures for the expansion of construction land, improve land utilization efficiency, and indirectly improve the regional quality of ecological habitats [69].

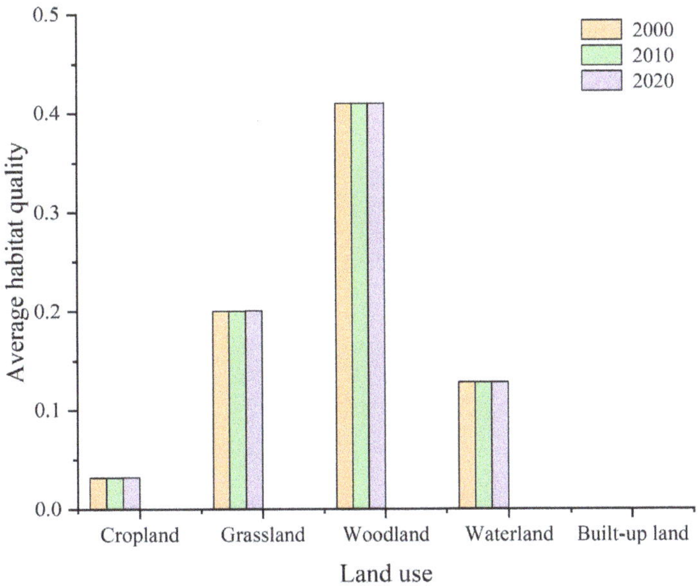

Figure 9. Average habitat quality index of different land use/cover types.

4.2. Changes in Habitat Quality and Landscape Pattern Indexes

Based on the MGWR model, a regression examination was conducted on the landscape pattern index and quality of ecological habitats of Guanling County. It was found that quality of ecological habitats and landscape pattern have a high degree of significance, and the changes in landscape pattern have a statistically significant impact on the quality of ecological habitats of Guanling County. The six landscape pattern indices have both positive and negative effects on quality of ecological habitats, indicating that the spatial heterogeneity in the Guanling karst mountainous area is intense, and there are statistically significant differences in the intensity of the influence of landscape patterns in different directions of the region on quality of ecological habitats [22]. This result is consistent with the study by Wang et al. [22] on the karst city of Guiyang, both confirming that the impact of landscape pattern indices on quality of ecological habitats exhibits spatial non-stationarity, but the driving mechanisms show statistically significant differentiation due to differences in regional ecological backgrounds. In this research effort, the compensation effect of forest expansion on quality of ecological habitats was statistically significantly higher than that in the study by Wang et al. [22] in karst mountain cities of Guizhou, which may be attributed to the fact that Wang et al. used 30 m land utilization data and 1000 m grid examination, while this research effort adopted 500 m grid examination. The ecological connectivity of forest patches is more easily captured at a fine scale. In terms of the social environment, the "Rocky Desertification Control Project" implemented in Guanling County in recent years has strengthened the positive effect of ecological land, while the study area of Wang et al. [22]. was more statistically significantly impacted by urbanization. Furthermore, the study by Hu et al. [29] on the rapidly urbanized areas of Nanjing also confirmed the advantage of the MGWR model in capturing spatial heterogeneity, and the positive effect of AI (Aggregation Index) on quality of ecological habitats was more statistically significant in ecological land. However, Hu et al. [29] found that the quality of ecological habitats in Nanjing presented a spatial configuration of "urban center decline-edge fluctuation", with the core driving force being construction land expansion, which differs from the pattern of "ecological core area aggregation-urban edge degradation" in Guanling County. In terms of data scale, Hu et al. [29]. used 700 m hexagonal grids to capture the landscape fragmentation in Nanjing, while the 500 m grid in this research effort is more suitable for the micro-geomorphological examination of karst peak-cluster depressions. The negative effect of the SHDI on quality of ecological habitats in Nanjing was mainly concentrated in the urban-rural transition zone, whereas in Guanling County, the negative effect of SHDI was more statistically significant in forest-concentrated areas, because the connectivity of forest patches in the karst ecosystem is more sensitive to landscape diversity. From the perspective of social environmental differences, habitat degradation in Nanjing was mainly driven by the rapid expansion of construction land; during the same period, Guanling County implemented the policy of returning farmland to forest, leading to large-scale forest expansion and a slowdown in the expansion rate of construction land, with quality of ecological habitats improvement mainly on the basis of ecological restoration projects.

PD, SHEI, and AI have a statistically significant positive impact on quality of ecological habitats, indicating that there are diverse patch types in the landscape of Guanling County, and these patches are relatively evenly distributed in space, which is conducive to supporting more species and maintaining the stability of the ecosystem. The persistence and migration of patches. In addition, such landscape patterns indicate that patches are aggregated to a certain extent, which helps to maintain the integrity and connectivity of habitats, facilitating the ecosystem of species to be in a relatively healthy state and providing good habitat conditions for various organisms [70–72]. The positive effects of these indices work together to form a healthy landscape pattern that can maintain biodiversity and promote ecosystem service functions. However, LPI, LSI, and SHDI have a statistically significant negative impact on quality of ecological habitats, which to a certain extent leads to the landscape of Guanling County being divided into many small patches, which increases the edge effect, reduces the effective area of internal habitats, and is not conducive to the survival of species; the landscape shape is relatively irregular, increasing the edge effect and reducing the quality of internal habitats; although diversity is usually beneficial, in some cases, too many patch types may lead to uneven resource allocation and thereby affect the survival of certain key species. These negative effects jointly indicate that the current landscape pattern is not conducive to certain specific species or ecological processes, and it is necessary to improve quality of ecological habitats through improved landscape planning and management. For example, the number of patches can be reduced, the shape of patches can be optimized, and the distribution of patch types can be adjusted to reduce the influence of LPI and LSI; the combination of patch types in the landscape can be optimized to improve SHDI, especially in areas with the distribution of forests and grasslands, to make them more suitable for the needs of target species. For instance, by establishing ecological protection corridors and enhancing the connectivity of ecological land, the quality of habitats in karst mountainous areas within the region can be directly improved and biodiversity can be maintained. As Liu et al. [73] pointed out in their study on the changes in quality of ecological habitats under land utilization in the Yellow River Delta, establishing a nature reserve in the Yellow River Delta can effectively restrict human development and construction activities, maintaining a high level of quality of ecological habitats within the protected area. In social and economic development, vigorously developing ecological parks, promoting ecological tourism, and ecological agricultural activities, or appropriately dividing ecological land in areas with concentrated construction and development of construction land, can thereby enhance the quality of ecological habitats of the region. Duan et al. [74] found in their study on the evolution of quality of ecological habitats under land utilization changes in the Jialing River Basin that in agricultural concentrated areas, encouraging the development of green agriculture, ecological tourism and other projects can minimize environmental degradation and improve quality of ecological habitats to the greatest extent.

4.3. Evaluation of the Importance of Habitat Quality Functions

The spatial distribution of the importance of quality of ecological habitats service functions in Guanling County and the examination of land utilization have a high correlation. Among them, the general important areas are mainly distributed in construction areas; the relatively important areas are mainly distributed in farmland areas; the medium important areas are mainly distributed in grassland areas; and the highly important areas are mainly distributed in forest areas. In 2020, the area of highly important areas accounted for more than 50%, which is an important area for ecosystem protection in Guanling County. This area is a widely distributed area of forest land in Guanling County and an important area for maintaining quality of ecological habitats and biodiversity of the ecosystem. Due to the influence of social economic development, urbanization, industrialization and other human activities, the area of relatively important areas has decreased year by year, but the implementation of ecological protection policies has provided corresponding compensation for highly important areas and medium important areas. While the importance of ecological service functions in some areas has declined, the importance of ecological service functions in other areas has been statistically significantly increased. The overall ecological benefit is good.

Ecological engineering can statistically significantly increase the carbon storage of terrestrial ecosystems [75]. Since the implementation of the reforestation and returning farmland to forest project in 2003, a large amount of farmland in Guanling County has been transformed into forest land. The expansion of forest land has effectively improved the quality of ecological habitats of the region and improved the habitat environment and increased biodiversity. However, due to the influence of rapid urbanization and industrialization, especially since 2006, Guizhou Province has actively promoted industrialization and urbanization development to achieve the goal of rapid social and economic development. Forest land, farmland and grassland have been extensively developed as construction land. As construction land for serving the social economy, construction land has promoted social and economic development while also causing a decline in quality of ecological habitats. The social and economic development has, in turn, promoted the expansion of construction land to forest land, grassland and farmland with higher quality of ecological habitats, thereby causing a further decline in quality of ecological habitats. Ecological construction projects have effectively improved the importance of ecological service functions and the ecological benefits of the region. However, to a certain extent, they will restrict social and economic development; while purely economic policies accelerate the process of urbanization and industrialization at the cost of sacrificing ecological benefits to meet the excessive demands of social and economic construction, the importance of ecological service functions has been statistically significantly reduced, and thus the ecological benefits are extremely low. In summary, only by taking into account both ecological benefits and social economic benefits at the same time can it be conducive to the harmonious coexistence of humans and nature and ultimately achieve the goal of sustainable development of human society.

4.4. Shortcomings and Prospects

The study established a 500 m × 500 m grid examination for the typical area of the Karst mountainous region in Guanling County to investigate the impact of landscape pattern changes on quality of ecological habitats. It achieved good simulation effects in regions with strong spatial heterogeneity. For small-scale areas, it has a certain degree of high simulation accuracy and calculation accuracy, but its generalization of the overall landscape pattern of Guanling County is limited. For example, the degree of forest fragmentation within a 500 m grid may be averaged, underestimating the actual habitat fragmentation risk; the spatial attenuation effect of threat sources may vary in different topographic areas, and the topographic barriers in karst peak-cluster areas may shorten the actual impact distance. Future research will expand the research scope and fully incorporate more natural and socio-economic factors to explore the overall landscape pattern of Guanling County and comprehensively analyze the impact of landscape pattern changes on quality of ecological habitats.

The study only explores the ecological environment and biodiversity service functions of the Karst mountainous region in Guanling County from the perspective of quality of ecological habitats. Although quality of ecological habitats is currently a hot topic in ecological research, there are still limitations in the research on the overall ecosystem service functions. Based on the above issues, future research will consider adding multiple modules including carbon storage, quality of ecological habitats, soil retention, and water yield to achieve a comprehensive study on the value trade-off and synergy of the entire ecosystem service value of the Southwest Karst region.

Relying on data from three discrete time points in 2000, 2010, and 2020 makes it difficult to capture the transient impacts of short-term ecological disturbances (such as the 2008 southern snow disaster and the mid-term effects of the 2015 rocky desertification control project) on quality of ecological habitats, thus leading to estimation errors in the trend of quality of ecological habitats changes. Future studies could integrate annual land utilization data to enhance temporal continuity.

5. Conclusions

Guanling County, as a core area of the karst mountainous region in the southwest, holds an important position in the regional ecological environment governance and protection. Habitat quality, as a key indicator for maintaining the health and stability of the karst mountainous region's ecological environment, plays a statistically significant role in the study of quality of ecological habitats and biodiversity in this research area. By combining ArcGIS 10.8 software and the InVEST model, the quality of ecological habitats of Guanling County from 2000 to 2020 was analyzed to explore the spatio-temporal evolution characteristics of quality of ecological habitats and its response to landscape pattern changes, and to analyze the importance of quality of ecological habitats service functions. This was achieved to conduct a comprehensive assessment of the quality of ecological habitats safety and biodiversity maintenance in the karst mountainous area of Guanling County. The main conclusions are as follows:

(1) Over the past 20 years, the spatial distribution of different levels of quality of ecological habitats in the karst mountainous area of Guanling has demonstrated a strong correlation with land utilization. Influenced by changes in land utilization and landscape pattern, regional quality of ecological habitats has undergone certain changes, which have had a statistically significant impact on the ecological environment quality of the karst mountainous area in Guanling County. Spatially, low-level quality of ecological habitats is concentrated in construction areas, medium-level quality of ecological habitats is concentrated in farmland areas, high-level quality of ecological habitats is concentrated in forest and water area, and high-level quality of ecological habitats is widely distributed in grassland areas. Expanding forest and grassland can, to a certain extent, counteract the negative impact of construction land on quality of ecological habitats. In high-high clustering areas, the "mountain closure for afforestation + artificial replanting" project should be implemented to improve regional quality of ecological habitats through forest expansion. In low-low clustering areas with hotspots of construction land expansion, the strategy of "compact development of construction land + construction of ecological buffer zones" should be adopted to mitigate habitat fragmentation. In areas with high fragmentation indices, ecological corridors should be planned along river corridors to connect isolated forest patches and enhance landscape connectivity.

(2) The quality of ecological habitats of the karst mountainous area in Guanling also has a high degree of spatial autocorrelation. However, the Moran's I value is decreasing, indicating that the spatial aggregation of the overall quality of ecological habitats in the karst mountainous area of Guanling County is showing a weakening trend. Compared with the OLS and GWR models, the MGWR model has better efficacy in data fitting. The six landscape indices have positive or negative effects on the quality of ecological habitats of different areas in the study area. The regression coefficients of each landscape index vary greatly, and the differences in the direction and intensity of their spatial effects are statistically significant, and their impact on quality of ecological habitats has statistically significant spatial heterogeneity.

(3) The landscape pattern of Guanling County is transforming from a relatively complete, highly diverse, and clustered state of patches to a fragmented one with complex shapes, reduced diversity, and more dispersed patch distribution. The spatial distribution of quality of ecological habitats in Guanling County is highly correlated with land utilization, and the changes in landscape pattern are of great significance for the spatio-temporal evolution of quality of ecological habitats. The landscape diversity of Guanling County has a negative impact on the quality of ecological habitats of ecological land utilizations such as forest land and grassland, while it has a positive impact on the quality of ecological habitats of social and economic development land utilizations such as construction land and cultivated land. The sensitivity of quality of ecological habitats to forest fragmentation can be reduced through returning farmland to forest and revegetation of abandoned mines. Patch connectivity can

be enhanced by "forest gap renovation + shrub replanting" to maintain ecosystem resilience. A smaller SHDI index indicates a stronger positive correlation between forest quality of ecological habitats and landscape diversity. Implementing the "patch integration" project to convert fragmented cropland into contiguous economic forests can statistically significantly improve quality of ecological habitats.

(4) The average area of the generally important, moderately important, moderately important, and highly important quality of ecological habitats zones from 2000 to 2020 was 28.84 km^2, 513.93 km^2, 112.62 km^2, and 812.62 km^2 respectively. The generally important, moderately important, and highly important quality of ecological habitats zones showed an increasing trend, and the importance of quality of ecological habitats service functions increased.

Author Contributions: All authors contributed to the manuscript. Conceptualization, Z.Z. and S.D.; methodology, S.D.; validation, D.H., F.D. and X.D.; formal analysis, S.D.; data curation, S.D. and D.H.; writing—original draft preparation, S.D.; writing—review and editing, S.D., Z.Z., D.H., F.D., X.D., Y.L.; Q.D. and Y.Y.; visualization, S.D. and D.H.; project administration, Z.Z.; funding acquisition, Z.Z. All authors have read and agreed to the published version of the manuscript.

Funding: Supported by Guizhou Provincial 2025 Central Government—Guided Local Science and Technology Development Fund Project, (Qian Ke He Zhong Yin Di [2025] 031); Guizhou Province Science and Technology Project: (Guizhou Science and Technology Cooperation Support 2023 General 211); Supported by Supported by Guizhou Provincial Key Laboratory Construction Project, (Qian Ke He Ping Tai [2025] 014); Supported by Guizhou Provincial [2023] Central Government—Guided Local Science and Technology Development Fund Project, (Qian Ke He Zhong Yin Di [2023] 005).

Institutional Review Board Statement: Not applicable.

Data Availability Statement: All the data used in this study are mentioned in Section 2 "Materials and Methods".

Acknowledgments: The authors gratefully acknowledge the financial support of Guizhou Normal University. We would also like to thank the editors and anonymous reviewers for their helpful and productive comments on the manuscript.

Conflicts of Interest: The authors declare no conflict of interest. The funders had a funding role in the design of the study; in the collection, analyses, or interpretation of data; in the writing of the manuscript; or in the decision to publish the results.

References

1. Chen, F.; Bai, X.; Liu, F.; Luo, G.; Tian, Y.; Qin, L.; Li, Y.; Xu, Y.; Wang, J.; Wu, L. Analysis long-term and spatial changes of forest cover in typical karst areas of China. *Land* **2022**, *11*, 1349. [CrossRef]
2. Wu, L.; Wang, S.; Bai, X.; Chen, F.; Li, C.; Ran, C.; Zhang, S. Identifying the multi-scale influences of climate factors on runoff changes in a typical karst watershed using wavelet analysis. *Land* **2022**, *11*, 1284. [CrossRef]
3. Liu, Q.; Wang, J.; Xiong, K.; Gong, L.; Chen, Y.; Yang, J.; Xiao, H.; Bai, J. Quantitative assessment of ecological assets in the world heritage karst sites based on remote sensing: With a special reference to South China Karst. *Herit. Sci.* **2024**, *12*, 129. [CrossRef]
4. Du, S.; Zhou, Z.; Huang, D.; Zhang, F.; Deng, F.; Yang, Y. The Response of Carbon Stocks to Land Use/Cover Change and a Vulnerability Multi-Scenario Analysis of the Karst Region in Southern China Based on PLUS-InVEST. *Forests* **2023**, *14*, 2307. [CrossRef]
5. Niu, L.; Shao, Q. Soil conservation service spatiotemporal variability and its driving mechanism on the Guizhou Plateau, China. *Remote Sens.* **2020**, *12*, 2187. [CrossRef]
6. Xie, B.; Zhang, M. Spatio-temporal evolution and driving forces of habitat quality in Guizhou Province. *Sci. Rep.* **2023**, *13*, 6908. [CrossRef]
7. Chen, J.; Yu, J.; Bai, X.; Zeng, Y.; Wang, J. Ecosystems, Environment, Fragility of karst ecosystem and environment: Long-term evidence from lake sediments. *Agric. Ecosyst. Environ.* **2020**, *294*, 106862. [CrossRef]
8. Pan, S.; Liang, J.; Chen, W.; Li, J.; Liu, Z. Gray forecast of ecosystem services value and its driving forces in Karst Areas of China: A case study in Guizhou Province, China. *Int. J. Environ. Res. Public Health* **2021**, *18*, 12404. [CrossRef]

9. Xiong, K.; Chi, Y. The problems in southern China karst ecosystem in southern of China and its countermeasures. *Ecol. Econ.* **2015**, *31*, 23–30.
10. Zhang, S.; Bai, X.; Zhao, C.; Tan, Q.; Luo, G.; Wu, L.; Xi, H.; Li, C.; Chen, F.; Ran, C. China's carbon budget inventory from 1997 to 2017 and its challenges to achieving carbon neutral strategies. *J. Clean. Prod.* **2022**, *347*, 130966. [CrossRef]
11. Li, Y.; Geng, H. Evolution of land use landscape patterns in karst watersheds of Guizhou Plateau and its ecological security evaluation. *Land* **2022**, *11*, 2225. [CrossRef]
12. Wang, Z.; Dai, L. Assessment of land use/cover changes and its ecological effect in karst mountainous cities in central Guizhou Province: Taking Huaxi District of Guiyang City as a case. *Acta Ecol. Sin.* **2021**, *41*, 3429–3440.
13. Du, C.; Bai, X.; Li, Y.; Tan, Q.; Zhao, C.; Luo, G.; Wu, L.; Chen, F.; Li, C.; Ran, C. Inventory of China's net biome productivity since the 21st Century. *Land* **2022**, *11*, 1244. [CrossRef]
14. Song, F.; Wang, S.; Bai, X.; Wu, L.; Wang, J.; Li, C.; Chen, H.; Luo, X.; Xi, H.; Zhang, S. A new indicator for global food security assessment: Harvested area rather than cropland area. *Chin. Geogr. Sci.* **2022**, *32*, 204–217. [CrossRef]
15. Deng, Y.; Jiang, Z.C. Characteristic of rocky desertification and comprehensive improving model in karst peak-cluster depression in Guohua, Guangxi, China. *Procedia Environ. Sci.* **2011**, *10*, 2449–2452. [CrossRef]
16. Xu, E.; Zhang, H.; Li, M. Mining spatial information to investigate the evolution of karst rocky desertification and its human driving forces in Changshun, China. *Sci. Total Environ.* **2013**, *458*, 419–426. [CrossRef]
17. Qi, X.; Wang, K.; Zhang, C. Effectiveness of ecological restoration projects in a karst region of southwest China assessed using vegetation succession mapping. *Ecol. Eng.* **2013**, *54*, 245–253. [CrossRef]
18. Peng, J.; Xu, Y.; Cai, Y.; Xiao, H. The role of policies in land use/cover change since the 1970s in ecologically fragile karst areas of Southwest China: A case study on the Maotiaohe watershed. *Environ. Sci. Policy* **2011**, *14*, 408–418. [CrossRef]
19. Chen, H.; Liu, J.; Wang, K.; Zhang, W. Spatial distribution of rock fragments on steep hillslopes in karst region of northwest Guangxi, China. *Catena* **2011**, *84*, 21–28. [CrossRef]
20. Yohannes, H.; Soromessa, T.; Argaw, M.; Dewan, A. Spatio-temporal changes in habitat quality and linkage with landscape characteristics in the Beressa watershed, Blue Nile basin of Ethiopian highlands. *J. Environ. Manag.* **2021**, *281*, 111885. [CrossRef]
21. Zhu, C.; Zhang, X.; Zhou, M.; He, S.; Gan, M.; Yang, L.; Wang, K. Impacts of urbanization and landscape pattern on habitat quality using OLS and GWR models in Hangzhou, China. *Ecol. Indic.* **2020**, *117*, 106654. [CrossRef]
22. Wang, Z.; Liu, S.; Su, Y. Spatiotemporal evolution of habitat quality and its response to landscape patterns in karst mountainous cities: A case study of Guiyang City in China. *Environ. Sci. Pollut. Res.* **2023**, *30*, 114391–114405. [CrossRef]
23. Li, Y.; Duo, L.; Zhang, M.; Wu, Z.; Guan, Y. Assessment and Estimation of the Spatial and Temporal Evolution of Landscape Patterns and Their Impact on Habitat Quality in Nanchang, China. *Land* **2021**, *10*, 1073. [CrossRef]
24. Gu, L.; Yan, J.; Li, Y.; Gong, Z. Evolution, Spatial–temporal evolution and correlation analysis between habitat quality and landscape patterns based on land use change in Shaanxi Province, China. *Ecol. Evol.* **2023**, *13*, e10657. [CrossRef]
25. Zhang, X.; Lyu, C.; Fan, X.; Bi, R.; Xia, L.; Xu, C.; Sun, B.; Li, T.; Jiang, C. Spatiotemporal variation and influence factors of habitat quality in Loess Hilly and Gully Area of Yellow River Basin: A case study of Liulin County, China. *Land* **2022**, *11*, 127. [CrossRef]
26. Fotheringham, A.S.; Yang, W.; Kang, W. Multiscale geographically weighted regression (MGWR). *Ann. Am. Assoc. Geogr.* **2017**, *107*, 1247–1265. [CrossRef]
27. Fotheringham, A.S.; Charlton, M.; Brunsdon, C. The geography of parameter space: An investigation of spatial non-stationarity. *Int. J. Geogr. Inf. Syst.* **1996**, *10*, 605–627. [CrossRef]
28. Brunsdon, C.; Fotheringham, S.; Charlton, M. Geographically weighted regression. *J. R. Stat. Soc. Ser. D (Stat.)* **1998**, *47*, 431–443. [CrossRef]
29. Hu, J.; Zhang, J.; Li, Y. Exploring the spatial and temporal driving mechanisms of landscape patterns on habitat quality in a city undergoing rapid urbanization based on GTWR and MGWR: The case of Nanjing, China. *Ecol. Indic.* **2022**, *143*, 109333. [CrossRef]
30. Tang, Z.; Li, R. Small watersheds are the best control and management unit for improving soil conservation services in karst areas. *Sci. Total Environ.* **2024**, *953*, 176162. [CrossRef]
31. Rong, Y.; Li, K.; Guo, J.; Zheng, L.; Luo, Y.; Yan, Y.; Wang, C.; Zhao, C.; Shang, X.; Wang, Z. Multi-scale spatio-temporal analysis of soil conservation service based on MGWR model: A case of Beijing-Tianjin-Hebei, China. *Ecol. Indic.* **2022**, *139*, 108946. [CrossRef]
32. Zhou, L.; Wu, T.; Xie, Z. Spatially heterogeneous relationships of PM_(2.5) concentrations with natural and land use factors in the Niger River Watershed, West Africa. *J. Clean. Prod.* **2023**, *394*, 136406. [CrossRef]
33. Kafy, A.-A.; Saha, M.; Fattah, M.A.; Rahman, M.T.; Duti, B.M.; Rahaman, Z.A.; Bakshi, A.; Kalaivani, S.; Rahaman, S.N.; Sattar, G.S. Integrating forest cover change and carbon storage dynamics: Leveraging Google Earth Engine and InVEST model to inform conservation in hilly regions. *Ecol. Indic.* **2023**, *152*, 110374. [CrossRef]
34. Li, Y.; Liu, Z.; Li, S.; Li, X. Multi-scenario simulation analysis of land use and carbon storage changes in changchun city based on FLUS and InVEST model. *Land* **2022**, *11*, 647. [CrossRef]

35. Dale, V.H.; Polasky, S. Measures of the effects of agricultural practices on ecosystem services. *Ecol. Econ.* **2007**, *64*, 286–296. [CrossRef]
36. Wen, B.; Liu, C.; Tian, X.; Zhang, Q.; Huang, S.; Zhang, Y. Interactive Effects of Ecological Land Agglomeration and Habitat Quality on Soil Erosion in the Jinsha River Basin, China. *Land* **2024**, *13*, 229. [CrossRef]
37. Wu, L.; Sun, C.; Fan, F. Estimating the characteristic spatiotemporal variation in habitat quality using the invest model—A case study from Guangdong–Hong Kong–Macao Greater Bay Area. *Remote Sens.* **2021**, *13*, 1008. [CrossRef]
38. Wang, B.; Cheng, W. Effects of land use/cover on regional habitat quality under different geomorphic types based on InVEST model. *Remote Sens.* **2022**, *14*, 1279. [CrossRef]
39. Wei, Q.; Abudureheman, M.; Halike, A.; Yao, K.; Yao, L.; Tang, H.; Tuheti, B. Temporal and spatial variation analysis of habitat quality on the PLUS-InVEST model for Ebinur Lake Basin, China. *Ecol. Indic.* **2022**, *145*, 109632. [CrossRef]
40. Xu, G.; Xiong, K.; Shu, T.; Shi, Y.; Chen, L.; Zheng, L.; Fan, H.; Zhao, Z.; Yang, Z. Bundling evaluating changes in ecosystem service under karst rocky desertification restoration: Projects a case study of Huajiang-Guanling, Guizhou province, Southwest China. *Environ. Earth Sci.* **2022**, *81*, 302. [CrossRef]
41. Wang, S.-J.; Li, Y.-B. Problems and development trends about researches on karst rocky desertification. *Adv. Earth Sci.* **2007**, *22*, 573.
42. Qiu, S.; Peng, J.; Zheng, H.; Xu, Z.; Meersmans, J. How can massive ecological restoration programs interplay with social-ecological systems? A review of research in the South China karst region. *Sci. Total Environ.* **2022**, *807*, 150723. [CrossRef]
43. Liu, J.; Kuang, W.; Zhang, Z.; Xu, X.; Qin, Y.; Ning, J.; Zhou, W.; Zhang, S.; Li, R.; Yan, C. Spatiotemporal characteristics, patterns, and causes of land-use changes in China since the late 1980s. *J. Geogr. Sci.* **2014**, *24*, 195–210. [CrossRef]
44. Sharp, R.; Tallis, H.; Ricketts, T.; Guerry, A.; Wood, S.; Chaplin-Kramer, R.; Nelson, E.; Ennaanay, D.; Wolny, S.; Olwero, N. *InVEST Version 3.2. 0 User's Guide. The Natural Capital Project*; The Nature Conservancy, and World Wildlife Fund, Stanford University, University of Minnesota: Minneapolis, MN, USA, 2015; Available online: https://scholar.google.com/scholar?hl=en&as_sdt=0,5&cluster=4950540898464922086#d=gs_cit&t=1752141390234&u=/scholar?q=info:5nkL07fas0QJ:scholar.google.com/&output=cite&scirp=0&scfhb=1&hl=en (accessed on 1 June 2025).
45. Xie, B.; Meng, S.; Zhang, M. Evolution of habitat quality and its response to topographic gradient effect in a Karst Plateau: A case study of the Key Biodiversity Conservation Project Area of Wuling Mountains. *Int. J. Environ. Res. Public Health* **2022**, *20*, 331. [CrossRef] [PubMed]
46. Gu, Y.-Y.; Huang, X.-F.; Zou, C.-X.; Ye, X.; Lin, N.-F.; Zhang, W.-M. Monitoring habitat quality changes in Yuanjiangyuan Nature Reserve based on Landsat images. *J. Ecol. Rural Environ.* **2019**, *35*, 764–772.
47. Zheng, H.; Li, H. Spatial–temporal evolution characteristics of land use and habitat quality in Shandong Province, China. *Sci. Rep.* **2022**, *12*, 15422. [CrossRef]
48. Li, S.; Hong, Z.; Xue, X.; Zheng, X.; Du, S.; Liu, X. Evolution characteristics and multi-scenario prediction of habitat quality in Yulin City based on PLUS and InVEST models. *Sci. Rep.* **2024**, *14*, 11852. [CrossRef]
49. Li, W.; Zhu, C.; Wang, H.; Xu, B. Multi-scale spatial autocorrelation analysis of cultivated land quality in Zhejiang province. *Trans. Chin. Soc. Agric. Eng.* **2016**, *32*, 239–245.
50. Cheng, Z. The spatial correlation and interaction between manufacturing agglomeration and environmental pollution. *Ecol. Indic.* **2016**, *61*, 1024–1032. [CrossRef]
51. Mi, Y.; Li, S.; Wang, Z. Spatial distribution and topographic gradient effects of habitat quality in the Chang-Zhu-Tan Urban Agglomeration, China. *Sci. Rep.* **2024**, *14*, 22563. [CrossRef]
52. Yang, Y. Evolution of habitat quality and association with land-use changes in mountainous areas: A case study of the Taihang Mountains in Hebei Province, China. *Ecol. Indic.* **2021**, *129*, 107967. [CrossRef]
53. Getis, A.; Ord, J.K. The analysis of spatial association by use of distance statistics. *Geogr. Anal.* **1992**, *24*, 189–206. [CrossRef]
54. Chen, Z.; Dong, H. Spatial and temporal evolution patterns and driving mechanisms of rural settlements: A case study of Xunwu County, Jiangxi Province, China. *Sci. Rep.* **2024**, *14*, 24342. [CrossRef] [PubMed]
55. Li, Z.; Fotheringham, A.S. Computational improvements to multi-scale geographically weighted regression. *Int. J. Geogr. Inf. Sci.* **2020**, *34*, 1378–1397. [CrossRef]
56. Yang, Y.; Li, H. Monitoring spatiotemporal characteristics of land-use carbon emissions and their driving mechanisms in the Yellow River Delta: A grid-scale analysis. *Environ. Res.* **2022**, *214*, 114151. [CrossRef]
57. Yu, L.; Li, Y.; Yu, M.; Chen, M.; Yang, L. Dynamic changes in agroecosystem landscape patterns and their driving mechanisms in karst mountainous areas of Southwest China: The case of Central Guizhou. *Sustainability* **2023**, *15*, 9160. [CrossRef]
58. Wu, L.; Zhou, J.; Xie, B. Comparative analysis of temporal-spatial variation on Mountain-Flatland landscape pattern in karst mountainous areas of southwest China: A case study of Yuxi city. *Land* **2023**, *12*, 435. [CrossRef]
59. Luo, Y.; Wang, Z.; Zhou, X.; Hu, C.; Li, J. Spatial-temporal driving factors of urban landscape changes in the karst mountainous regions of southwest China: A case study in central urban area of guiyang city. *Sustainability* **2022**, *14*, 8274. [CrossRef]

60. Zhang, D.; Wang, J.; Wang, Y.; Xu, L.; Zheng, L.; Zhang, B.; Bi, Y.; Yang, H. Is there a spatial relationship between urban landscape pattern and habitat quality? Implication for landscape planning of the yellow river basin. *Int. J. Environ. Res. Public Health* **2022**, *19*, 11974. [CrossRef]
61. Chen, Y.; Wu, R.; Ma, Y.-W.; Feng, X.-H.; Pan, J.-F.; Chen, Y.-M. Spatial and temporal differentiation and simulation of habitat quality in typical karst areas. *J. Ecol. Rural Environ.* **2022**, *38*, 1593–1603.
62. Bai, L.; Xiu, C.; Feng, X.; Liu, D. Influence of urbanization on regional habitat quality: A case study of Changchun City. *Habitat Int.* **2019**, *93*, 102042. [CrossRef]
63. Xie, Y.; Gong, J.; Zhang, S.; Ma, X.; Hu, B. Spatiotemporal change of landscape biodiversity based on InVEST model and remote sensing technology in the Bailong river watershed. *Sci. Geogr. Sin.* **2018**, *38*, 979–986.
64. Li, W.; Wang, Y.; Xie, S.; Cheng, X. Coupling coordination analysis and spatiotemporal heterogeneity between urbanization and ecosystem health in Chongqing municipality, China. *Sci. Total Environ.* **2021**, *791*, 148311. [CrossRef] [PubMed]
65. Wei, W.; Wang, Z.; Bao, Y.; Chen, X.; Sun, Y.; Zeng, M.; Mo, Y. *Response of Habitat Quality to Land Use Change in Urban Built-Up Area of Karst Mountainous City*; Research Square: Durham, NC, USA, 2022.
66. Zhang, S.; Xiong, K.; Qin, Y.; Min, X.; Xiao, J. Evolution and determinants of ecosystem services: Insights from South China karst. *Ecol. Indic.* **2021**, *133*, 108437. [CrossRef]
67. He, J.; Huang, J.; Li, C. The evaluation for the impact of land use change on habitat quality: A joint contribution of cellular automata scenario simulation and habitat quality assessment model. *Ecol. Model.* **2017**, *366*, 58–67. [CrossRef]
68. Liu, Z.; Tang, L.; Qiu, Q.; Xiao, L.; Xu, T.; Yang, L. Temporal and spatial changes in habitat quality based on land-use change in Fujian Province. *Acta Ecol. Sin.* **2017**, *37*, 4538–4548.
69. Wang, H.; Tang, L.; Qiu, Q.; Chen, H. Assessing the impacts of urban expansion on habitat quality by combining the concepts of land use, landscape, and habitat in two urban agglomerations in China. *Sustainability* **2020**, *12*, 4346. [CrossRef]
70. Liu, F.; Wang, W.; Wang, J.; Zhang, X.; Ren, J.; Liu, Y. Multi-scale analysis of the characteristics of the changing landscape of the typical mountainous region of Southwest China over the past 40 years. *PeerJ* **2021**, *9*, e10923. [CrossRef]
71. Ntshanga, N.K.; Procheş, S.; Slingsby, J.A. Assessing the threat of landscape transformation and habitat fragmentation in a global biodiversity hotspot. *Austral Ecol.* **2021**, *46*, 1052–1069. [CrossRef]
72. Zheng, L.; Wang, Y.; Li, J. Quantifying the spatial impact of landscape fragmentation on habitat quality: A multi-temporal dimensional comparison between the Yangtze River Economic Belt and Yellow River Basin of China. *Land Use Policy* **2023**, *125*, 106463. [CrossRef]
73. Yubin, L.; Mei, H.; Min, W.; Chao, F.; Hang, Z. Habitat Quality Assessment in the Yellow River Delta Based on Remote Sensing and Scenario Analysis for Land Use/Land Cover. *Sustainability* **2022**, *14*, 15904. [CrossRef]
74. Duan, X.; Chen, B.; Zhang, T.; Guan, Y.; Zeng, K. Habitat Quality Evolution and Multi-Scenario Simulation Based on Land Use Change in the Jialing River Basin. *Sustainability* **2024**, *16*, 6968. [CrossRef]
75. Tong, X.; Brandt, M.; Yue, Y.; Horion, S.; Wang, K. Increased vegetation growth and carbon stock in China karst via ecological engineering. *Nat. Sustain.* **2018**, *1*, 44–50. [CrossRef]

land

MDPI

Article

Nature-Based Solutions for Landscape Performance Evaluation—Handan Garden Expo Park's "Clear as a Drain" Artificial Wetland as an Example

Jiaju Liu [1,2,3], Yujia Guo [4], Jingyi Han [3], Feng Yang [1,2], Nan Shen [1,2], Fei Sun [1,2], Yanjie Wei [1,2], Peng Yuan [1,2,*] and Jiawei Wang [5,*]

1 State Key Laboratory of Environmental Criteria and Risk Assessment, Chinese Research Academy of Environmental Sciences, Beijing 100012, China; ljj19910213@pku.edu.cn (J.L.); yang.feng@craes.org.cn (F.Y.); ecoshennan@craes.org.cn (N.S.); sunfei@craes.org.cn (F.S.); yanjiewei@craes.org.cn (Y.W.)
2 Research Center for Integrated Control of Watershed Water Pollution, Chinese Research Academy of Environmental Sciences, Beijing 100012, China
3 Architecture and Landscape Architecture, Peking University, Beijing 100871, China; hanjingyi@pku.edu.cn
4 College of Civil Engineering, Hebei University of Water Resources and Electric Power, Cangzhou 061016, China; guoyujia@hbwe.edu.cn
5 Hebei Key Laboratory of Water Quality Engineering and Comprehensive Utilization of Water Resources, Hebei University of Architecture, Zhangjiakou 075000, China
* Correspondence: 2006391290@pku.edu.cn (P.Y.); 18632312133@163.com (J.W.)

Abstract: As a technology for water landscape performance that considers landscape, ecological, and social effects, nature-based solutions play a crucial role in enhancing the functionality of integrated ecosystem services on the micro-scale. This study conducted a systematic investigation into the landscape performance of the "Clear as a Drain" composite sponge facility at Handan Garden Expo Park. The following conclusions were drawn: (1) In terms of ecological restoration support services, the "Clear as a Drain" artificial wetland exhibited diverse habitat types, a rich variety of plant species specific to the site's region, and high plant diversity indices for shrubs (1.776) and herbaceous aquatic plants (3.352). Reclaimed water reused in the artificial terraced wetland promoted plant growth and diversity while contributing to site self-rehabilitation; plants also significantly contributed to carbon fixation, oxygen release, and carbon emission reduction. (2) Regarding ecological restoration regulation services, the artificial wetland effectively purified reclaimed water with substantial improvements observed in incoming water quality during spring, summer, and autumn—particularly notable purification effects were observed during the summer months. Pollutant reduction rates for COD, BOD_5 ammonia nitrogen, TP, and TN reached 75.8%, 72.1%, 93.8%, 96.7%, and 90.3%, respectively; different independent subsystems within the wetland demonstrated distinct advantages in pollutant removal; park plants displayed strong air purification capabilities; annual energy savings from park plants could fully cover daily energy consumption for nearby residents. This case could serve as guidance for scientific management and design parameter optimization of other composite sponge facilities.

Keywords: nature-based solutions; composite sponge facilities; landscape performance effectiveness; "Clear as a Drain" artificial wetland

1. Introduction

Nature-based solutions are actions aimed at protecting, sustainably managing, restoring, or improving nature and ecosystems to efficiently solve societal challenges, such as climate change, food security, water security, human health, natural disasters, and social and economic development, while also aiming to provide benefits for human well-being and biodiversity. Among these, the use of nature-based techniques is widely applied in water ecosystem restoration [1].

With the promotion of the concept of nature-based solutions, composite sponge facilities such as sunken green spaces, ecological ponds, surface-flow artificial wetlands, and submerged artificial wetlands are widely used in urban stormwater management [2]. Through the continuous tracking and monitoring of various parameters of completed projects, a comprehensive evaluation of various ecosystem services provided by sponge facilities is carried out, and then the coupling law of various factors is grasped, and thus the optimal and sustainable operation parameters of sponge facilities are proposed. This has a positive effect on broadening the theoretical approach to water landscape performance.

Currently, LEED, SITES, and LPS are the three most influential systems for evaluating landscape performance. These systems are associated with the assessment of objects and factors related to ecological factors such as land, water, habitat, energy utilization, and raw materials. The social benefit index primarily focuses on leisure tourism and similar aspects, while economic research is relatively limited [3,4]. A series of experts and scholars at home and abroad have carried out relevant monitoring studies for different types (ecological pond type, submerged artificial wetland type, composite type, etc.) of completed sponge facilities [5,6]. Specifically, they have included research on the reduction rate of water inlet and outlet of sponge facilities, research on seasonal purification effects, and research on purification effects along the way. The United States government has carried out monitoring and evaluation of ecological restoration in Chesapeake Bay for 40 years, accumulating a large amount of research data, which provide data support for the evaluation of restoration effects, the selection of restoration programs, the selection of restoration technologies, and decisions regarding government-designed impervious ground areas and water purification efficiency to measure the effectiveness of performance-based payment contracts. At the same time, related economic benefits such as increasing employment have begun to be included in the scope of collaborative benefits of ecological restoration work [7]; Seattle's Thornton Creek Water Quality Channel (TCWQC) was monitored for three years after the project was completed in accordance with the project's quality assurance plan, including water quality, flow, sediment, vegetation, etc., to demonstrate that it can provide multiple ecological benefits in highly urbanized environments [8]. Chang [9] conducted monitoring experiments on the terraced submerged wetland built in Fengxiang Park in Haikou City and used a multivariate statistical regression model to study the impact of different elements on the pollutant removal rate, and then obtained the influence of each design parameters on the water quality purification effects. Gong proposed reconnaissance verification methods, monitoring methods, and joint monitoring and simulation assessment methods [10], but post-use evaluation and the evaluation of sustainability characteristics have been relatively lacking in terms of actual observation and study of the service quality of these built artificial ecosystems to establish the relationship between design elements, system function and post occupancy evaluation. This is also an organic part of design ecology, which has theoretical and practical significance for the promotion and continuous optimization of design.

At this stage, there are deficiencies in evaluation based on natural technology in the following aspects: insufficient continuous monitoring of the operational effects and water quality evaluation; lack of monitoring of long-term scale and seasonal changes in water quality of built projects; emphasis on a single evaluation of water quality purification effects; and lack of a comprehensive evaluation of the operational status of the composite sponge facilities. It is urgent to establish a comprehensive and systematic evaluation system framework to comprehensively evaluate the comprehensive and sustainable services of the sponge facilities [11–15]. This study took the "Clear as a Drain" artificial wetland in Handan Expo Park as the research object, which is a composite sponge facility designed for water purification and has been standing and operating for a certain period of time, with stable landscape effects. Based on water quality monitoring, plant diversity surveys, model calculation, etc., the ecological restoration performance of the "Clear as a Drain" composite sponge facilities was systematically studied, including landscape support services, landscape regulation services, and influencing factors affecting ecological restoration performance. The results were used to guide the scientific management and optimization of other nature-based composite sponge facilities.

2. Overview of the Garden Expo Park and the Artificial Wetland

Handan Garden Expo Park is located in the Fuxing District of Handan City. It is the source of the Qin River, the mother river of the Handan, which flows into Handan City. Due to urban expansion and industrial development, it became an industrial wasteland. The industrial brownfield area in the park reached 3 square kilometres and caused serious damage to the local air, soil, rivers, habitats, etc. The original "Clear as a Drain" artificial wetland site had a large amount of industrial waste. The surface was mostly hard concrete and bare soil. The plant condition was poor, and the flood storage capacity was weak. The terrain within the site was undulating, with large vertical, with a maximum height difference was about 30 m. The site is surrounded by three natural villages, each housing approximately 700 households within a 500-m radius.

The landscape design of the Garden Expo Park made full use of the existing terrain and water systems to sort and connect the original Qin River waterways, forming a water network that effectively connected the urban ecology and achieving the systematic restoration of the Qin River. The ecological design concept of minimal intervention was adopted for the existing water bodies. The concept was to retain the original wetland texture and replant vegetation appropriately in the wetland edge areas to provide habitat for birds and other creatures; for the steeply sloped areas with large height differences, the terrace texture was formed by cascading fill, and for the surface–flow artificial wetland, submerged artificial wetland and ecological ponds were set up on the terrace. For the problem of soil pollution, the methods of depression filling, soil replacement, and water purification were chosen. Existing slags were excavated, part of the soil was replaced, or wetland water systems were introduced and natural water systems were ecologically purified. Through the above landscape performance strategies, the industrial wasteland was restored to vitality, a complete ecosystem was formed, and natural self-regulation and purification capabilities were utilized and ultimately enabled the Garden Expo Park site to achieve the reproduction and regeneration of the natural landscape (Figure 1).

Figure 1. Comparison of Handan Garden Expo Park before and after construction.

3. Materials and Methods

3.1. Monitoring of Water Quality Purification Effect

Based on the research purpose, representative wetland single ponds were selected for sampling. Three time points were selected for monitoring: November 2020 (autumn), April 2021 (spring), and July 2021 (summer). Sampling points (water inlet, T1–T5) were set up at the water inlet and 5 independent subsystem water outlets of the "Clear as a Drain" composite sponge facility area to characterize the water quality removal performance of each subsystem.

The principle of selecting continuous series of single-pond wetlands was to start monitoring from the single-pond wetland at the uppermost source of the wetland, to maintain non-crossing series of wetlands that were as long as possible, and to try to encompass each wetland type into the monitoring sampling points: 14 sampling points were set up in the T2 independent subsystem of the terraced flow with better operating conditions and landscape effects (T2-water inlet, T2-1, T2-4, T2-7, T2-9, T2-12, T2-13, T2-16, T2-19, T2-20, T2-21, T2-25, T2-27, T2-water outlet); 11 sampling points were set up in the T3 independent subsystems of the terraced flow with better operating conditions and landscape effects (T3-water inlet, T3-5, T3-13, T3-19, T3-21, T3-22, T3-25, T3-27 T3-30, T3-35, T3-37). The locations of wetland sampling points and observation points at all levels of the "Clear as a Drain" wetland area are shown in Figure 2.

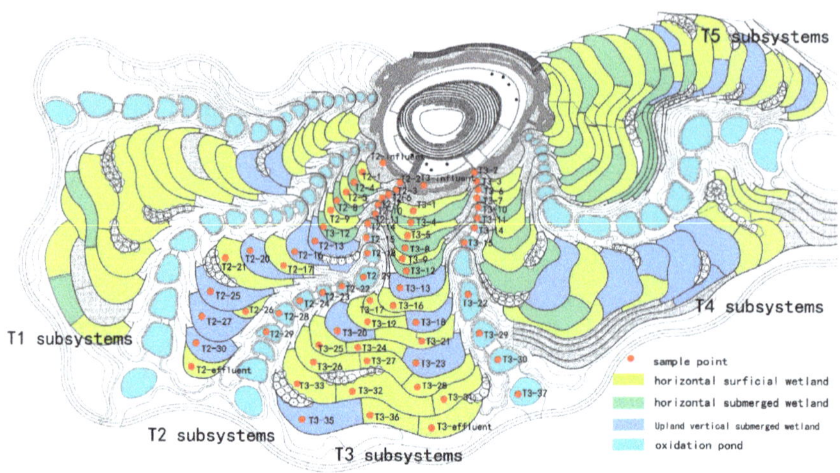

Figure 2. Location of sampling points and distribution of sponge facilities in the Clear Drainage Rush.

The collected water samples were tested for total phosphorus (TP), total nitrogen (TN), ammonia nitrogen (NH_3-N), chemical oxygen demand (COD), biochemical oxygen demand (BOD_5), and suspended solids (SSs) in the laboratory.

3.2. *Biodiversity Maintenance Investigation*

Through field investigation, measurement data such as habitat types and plant species were obtained, followed by data analyses and comparative studies, which were compared with the original plant design parameters of the park. The investigation was conducted in the area of "Clear as a Drain" wetland area. The habitat types were investigated via visits, photography, and classification of the plant habitats in the park according to land use types and community type classification. On this basis, the "dominance analysis" method [16] was used to analyze the structures, dominant species, and understory plants of each habitat type, in which the dominant species were determined by calculating the basal area. For plant species, the plant quadrat method was adopted for the investigation and compared with the original plants design parameters. The location distribution of each quadrat is shown in Figure 2. The total area of quadrats for trees and shrubs accounted for 1% of the built-up area of the "Clear as a Drain" wetland area, using a unit area of 10 × 10 m; the total area of herbaceous aquatic plants quadrats accounted for 0.2% of the built-up area of the "Clear as a Drain" wetland area, using a unit area of 2 × 2 m. The site area of the "Clear as a Drain" wetland area was about 103,834 m^2, the calculated number of tree and shrub quadrats was 10, and the number of herbaceous and aquatic plants quadrats was 50. Within the tree and shrub quadrats, we recorded the species name, quantity, diameter at breast height, crown width, tree height, height under branches, height under leaves, and base diameter (the shrubs were recorded in the same way as the herbaceous aquatic plants); within the herbaceous aquatic plant quadrats, we recorded the species names, height, and covered area.

The diversity index of trees, shrubs, and herbaceous aquatic plants at the site was calculated using the following formula:

$$H = -\sum_{i}^{S}(P_i \times \ln P_i) \tag{1}$$

In the formula: S represents the total number of species within the site, i represents the species, and P_i is the relative importance value of species i.

3.3. Modelling Calculations

The study took the original plant design parameters of Handan Expo Park as the basic data source and used the tree benefit calculator NTBC to calculate the ecological benefits of trees in the park in terms of carbon sequestration and oxygen release; when calculating, the corresponding climate zone should be selected, and the variables to input consist of the species, diameter at breast height, and land types of trees in the park to obtain the amount of carbon sequestration, oxygen release, and carbon reduction per plant of the park's various species and multiplied by the number of individuals of each tree species, and finally the relevant reference data are combined to further calculate the economic value of the trees (the calculation process was the same as for the benefits evaluation of ecological restoration regulation services of air purification and energy conservation).

The ecological restoration regulation services benefits of air purification and energy conservation refer to the official landscape performance evaluation website [17] and existing research results. The ecological restoration regulation service benefits of air purification and energy saving was evaluated in terms of the reduction of four air pollutants, namely, O_3, NO_2, SO_2 and PM10, and the amount of energy savings in electricity, petroleum, and natural gas, as well as their corresponding economic values. In addition to the combination of construction drawings and field surveys, data were recorded regarding the types, quantities, and diameters breast height of plants in the park. NTBC was used to conduct simulate simulations for each tree to obtain the corresponding functional quantities of each tree, and then the corresponding economic values were converted based on the conversion standards.

4. Evaluation of Landscape Performance in Handan Garden Expo Park

4.1. Assessment of Ecological Restoration Support Services

4.1.1. Habitat Types

Through investigation, it was found that the plant habitats in the "Clear as a Drain" wetland area could be divided into four types, namely deciduous broad-leaved forests, shrubs, grasses, and shallow swamps (Figure 2). At the same time, it was found that the habitat structures, dominant species, and lower-level plants of different habitats in the site were different, with significant differences (Table 1). The habitat structures within the site belonged to the "tree–shrub–grass" three-layer plant structure, but the shrub layer was relatively uniform, mainly composed of *Amorpha fruticosa*. The dominant species at the site, such as *Amorpha fruticosa* and gladiolus, showed strong water adaptability, tolerance to barrenness, and salt/alkali conditions, thus achieving the design goals of soil improvement, water quality purification, and ecological restoration; 17 naturally growing wild native plants of 17 genera and 9 families, including *Echinochloa crusgali*, *Setaria viridis*, and *Kochia scoparia*, had appeared in the grass and shallow swamp habitats (Table 2).

Table 1. Analysis of habitat types in "Clear as a Drain" wetland area.

Habitat Types	Habitat Structures	Dominant Species	Lower Level Plants
Deciduous broad-leaved forest	Mainly composed of deciduous broad-leaved forests, with a combination of tree and shrub layers and herbs	*Melia azedarach* L. *Euonymus maackii Rupr.*	*Acer negundo* L., *Pinus*, *Ligustrum compactum* (Wall. ex G. Don) Hook. f., etc.
Shrubs	Combination of shrubs and herbs	*Amorpha fruticosa* Linn.	*Rosa xanthina* f.

Table 1. *Cont.*

Habitat Types	Habitat Structures	Dominant Species	Lower Level Plants
Grass	No tree layer	*Setaria viridis*, *Axonopus compressus Iris tectorum* Maxim. chickweed chaparral *Conyza canadensis* (L.) Cronq., *Salvia japonica* Thunb., *Echinochloa crusgalli* (L.) Beauv	*Sonchus asper* (L.) Hill, *Ophiopogon bodinieri* Levl., *Humulus scandens* (Lour.) Merr. (Humulus *scandens*), *Physostegia virginiana* Benth., *Beckmannia syzigachne* (Steud.) Fern., *Artemisia carvifolia* Buch.-Ham. ex Roxb. Hort. Beng.
Shallow water swamp	A large number of emergent plants planted in the shallow water area of the wetland	*Acorus calamus* L., *Phragmites australis* (Cav.) Trin. ex Steud.	*Liatris spicata* (L.) Willd., *Pontederia cordata* L., *Canna indica* L., et al.

Table 2. Wild Native Plants in "Clear as a Drain".

The Name of the Plant	Family	Genus
Echinochloa crusgalli (L.) Beauv.	Poaceae	Echinacea
Kochia scoparia (L.) Schrad.	Chenopodiaceae	Kochia
Impatiens balsamina L.	Impatiensaceae	Impatiens
Setaria viridis (L.) Beauv.	Poaceae	Setaria
Cirsium japonicum Fisch. ex DC.	Asteraceae	Thistle
Pogonatherum crinitum (Thunb.) Kunth	Poaceae	Goldengrass
Cichorium endivia L.	Asteraceae	Chicory
Sonchus oleraceus L.	Asteraceae	Gesneria
Chenopodium album L.	Chenopodiaceae	Chenopodium
Solanum nigrum L.	Solanaceae	Solanum
Humulus scandens (Lour.) Merr.	Sanko	Humulus
Portulaca oleracea L.	Portulaceae	Portulaca
Digitaria sanguinalis (L.) Scop.	Poaceae	Crabgrass
Artemisia carvifolia Buch.-Ham. ex Roxb.	Asteraceae	Artemisia
Rumex acetosa L.	Polygonaceae	Sorrel
Beckmannia syzigachne (Steud.) Fern.	Poaceae	Brassica
Cyperus rotundus L.	Poaceae	Cyperus
Total: 17	9 family	17 Genus

4.1.2. Plant Diversity

(1) Analysis of the basic situation and changes of plant species

There were a total of 29 families, 49 genera, and 54 species of plant in the artificial wetlands. Compared with the original design parameters (Figure 3), there were 11 families, 26 genera, and 28 species added. The main community structure was the tree layer and herbaceous layer, and the regional characteristics of the plant community were more obvious. In terms of dominant species, the originally planted vegetation such as *Acer miyabei Ligustrum lucidum* and other vegetation still accounted for a relatively high proportion; *Melia azedarach*, *Euonymus maackii*, *Amorpha fruticosa*, *Setaria viridis*, *Iris tectorum*, *Axonopus compressus*, and other plants had become new dominant species.

Figure 3. Locations, habitat types, and dominant species in the "Clear as a Drain" quadrat site.

(2) Analysis of plant diversity index and its changes

The overall diversity index of trees and shrubs was 1.776, and the overall diversity index of herbaceous aquatic plants was 3.352, which is much higher than the diversity index of trees and shrubs, and the diversity index of each quadrat is shown in Table 3. The current situation is compared with the original plant design parameters of the exhibition garden. From a longitudinal perspective, 50% of the quadrats exhibited a Shannon index of 0 at the inception of the garden design; from a horizontal comparison, the Shannon index of trees and shrubs were all in the same range, and plant diversity had almost no change. In terms of herbaceous aquatic plants, from a longitudinal perspective, at the beginning of the garden design, the plant quadrats were all of a single species, accounting for 100%, and the plant types were uniform. While the current plant quadrats with Shannon index > 0 accounted for 58%, and the plant diversity was relatively high. In terms of horizontal comparison, the Shannon index of all the quadrats at the beginning of the design was 0, but the Shannon index scores in the current area increased significantly at each interval. After the completion of the park, the habitat in this area was better, and plant diversity was improved. Among these plants, the diversity of the surrounding areas of the terraced wetlands had increased more significantly. Through the establishment of optimal habitats, there will be a sustained enhancement in plant diversity and ecosystem services for the foreseeable future.

From a spatial perspective, in the "Clear as a Drain" wetland area, there was relatively low diversity in the terraced wetlands mainly planted with artificial large-area water purification or soil improvement plants, while the quadrats with higher Shannon–Weiner index were mostly located around the terraced wetlands and in the northern forest land, and at the same time, the plant diversity in the dry terraced wetlands was also higher. The above-mentioned areas with higher diversity are mostly naturally regenerated herbaceous plants, showing a good growth trend, such as *Setaria viridis*, *Setaria viridis*, *Conyza canadensis*, and many other new local herbaceous plant species. Meanwhile, it could be found that plant diversity index increased mainly in the above three areas, namely, around the terraced wetlands, the northern forest land, and the dry terraced wetlands.

Table 3. Comparison of Shannon–Weiner index of quadrat plants after the constructed artificial wetland and at the beginning of the design.

Plant Types	**Shannon–Weiner Index**				
Arbor and bush	Quadrat 1 0.937 (0.637)	Quadrat 2 0 (zero)	Quadrat 3 0.849 (0)	Quadrat 4 0.673 (0.673)	Quadrat 5 0 (0.673)
	Quadrat 6 0 (0.673)	Quadrat 7 0 (0.683)	Quadrat 8 0 (0)	Quadrat 9 0.562 (0)	Quadrat 10 1.011 (0)
Grass	Quadrat 1 0 (0)	Quadrat 2 0.562 (0)	Quadrat 3 0 (0)	Quadrat 4 0.410 (0)	Quadrat 5 0 (0)
	Quadrat 6 0 (0)	Quadrat7 1.674 (0)	Quadrat 8 0.760 (0)	Quadrat 9 0.500 (0)	Quadrat 10 0.325 (0)
	Quadrat 11 1.474 (0)	Quadrat 12 0.693 (0)	Quadrat 13 0 (0)	Quadrat 14 0.637 (0)	Quadrat 15 0 (0)
	Quadrat 16 0.693 (0)	Quadrat 17 0.673 (0)	Quadrat 18 1.560 (0)	Quadrat 19 1.010 (0)	Quadrat 20 0 (0)
	Quadrat 21 0.325 (0)	Quadrat 22 0 (0)	Quadrat 23 0 (0)	Quadrat 24 0 (0)	Quadrat 25 0.758 (0)
	Quadrat 26 0.802 (0)	Quadrat 27 0.858 (0)	Quadrat 28 0.802 (0)	Quadrat 29 0 (0)	Quadrat 30 0 (0)
	Quadrat 31 0.950 (0)	Quadrat 32 0.673 (0)	Quadrat 33 1.089 (0)	Quadrat 34 0.950 (0)	Quadrat 35 0.693 (0)
	Quadrat 36 1.540 (0)	Quadrat 37 0.218 (0)	Quadrat 38 0 (0)	Quadrat 39 0 (0)	Quadrat 40 0.191 (0)
	Quadrat 41 0.245 (0)	Quadrat 42 0 (0)	Quadrat 43 0 (0)	Quadrat 44 0 (0)	Quadrat 45 0.898 (0)
	Quadrat 46 0 (0)	Quadrat 47 0 (0)	Quadrat 48 0.509 (0)	Quadrat 49 0 (0)	Quadrat 50 0 (0)

Note: The values in parentheses represent the Shannon–Weiner index at the beginning of the design.

4.1.3. Carbon Sequestration and Oxygen Release

Plants have significant functions in sequestering carbon and releasing oxygen, as well as reducing carbon emissions. In recent years, their functions of sequestering carbon and releasing oxygen to improve the environment in cities has gradually received more attention. Plants mainly reduce atmospheric carbon in two ways: firstly, they absorb CO_2 from the air through photosynthesis during growth, convert it into carbohydrate storage bodies, and release O_2, directly reducing the CO_2 content in the atmosphere; secondly, trees can regulate urban climate environments and reduce energy consumption, thereby indirectly reducing carbon emissions from the use of fossil fuels.

After calculation, Handan Garden Expo Park could sequester 748.76 t of carbon, release 545.79 t of oxygen, and reduce carbon emissions by 965.90 t in a year. According to the data provided by NTBC, an ordinary medium-sized car traveling about 20,000 kilometres per year will produce about 5 t of carbon dioxide. The fixed annual CO_2 emissions of Handan Garden Expo Park will be about 150 medium-sized cars, and this value will be even more significant in the future with the growth of the park's trees. Meanwhile, referring to the unit price of 251.40 yuan/t for carbon sequestration and carbon sequestration functions in China's afforestation cost method, the CO_2 fixation amount and CO_2 emission reduction in the park will be converted into the corresponding economic value of CNY 431,066.27 (Table 4).

Table 4. Oxygen sequestration by vegetation in Handan Garden Expo Park.

Carbon Sequestration (t/Year)	Oxygen Release (t/Year)	Reduction of CO_2 Emissions (t/Year)	Total Atmospheric CO_2 Reduction (t/Year)	Economic Value ($/Year)
748.76	545.79	965.90	1714.66	431,066.27

4.2. Evaluation of Ecological Restoration Regulation Services

4.2.1. Water Purification

At the same time, the design parameters of each single pond wetland along the route were recorded to study the effects of different types of wetlands and design parameters on water quality purification. The locations of wetland sampling points and observation points at all levels of the "Clear as a Drain" wetland area are shown in Figure 2.

The detection results of different subsystems were sorted and combined according to the serial order of the wetlands, and the change diagram of the concentration of each pollutant was obtained. In Figure 2, the abscissa represents the number of the wetland group; for example, T2-1 refers to the first level wetland through which the inlet water flows in the T2 subsystem, and the vertical axis represents the concentration of pollutants. Thus, research on the seasonal purification effects of terraced wetlands was carried out.

(1) Pollutant Removal Effects of "Clear as a Drain" Artificial Wetland in Spring

The average values of COD, BOD_5, ammonia nitrogen, TP, and TN at the sampling points in spring were 17.9 mg/L, 4.2 mg/L, 0.07 mg/L, 0.01 mg/L, and 0.32 mg/L, respectively; the pollutant removal rates were 45.9%, 46%, 40.9%, 85%, and 66.7%, respectively (Figure 4). The purification effect was more obvious during the closure period of the Garden Expo Park in spring, when the plant maintenance in "Clear as a Drain" was poor, and the number of blue-green algae in the water body increased, resulting in significant fluctuations in dissolved oxygen within a day. At the same time, some plants withered and died without timely harvesting, causing nutrients such as nitrogen and phosphorus to re-enter the water body.

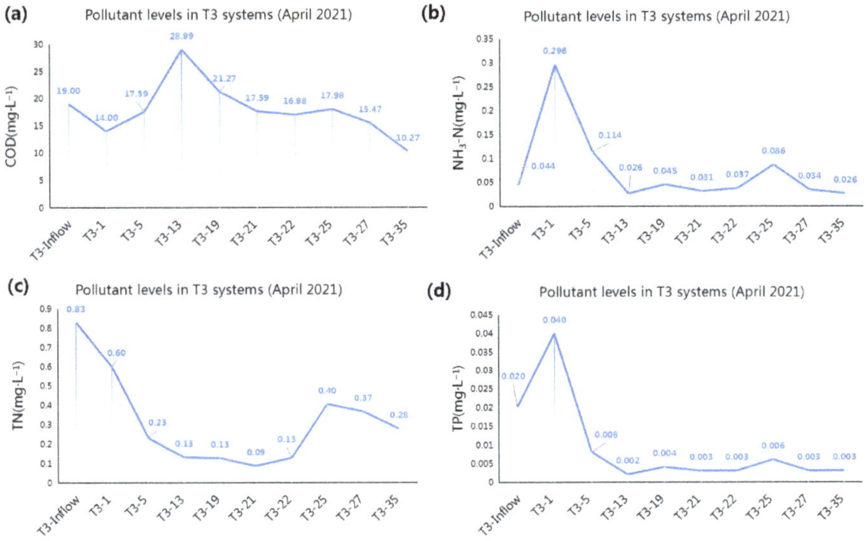

Figure 4. Changes in pollutant concentrations in wetlands at all levels of the T3 system during the spring of 2021. (**a**) COD (**b**) NH_3-N (**c**) TN (**d**) TP.

(2) The Removal Effects of Pollutants in "Clear as a Drain" Artificial Wetland in Summer

The average value of COD, BOD_5, ammonia nitrogen, TP, and TN at summer sampling points were 22.1 mg/L, 6.77 mg/L, 0.61 mg/L, 0.08 mg/L, and 0.95 mg/L, respectively; the pollutant removal rates were 57.2%, 33.9%, 95.8%, 96.3%, and 73%, respectively (Figure 5). The water quality changed from Class V incoming water to Class II water, and the water quality standard reached the design expectations. The research results show that the dissolved oxygen concentration in the "Clear as a Drain" artificial wetland increased, and COD and BOD_5 concentrations showed the opposite trend, with significant negative correlations. With the use of a large number of aquatic plants to sequester carbon and release oxygen, and ecological ponds and other falling water aeration facilities, the dissolved oxygen concentration gradually increased, and the pollutant purification effect gradually became apparent. Both water quality indicators and reduction rates performed best in summer [18]. This shows that the temperature and plant growth in summer have a significant impact on water purification.

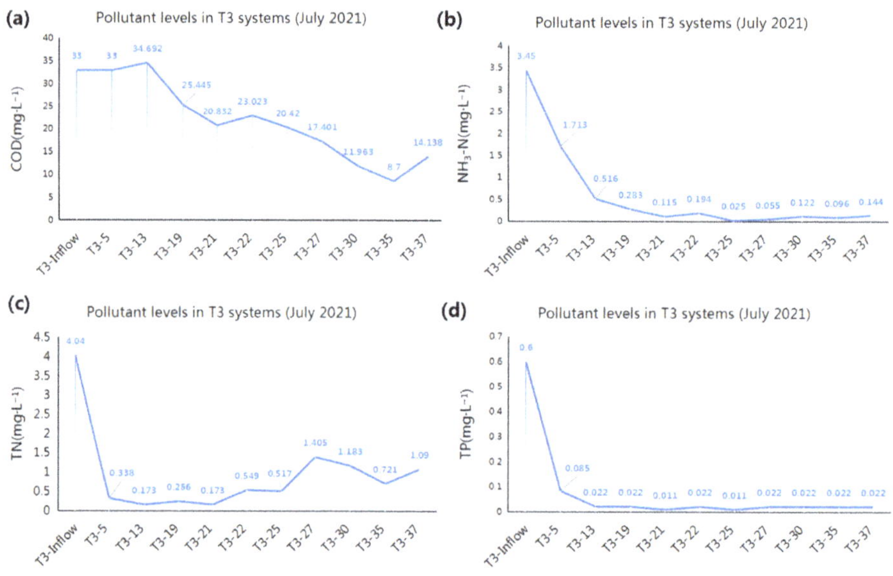

Figure 5. Changes in pollutant concentrations in wetlands at all levels of the T3 system during the summer of 2021. (**a**) COD (**b**) NH_3-N (**c**) TN (**d**) TP.

(3) The Removal Effects of Pollutants in "Clear as a Drain" Artificial Wetland in Autumn

The average values of COD, BOD_5, ammonia nitrogen, TP, and TN at the autumn sampling points were 18.01 mg/L, 7.15 mg/L, 0.27 mg/L, 0.03 mg/L, and 5.46 mg/L, respectively, and the removal rates were 51.8%, 35%, 42.8%, 72%, and 4.8%, respectively (Figure 6). Due to the obvious aerobic response of composite sponge facilities such as surface flow constructed wetlands and ecological ponds, the ammonia and nitrogen removal rate of sponge facilities was relatively high, reaching 43%, and the changes along the process were obvious. However, the removal effect of the composite sponge facility on TN was not obvious, and no obvious changes were seen in levels of TN (the outlet water quality did not meet the standard). It is possible that the system was not stable in the early plant growth stage and required long-term monitoring. In addition, the C/N ratio of the incoming water was relatively low, and denitrification would consume a large amount of carbon, affecting the denitrification of TN [19].

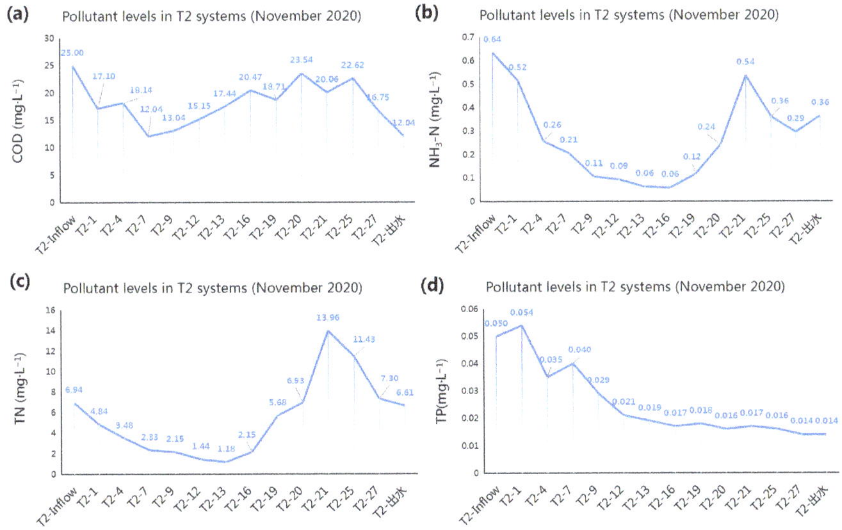

Figure 6. Changes in pollutant concentrations in wetlands at all levels of the T3 system during the autumn of 2021. (**a**) COD (**b**) NH_3-N (**c**) TN (**d**) TP.

From the above research results, it can be seen that the reduction rates of water pollutants in the inlet and outlet water in the three seasons were relatively significant. The purification efficiency of wetland water in summer was significantly higher than that in spring and autumn, and the water quality changed from Class V inflow to Class II. This shows that this composite constructed wetland had good water purification effects, and the water quality standard met the design expectations. It is worth noting that due to many uncontrollable influencing factors during actual site operation, the concentration of pollutants increased instead of decreasing. There are various reasons for this; it may be due to the poor vegetation growth of the single-pond wetland, resulting in an increase in pollutants, or due to local blockage of the wetland, leading to the accumulation of pollutants, or possibly due to the instability of the pollutants themselves, which were not fully decomposed in reversible reactions [20,21].

4.2.2. Air Purification

Using NTBC to analyze and calculate the plants in the park, the annual absorption of ozone, nitrogen dioxide, sulfur dioxide, and particulate matter by the vegetation in the park was obtained [22]. Then, based on the amount of air purification, the economic value brought by the vegetation's absorption of various pollutants could be converted. It could be concluded that the total amount of air purified by the vegetation in Handan Garden Expo Park throughout the whole year is about 13,836.56 kg, including 2181.73 kg of purified ozone, 4481.32 kg of nitrogen dioxide, 3422.85 kg of sulfur dioxide, and 3750.66 kg of particulate matter. It was found that the vegetation in Garden Expo Park had the strongest purification ability for nitrogen dioxide, followed by the purification ability for sulfur dioxide and particulate matter, and the purification ability for ozone was the weakest.

Due to the lack of a unified and clear economic conversion standard for ecological benefits in China, this study converted the economic benefits of plant purification of air pollutants based on the standards of the National Public Service Commission in the United States. The conversion criteria and economic benefits are as follows (Table 5).

Table 5. Price conversion criteria for the economic value of vegetation purification of atmospheric pollutants.

Atmospheric Pollutants	Relevant Economic Value (CNY/kg)	Amount of Air Purified (kg/year)	Economic Value (USD/Year)
O_3	54.23	2181.73	118,318.35
NO_2	54.18	4481.32	243,067.55
SO_2	23.22	3422.85	79,478.68
PM10	36.13	3750.66	135,511.75
Overall amount		13,836.56	576,376.33

4.2.3. Energy Saving

Buildings such as venues, exhibition halls, toilets, and shops in the park would generate energy consumption [23]. Reasonable plant configurations could generate certain economic benefits in terms of energy consumption. For example, in hot summer weather, the shading effect of plants lowered the indoor temperature of the building, thereby reducing electricity consumption; in cold winter weather, the design of plant groups blocked the cold wind and played a role in keeping warm, reducing the consumption of natural gas [24]. In addition, reasonable vegetation design was conducive to creating a good environment, improving local microclimate, and saving related electricity and air purification costs for nearby residents [25].

This study calculated and evaluated the energy saving benefits of plants in the park with the help of NTBC, and found that the total amount of electricity saved by the trees in Handan Garden Expo Park over the year was 812,665 kw·h, and the amount of oil or natural gas reduced was 851,402.67 m^3. Based on a monthly electricity consumption of 300 kw·h per household, the annual energy saved by plants in the park was enough for 100 households to use in 2.3 years; based on a monthly use of 30 m^3 of natural gas by each household, the annual amount of natural gas saved by the park's vegetation was enough for 100 households for 23.7 years.

Since Handan residents have implemented time-of-use electricity prices, 8 a.m. and 10 p.m. were used as the dividing point; the peak period was 8:00–22:00 (electricity price of CNY 0.55), and the valley period was 22:00–8:00 (electricity price of CNY 0.30). This study used the standard of 0.55 to convert the economic value of electricity. Oil and natural gas were according to the latest sales pricing of 2.68 yuan/cubic meter adjusted by the Handan Municipal Development and Reform Commission. It was calculated that the total economic value of energy saving by plants in the park in 2021 reached 2,646,556 yuan (Table 6).

Table 6. Energy saving status of vegetation in Handan Garden Expo Park.

	Power Cooling	Oil/Gas Reduction	Total Value
Energy savings	812,665 (kw·h/year)	300,849 (therm/year)	
Energy savings	812,665 (kw·h/year)	851,402.67 (m^3/year)	
Economic value ($/year)	364,796.85	2,281,759.15	2,646,556.00

5. Conclusions

This study was guided by concepts of water control in the new era and uses nature-based technology to solve China's present water problems. It aimed to transition from simple water conservancy project construction and water quality standards to comprehensively improving the service functions of water ecosystems and restoring the natural virtuous cycle of water bodies. This study focused on the actual landscape performance and influencing factors of the green technology that comprehensively enhance the ecosystem service function of composite sponge facilities. Empirical research was conducted using the example of the "Clear as a Drain" artificial wetland in Handan Garden Expo Park. The results showed that constructed wetlands play an important role in ecosystem support services and ecosystem regulation services, especially in water purification and biodiversity enhancement in summer. Relevant conclusions are as follows:

(1) The "Clear as a Drain" artificial wetland in Handan Garden Expo Park had a variety of habitat types, maintaining material circulation and the normal operation of the ecosystem. The dominant species presented adaptive landscape characteristics of purifying water quality and improving soil, and had strong ecological restoration functions. The rich plant species in the site has accelerated the natural succession process of native plants, with an increasing trend in plant species, especially the addition of many new local plants, which grew well and provided a good habitat for other organisms, demonstrating the good sustainability of the site. Meanwhile, plant diversity was relatively high, especially in the area around the terraced wetland, the northern forest land, and the dry terraced wetlands, where plant diversity had significantly increased compared with before. In addition, the ecological benefits of the site plants in carbon sequestration, oxygen release, and reducing carbon emissions were also significant. In short, the scientific and reasonable plant design of Handan Garden not only met the aesthetic functional requirements, but also provided good conditions for the self-restoration and regeneration of the site's habitats, thereby providing sustainable natural services for the city and its residents.

(2) Handan Expo Park had played a positive role in ecosystem regulation. Among them, the "Clear as a Drain" artificial wetlands had a more significant effect on purifying recycled water. The inflow water quality significantly improved in spring, summer, and autumn, with the highest purification effect in summer. Therefore, during the operation of Garden Expo Park, the treatment volume in summer should be increased and the proportion of deep treatment of recycled water in summer should be increased throughout the year. The water quality gradually became better as the steps increased, and the reduction rate was obvious, which effectively conformed to the basic principle of first-order kinetic reaction. There was an obvious negative correlation between pollutant concentration and dissolved oxygen, and the effect of oxygen explosion facilities was obvious. The wetland purification effects of different independent subsystems had their own advantages in the removal of pollutants, and the length of the water flow path was positively correlated with the pollutant removal rate; the environment and air quality of the park had been greatly improved, and the plants in the park had shown strong air purification functions; the vegetation in the park was like a natural sponge, absorbing and digesting rainwater as a precious resource; the annual energy savings of plants in the park can fully cover the daily energy consumption of nearby residents in Qi Village.

(3) At the beginning of the planning of the constructed wetland, the sustainability of the ecosystem was fully considered. In the construction process, local renewable and recyclable materials, including steel slag, plant materials, wood, etc. were given priority to reduce the "ecological footprint" and "life cycle cost". In terms of plant selection, a large number of native plants are selected, which can effectively adapt to local climate and soil conditions while reducing maintenance costs and water consumption. The establishment of good habitats helps to support the sustainable provision of services in local ecosystems. In the future, with the continuous investment in operation and maintenance, the artificial wetland will provide more systematic and integrated ecosystem services, such as stormwater management (reducing regional water security risks), aesthetic enlightenment (promoting regional culture), and so on.

Through empirical research on the "Clear as a Drain" artificial wetland in Handan Garden Expo Park, a quantitative study on the multiple comprehensive values of ecological restoration was carried out based on a comprehensive evaluation framework. This research results can provide support for the operation and maintenance of the "Clear as a Drain" artificial wetland in the Garden Expo Park and the subsequent design of composite sponge facilities. In the future, the project should be continuously tracked and studied, and the time dimension should be integrated into the open system for comprehensive analysis, in order to reflect the sustainability of the ecological restoration performance of composite sponge facilities.

Author Contributions: Conceptualisation, J.L. and J.W.; methodology, J.H. and Y.G.; validation, F.Y.; formal analysis, N.S. and Y.G.; investigation, J.L.; resources, P.Y. and Y.W.; data P.Y.; writing—review and editing, J.L. and J.W.; visualisation, F.S.; supervision, J.W.; project administration, J.L.; administration, J.W. All authors have read and agreed to the published version of the manuscript.

Funding: The research was funded by National Key Research and Development Programme, funding number 2016YFC0401108; Key research and development plan of Ministry of Science and Technology, funding number 2021YFC3201504.

Institutional Review Board Statement: Not applicable.

Informed Consent Statement: Not applicable.

Data Availability Statement: Data can be requested from J.L. ljj19910213@pku.edu.cn.

Acknowledgments: All the authors wish to thank all who assisted in conducting this work.

Conflicts of Interest: The authors declare no conflict of interest.

References

1. Lau, C. Nature-based solutions. *Planning* **2018**, *84*, 34.
2. Jansson, R.; Backx, H.; Boulton, A.J.; Dixon, M.; Dudgeon, D.; Hughes, F.M.R.; Nakamura, K.; Stanley, E.H.; Tockner, K. Stating mechanisms and refining criteria for ecologically successful river restoration: A comment on Palmer et al. (2005). *J. Appl. Ecol.* **2005**, *42*, 218–222. [CrossRef]
3. Ndubisi, F.; Whitlow, H.; Deutsch, B. Landscape Performance: Past, Present, and Future. *Landsc. Archit.* **2015**, *1*, 40–51.
4. Keeton, J.M. The Road to Platinum: Using the USGBC's LEED-EB Green Building Rating System to Retrofit the U.S. Environmental Protection Agency's Region 10 Park Place Office Building. *J. Green Build.* **2010**, *5*, 55–75. [CrossRef]
5. Palmer, M.A.; Bernhardt, E.S.; Allan, J.D.; Lake, P.S.; Alexander, G.; Brooks, S.; Carr, J.; Clayton, S.; Dahm, C.N.; Follstad Shah, J.; et al. Standards for ecologically successful river restoration. *J. Appl. Ecol.* **2005**, *42*, 208–217. [CrossRef]
6. Saiyid, A.H. States Have EPA-Developed Templates to Write Water Standards for Wetlands. *Environ. Rep.* **2016**, *47*, 1565.
7. Seattle Public Utilities, Restore Our Waters. Thornton Creek Water Quality Channel Final Report. 2009. Available online: http://www.seattle.gov/util/cs/groups/public/documents/webcontent/spu01_006146.pdf (accessed on 2 December 2018).
8. Giraldo, G.; Lo, M.; Davies, M. Thornton Creek Water Quality Channel, Urban Water Quality, and Environmental Benefits. In *Low Impact Development: New and Continuing Applications*; ASCE: Reston, VI, USA, 2008.
9. Chang, H.X. *Design Parameters of Cascade Subsurface Flow Wetland and Their Effects on Water Purification: A Case Study of Fengxiang Park in Haikou City*; Peking University: Beijing, China, 2018. (In Chinese)
10. Gong, Y.W.; Liu, C.; Li, J.Q. Exploration of Acceptance and Assessment Methods for the Main Objectives of Sponge City Construction. *China Water Supply Drain.* **2015**, *31*, 114–117. (In Chinese)
11. Aya, S.; Aka, L. Treatment of wastewater by natural systems. *Environ. Int.* **2001**, *26*, 189–195.
12. Sim, C.H.; Yusoff, M.K.; Shutes, B.; Ho, S.C.; Mansor, M. Nutrient removal in a pilot and full scale constructed wetland, Putrajaya city, Malaysia. *J. Environ. Manag.* **2008**, *88*, 307–317. [CrossRef]
13. Vitturi, M.; Iannelli, R.; Giraldi, D.; de Michieli Vitturi, M.; Iannelli, R. FITOVERT: A dynamic numerical model of subsurface vertical flow constructed wetlands. *Environ. Model. Softw.* **2010**, *25*, 633–640.
14. Wallace, S.; Parkin, G.; Cross, C. Cold climate wetlands: Design and performance. *Water Sci. Technol.* **2001**, *44*, 259–265. [CrossRef]
15. Werker, A.; Dougherty, J.M.; Mchenry, J.L.; Van Loon, W.A. Treatment variability for wetland wastewater treatment design in cold climates. *Ecol. Eng.* **2003**, *19*, 1–11. [CrossRef]
16. Ouyang, Y.; Jia, E.; Zhang, J.E.; Lin, D.; Liu, G.D. A STELLA model for the estimation of atrazine runoff, leaching, adsorption, and degradation from an agricultural land. *J. Soil Sediments* **2010**, *10*, 263–271. [CrossRef]
17. Qing, C.; Honghu, Z.; Yanpeng, L.; Chen, Q.; Zeng, H.; Liang, Y.; Qin, L.; Peng, G.; Huang, L.; Song, X. Purification Effects on β-HCH Removal and Bacterial Community Differences of Vertical-Flow Constructed Wetlands with Different Vegetation Plantations. *Sustainability* **2021**, *13*, 13244. [CrossRef]
18. Faulwetter, J.L.; Gagnon, V.; Sundberg, C.; Chazarenc, F.; Burr, M.D.; Brisson, J.; Camper, A.K.; Stein, O.R. Microbial processes influencing performance of treatment wetlands: A review. *Ecol. Eng.* **2009**, *35*, 987–1004. [CrossRef]
19. Tomei, C.M.; Carozza, A.N.; Angelucci, M.D. Post-aerobic digestion of waste sludge: Performance analysis and modelling of nitrogen fate under alternating aeration. *Int. J. Environ. Sci. Technol.* **2016**, *13*, 21–30. [CrossRef]
20. Jizheng, P.; Houhu, Z.; Xuejun, L.; Pan, J.; Zhang, H.; Lu, X.; Li, Y.; Zhao, M.; Xu, H. Enhanced nitrogen removal by the integrated constructed wetlands with artificial aeration. *Environ. Technol. Innov.* **2019**, *14*, 100362. [CrossRef]
21. Linares, G.; Pulgar, M.; López, S.M.; Gilabert, C.V. A Territorial Strategy for the Activation of Tourism in Low Population Density Heritage Landscapes. *Land* **2024**, *13*, 574. [CrossRef]
22. Hamidova, E.; Bosino, A.; Franceschi, L.; De Amicis, M. Nature-Based Solution Integration to Enhance Urban Geomorphological Mapping: A Methodological Approach. *Land* **2024**, *13*, 467. [CrossRef]

23. Brom, P.; Engemann, K.; Breed, C.; Pasgaard, M.; Onaolapo, T.; Svenning, J.-C. A Decision Support Tool for Green Infrastructure Planning in the Face of Rapid Urbanization. *Land* **2023**, *12*, 415. [CrossRef]
24. Babí Almenar, J.; Petucco, C.; Navarrete Gutiérrez, T.; Chion, L.; Rugani, B. Assessing Net Environmental and Economic Impacts of Urban Forests: An Online Decision Support Tool. *Land* **2023**, *12*, 70. [CrossRef]
25. Stašek, J.; Krása, J.; Mistr, M.; Dostál, T.; Devátý, J.; Středa, T.; Mikulka, J. Using a Rainfall Simulator to Define the Effect of Soil Conservation Techniques on Soil Loss and Water Retention. *Land* **2023**, *12*, 431. [CrossRef]

land

Article

Ecoparque: An Example of Nature-Based Solutions Implementation at Tijuana a Global South City

Lina Ojeda-Revah and Gabriela Muñoz-Meléndez *

Departamento de Estudios Urbanos y del Medio Ambiente, El Colegio de la Frontera Norte, Km 18.5 Carretera Tijuana-Ensenada, San Antonio del Mar, Tijuana B.C. 22560, Mexico; lojeda@colef.mx
* Correspondence: gmunoz@colef.mx; Tel.: +52-(664)-6316310

Abstract

Nature-Based Solutions (NBS) are recognized as urban strategies to face environmental degradation and climate change vulnerability to address social challenges. However, NBS are context-dependent and must be based on evidence. Thus, this document details the NBS implementation in a global south city such as Tijuana, a semiarid city at the Mexico–USA border, which has rapidly grown under poor urban planning, widespread irregular settlements, increase in air and water pollution, and limited green spaces. In response, six hectares of a severely eroded slope have been transformed by El Colef into Ecoparque, an Urban Resilience Laboratory. This academic initiative aims to enhance residents' quality of life by analyzing environmental problems, raising awareness, and engaging the community, in addition to identifying opportunities for implementing NBS. This paper presents the 32 years' experience of implementing NBS at Ecoparque, such as a constructed wetland as part of a wastewater treatment, reforestation with native plants grown in an in situ nursery, soil restoration using its own-produced compost, and urban ecosystem rehabilitation. Moreover, main challenges and upscaling opportunities are identified to adopt NBS in a Global South city. Results showed that the most relevant problems have been insufficient human and financial resources, as well as the lack of a proper legal framework. This study provides an analytical significance that could be useful to apply under similar contexts.

Keywords: NBS; urban-resilience laboratory; urban park; urban sustainability; Tijuana; Mexico

1. Introduction

By 2023, more than half of the population worldwide lived in cities and towns; this number is expected to increase to about 5 billion in 2030 and to 2 billion by 2050 [1]. This increase of urbanites is worrisome given that cities are both victims of climate change and among its worst perpetrators; the former because metropolises are disproportionately exposed to climate impacts, the latter because they are also responsible for generating a large share of global emissions [2]. Cities are responsible for 80% of greenhouse gas emissions and consume 80% of the world's resources [3]. Furthermore, with urbanization, the consumption of energy and water resources has increased, causing environmental problems that impact human wellbeing [4], such as the pollution of air and water and generation of waste disposal [5].

The increasing global urbanization and its environmental impacts are hindering urban sustainability and resilience currently but are likely to continue and intensify in the future [6]. In this context, NBS are being advocated for as a strategy for cities dealing with

https://doi.org/10.3390/land15010089

the degradation of natural resources and ecosystem services and vulnerability to climate change and extreme weather events, as well as incertitude about citizens' health and wellbeing. Inspired by nature, NBS are "actions to protect, conserve, restore, sustainably use and manage natural or modified terrestrial, freshwater, coastal and marine ecosystems which address social, economic and environmental challenges effectively and adaptively, while simultaneously providing human wellbeing, ecosystem services, resilience and biodiversity benefits" [7].

NBS serve as a transdisciplinary umbrella term that brings together concepts like ecosystem services, green infrastructure, and natural capital, to address threats to society [8] and for maintaining human well-being [9]. A literature review on NBS reveals they are site specific in function of the urban context and landscape scale, require a systematic understanding between nature conservation and societal needs, benefit both people—including equity—and biodiversity, empower governance processes that promote broad participation and foster communication, are practical and viable economically, and, most importantly, are challenge oriented in ecological and social terms [8,10–12].

However, our NBS literature review also highlights several significant challenges hindering their effective implementation. These include a lack of legal, policy frameworks, and guidance and tools; weak law enforcement; and limited stakeholder engagement. Also mentioned are organizational issues such as inflexible and fragmented structures, sectoral silos, and limited collaborative governance, as well as insufficient public awareness and engagement. Another set of issues relate to deficient monitoring and assessment leading to shortages of reliable data, appropriate indicators, and clear measures of effectiveness, which create uncertainties regarding the added value and benefits. Finally, there is a dearth of public or private financial resources, funding instruments (e.g., maintenance and human resources), low private sector engagement, and power dynamics tied to property interests [13–15].

As in many countries of the Global South, in Latin America, although urban ecosystem services research related to NBS has recently increased, its integration into planning is still limited [16]; in Mexico even greenspace planning and management face the very same problem [17]. But, going back to NBS research, most focus on using them to ensure water supply, enhance quality, lower landslide risk, and control flooding [18,19]. Yet, there are prejudices to overcome; for example, Latin American participants of a symposium to deepen understandings of nature-based thinking noted that putting nature ahead of human needs represents a privileged viewpoint, since tackling poverty and inequality must come before environmental efforts [18].

Within this framework, the purpose of this document is to examine the NBS implemented at an urban park in a city of the Global South, as well as their performance, challenges, and opportunities of upscaling. This manuscript draws from a case study methodology to know the local context and main environmental problems of the city of Tijuana, Baja California, Mexico, and the local park "Ecoparque" and its NBS adoption through historical and technical data; difficulties and opportunities identified are described later. Then, obstacles during achievement and enhancement strategies are debated. Finally, conclusions are given.

2. Materials and Methods

This academic work was carried out as case study research, a methodology of the social sciences, that has been defined as an empirical method that investigates a contemporary phenomenon in depth and within its real-world context, when the boundaries between phenomenon and context may not be clearly evident [20]. This research initiative uses a clear-cut focus on a unit of analysis, applying an empirical approach to deal with complex

https://doi.org/10.3390/land15010089

research questions, in processes or objects with multiple not obvious variables, utilizing qualitative, quantitative, and mixed methods applied either in a positivist or interpretivist mode, presented in a narrative way [21] that can provide important evidence for the argument [22].

According to their epistemological function, case studies could be exploratory, descriptive, or explicative, while categorized by their characteristics, could be critical, unusual, common, revealing, and longitudinal [20]. It must be added that case studies are not samples, and, in that sense, results from their characterization could not be generalized in a statistical sense but in a logical and analytical fashion; moreover, case studies could impact theories and key concepts and shed light on the understanding of a given phenomenon [22].

Here, the case study research begins with a brief description of Tijuana to set the urban context, then the case study of Ecoparque is presented in a longitudinal way, as seen in Figure 1. For 27 years, interventions were carried out more in a reactive way; noticeably, it was 2013 when a committee of specialists was established, composed of civil, biochemical, and chemical engineers, and a biologist. All of them are assigned to El Colef, in addition two independent external volunteers joined such committee, one specialized in native plants and the other in urban agriculture.

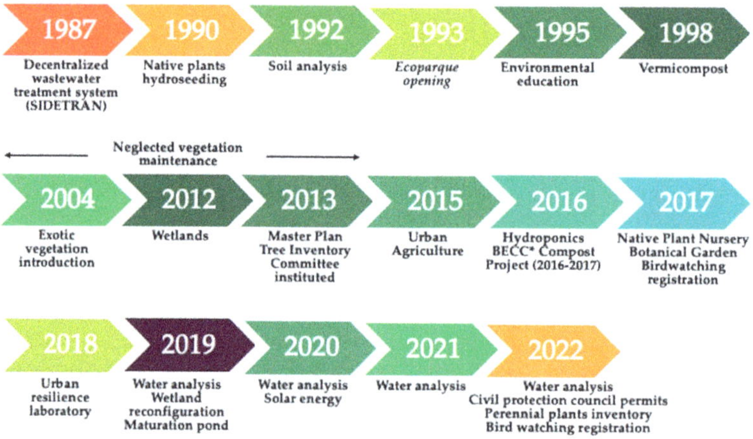

Figure 1. Ecoparque's timeline. Source: Authors' own elaboration.

From then until now, this committee has been in charge of the site planning and decision making, as depicted in Figure 2. It should be mentioned that planned actions have been developed under financial constraints, and they have been prioritized under opportunity. When actions are planned, depending on the program, a set of appropriate indicators are drafted to identify interactions and overlaps, data sources, and monitoring details to know if goals are achieved, and after the intervention ends the committee analyzes it, registering lessons. Common indicators used for repetitive interventions are

- Vegetation inventories (species' richness and abundance, natives, and exotic)
- Water quality (bacteriological, organic, inorganic, and nutrient content)
- Birdwatching (species)
- Compost quality (NPK content)
- Solar energy generation (kwh per time)
- Risk (firewalls, security measures, and number of accidents)
- Environmental education (number of visitors per educational level, workshops, tours, volunteering, and consulting)

 https://doi.org/10.3390/land15010089

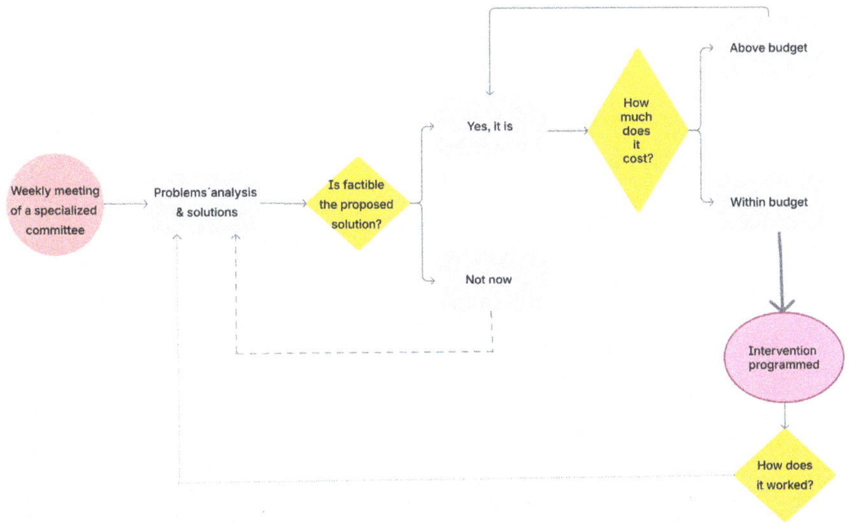

Figure 2. Flow diagram of decision making at Ecoparque. Source: Authors' own elaboration.

2.1. Tijuana's Urban Context and Challenges

Tijuana is in the far northwest of Mexico, in the state of Baja California, bordering with the USA, south of San Diego County, California, and with the Pacific Ocean to the west (Figure 3). The city lies on a terrain with a rugged topography, ranging from sea level in the west to more than 1800 m above sea level in the eastern mountain range. The soil is unstable and saturates quickly, thus it is prone to floods, erosion, and landslides. The Mediterranean-type climate is semiarid and temperate, with hot dry summers and rainy winters, with an average annual temperature of 17.7 °C and scarce and irregular rainfall (237 mm annual average) [23]. The city is in the California Floristic Province, which has great biodiversity and endemic species, and is considered one of the world's biodiversity hot spots, threatened by rapid urbanization [24]. The main ecosystems present are chaparral, coastal scrubland, and riparian vegetation with some trees along intermittent streams [25].

Tijuana has a demographic dynamic greatly influenced by migration, attracted in large numbers by being a commercial center with services and industrialization, given its proximity to the USA. By 2020, the city's population reached 1,922,523 inhabitants [26]. The city's urban area expanded from 7789 hectares in 1972 to 30,626 hectares in 2017 (on average 507 hectares per year) [27], reducing and fragmenting natural habitats and affecting species diversity [25].

Rapid growth and poor urban planning have led to capacity mismatches to meet the needs of the population, aggravated by the irregular origin of 43% of human settlements [28] by a progressive occupation of hillsides with slopes unsuitable for construction [29]. Consequently, Tijuana has become an area vulnerable to flooding and landslides, due to modifications made to a landscape with abrupt topography, the increase in impermeable surfaces, the removal of the natural vegetation, and an inefficient network of storm drains, which destabilize the slopes, reduces the infiltration of rainwater, and accelerates its runoff. Moreover, the city has only 1.44 m^2 of public green areas per inhabitant unevenly distributed [30].

https://doi.org/10.3390/land15010089

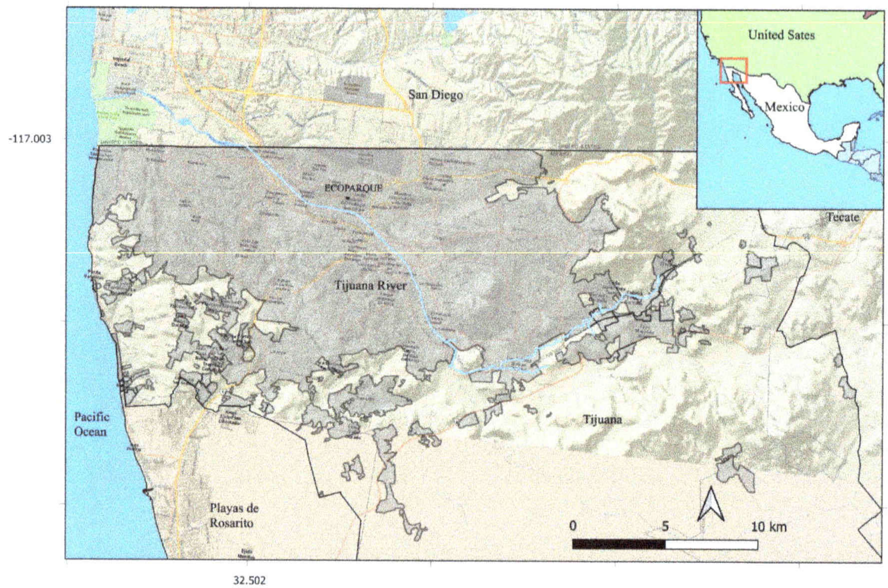

Figure 3. Study area. Source: Authors' own elaboration.

On the other hand, the main water source to supply Tijuana comes from the Colorado River, which is transported via an aqueduct that runs 125 km, from the Mexicali Valley to Tijuana, overcoming a height of up to 1061 m [31]. During 2021, 166,509,443 m^3 of water was moved through the Colorado River—Tijuana Aqueduct using 658,134,925 KWh for pumping at a cost of MXN 1,293,772,489 (USD 79 million at the value of 2025); 92% of this water was consumed at Tijuana [32]. Due to energy consumption to operate this aqueduct, 68,045 tons of CO_2, 409 tons of SO_2, 339 tons of NO_x, and 90 tons of PM_{10} were estimated to be generated [33].

These expensive and energy-embedded waters are used once (a mere 2.8% is reused), treated, and discharged (88% of population has access to sewerage system) into the Tijuana River, whose waters drain to the Pacific Ocean passing across Imperial Beach, CA, USA. Tijuana River waters are treated at the South Bay International Wastewater Treatment Plant just across the border; however, its treatment capacity of 9.5 million l/day has been regularly exceeded since 2014 [34], especially during the rainy season, causing tensions between Mexico and the USA.

Along with water dependency, the accelerated urbanization in Tijuana has increased between 1950 and 2010, an increment of 235 times of construction materials, except for wood, adobe, and waste material, the increase of which was eleven, one, and nine times, respectively. For its part, the consumption of electricity increased nearly 18 times from 1980 to 2010. Additionally, all urbanization activities have deteriorated the air quality. In particular, ozone and PM_{10} have been historically above maximum permitted levels. Another impact related to air quality are wildfires (associated with the Mediterranean climate in the region) that have increased due to urban waste fires (from 2010 to 2015, solid waste generation has increased to 28.5 thousand tons), caused by electric shocks, cigarette butts, and bonfires [35]. Additionally, it is expected that climate change conditions will increase climate sensitivity, such as water scarcity and wildfires [36].

https://doi.org/10.3390/land15010089

2.2. *Ecoparque, an Urban Resilience Laboratory*

In 1987, El Colef created Ecoparque, a project focused on environmental education, academic outreach, and applied research, to look for answers to environmental problems caused by Tijuana's accelerated urbanization. From then until now, Ecoparque has transformed into an Urban Resilience Laboratory, where the conditions of environmental management in the city are analyzed, experiments are designed, solutions are provided in technical and scientific terms, and areas of opportunity are identified through applied research, all with the aim to improve the quality of life of Tijuana's inhabitants.

Ecoparque is found within the urban area of Tijuana, and it extends over six hectares of a hillside with an elevation of around 80 m above the riverbed level (see Figure 4). Activities carried out in the park are organized into five interrelated multipurpose programs: water, botanical garden with a native plant nursery, urban agriculture, energy, environmental education, and a soil amendment subprogram (compost and vermicompost) (Figure 5). As observed, colored lines show interactions among programs, for example, soil restoration is intrinsically associated with urban agriculture, reforestation, and environmental education. In the Supplementary Materials a short video of the park can be downloaded.

Figure 4. Panoramic view of Ecoparque. Source: Guillermo Arias.

2.2.1. Constructed Wetlands

In 1993, a secondary DEWATs started to operate in Ecoparque with 5 lps of installed capacity. This system treats domestic wastewater that arrives by gravity from a settlement of 6193 inhabitants distributed in 2090 homes and 153 small businesses. Wastewater is received by an inlet grid and micro sieve (primary stage) to pass to a biofilter and sedimentation tank (secondary stage). By 2012, two artificial wetlands' horizontal subsurface were added to DEWATs. This addition aimed to improve water quality to reduce the high concentrations of fecal coliforms, N and P. The installed system was composed of

- Unit 1. Surface ≈ 533 m^2 (width = 13 m, length = 41 m), depth of 0.52 m at 1.06 m (slope 1%)
- Unit 2. Surface ≈ 560 m^2 (width = 13 m, length = 43 m), depth of 0.90 m at 1.25 m (slope 1%)

 https://doi.org/10.3390/land15010089

- Water tanks ≈ 60 m^2 (width = 5 m, length = 12 m) depth = 2.5 m

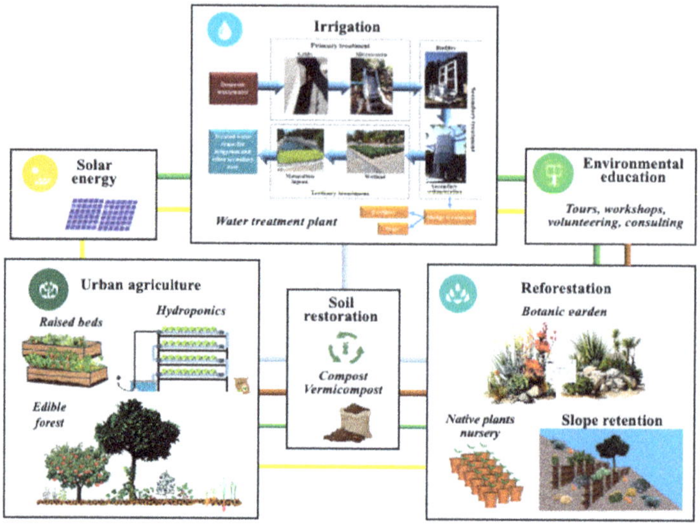

Figure 5. Ecoparque's programs, their main functions, and interactions. Source: Authors' own elaboration.

The substrate layer comprises gravel of 4″ to 6″ at the inlet and outlet, with some gravel of 1 $\frac{1}{2}$″. The rest of the wetland has gravel of $\frac{3}{4}$″ in diameter. On the other hand, a variety of vegetation was planted. For example, *Schoenoplectus californicus* (tule) is a perennial grass-like herb, native to Baja California, that grows in dense stands in marshes and along pond margins [37]. After several years of unsuccessfully trying for better water quality, by late 2019, unit 1 was kept as a wetland with the addition of an Anaerobic Baffled Reactor (ABR) before the inlet and reorganization and cleaning of the substrate layer. This wetland is currently planted with *Canna hybrid*. Unit 2 was transformed into a maturation pond. Modifications were achieved to comply with water quality standards. Water flows at the two units are 1 L per second (Q = 1 lps), while the hydraulic retention time (HRT) is around six days per unit.

The quality of the water produced is periodically monitored to ensure compliance with the NOM-001-SEMARNAT-1997 [38]. This Mexican regulation was updated in 2021 to NOM-001-SEMARNAT-2021 [39]. More than 30% of the treated water is reused in situ by means of a (1) Gravity irrigation system using two elevated tanks of ten m^3 of capacity, (2) Pressurized irrigation system using a seven and a half HP centrifugal pump, and (3) two and a half km irrigation network. About four hectares of green areas are irrigated, including the edible forest and native plants established.

2.2.2. Reforestation with Native Plants

Reforestation started in 1990 by means of hydroseeding, a technology that uses water pressure to embed seeds mixed with environmentally friendly fertilizers into the steep slopes. However, due to changes in management scope, many introduced species were planted over several years, without a design and without proper maintenance. In 2013, a diagnosis was carried out, accounting for 53 exotic species (trees and shrubs); then, a master plan was developed (see Figure 6).

https://doi.org/10.3390/land15010089

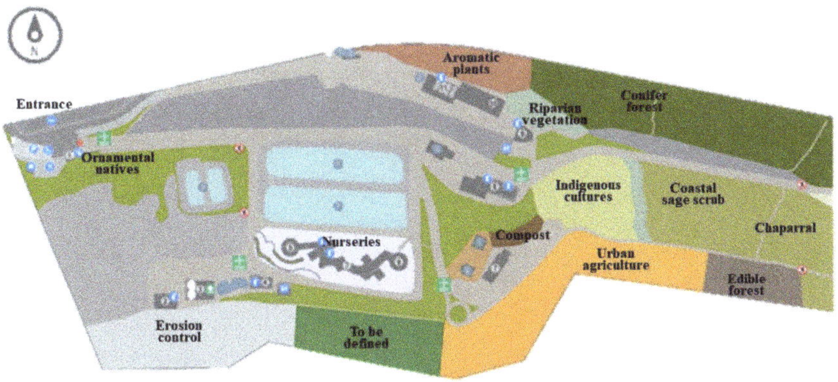

Figure 6. Botanic garden master plan. Source: Authors' own elaboration.

By 2016, the park was registered as a facility that manages confined wildlife outside of its natural habitat (No. SPA-PIMVS-JB-0009-BC/17), with the environmental authorities (SMADS/SEMARNAT), and a Botanic Garden was constructed with documented collections of living plants along with a native plant nursery. According to our 2022 perennial plant inventory, the park houses 1675 plants, belonging to 141 different trees and shrub species, of which 35.46% are native. The nursery has propagated 30 species of native plants.

2.2.3. Soil Restoration Using Compost and Vermicompost

From 2 May 2016 to 31 March 2017, a composting pilot program using forced aeration with a blower was carried out in Ecoparque. Up to 118 households were enrolled in this program, of which 77% actively participated. A total of 17,624 kg of organic waste was collected, and 4000 kg of compost was produced during around 100 days; analysis according to national standards [40] on the compost quality render it as a soil amendment due to its N-P-K composition, pH, and organic content percentage. The compost produced at Ecoparque could be used for ecological agriculture and reforestation as well as for urban green areas.

After the composting program ended, a vermicomposting program was initiated using California worms (*Eisenia* sp.). Compost has been utilized for ornamental plants propagation for the environmental education program, for the seeding of plants, as substrate and breeder in elevated beds of the urban garden in the urban agriculture program, and has been employed to improve soil in Ecoparque's slopes.

2.2.4. Ecosystem Restoration

After decades of reforestation and restoration within Ecoparque, to date, 35 bird species have been recorded, of which the majority are migratory residents (43%), followed by residents (30%) and migratory (19%). One of them, the red-shouldered hawk (*Buteo lineatus*), is considered by the International Union for Conservation of Nature (IUCN) under the category of "least concern", but under NOM-059 Mexican legislation it has a "special protection" category [41].

At the southern end of the park, on a very steep slope of approximately 7000 m^2, a native vegetation patch with 20 native species has been preserved. This is an important area because it proves that native vegetation can survive in the city's harsh environment.

2.2.5. Edible Forest

This multi-stratum agroforestry system with six strata: large, medium, and small crown trees, shrubs, vines, and herbs. Initiated in 2015, by 2025 this area comprises

https://doi.org/10.3390/land15010089

31 different edible species of which 14 are trees, one is a vine, six are shrubs, and ten vegetables and tender herbs.

3. Results as a Summary of Key Experiences and Lessons Learned

Decades of experience at implementing and operating NBS in Ecoparque have left valuable lessons, and these are summarized in Table 1.

Table 1. NBS at Ecoparque: problems, challenges, solutions, and environmental problems addressed.

NBS	Problems in Ecoparque	Solutions	Environmental Problems Addressed
Constructed wetland	"Siphon effect at wetland" (vacuum effect)	A TEE PVC piece venting to the atmosphere was installed.	Water scarcity by treating wastewater and reusing it in irrigation of greenspace
	Substrate clogging with sediments	An ABR with plastic filers were installed before the inlet. Such filters are periodically cleaned.	
	Lack of wastewater inlets due to clogging	Contact the municipal provider of wastewater. The installation of a system of pipelines and pumps to recirculate water in the DEWAT system (biofilter, sedimentation tank, wetland, and maturation pond)	
Reforestation with native plants	Introduction of exotic species	Long-term program to replace them with native plants	Biodiversity loss due to urbanization by conservation
	Adaptation of native plants to the urban environment	Research on conditions needed by native plants to adapt to urban environments	Native plants propagation
	Steep slopes	Development of different slope retention techniques	Erosion and landslide risk
	Pest and disease control	Use of organic pesticides, pest control plant species and promoting plant biodiversity	
	Native species propagation	Research to develop plant propagation protocols, mainly by seed to ensure genetic variability	Lack of urban greenspace
	Social preference for lustrous vegetation of humid places	Promote awareness to appreciate native brown–green vegetation with attractive landscape designs	Social awareness of ecosystem conservation
Compost producing	Quality of residues to be composted	Flyers informing on the "separation at source", and acceptable residues were distributed to participating households.	Soil erosion and degradation by restoring moisture retention
	Household enrolment	Incentives such as ornamental plants and 5 kg compost packages were given out.	Natural soil fertilization
	Rodents at the composting beds	Humidity and temperature control	
	Supersaturation of composting piles during the rainy season	Piles were covered with black sturdy plastic sheets.	
Native ecosystems restoration	Lack of personnel for bird watching	Volunteerism of specialized personnel	Biodiversity loss by: -Providing habitat for birds and other fauna -Native vegetation conservation
	Lack of personnel to monitor the natural vegetation patch	Research as master thesis topic	
Edible forest	Thefts of tools and urban agricultural production	Sturdy lockers were bought to store tools	Biodiversity loss
	Adaptation to local urban and site conditions	Seed selection of cultivars that thrive in urban conditions	Soil health improvement

https://doi.org/10.3390/land15010089

As seen in Table 1, NBS at Ecoparque have the potential for addressing environmental problems in the city as they have already showed their effectiveness in situ, see graphical abstract examples depicted in Figure 7. As observed in Figure 7a, the treated water can sustain vegetation, native plants included (see Figure 7b), in Ecoparque.

Figure 7. Graphical abstract examples of NBS in Ecoparque. Source: Authors' own preparation.

In addition to risk management, the urban park is an example of how to respond to the lack of urban greenspace (as noted in Figure 7c). Ecoparque is the fourth most extensive public green space in Tijuana, only behind three parks managed by the municipal government.

On the other hand, Figure 7d presents the compost used to restore and fertilize the soil. Moreover, this urban park has been, for decades, a space of environmental awareness for society by receiving visitors (Figure 7e).

Biodiversity loss due to urbanization has been handled by the propagation of native plants in our own nursery as shown in Figure 7f, while Figure 7g exhibits a technique to reduce erosion and landslide risk installed in Ecoparque few years ago.

Finally, all the work developed for more than three decades is paying back by restoring biodiversity. Our park provides a habitat for birds and other fauna, as demonstrated in Figure 7h.

4. Discussion on NBS Principles and Challenges at Ecoparque

4.1. General Challenges for NBS in Ecoparque

In the introduction of this document, a literature review was carried out identifying NBS challenges [13–15], as well as green infrastructure implementation [42], and greenspace planning and management in Mexico [17]. Although with different classifications, authors identified similar problems. In Tijuana and in Ecoparque most complications have been presented eventually, some of the most excruciating ones are detailed below.

4.2. During Implementation and Maintenance

Ecoparque is a complex project that has been operating for 32 years. During this period, successes and failures have been experienced at implementing and maintaining actions, those faced by specific NBS have been detailing in Table 1. Categorizing these by principles and challenges reported in the literature, major obstacles could be classified as

https://doi.org/10.3390/land15010089

Challenge oriented principle related to social awareness comprise the following:

- Social preference for lustrous vegetation of humid places
- Quality of residues to be composted
- Household enrolment

Systematic understanding, practical and economic viability, and challenged oriented principles, are associated with

- Insufficient funds
- Limited specialized human resources

Site specific principle together with external factor difficult monitoring and evaluation assessment was

- Thefts of tools and urban agricultural production

Furthermore, a different set of challenges due to technical issues determined by site specific principle were identified:

- "Siphon effect at wetland" (vacuum effect)
- Substrate clogging with sediments
- Lack of wastewater inlets due to clogging
- Steep slopes
- Supersaturation of composting piles during the rainy season

In addition, other technical issues determined by site specific conditions and social challenges principles were

- Pest and disease control
- Native species propagation
- Rodents at the composting beds
- Adaptation to local urban and site conditions

Finally, a technical issue under specific social challenges and systemic understanding principles was

- Adaptation of native plants to the urban environment

More importantly, Ecoparque, as part of El Colef, belongs to a network of CPI whose financial resources entirely come from the federal government; such resources tend to vary annually. Furthermore, funding restrictions applied to CPIs such as co-financing, private–public partnerships, and tax deduction are tools that cannot be used to look for additional funds. At last, while technical issues have not been reported as a challenge in our literature, they may be linked to the presence of specialized personnel. At Ecoparque, these challenges have been addressed through collaborations with expert volunteers.

4.3. Scaling up NBS

There is no experience of scaling up NBS projects developed in Ecoparque, besides a short-lived purple line project to reuse treated water to irrigate green spaces in Tijuana. The reasons for falling to advance NBS at the city level could be due to problems discussed before. Yet, of special mention is that in Mexico there are laws that offer a framework for NBS (greenspace and green infrastructure) development, but there is a lack of clear definitions, roles, and functions [43]. This is partly due to the low relevance given in the legislation to environmental services conservation [44], including urban greenspace planning focused on cultural services [17].

Nonetheless, there are national examples that have shown that political will could defeat challenges; for example, the experience of Xalapa's floods and landslides showed how sustained political will, in close collaboration with different organizations, a long-term

https://doi.org/10.3390/land15010089

vision, and gradual implementation can drive change. As established in the Xalapa Municipal Climate Plan 2013, this city has introduced various adaptation strategies, including NBS and green infrastructure [45]. Another successful example is the Green Infrastructure Network Special Program implemented in Mexico City [46].

In addition, a recently published federal rule [47] establishes guidelines to reinforce the territorial system, to withstand, adapt to, and recover from natural hazards and climate change through spatial planning. It highlights green infrastructure as an essential component of land-use planning and management for promoting resilience against climate change and to improve quality of life.

5. Conclusions

Ecoparque is an urban resilience laboratory that promotes a culture of experimentation with a safe-to-fail approach, to stimulate innovation for addressing urban environmental problems. With an integrated resource management vision, NBS are developed to try to reduce the need for external inputs. In a semi-arid area, treated water is used for irrigation, reforestation, with native plants for conservation, while reducing erosion, enhancing slope stabilization, and increasing green space. Compost is also produced to improve the soil and fertilize plants, and energy consumption is reduced by taking advantage of the slope to treat water and using solar energy. Through the participation of society, an experiential approach to caring for the urban environment is generated and transmitted to the community. Urban agriculture is also promoted to improve healthy diets and food security.

Maintenance and implementation of NBS at Ecoparque have faced significant challenges mainly due to a lack of personnel and financial resources. Most importantly, scaling up of NBS developed in Ecoparque at Tijuana face the absence of a proper legal framework, although this can be overcome by political will; however, the lack of this has the potential to end years of efforts and united will of transdisciplinary and interdisciplinary work on this side of the world. Despite difficulties and obstacles in environmental problem-ridden cities of the Global South, projects such Ecoparque have demonstrated that this can work and should be included in urban planning procedures carried out by corresponding levels of government; furthermore, it provides solid-based indicators for longitudinal monitoring in decades-long projects. And, last but not least, Ecoparque has endured mainly due to the team's long-term commitment and creativity.

Supplementary Materials: The following short video can be downloaded at https://www.youtube.com/watch?v=f5O46tSg4Ug (accessed on 20 December 2025)

Author Contributions: Conceptualization, L.O.-R. and G.M.-M.; formal analysis, L.O.-R. and G.M.-M.; investigation, L.O.-R. and G.M.-M.; resources, L.O.-R.; data curation, L.O.-R. and G.M.-M.; writing—original draft preparation, G.M.-M.; writing—review and editing, L.O.-R. and G.M.-M.; visualization, L.O.-R. and G.M.-M. All authors have read and agreed to the published version of the manuscript.

Funding: This research received no external funding.

Data Availability Statement: The original contributions presented in this study are included in the article/Supplementary Materials. Further inquiries can be directed to the corresponding author.

Acknowledgments: We would like to acknowledge the help received from the Ecoparque team members Xiomara Delgado, Armando Arteaga, Samuel Pérez, Ignacio Martinez, and Tania Bermudez. Thanks also to Fernando Jauregui for making the map.

Conflicts of Interest: The authors declare no conflicts of interest.

https://doi.org/10.3390/land15010089

Abbreviations

The following abbreviations are used in this manuscript:

El Colef	El Colegio de la Frontera Norte
CPI	Centro Público de Investigación
NBS	Nature-based solutions
CA	California
USA	United States of America
KWh	Kilo Watt hour
DEWATs	Decentralized Wastewater Treatment Plant
HP	Horse Power
SMADS	Secretaría de Medio Ambiente y Desarrollo Sustentable
SEMARNAT	Secretaría de Medio Ambiente y Recursos Naturales
UNPFA	United Nations Population Fund
UN-Habitat	United Nations Human Settlements Programme
IMPLAN	Instituto Metropolitano de Planeación de Tijuana
INEGI	Instituto Nacional de Geografía y Estadística
CEABC	Comisión Estatal del Agua de Baja California
SPA	Secretaria de Protección al Ambiente
SEDEMA	Secretaría del Medio Ambiente
SIDETRANS	Sistema Descentralizado de Tratamiento y Reúso de Aguas Negras en Zonas Urbanas
LOR	Lina Ojeda-Revah
GMM	Gabriela Muñoz-Meléndez

References

1. United Nations Population Fund (UNPFA). Urbanization. Available online: https://www.unfpa.org/urbanization (accessed on 7 March 2025).
2. United Nations Human Settlements Programme (UN-Habitat). *Cities and Climate Change Action. World Cities Report 2024*; United Nations Human Settlements Programme (UN-Habitat): Nairobi, Kenya, 2024. Available online: https://unhabitat.org/world-cities-report-2024-cities-and-climate-action (accessed on 12 December 2025).
3. Yigitcanlara, T.; Kamruzzamanf, M.; Buysb, L.; Giuseppe, I.; Sabatini-Marques, J.; Moreira da Costa, E.; Yune, J.J. Understanding 'Smart Cities': Intertwining Development Drivers with Desired Outcomes in a Multidimensional Framework. *Cities* **2018**, *81*, 145–160. [CrossRef]
4. Kong, L.; Liu, Z.; Wu, J. A Systematic Review of Big Data-Based Urban Sustainability Research: State-of-the-Science and Future Directions. *J. Clean. Prod.* **2020**, *273*, 123142. [CrossRef]
5. Bibri, S.E.; Krogstie, J. Smart Sustainable Cities of the Future: An Extensive Interdisciplinary Literature Review. *Sustain. Cities Soc.* **2017**, *31*, 183–212. [CrossRef]
6. Almenar, J.B.; Elliot, T.; Rugani, B.; Philippe, B.; Navarrete Gutierrez, T.; Sonnemann, G.; Geneletti, D. Nexus between Nature-Based Solutions, Ecosystem Services and Urban Challenges. *Land Use Policy* **2021**, *100*, 104898. [CrossRef]
7. European Commission. Brief Me on Nature-Based Solutions. Available online: https://knowledge4policy.ec.europa.eu/biodiversity/brief-me-nature-based-solutions_en (accessed on 8 March 2025).
8. Albert, C.; Brillinger, M.; Guerrero, P.; Gottwald, S.; Henze, J.; Schmidt, S.; Ott, E.; Schröter, B. Planning Nature-Based Solutions: Principles, Steps, and Insights. *Ambio* **2021**, *50*, 1446–1461. [CrossRef] [PubMed]
9. *Millenium Ecosystem Assessment. Ecosystems and Human Well-Being: Synthesis*; Island Press: Washington, DC, USA, 2005.
10. Cohen-Shacham, E.; Andrade, A.; Dalton, J.; Dudley, N.; Jones, M.; Kumar, C.; Maginnis, S.; Maynard, S.; Nelson, C.R.; Renaud, F.G.; et al. Core Principles for Successfully Implementing and Upscaling Nature-Based Solutions. In *Environmental Science and Policy*; Elsevier Ltd.: Amsterdam, The Netherlands, 2019; pp. 20–29. [CrossRef]
11. International Union for Conservation of Nature. *IUCN Global Standard for Nature-Based Solution. A User-Friendly Framework for the Verification, Design and Scaling up of NbS*; International Union for Conservation of Nature: Gland, Switzerland, 2020.
12. Kabisch, N.; Frantzeskaki, N.; Hansen, R. Principles for Urban Nature-Based Solutions. *Ambio* **2022**, *51*, 1388–1401. [CrossRef] [PubMed]
13. Dorst, H.; van der Jagt, A.; Toxopeus, H.; Tozer, L.; Raven, R.; Runhaar, H. What's behind the Barriers? Uncovering Structural Conditions Working against Urban Nature-Based Solutions. *Landsc. Urban. Plan.* **2022**, *220*, 104335–104348. [CrossRef]

https://doi.org/10.3390/land15010089

14. Sarabi, S.; Han, Q.; Romme, A.G.L.; de Vries, B.; Valkenburg, R.; Den Ouden, E.; Zalokar, S.; Wendling, L. Barriers to the Adoption of Urban Living Labs for NbS Implementation: A Systemic Perspective. *Sustainability* **2021**, *13*, 13276. [CrossRef]
15. Seddon, N.; Chausson, A.; Berry, P.; Girardin, C.A.J.; Smith, A.; Turner, B. Understanding the Value and Limits of Nature-Based Solutions to Climate Change and Other Global Challenges. In *Philosophical Transactions of the Royal Society B: Biological Sciences*; Royal Society Publishing: London, UK, 2020. [CrossRef]
16. Romero-Duque, L.P.; Trilleras, J.M.; Castellarini, F.; Quijas, S. Ecosystem Services in Urban Ecological Infrastructure of Latin America and the Caribbean: How Do They Contribute to Urban Planning? *Sci. Total Environ.* **2020**, *728*, 138780. [CrossRef] [PubMed]
17. Ojeda-Revah, L.; Ochoa González, Y.; Vera, L. Fragmented Urban Greenspace Planning in Major Mexican Municipalities. *J. Urban Plan. Dev.* **2020**, *146*, 04020019. [CrossRef]
18. Mercado, G.; Wild, T.; Hernandez-Garcia, J.; Baptista, M.D.; van Lierop, M.; Bina, O.; Inch, A.; Ode Sang, Å.; Buijs, A.; Dobbs, C.; et al. Supporting Nature-Based Solutions via Nature-Based Thinking across European and Latin American Cities. *Ambio* **2024**, *53*, 79–94. [CrossRef] [PubMed]
19. Ozment, S.; Gonzalez, M.; Schumacher, A.; Oliver, E.; Silva, M.; Grunwald, A.; Watson, G. *Nature-Based Solutions in Latin America and the Caribbean Regional Status and Priorities for Growth*; World Resources Institute: Washington, DC, USA, 2021.
20. Yin, R.K. *Case Study Research and Applications*, 6th ed.; AGE Publications, Inc.: London, UK, 2018.
21. Remenyi, D. *Case Study Research: The Quick Guide Series*; van Zyl, W., Ed.; Research Essentials; UJ Press: Johannesburg, South Africa, 2022. [CrossRef]
22. Gerring, J. *Case Study Research: Principles and Practices*, 1st ed.; Cambridge University Press: Cambridge, UK, 2006.
23. Instituto Metropolitano de Planeación de Tijuana. *Programa de Desarrollo Urbano del Centro de Población de Tijuana 2020–2040*; Instituto Metropolitano de Planeación de Tijuana: Tijuana, Mexico, 2019.
24. Mittermeier, R.A.; Rylands, A.B. Biodiversity Hotspots. In *Encyclopedia of the Anthropocene: Volumes 1-5*; Elsevier: Amsterdam, The Netherlands, 2017; Volume 3, pp. 67–75. [CrossRef]
25. Eaton-Gonzalez, R.; Mellink, E. One Shared Region and Two Different Change Patterns: Land Use Change in the Binational Californian Mediterranean Region. *Land* **2015**, *4*, 1138–1154. [CrossRef]
26. Instituto Nacional de Estadística y Geografía. Censo de Población y Vivienda (CPV). 2020. Available online: https://www.inegi.org.mx/programas/ccpv/2020/default.html#Microdatos (accessed on 21 March 2025).
27. Sanchez-Rodriguez, R.A.; Morales-Santos, A.E. Vulnerability Assessment to Climate Variability and Climate Change in Tijuana, Mexico. *Sustainability* **2018**, *10*, 2352. [CrossRef]
28. Alegria-Olazabal, T.; Ordoñez-Barba, G.M. *Legalizando La Ciudad. Asentamientos Informales y Procesos de Regulación En Tijuan*; El Colegio de la Frontera Norte: Tijuana, Mexico, 2016.
29. Bringas-Rabago, N.L.; Sanchez-Rodriguez, R.A. Social Vulnerability and Disaster Risk in Tijuana: Preliminary Findings. In *Equity and Sustainable Development. Reflection from the U.S.-Mexico Border; Center for U.S.-Mexican Studies*; UCSD: La Joya, TX, USA, 2006; pp. 150–172.
30. Ojeda-Revah, L.; Ochoa-Gonzalez, Y. Infraestructura Verde Para Tijuana, México. In *Desarrollo Sostenible en la Frontera Norte de México: Reflexiones Para una Agenda de Acción*; El Colegio de la Frontera Norte: Tijuana, Mexico, 2018; pp. 21–48.
31. Comision Estatal del Agua. *Sabes Cómo Llevamos el Agua Desde el Río Colorado Hasta las Ciudades de Tecate, Tijuana y Playas de Rosarito?* Gobierno del Estado de Baja California: Playas de Rosarito, BC, Mexico, 2025. Available online: http://www.cea.gob.mx/documents/arct/perfil.pdf (accessed on 8 March 2025).
32. Operadores Sistemas de Agua Potable y Alcantarillado. *Indicadores de Gestión. Organismos Operadores de Sistemas de Agua Potable y Alcantarillado*; Gobierno del Estado de Baja California: Mexicali, Mexico, 2022.
33. Munoz-Melendez, G. La habitabilidad en una región de emergencia ambiental: El caso de la cuenca del río Tijuana. In *Visiones Sobre La Habitabilidad Terrestre Y Humana Frente Al Cambio Climático. Una Primera Aproximación Epistemológica*; Cervates-Nuñez, S., Ed.; UNAM: Ciudad de Mexico, Mexico, 2021; pp. 169–186.
34. Minute 320 Binational Technical Team. *Report of Transboundary Bypass Flows into the Tijuana River*; International Boundary and Water Commission United States Mexico: San Diego, CA, USA, 2017.
35. Munoz-Melendez, G. Capítulo 6. El vínculo agua-energía-desarrollo urbano en ciudades de la frontera norte. In *Gestion del Agua en Mexico: Sustentabilidad y Gobernanza*; Aguilar-Benitez, I., Ed.; El Colegio de la Frontera Norte: Tijuana, Mexico, 2020; pp. 197–222.
36. Secretaria de Proteccion al Ambiente. *Programa Estatal de Accion ante el Cambio Climatico de Baja California*; Secretaria de Medio Ambiente y Recursos Naturales (SEMARNAT), Gobierno del Estado de Baja California (GobBC) & Instituto Nacional de Ecologia (INE): Mexicali, Mexico, 2012.
37. Rebman, J.P.; Roberts, N.C. *Baja California Plant Field Guide*; San Diego Natural History Museum: San Diego, CA, USA, 2012.
38. *NOM-001-ECOL-1996*; Que Establece los Límites Máximos Permisibles de Contaminantes en las Descargas de Aguas Residuales en Aguas y Bienes Nacionales. Diario Oficial de la Federación: Ciudad de Mexico, Mexico, 1997; p. 35.

 https://doi.org/10.3390/land15010089

39. *NOM-001-SEMARNAT-2021*; Que Establece los Límites Permisibles de Contaminantes en las Descargas de Aguas Residuales en Cuerpos Receptores Propiedad de la Nación. Diario Oficial de la Federación: Ciudad de Mexico, Mexico, 2021; p. 18.
40. *NADF-020-AMBT-2011*; QUE ESTABLECE LOS REQUERIMIENTOS MÍNIMOS PARA LA PRODUCCIÓN DE COMPOSTA A PARTIR DE LA FRACCIÓN ORGÁNICA DE LOS RESIDUOS SÓLIDOS URBANOS, AGRÍCOLAS, PECUARIOS Y FORESTALES, ASÍ COMO LAS ESPECIFICACIONES MÍNIMAS DE CALIDAD DE LA COMPOSTA PRODUCIDA Y/O DISTRIBUIDA EN EL DISTRITO FEDERAL. Gaceta Oficial del Distrito Federal: Ciudad de Mexico, 2012; p. 14.
41. *NOM-059-SEMARNAT-2010*; Protección Ambiental-Especies Nativas de México de Flora y Fauna Silvestres-Categorías de Riesgo y Especificaciones Para su Inclusión, Exclusión o Cambio-Lista de Especies en Riesgo. Diario Oficial de la Federación: Ciudad de Mexico, Mexico, 2010.
42. Zuniga-Teran, A.A.; Staddon, C.; de Vito, L.; Gerlak, A.K.; Ward, S.; Schoeman, Y.; Hart, A.; Booth, G. Challenges of Mainstreaming Green Infrastructure in Built Environment Professions. *J. Environ. Plan. Manag.* **2020**, *63*, 710–732. [CrossRef]
43. Salgado-López, J.A.; Galván-Benítez, R. Infraestructura verde. Conceptualización y análisis normativo de México Green Infrastructure. Conceptualization and legal analysis of Mexico. *Quivera Rev. Estud. Territ.* **2022**, *24*, 105–128.
44. Fernández Mesta, M.E. Servicios ambientales: Elementos para el desarrollo de un marco jurídico. *Terra Latinoam.* **2016**, *34*, 155–166.
45. Balderas Torres, A.; Angón Rodríguez, S.; Sudmant, A.; Gouldson, A. *Adapting to Climate Change in Mountain Cities: Lessons from Xalapa, Mexico*; Coalition for Urban Transitions: Washington, DC, USA, 2021. Available online: https://urbantransitions.global/publications (accessed on 20 December 2025).
46. Secretaria de Medio Ambiente, de la Ciudad de México. Infraestructura Verde, Programa Especial de la Red de Infraestructu-ra Verde (PERIVE); Secretaria de Medio Ambiente, de la Ciudad de México, 2023. Available online: https://www.sedema.cdmx.gob.mx/programas/programa/infraestructura-verde (accessed on 20 December 2025).
47. *NOM-003-SEDATU-2023*; Que Establece los Lineamientos Para el Fortalecimiento del Sistema Territorial Para Resistir, Adaptarse y Recuperarse Ante Amenazas de Origen Natural y del Cambio Climático a Través del Ordenamiento Territorial. Diario Oficial de la Federación: Ciudad de Mexico, Mexico, 2024.

https://doi.org/10.3390/land15010089

land

MDPI

Article

Urban Habitat Quality Enhancement and Optimization under Ecological Network Constraints

Yanhai Zhou *, Jianwei Geng and Xingzhao Liu

College of Landscape Architecture and Art, Fujian Agriculture and Forestry University, Fuzhou 350002, China; 62419052015@fafu.edu.cn (J.G.); xzliu@fafu.edu.cn (X.L.)
* Correspondence: zhouyanhai@fafu.edu.cn; Tel.: +86-130-6745-5200

Abstract: The process of urbanization leads to the rapid expansion of construction land and brings a series of ecological and environmental problems. The ecological network, as a linear landscape element, is of great significance to improve the quality of the regional ecological environment. In this study, the Morphological Spatial Pattern Analysis (MSPA) and the model of Minimum Cumulative Resistance (MCR) were used to construct the ecological corridors in the central city of Fuzhou, and the land use pattern under the constraints of the ecological network was simulated and quantified by the patch-level land use simulation (PLUS) tool with the results of the identification of ecological corridors. Meanwhile, with the help of InVEST habitat quality model, the regional habitat quality under different development scenarios was compared. The results show that (1) 19 ecological sources and 35 ecological corridors were identified; (2) under the constraints of ecological corridors, the area of forested land in the study area in 2027 was increased by 1.57% and the area of built-up land was reduced by 0.55% compared with that in 2022; (3) and under the constraints of ecological corridors, the mean value of habitat quality in Fuzhou City improved by 0.0055 and 0.0254 compared with 2022 and 2027 natural evolution scenarios, respectively. The study provides decision-making assistance for the construction of ecological corridors from the perspective of land use planning.

Keywords: ecological corridor; PLUS model; land use planning; InVEST model; habitat quality

1. Introduction

With the accelerating urbanization process in China, the land use structure in the central urban area has changed dramatically in the past 20 years [1,2]. The phenomenon is mainly characterized by the reduction in ecological land use such as woodland and shrubs and the massive expansion and erosion of impervious surfaces. As a result, the quality of urban ecological environment has been affected, causing ecological and social problems such as urban heat island, noise pollution, and traffic congestion [3–5]. Relevant studies have shown that ecological corridors, as a linear landscape element, play an important role in maintaining regional biodiversity and enhancing the regional ecological environment by connecting large green patches in cities [6,7]. However, how does the construction of ecological networks guide the rational planning of real land use? How to evaluate and quantify the ecological benefits generated after planning is a difficult problem.

In terms of ecological corridor modeling, the predominant practice is based on landscape ecology, utilizing a framework system that includes patch resistance and surface corridors. The extraction of ecological patches is carried out through, for example, Morphological Spatial Pattern Analysis (MSPA) [8], or by utilizing the landscape pattern index, combined with the particle size backpropagation method [9]. Meanwhile, the connectivity index was utilized for the selection of patches, which in turn led to the final determination of suitable ecological source sites [10,11]. Meanwhile, related scholars have utilized the InVEST model and RSEI model to quantify the regional ecosystem service function and ecological environment quality and then make the determination of source sites [12,13]. As for the construction of resistance surfaces, most scholars adopted factors such as land use type, elevation, and slope as the key influencing factors affecting the direction of ecological corridors, and the Analytic Hierarchy Process (AHP) evaluation method was used for the determination of weights [14]. Minimum Cumulative Resistance (MCR), Least Cost Path (LCP), and current theory models are often used in the construction and optimization of ecological corridors, and the MCR model can visually express the layout of the corridor, while the current theory is used to judge the degree of connectivity of the overall landscape from the perspective of the overall landscape [15,16]. In terms of the prediction of future land use simulation, relevant studies have mainly used the CA_Markov model [17], FLUS model [18], and PLUS model [19]. It is mainly applied at the city scale, such as Nanjing [20], Shenzhen [21], and Wuhan [22] in China. All these models show good simulation results. Their general principle is to first quantitatively determine the future land use through Markov chain or linear regression model, and at the same time, under the quantitative constraints, combine with algorithms such as CA model to carry out spatial simulation of like elements, so as to realize the spatial quantification of land use in the future period. With the continuous deepening of research, the simulation of land use is not limited to the natural scenarios based on history, and more and more scholars began to carry out a certain degree of human "interference" to explore the future land use change rules under different scenarios, such as ecological protection scenarios, urban development scenarios, and synergistic co-development scenarios [23,24]. At the same time, it is the main research trend to combine relevant ecological models, such as the InVEST model, to explore the evaluation of ecosystem service function under different urban development scenarios [25–27].

Although, there are more studies on the construction of ecological corridors and land use simulation, there are still some shortcomings. For example, MSPA and other methods only carry out the determination of ecological sources from the perspective of the landscape structure, ignoring the attributes of the ecological source quality. Relevant studies combined MSPA and RSEI to perform ecological source determination from the perspective of structure and function, which more accurately identified the ecological source area [13], but it was also based on the current static construction idea, which ignored the influence of dynamic changes in the ecological source area. Second, in the evaluation and optimization of ecological corridors, most of the studies were conducted by reselecting source sites, identifying barrier and pinch points, or structurally grading and evaluating the network [28,29]. These processes are not sufficient to show the ecological benefits and ecological functions that ecological corridors can produce, which challenges the necessity of ecological corridor construction. Meanwhile, related scholars have quantified the future ecological corridor pattern of cities by simulating the future land use structure and realizing the future ecological corridor pattern on this basis [30,31]. In fact, this is still a result constructed based on a static period of time, which still restricts the rational planning and utilization of land in reality. In addition, related scholars have modified the transfer matrix of the PLUS model, combined with the ecological protection red line, urban development boundary, and so on, to carry out ecological priority land use simulation. However, these policy documents are difficult to obtain, and from the results, most of the studies are based on the model of reducing construction land and thus improving the quality of ecological environment [32,33]. Obviously, this is not in line with the current urban development law.

The purpose of our study is to explore how the construction of ecological corridors can guide the rational use of real land and conduct impact assessments. To this end, our study starts from the basic theoretical basis that the construction of ecological corridors changes land use types and land use types affect habitat quality. This study takes the central urban area of Fuzhou City as the study area, and based on the results of the land use classification in 2012, 2017, and 2022, quantifies the landscape types of each period and extracts the core patches by using MSPA, and combines the landscape connectivity analysis with the MCR model to construct the ecological corridor in the central city of Fuzhou. On this basis, using the PLUS model, the ecological corridor is taken as one of the driving factors affecting land use change, and the land use under future scenarios with ecological corridor constraints is simulated, while the InVEST model is used to explore the changing patterns of regional habitat quality and landscape pattern under ecological corridor constraints and natural development scenarios. The main contents of this study are as follows: (1) the construction of ecological corridor in Fuzhou City; (2) a land use simulation prediction based on PLUS model; (3) and a land use evaluation based on InVEST habitat quality. The study provides decision-making assistance for the construction of ecological corridors in Fuzhou City from the perspective of land resource planning and utilization.

2. Materials and Methods

2.1. Study Area

The downtown area of Fuzhou City is located on the southeast coast of China (25°15′–26° 39′ N~118° 08′–120° 31′ E), with a total area of about 2212.94 km^2. The climate is warm and humid, with an average annual temperature of 16–20 °C. It has a typical subtropical monsoon climate. The main city of Fuzhou is surrounded by mountains, with an elevation of 600–1000 m above sea level, which is a typical estuarine basin. With the continuous expansion of urbanization scale in recent years and the barrier of the mountains, the trend of its urban expansion is along the river to the sea, eastward, and southward (Figure 1).

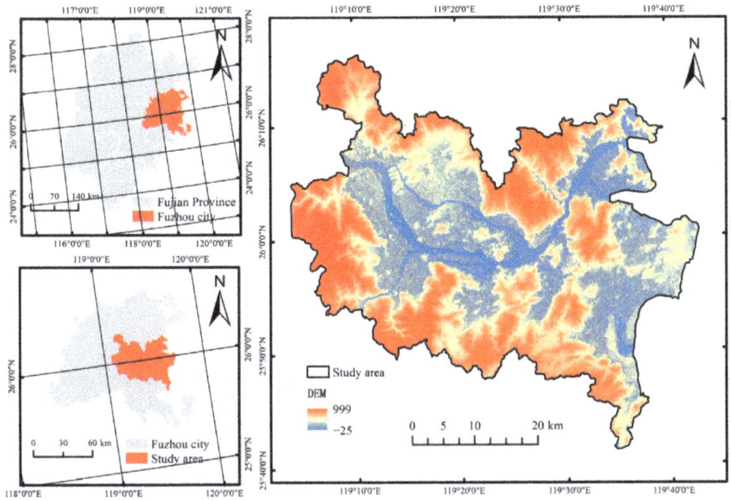

Figure 1. Study area location and DEM.

2.2. Data Sources

The 2012–2022 land use data were obtained from the open-source dataset published by Wuhan University [34], which categorizes China's land use types into nine major categories: farmland, forest, shrubland, grassland, water bodies, snow and ice, bare land, impervious surface, and wetland. The spatial resolution of the data was 30 m. The DEM data of the study area were obtained from the Distributed Activity Archive of the Alaska Satellite Facility (ASF) in the United States with a spatial resolution of 12.5 m, and were used to calculate the elevation and slope factors. Vector data of urban roads, highway roads, railroads, and POIs were downloaded from open street map. The population density of the study area was obtained from the world population dataset, and the average annual precipitation data were obtained from the National Center for Earth System Data Science. The Gross Domestic Product (GDP) raster data were obtained from the Center for Resource and Environmental Science and Data, Chinese Academy of Sciences. Please see Table 1 for details.

Table 1. Data types, formats, and sources.

Data Type	Data Format	Resolution	Source
2012–2022 Land use type data	.tif	30 m	https://zenodo.org/records/8176941, accessed on 7 August 2024
Boundary of the study area	.shp	/	http://zygh.fuzhou.gov.cn/, accessed on 7 August 2021
DEM	.tif	12.5 m	https://asf.alaska.edu/, accessed on 7 August 2024
City roads, railroads, highways, etc.	.shp	/	https://www.openstreetmap.org/, accessed on 7 August 2024
Population density	.tif	100 m	https://www.worldpop.org/, accessed on 7 August 2024
GDP	.tif	1000 m	https://www.resdc.cn/, accessed on 7 August 2024
Average annual precipitation	.tif	1000 m	http://www.geodata.cn/, accessed on 7 August 2024

2.3. Methodology

2.3.1. Identification of Ecological Source Areas

Morphological Spatial Pattern Analysis (MSPA) is a model based on mathematical morphology algorithms for processing binary images and is often used to determine forest geometry and extract ecological source sites. It is able to classify land classes with ecological value, such as woodland and grassland, into seven landscape structures, namely Core, Islet, Perforation, Edge, Loop, Bridge, and Branch [35,36]. In our study, woodland, shrubland, and grassland in 2012, 2017, and 2022 were selected as foreground data and pre-processing of the data was carried out using GuidoToolbox3.0 analysis software with 8-domain method and the edge width set to 1, which is the smallest unit of the raster, and 30 m. At the same time, the most ecologically valuable core areas of the seven landscape types were extracted, and the results of the three years were intersected to obtain the preliminary core area patches that were more stable in 10 years.

Landscape connectivity indices are often used to determine the degree of interaction between patches, of which the most commonly used index is the patch importance index (dPC). The formula is as follows:

$$\mathrm{PC} = \frac{\sum_{i=1}^{n}\sum_{j=1}^{n} a_i a_j p_{ij}^{*}}{A_L^2} \tag{1}$$

$$\mathrm{dPC} = \frac{\mathrm{PC} - \mathrm{PC}_{\mathrm{remove}}}{\mathrm{PC}} \times 100\% \tag{2}$$

where A_L is the total area of the landscape; n denotes the number of patches; a_i and a_j are the areas of patches i and j; and p_{ij} denotes the maximum probability that a patch is connected between i and j. PC refers to the connectivity index, and takes the value in the range of 0 to 1. The higher the value of dPC, the better and more useful it is on behalf of the connectivity of the patches. PC_{remove} is the value of the probability of connectivity of a particular patch after removing it.

After obtaining the preliminary ecologically valuable core areas based on MSPA, we selected the patches with an area of more than 0.5 km^2 and a dPC index of more than 0.1 as the ecological source sites for this study.

2.3.2. Construction of Integrated Resistance Surfaces and Ecological Corridors

The spatial movement of ecological material is affected by a variety of resistances. These include land use type, elevation, slope, and landscape pattern. Among them, ecological land such as woodland, shrubland, and grassland is more conducive to the construction of ecological corridors, and therefore the resistance value is lower. On the contrary, impervious surfaces are more difficult to construct and have more human interference, so the resistance value should be set higher. Landscape type as a further categorization of ecological land can distinguish the detailed parts of ecological land more precisely. For example, the core, as a large area of dominant ecological sources, should have a lower resistance value than ecological patches such as islet and edge. Similarly, the resistance effect of elevation is also obvious, for example, low elevation and low slope areas are more suitable for the construction of ecological corridors, so the resistance value should be set lower. We superimpose these resistance factors with the same weights, and the resistance value setting of each factor is shown in Table 2 [37–39].

Table 2. Resistance factors and their resistance value settings.

Type of Resistance	Resistance Factor	Resistance Value
Land use type	Cropland	3
	Forest, Shrub, Grassland	1
	Water	5
	Barren	7
	Impervious	9
MSPA landscape types	Core	1
	Bridge, Loop	3
	Branch, Islet	5
	Edge, Perforation	7
	Background	9
Slope	<5°	1
	5–15°	3
	15–25°	5
	25–35°	7
	>35°	9
DEM	<200	1
	200–300	3
	300–400	5
	400–600	7
	>600	9

The Minimum Cumulative Resistance (MCR) model is often used for the construction of ecological corridors [40], and LinkageMapper, an open-source ArcGIS toolkit, is able to quantify the ecological corridors in the study area quickly and efficiently [41]. Based on the constructed integrated resistance surface, the quantification of ecological corridors is achieved by calculating the minimum cost to be spent between ecological sources and thus the quantification of ecological corridors. The equation of the model can be written as

$$MCR = f_{min} \sum_{j=n}^{i=m} (D_{ij} \times R_i) \tag{3}$$

where MCR is the Minimum Cumulative Resistance, f is the positive correlation between the Minimum Cumulative Resistance and the ecological process, D_{ij} refers to the spatial distance from patch i to j, and R_i denotes the resistance coefficient of spatial expansion of patch i.

2.3.3. PLUS Model

The PLUS model is a cellular automata (CA) model based on raster data that can be used for the simulation of land use changes at the patch scale. It mainly extracts the transformation probability between the land use datasets of two periods, randomly extracts sampling points in the areas where each class changes for analysis, uses the random forest algorithm to excavate the driving mechanism of land use change, and then combines with the generation of random plaques and the setting of transfer matrices to ultimately realize the simulation of the future land use [19,42]. The PLUS model is mainly composed of the following modules:

(1) Land Expansion Analysis Strategy (LEAS)

By extracting the expansion portion of each class of land in different periods and sampling from the increase portion, the expansion and driving mechanism of each class is subsequently mined using the random forest algorithm to obtain the development probability of each class and the degree of contribution of the driving factors, which is shown in Equation (4):

$$P^{d}_{i,k(X)} = \frac{\sum_{n=1}^{M} I[h_n(X) = d]}{M} \quad (4)$$

where d takes the value of 0 or 1, the value of 1 indicates the shift of other land use types to the land use type k, and the value of 0 indicates no shift. X is a vector consisting of driving factors, M is the number of decision trees, and $h_n(X)$ is the predicted type of land use computed when the decision tree is n. I is the indicator function of the decision tree; $P^{d}_{i,k(X)}$ is the probability of growth of the land use type k at the spatial unit, and i is the probability of the growth of the k land use types at the place.

(2) CA model based on multi-class randomized patch seeding (CARS)

The CARS module performs land use simulation by inputting land use data and development probabilities for each type of expansion, which is the result calculated by LEAS above, and by setting the land use conversion matrix and domain weights. The development zone, as an essential component of the CARS module, aims to quantify the impact of land use simulation after the inclusion of planned development zones. In our study, by incorporating data on ecological corridor buffer zones and setting relevant parameters, it is possible to simulate and predict future land use under the constraints of ecological corridors. Land use type k can be expressed by the following equation:

$$P^{=1,t}_{O,i,k} = P^{d=1}_{i,k} \times \Omega^{t}_{i,k} \times D^{t}_{k} \quad (5)$$

where $P^{=1,t}_{O,i,k}$ is the growth probability of land use type k on tuple i, $\Omega^{t}_{i,k}$ is the domain effect of tuple i, and D^{t}_{k} is the effect of future demand for land use type k. The formula can be written as

$$\Omega^{t}_{i,k} = \frac{con\left(c^{t-1}_{i} = k\right)}{n \times n - 1} \times \omega k \quad (6)$$

$$D^{t}_{k} = \begin{cases} D^{t-1}_{k}\left(\left|D^{t-1}_{k}\right| \leq \left|D^{t-2}_{k}\right|\right) \\ D^{t-1}_{k} \times \frac{D^{t-2}_{k}}{D^{t-1}_{k}}\left(D^{t-1}_{k} < D^{t-2}_{k} < 0\right) \\ D^{t-1}_{k} \times \frac{D^{t-1}_{k}}{D^{t-2}_{k}}\left(0 < D^{t-2}_{k} < D^{t-1}_{k}\right) \end{cases} \quad (7)$$

where con is the total number of grid cells occupied by the kth land use type in the last iteration in the window and $n \times n$. ω is the weight between different land use types. D is the difference between the current and future demands of land use type k at $t-1$ and $t-2$ iterations, respectively.

(3) Land use drivers

By referring to the results of previous research, the study selected a total of 13 topographic and geomorphological features, including socio-economic and demographic factors, transportation elevation, slope, precipitation, population density, GDP, distance from water bodies, distance from railroads, distance from highways, distance from other roads in the city, distance from railroad stations, distance from the government, and distance from ecological sources and ecological corridors, as the set of driving factors for simulating land use in 2027 (Figure 2). All data were resampled to 30 m spatial resolution and projected to the same coordinate system. Among them, in the natural state, the driving factors do not consider the influence of ecological sources and ecological corridors when we simulate the change in land use types in 2027 in downtown Fuzhou (Figure 2a–k). In contrast, as a control, under the ecological corridor constraint scenario, we consider all the driving factors (Figure 2a–m).

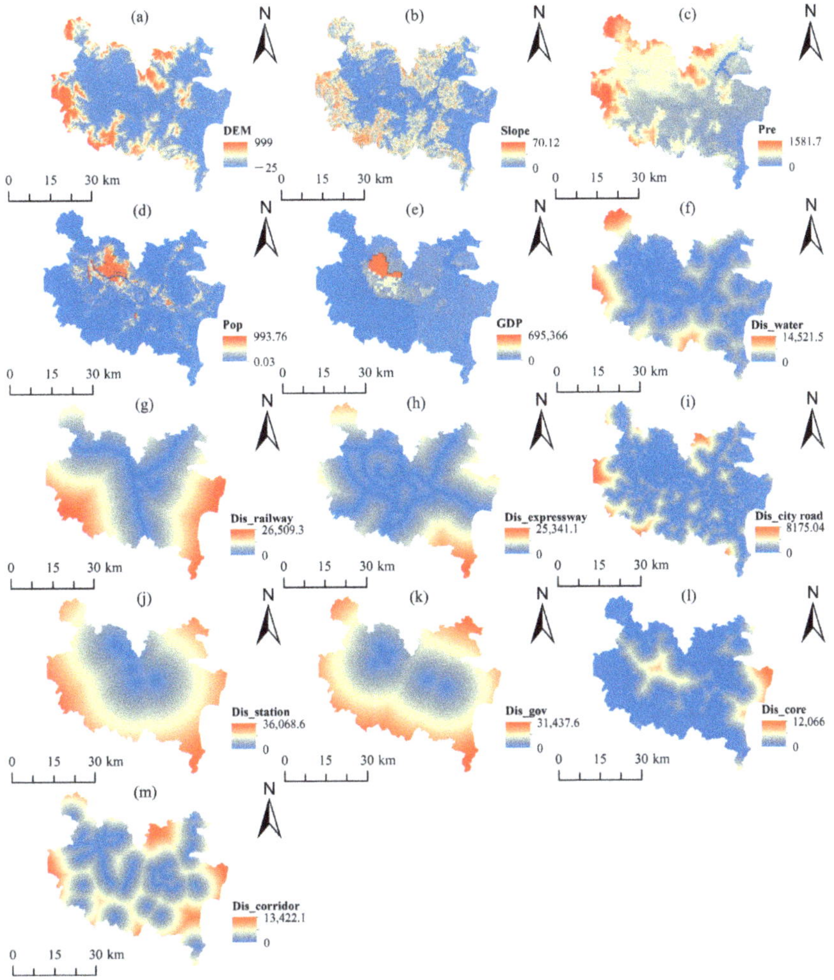

Figure 2. Land use driving factors. (**a**) DEM; (**b**) slope; (**c**) precipitation; (**d**) POP; (**e**) GDP; (**f**) Dis_water; (**g**) Dis_railway; (**h**) Dis_expressway; (**i**) Dis_city road; (**j**) Dis_station, (**k**) Dis_government; (**l**) Dis_Core; (**m**) Dis_corridor.

(4) Parameter setting

Since the construction of the ecological corridor will inevitably lead to the transformation of the land use type, in the development zone of the CARS module, we take the buffer zone of the ecological corridor of 150 m as the planned development zone and the extracted patches of the core zone as the transformation restriction zone to maintain the stability of the ecological source land. In the development type, we choose the ID value of woodland, and the development weight is set to 0.8 to emphasize the important role of woodland construction in the buffer zone of ecological corridor, the patch generation threshold is set to 0.1, and the expansion coefficient is set to 0.9. The transfer matrix is set to 0.1. The transfer matrix and neighborhood weight were set as shown in Tables 3 and 4 below. At the same time, we also simulate the future land use structure change under the natural development state for comparison and verification, and the transfer matrix and neighborhood weight under the natural state are both at default.

Table 3. Land use transfer matrix setup under ecological corridor constraints.

	Cropland	Forest	Shrub	Grassland	Water	Barren	Impervious
Cropland	1	1	1	1	0	1	1
Forest	0	1	0	0	0	0	0
Shrub	0	1	1	1	0	0	1
Grassland	0	1	1	1	0	0	1
Water	0	1	0	0	1	0	0
Barren	1	1	1	1	1	1	1
Impervious	1	1	0	0	0	0	1

Table 4. Domain weights for each category under ecological corridor constraints.

	Cropland	Forest	Shrub	Grassland	Water	Barren	Impervious
Neighborhood Weight	0.2	0.7	0.2	0.1	0.1	0.1	0.7

(5) Accuracy test

Accuracy testing is a tool used to evaluate the reliability of PLUS results. The accuracy is obtained by calculating the proportion of correctly predicted samples to the total number of samples. We input the land use and driving factors of the study area in 2012 and 2017 into the CARS module. We simulate the land use structure of the study area in 2022 through a Markov chain to obtain the spatial distribution of land use in 2022 based on the PLUS model and compare it with the real land use in 2022. The results show that the Kappa coefficient reaches 91.82% and the overall accuracy reaches 94.52%, which meets the research needs.

2.3.4. Habitat Quality Model

The InVEST habitat quality model is mainly based on land use data and is used for the quantitative assessment of regional ecological environment quality by combining the ecological sensitivity of different feature types, the degree of coercion, and the mode of impact of threat sources [3,43]. Due to its high efficiency and good visualization, it is often used for the evaluation of regional ecological environment quality. In this regard, threat sources can be understood as non-ecological land uses in land use, such as impervious surfaces, bare ground, and transportation roads. The formula of the model is as follows:

$$D_{xj} = \sum_{r=1}^{R}\sum_{y=1}^{Y_r}\left(\frac{w_r}{\sum_{r=1}^{R} w_r}\right) r_y i_{rxy} \beta_x S_{jr} \quad (8)$$

where Y_r is the total number of raster cells for stressor r; R is the number of stressors; w_r is the normalized weight value; r_y is the number of stressors on the raster cell; β_x is the approachable level of raster x; S_{jr} denotes the sensitivity of the landscape j to stressor r; and i_{rxy} is the maximum distance of influence of the stressor.

$$Q_{xj} = H_j\left(1 - \left(\frac{D_{xj}^z}{k^z + D_{xj}^z}\right)\right) \quad (9)$$

where Q_{xj} is the habitat quality of patch group x in landscape type j; H_j is the habitat suitability score for landscape type j; z is the scaling factor, generally taken as 2.5; and k is the half-saturation constant, taken as 0.05.

The data to be inputted into the model mainly include the land use type data, the weights of the threat source factors and their attenuation distances, and the sensitivity table between landscape types and threat sources. We set the model-related parameters by referring to the related literature [43,44] and the InVEST user manual (Tables 5 and 6).

Table 5. Threat source impact distance and weights.

THREAT	MAX_DIST	WEIGHT	DECAY
Cropland	2	0.6	linear
Barren	3	0.5	exponential
Impervious	8	1	exponential

Table 6. Habitat sensitivity by category.

LULC	HABITAT	Cropland	Barren	Impervious
Cropland	0.3	0	0.1	0.8
Forest	1	0.3	0.2	0.6
Shrub	0.8	0.6	0.3	0.9
Grassland	0.6	0.8	0.4	0.8
Water	0.8	0.1	0.1	0.2
Barren	0.2	0.7	0	0.9
Impervious	0	0.1	0	0

3. Results

3.1. Landscape Type Dynamics and Ecological Source Extraction Results

By extracting woodland, shrubland, and grassland with ecological value, the landscape types were extracted by using the MSPA method (Figure 3, Table 7). The results showed that the landscape types in the center of Fuzhou City were mainly dominated by Core, followed by Edge. However, with the advancement of urbanization, the areas of Core, Islet, Loop, Bridge, and Branch landscape types all decreased, while Perforation and Edge increased by 31.86% and 10.80%, respectively, during the 10-year period, indicating that the edge areas between the patches in the study area increased, and the integrity and connectivity deteriorated. In terms of spatial distribution, the core area is mainly distributed near the surrounding mountains in the study area, while the smaller core area in the center has been gradually affected by urban expansion in recent years, and its area has been gradually reduced (Figure 4). After overlaying the core areas of the three phases, we extracted a total of 19 ecological source sites, of which the largest ecological source site was No. 12, with an area of 118.22 km^2, while the smallest source site was No. 3, with an area of 0.5049 km^2 (Figure 5).

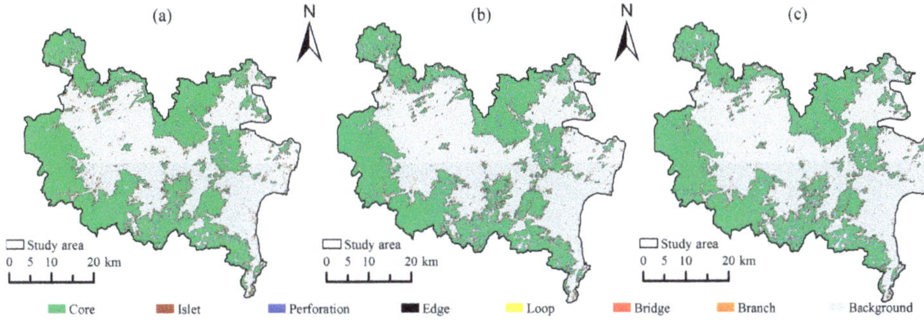

Figure 3. Spatial distribution of MSPA landscape types. (**a**) 2012; (**b**) 2017; (**c**) 2022.

Table 7. 2012–2022 landscape area and rate of change by landscape type.

	2012 (km^2)	2017 (km^2)	2022 (km^2)	Rate of Change 2012–2022
Core	949.35	872.19	865.54	−8.83%
Islet	7.98	7.72	5.89	−26.10%
Perforation	20.11	26.08	26.51	31.86%
Edge	53.83	58.63	59.64	10.80%
Loop	9.92	10.62	7.99	−19.48%
Bridge	6.51	5.82	4.25	−34.68%
Branch	15.02	16.55	13.76	−8.44%

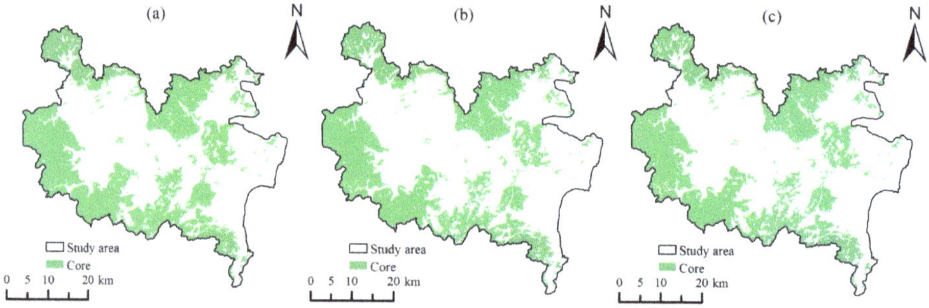

Figure 4. Spatial distribution of the core area. (**a**) 2012; (**b**) 2017; (**c**) 2022.

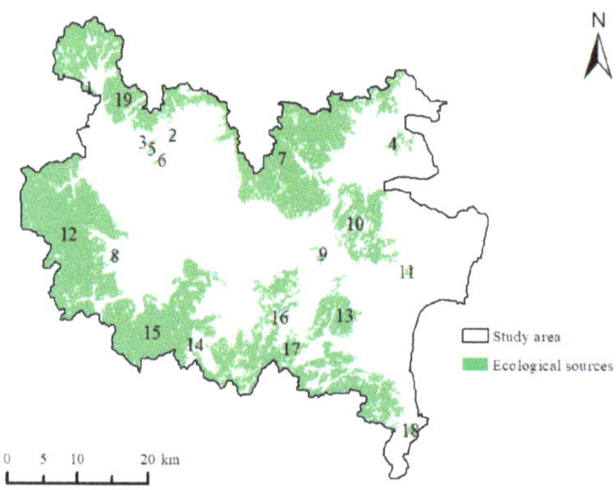

Figure 5. Spatial distribution of ecological source.

3.2. *Resistive Surface Construction and Spatial Quantification of Ecological Corridors*

We obtained the integrated resistance surface by superimposing the resistance factors (Figure 6), and the results showed that the areas with high resistance values were mainly concentrated in the city, while the areas with low resistance were mainly concentrated in the peripheral boundaries of the city. Combined with the identification results of ecological source sites, we used the LinkageMapper toolkit to realize the quantification of the spatial pattern of ecological corridors in the study area, and a total of 35 ecological corridors were identified, with an average length of 3314.31 m. Among them, the longest corridor reaches 14,958.51 m, which connects the source sites 6–15, and the shortest is only 42.43 m, which connects the source sites 7–19. In terms of spatial distribution, it is mainly concentrated in the northwestern and eastern parts of the study area (Figure 7).

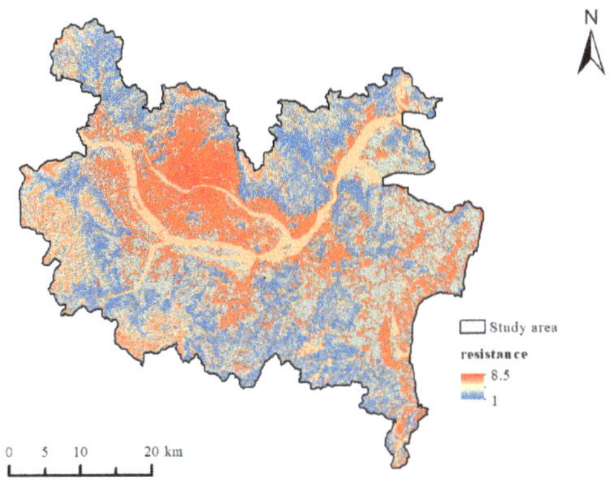

Figure 6. Spatial distribution of integrated resistance surfaces.

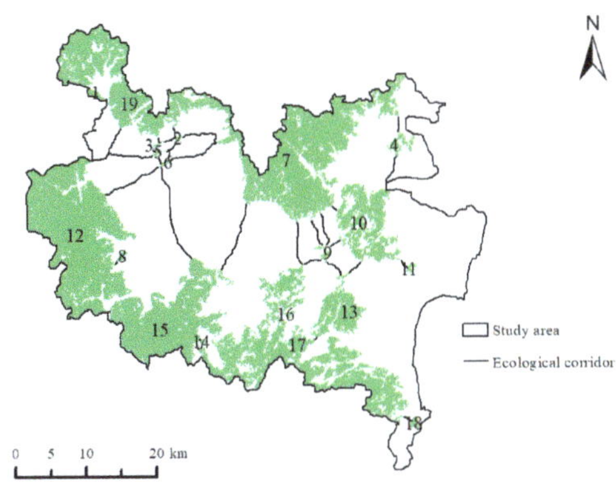

Figure 7. Spatial distribution of ecological corridors.

3.3. *Land Use Modeling and Prediction*

Using the PLUS model, the study simulated the land use of Fuzhou City center in 2027 under the natural state and ecological constraints by combining the historical evolution patterns of the three phases of 2012, 2017, and 2022 as well as the constraints of the ecological sources and ecological corridors (Figure 8 and Table 8). The results show that under different simulation scenarios, compared with 2022, cropland, shrub, grassland, and barren in 2027 are reduced because these land categories themselves have a smaller area advantage and are easily affected by other land categories; the second reduction in cropland is due to urban construction and the centralized remediation and planning of cropland. The changes in forest and impervious are not the same. It can be seen that compared to 2022, the area of forest in its natural state has decreased in 2027 by 1.47%, while impervious has increased by 7.27%. It can be seen that the natural scenario based on historical patterns will see further urban expansion accompanied by a decrease in forest. In contrast, among the land use types under the ecological corridor constraints, the area of forest has increased, and the area of impervious is also basically the same as in 2022. This also reflects the constraining effect of the overall ecological network.

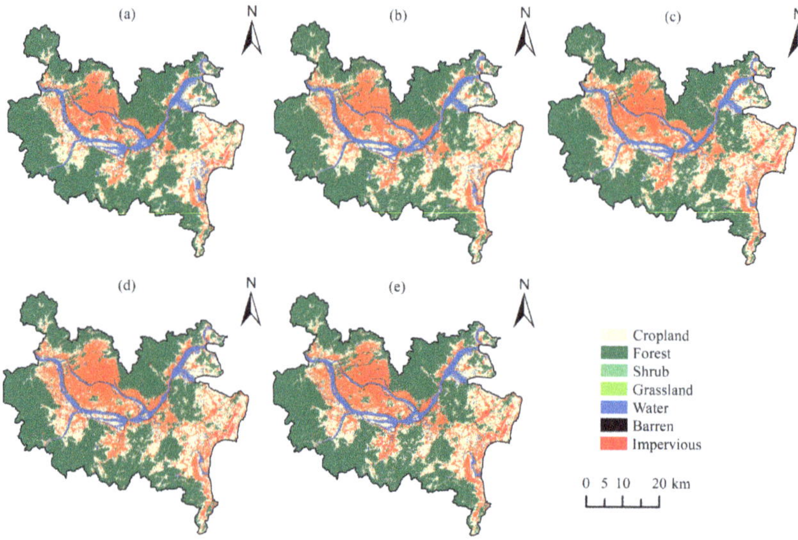

Figure 8. Land use maps by period. (**a**) 2012; (**b**) 2017; (**c**) 2022; (**d**) 2027 state of nature; (**e**) 2027 corridor constraints.

Table 8. Land use area statistics by period.

Land Use Type	2012 (km^2)	2017 (km^2)	2022 (km^2)	2027 State of Nature (km^2)	2027 Corridor Constraints (km^2)
Cropland	543.29	568.38	562.12	555.10	547.54
Forest	1061.60	997.09	983.31	968.87	998.79
Shrub	0.04	0.03	0.03	0.02	0.02
Grassland	1.09	0.51	0.24	0.16	0.13
Water	163.48	152.65	141.70	131.66	143.45
Barren	0.72	0.46	0.35	0.27	0.22
Impervious	352.75	403.84	435.20	466.86	432.80

3.4. Land Use Evaluation Based on Habitat Quality

Based on the data of land use types in each period, we further clarified the dynamics of its ecological quality by quantifying its habitat quality (Figure 9). The results show that the mean values of habitat quality in each period are 0.4542, 0.3991, and 0.3754, respectively, which shows that the habitat quality decreases in the 10-year period from 2012 to 2022, and further decreases to 0.3555 in 2027 under the natural evolution scenario based on the historical pattern, while the habitat quality after the ecological corridor constraints is improved, with a mean value of 0.3809. Meanwhile, we reclassified the habitat quality according to the equal spacing method (Table 9), and it can be seen that the area with a low habitat quality rating continued to increase during the 10-year period, from 688.78 km^2 in 2012 to 897.14 km^2 in 2022, whereas the area with a high habitat quality area gradually decreased, from 455.45 km^2 to a smaller area. The overall pattern still maintains the spatial distribution pattern of low in the center and high in the surroundings. In 2027, the area of low habitat quality under natural conditions will further expand, while the habitat quality under ecological corridor constraints will improve in 2027, which is mainly reflected in the reduction in the area of low habitat quality and the ecological restoration of high habitat quality areas.

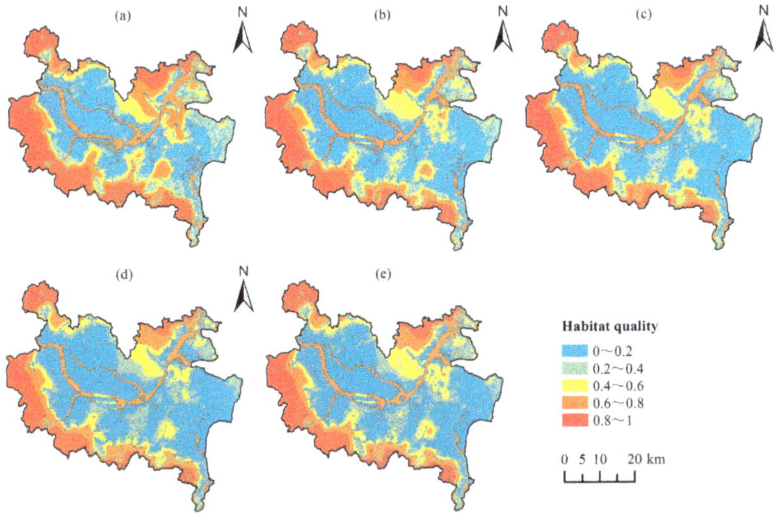

Figure 9. Habitat quality by period. (**a**) 2012; (**b**) 2017; (**c**) 2022; (**d**) 2027 natural state; (**e**) 2027 corridor constraints.

Table 9. Changes in habitat quality rating by period.

Habitat Quality Rating	2012	2017	2022	2027 State of Nature	2027 Corridor Constraints
Lowest (0~0.2)	688.78	836.41	897.14	950.28	878.83
Lower (0.2~0.4)	334.19	316.10	316.86	320.56	317.29
Medium (0.4~0.6)	220.00	259.40	253.86	236.55	260.04
Higher (0.6~0.8)	424.55	358.38	342.13	334.30	346.68
Highest (0.8~1)	455.45	352.68	312.98	281.28	320.13

4. Discussion

4.1. Impact of the Construction of Ecological Corridors on Land Use Modeling

In contrast to previous research ideas, related studies first construct ecological corridors for a certain period of time in the future by modeling future land use, and this approach neglects the impact of ecological corridor construction on land use types [45]. Starting from the basic theoretical basis that ecological corridor construction changes land use types and land use types affect regional ecological quality, our study proposes a modeling framework that integrates the ecological corridor, land use, and habitat quality, which provides a basis for corroborating the positive effects of corridor construction. This modeling framework integrating the ecological corridor, land use, and habitat quality is proposed, which provides a basis for proving the positive effects of corridor construction. The advantage of this framework is that it integrates the construction of ecological corridors into land use simulation and prediction from a dynamic perspective, which enriches the connotation of land use simulation and prediction and ecological corridor construction. Relevant scholars have strictly limited the transfer of forest land, reduced the expansion of construction land, and delineated the red line of ecological protection on the outskirts of the city in order to perform the land use simulation in an ecological context. Although this approach restores forest patches and improves the overall habitat quality, the improvement is still in the periphery of the city, and it is still difficult to restore and improve the habitat quality in the city. From the results of our study (Figure 10), it can be seen that the impervious surfaces of the city remain highly intensified under natural development scenarios based on historical patterns. Under the constraining effect of ecological corridors, more new woodland patches were generated around the corridors, especially around the impervious surfaces. Obviously, the constraining effect of ecological corridors has played a certain effect on the change in land use types. At the same time, ecological corridors in the city also enhance the quality of habitats within the city.

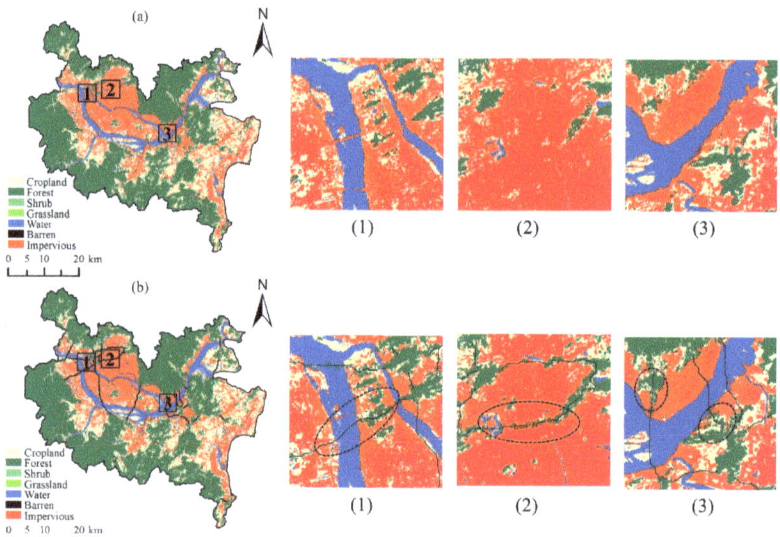

Figure 10. 2027 land use simulation projections. (**a**) Natural state; (**b**) corridor constraints.

4.2. Strategies for Planning and Construction of Ecological Corridors in Fuzhou City

Our study classified the ecological corridors and ecological source areas in the central city of Fuzhou into three levels according to the natural discontinuity point method by using the current theory through the LinkageMapper module. Among them, there are six primary corridors, fourteen secondary corridors, and fifteen tertiary corridors and six primary source sites, six secondary source sites, and seven tertiary source sites. At the same time, the study intersected the ecological corridors with urban roads, highways, and railways in order to identify key nodes that impede the corridor connectivity, which resulted in the identification of a total of two hundred thirty-nine nodes of urban roads, ninety-one nodes of highways, and eight nodes of railways (Figure 11). In terms of spatial distribution, the primary corridor mainly connects the primary source area and part of the secondary source area, mainly distributed in the edge of the city, with an average length of 436.98 m. In the actual planning and design, the natural restoration should be taken as the main way to promote the integration between the adjacent primary source areas. As an important corridor connecting large ecological sources at the border and small sources within the city, the average length of the secondary corridor is 1150.80 m, and the integrity of the corridor should be guaranteed in the planning and design. For the nodes of some secondary corridors and road networks, the corridor design should adopt the sunken design to avoid the interference and destruction of road networks. The average length of the tertiary corridor is 6484.51 m, which plays the role of connecting the north and south ecological sources, and because of its long length, it has the most road nodes with the urban area. In the planning and design, the tertiary corridor should be attached to the city road, belt green space, and the river areas along the Min River and Wulong River as much as possible. For corridors that are too long, existing green space resources should be integrated, and the role of urban parks and street green spaces as stepping stones should be utilized to enhance the overall connectivity of the ecological corridor. For areas that are difficult to change, designs such as vertical greening and rooftop gardens can be implemented to reduce the negative impact of construction land on the ecological corridor. At the same time, ecological corridors can be organized and connected through landscape green belts, plant landscape belts, and bicycle and walking paths. Finally, for the ecological source areas in the downtown area of Fuzhou City, the red line of the 19 ecological source areas will be strictly regulated, and the internal quality of the source areas will be improved by increasing the vegetation cover and enriching the three-dimensional structure of the vegetation canopy. Protective buffer zones should be set up in the bridging and edge areas around the ecological source areas to reduce the negative impacts caused by human activities.

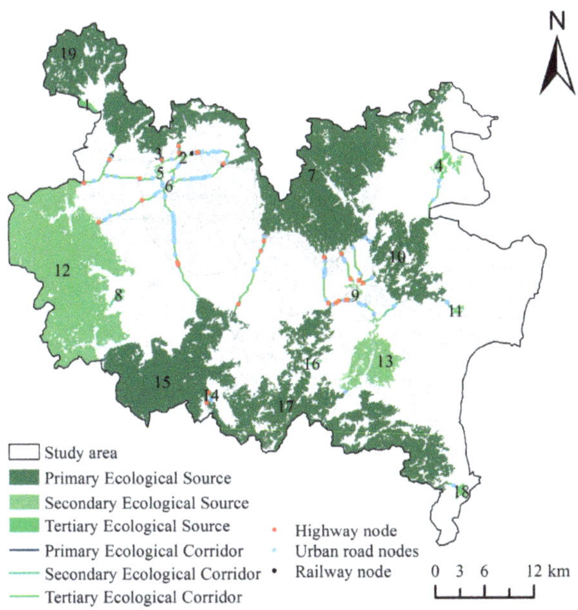

Figure 11. Ecological corridor classification and spatial distribution of nodes.

4.3. Shortcomings and Prospects

The study simulates and predicts the future land use under the ecological situation by constructing the ecological corridor in the central city of Fuzhou and taking it as the key factor to constrain the development of the city. However, there are still some deficiencies in the research process; for example, in the construction of the ecological corridor resistance surface, we used the common land use types, MSPA landscape types, and elevations and slopes and did not take into account the differences caused by different land use properties and development intensities in the city. Meanwhile, in the simulation prediction, the study took 150 m as the area of future corridor development restriction. The ecological benefits brought by different width thresholds of the corridor and the trade-off relationship with the construction cost of different width thresholds of the corridor should be further determined in the future.

5. Conclusions

The study begins with the fundamental theoretical premise that the construction of ecological corridors alters the land use type, which in turn affects the quality of the ecological environment. It then explores the impact of the constraining effect of ecological corridors on the simulation prediction of land use, with a view to provide new ideas for the planning and construction of urban ecological corridors. The principal findings of the study are as follows: (1) This study identifies a total of 19 ecological sources and 35 ecological corridors in the central city of Fuzhou based on the MSPA and MCR models and (2) the PLUS model is used to simulate and predict the land use types in the year 2027 under the historical evolution scenario and ecological constraints. The results show that The forested area in the study area increased by 1.57% compared to the area in 2022. This was due to an increase in improved areas of 1.57% and a decrease in impervious surfaces of 0.55%. The forested land area exhibited a further decrease of 1.47% under the historical evolution scenario, while the impervious surface area increased by 7.27% in comparison to 2022. (3) The ecological environment in the study area demonstrated a deterioration between 2012 and 2022, with the mean values of habitat quality reaching 0. The mean values of habitat quality were 0.4542, 0.3991, and 0.3754, respectively. In contrast, the historical evolution scenario resulted in a further decrease in mean habitat quality to 0.3555. However, under the constraining effect of ecological corridors, the habitat quality of the study area improved to 0.3809.

Author Contributions: Conceptualization, Y.Z.; methodology, Y.Z., J.G. and X.L.; software, J.G. and X.L.; validation, Y.Z. and J.G.; formal analysis, J.G. and X.L.; investigation, J.G. and X.L.; resources, Y.Z.; data curation, Y.Z. and J.G.; writing—original draft preparation, Y.Z.; writing—review and editing, J.G. and X.L.; visualization, J.G., X.L. and Y.Z.; supervision, Y.Z.; project administration, Y.Z.; funding acquisition, Y.Z. All authors have read and agreed to the published version of the manuscript.

Funding: This work was supported by National Natural Science Foundation of China, Grant No. 31800401. The Response of Plant Carbon-Phosphorus Stoichiometry Characteristics to Urban-Rural Temperature Differences and Its Mechanism.

Data Availability Statement: The data that support the findings of this study are available from the author upon reasonable request.

Conflicts of Interest: The authors declare no conflicts of interest.

References

1. Zhang, W.; Chang, W.J.; Zhu, Z.C.; Hui, Z. Landscape ecological risk assessment of Chinese coastal cities based on land use change. *Appl. Geogr.* **2020**, *117*, 102174. [CrossRef]
2. Wu, W.; Zhao, S.; Zhu, C.; Jiang, J. A comparative study of urban expansion in Beijing, Tianjin and Shijiazhuang over the past three decades. *Landsc. Urban Plan* **2015**, *134*, 93–106. [CrossRef]
3. Li, M.; Zhou, Y.; Xiao, P.; Tian, Y.; Huang, H.; Xiao, L. Evolution of habitat quality and its topographic gradient effect in northwest hubei province from 2000 to 2020 based on the invest model. *Land* **2021**, *10*, 857. [CrossRef]
4. Wu, L.; Sun, C.; Fan, F. Estimating the characteristic spatiotemporal variation in habitat quality using the invest model—A case study from Guangdong–Hong Kong–Macao Greater Bay Area. *Remote Sens.* **2021**, *13*, 1008. [CrossRef]
5. Yang, L.; Yu, K.; Ai, J.; Liu, Y.; Yang, W.; Liu, J. Dominant factors and spatial heterogeneity of land surface temperatures in urban areas: A case study in Fuzhou, China. *Remote Sens.* **2022**, *14*, 1266. [CrossRef]
6. Dong, J.; Peng, J.; Liu, Y.; Qiu, S.; Han, Y. Integrating spatial continuous wavelet transform and kernel density estimation to identify ecological corridors in megacities. *Landsc. Urban Plan* **2020**, *199*, 103815. [CrossRef]
7. Peng, J.; Zhao, H.; Liu, Y. Urban ecological corridors construction: A review. *Acta Ecologica Sinica.* **2017**, *37*, 23–30. [CrossRef]
8. Hu, C.; Wang, Z.; Wang, Y.; Sun, D.; Zhang, J. Combining MSPA-MCR model to evaluate the ecological network in Wuhan, China. *Land* **2022**, *11*, 213. [CrossRef]
9. Guan, D.; Jiang, Y.; Cheng, L. How can the landscape ecological security pattern be quantitatively optimized and effectively evaluated? An integrated analysis with the granularity inverse method and landscape indicators. *Environ. Sci. Pollut. Res.* **2022**, *29*, 41590–41616. [CrossRef]
10. Lalechère, E.; Bergès, L. A validation procedure for ecological corridor locations. *Land* **2021**, *10*, 1320. [CrossRef]
11. Wang, S.; Wu, M.; Hu, M.; Fan, C.; Wang, T.; Xia, B. Promoting landscape connectivity of highly urbanized area: An ecological network approach. *Ecol. Indic.* **2021**, *125*, 107487. [CrossRef]
12. Dai, Y. Identifying the ecological security patterns of the Three Gorges Reservoir Region, China. *Environ. Sci. Pollut. Res.* **2022**, *29*, 45837–45847. [CrossRef] [PubMed]
13. Geng, J.; Yu, K.; Sun, M.; Xie, Z.; Huang, R.; Wang, Y.; Zhao, Q.; Liu, J. Construction and Optimisation of Ecological Networks in High-Density Central Urban Areas: The Case of Fuzhou City, China. *Remote Sens.* **2023**, *15*, 5666. [CrossRef]
14. Li, Y.; Zhang, Y.; Jiang, Z.; Guo, C.; Zhao, M.; Yang, Z.; Guo, M.; Wu, B.; Chen, Q. Integrating morphological spatial pattern analysis and the minimal cumulative resistance model to optimize urban ecological networks: A case study in Shenzhen City, China. *Ecol. Process.* **2021**, *10*, 63. [CrossRef]

15. Huang, K.; Peng, L.; Wang, X.; Deng, W. Integrating circuit theory and landscape pattern index to identify and optimize ecological networks: A case study of the Sichuan Basin, China. *Environ. Sci. Pollut. Res.* **2022**, *29*, 66874–66887. [CrossRef] [PubMed]
16. Wang, Y.; Qu, Z.; Zhong, Q.; Zhang, Q.; Zhang, L.; Zhang, R.; Yi, Y.; Zhang, G.; Li, X.; Liu, J. Delimitation of ecological corridors in a highly urbanizing region based on circuit theory and MSPA. *Ecol. Indic.* **2022**, *142*, 109258. [CrossRef]
17. Hamad, R.; Balzter, H.; Kolo, K. Predicting land use/land cover changes using a CA-Markov model under two different scenarios. *Sustainability* **2018**, *10*, 3421. [CrossRef]
18. Liu, X.; Liang, X.; Li, X.; Xu, X.; Ou, J.; Chen, Y.; Li, S.; Wang, S.; Pei, F. A future land use simulation model (FLUS) for simulating multiple land use scenarios by coupling human and natural effects. *Landsc. Urban Plan* **2017**, *168*, 94–116. [CrossRef]
19. Liang, X.; Guan, Q.; Clarke, K.C.; Liu, S.; Wang, B.; Yao, Y. Understanding the drivers of sustainable land expansion using a patch-generating land use simulation (PLUS) model: A case study in Wuhan, China. *Comput. Environ. Urban Syst.* **2021**, *85*, 101569. [CrossRef]
20. Gao, L.; Tao, F.; Liu, R.; Wang, Z.; Leng, H.; Zhou, T. Multi-scenario simulation and ecological risk analysis of land use based on the PLUS model: A case study of Nanjing. *Sust. Cities Soc.* **2022**, *85*, 104055. [CrossRef]
21. Tang, W.; Cui, L.; Zheng, S.; Hu, W. Multi-scenario simulation of land use carbon emissions from energy consumption in Shenzhen, China. *Land* **2022**, *11*, 1673. [CrossRef]
22. Wang, Z.; Zeng, J.; Chen, W. Impact of urban expansion on carbon storage under multi-scenario simulations in Wuhan, China. *Environ. Sci. Pollut. Res.* **2022**, *29*, 45507–45526. [CrossRef] [PubMed]
23. Cao, M.; Chang, L.; Ma, S.; Zhao, Z.; Wu, K.; Hu, X.; Gu, Q.; Lü, G.; Chen, M. Multi-scenario simulation of land use for sustainable development goals. *IEEE J. Sel. Top. Appl. Earth Observ. Remote Sens.* **2022**, *15*, 2119–2127. [CrossRef]
24. Xiong, N.; Yu, R.; Yan, F.; Wang, J.; Feng, Z. Land use and Land cover changes and prediction based on Multi-Scenario Simulation: A case study of Qishan County, China. *Remote Sens.* **2022**, *14*, 4041. [CrossRef]
25. Li, Y.; Liu, Z.; Li, S.; Li, X. Multi-scenario simulation analysis of land use and carbon storage changes in changchun city based on FLUS and InVEST model. *Land* **2022**, *11*, 647. [CrossRef]
26. Li, X.; Liu, Z.; Li, S.; Li, Y. Multi-scenario simulation analysis of land use impacts on habitat quality in Tianjin based on the PLUS model coupled with the InVEST model. *Sustainability* **2022**, *14*, 6923. [CrossRef]
27. Liu, Y.; Jing, Y.; Han, S. Multi-scenario simulation of land use/land cover change and water yield evaluation coupled with the GMOP-PLUS-InVEST model: A case study of the Nansi Lake Basin in China. *Ecol. Indic.* **2023**, *155*, 110926. [CrossRef]
28. Jiang, H.; Peng, J.; Zhao, Y.; Xu, D.; Dong, J. Zoning for ecosystem restoration based on ecological network in mountainous region. *Ecol. Indic.* **2022**, *142*, 109138. [CrossRef]
29. Lv, L.; Zhang, S.; Zhu, J.; Wang, Z.; Wang, Z.; Li, G.; Yang, C. Ecological restoration strategies for mountainous cities based on ecological security patterns and circuit theory: A case of Central Urban Areas in Chongqing, China. *Int. J. Environ. Res. Public Health* **2022**, *19*, 16505. [CrossRef]
30. Xu, X.; Wang, S.; Rong, W. Construction of ecological network in Suzhou based on the PLUS and MSPA models. *Ecol. Indic.* **2023**, *154*, 110740. [CrossRef]
31. Li, L.; Huang, X.; Wu, D.; Yang, H. Construction of ecological security pattern adapting to future land use change in Pearl River Delta, China. *Appl. Geogr.* **2023**, *154*, 102946. [CrossRef]
32. Liu, J.; Liu, B.; Wu, L.; Miao, H.; Liu, J.; Jiang, K.; Ding, H.; Gao, W.; Liu, T. Prediction of land use for the next 30 years using the PLUS model's multi-scenario simulation in Guizhou Province, China. *Sci. Rep.* **2024**, *14*, 13143. [CrossRef] [PubMed]
33. Wang, S.; Ge, Y. Ecological quality response to multi-scenario land-use changes in the Heihe River Basin. *Sustainability* **2022**, *14*, 2716. [CrossRef]
34. Yang, J.; Huang, X. The 30 m Annual Land Cover Datasets and Its Dynamics in China from 1985 to 2022. *Earth Syst. Sci. Data* **2023**, *13*, 3907–3925. [CrossRef]
35. Soille, P.; Vogt, P. Morphological segmentation of binary patterns. *Pattern Recognit. Lett.* **2009**, *30*, 456–459. [CrossRef]
36. Vogt, P.; Ferrari, J.R.; Lookingbill, T.R.; Gardner, R.H.; Riitters, K.H.; Ostapowicz, K. Mapping functional connectivity. *Ecol. Indic.* **2009**, *9*, 64–71. [CrossRef]
37. Tang, Y.; Gao, C.; Wu, X. Urban ecological corridor network construction: An integration of the least cost path model and the InVEST model. *ISPRS Int. J. Geo-Inf.* **2020**, *9*, 33. [CrossRef]
38. Hou, W.; Zhou, W.; Li, J.; Li, C. Simulation of the potential impact of urban expansion on regional ecological corridors: A case study of Taiyuan, China. *Sust. Cities Soc.* **2022**, *83*, 103933. [CrossRef]
39. Li, Q.; Zhou, Y.; Yi, S. An integrated approach to constructing ecological security patterns and identifying ecological restoration and protection areas: A case study of Jingmen, China. *Ecol. Indic.* **2022**, *137*, 108723. [CrossRef]
40. Knaapen, J.P.; Scheffer, M.; Harms, B. Estimating habitat isolation in landscape planning. *Landsc. Urban Plan* **1992**, *23*, 1–16. [CrossRef]
41. McRae, B.H.; Kavanagh, D.M. *Linkage Mapper Connectivity Analysis Software*; The Nature Conservancy: Seattle, WA, USA, 2011.
42. Xu, L.; Liu, X.; Tong, D.; Liu, Z.; Yin, L.; Zheng, W. Forecasting urban land use change based on cellular automata and the PLUS model. *Land* **2022**, *11*, 652. [CrossRef]
43. Chen, C.; Liu, J.; Bi, L. Spatial and temporal changes of habitat quality and its influential factors in China based on the InVEST model. *Forests* **2023**, *14*, 374. [CrossRef]

44. Terrado, M.; Sabater, S.; Chaplin-Kramer, B.; Mandle, L.; Ziv, G.; Acuña, V. Model development for the assessment of terrestrial and aquatic habitat quality in conservation planning. *Sci. Total Environ.* **2016**, *540*, 63–70. [CrossRef] [PubMed]
45. Luo, X.; Le, F.; Zhang, Y.; Zhang, H.; Zhai, J.; Luo, Y. Multi-scenario analysis and optimization strategy of ecological security pattern in the Weihe river basin. *J. Environ. Manag.* **2024**, *366*, 121813. [CrossRef] [PubMed]

sustainability

MDPI

Article

Exploring the Relationship between Key Perceptual Elements of Urban Secondary Wilderness and Its Restorative Benefits

Keyan Chen, Yan Xu, Kaiyuan Zhan, Yangshuo Gao and Xiangcai Xie *

College of Landscape Architecture and Art, Fujian Agriculture and Forestry University, Fuzhou 350002, China; 1221775009@fafu.edu.cn (K.C.); 1221775008@fafu.edu.cn (Y.X.); 1221775028@fafu.edu.cn (K.Z.); 1221775005@fafu.edu.cn (Y.G.)

* Correspondence: fjlxyxxc@126.com; Tel.: +86-15860277990

Abstract: Urban secondary wilderness (USWs) is a near-natural place in cities and is an important link to reconnect humans with the natural world. Assessing the perceptual elements of USWs not only provides insight into public perceptions and preferences, but also helps to further explore its relevance to experiential values such as environmental restoration benefits. In this study, we selected three USWs cases located in Fuzhou, the capital city of Fujian Province, China, and constructed an evaluation system containing two types of dimensions and seventeen perceptual elements. Based on the public's evaluation of the importance of and satisfaction with the perceived elements, we extracted the key perceived elements and analyzed their relationship with environmental restoration benefits. The results indicated that all three USWs cases scored over 4.9 on the Perceived Restoration Scale (PRS). Plants, Landscape color, Water, Landform, Climate, Freedom, and Naturalness are the key perceptual elements of the USWs, and the correlation coefficients between the satisfaction with the above seven perceptual elements and the scores of the PRS were all positive. In summary, USWs is an ideal place for residents to carry out restoration activities, and its key perceptual elements play a driving role.

Keywords: urban secondary wilderness (USW); perceptual elements; restorative benefits; correlation studies

1. Introduction

In recent years, people living in the city have realized that they have fewer opportunities to interact with nature [1–3], a phenomenon known as "experiential extinction" [4]. Research has demonstrated that "experiential extinction" not only negatively affects the physical and mental health and social well-being of residents [5–7], but also undermines their positive feelings, attitudes, and behaviors towards ecological conservation [8–10]. Therefore, in the context of rapid global urban expansion, meeting the needs of urban residents to experience nature and reconnecting people with nature are becoming key issues of the times [11,12].

For this reason, scholars have proposed the concept of urban wilderness (UWs) based on natural wilderness research, emphasizing the feasibility of accessing wild nature in cities [13–15]. UWs refers to land dominated by natural processes rather than by humans in or near urban areas, includes three types: urban primary wilderness (UPWs), urban secondary wilderness (USWs), and urban-like wilderness (ULWs) [16,17]. All three types have the potential to increase the connection between human beings and the natural world [18,19], but UPWs emphasizes the natural retention properties of the site, and the scarcity of wilderness resources in cities limits the development of related research [14]; ULWs creation involves species introduction and other ecological modifications, and some scholars are concerned about the risk of damaging fragile urban ecosystems [20,21]; however, taking USWs as the object of research may avoid the abovementioned research shortcomings. USWs refers to the wilderness environment that is re-formed by the natural law after persistent and high-intensity anthropogenic control or influence has ceased [22,23]. Nowadays, the increasing number of abandoned green spaces in cities not only provides sufficient samples for the study of USWs, but also effectively avoids the potential ecological risks of artificial intervention [24–26].

In the past, based on case studies of abandoned woodlands, riverbanks, and parks, scholars have found that USWs areas are distinguished from other types of urban environments, and usually contain more natural and wild environmental elements [27–29], such as unpruned plants, abundant wildlife, undulating terrain, and a natural boundary. At the same time, according to the framework of restorative environment research constituted by stress reduction theory (SRT) and attention restoration theory (ART), these environmental elements are viewed as key to supporting and contributing to the restorative benefits of USWs [30–33]. On this basis, a large number of studies have focused on the experiential perception of USWs, confirming on the one hand that USWs, as a near-natural place, has the potential to alleviate daily pressures [34,35], inhibit negative emotions [36,37], and increase life satisfaction [38]; on the other hand, they have found that there are differences in the public's perceptions of and individual preferences for the objective characteristics of USWs and have emphasized that the evaluation of perceptions is a key perspective for recognizing USWs [39,40].

Perception refers to the human subject's subjective view of external object things [41]. Drawing on past theoretical models, the perceptual elements of can be divided into physiological and psychological dimensions [42,43]. The physiological dimension refers to the perception of the objective environmental elements of the USWs based on the five senses, including visual elements such as organisms, landscapes, facilities, water, and landform [44,45], auditory elements such as natural and biological sounds [46], olfactory elements such as air freshness [47], and somatosensory elements such as temperature and humidity [48]. And the psychological dimension is the experiential evaluation of the USWs environment, including feelings of fun, comfort, freedom, and naturalness [49–52]. However, existing studies have mostly assessed only a few perceptual elements from a single evaluation dimension of physiology or psychology, and it is still unclear about the public's cognition and preference for the perceptual elements of USWs areas, and even less is known about the perceptual elements that play a key role in them. On this basis, it is difficult to further confirm the correlation between experiential benefits such as environmental restoration of USWs and human subjective perceptions.

Therefore, the study aimed to scientifically and comprehensively evaluate the perceptual elements of USWs and to explore the association between key perceptual elements and experiential benefits such as environmental restoration. Based on the cases of USWs in Fuzhou City, Fujian Province, China, the study tried to construct an evaluation system for USWs perception based on the results of related studies and field research. Then, we extracted the perceptual elements that have a key impact on the experience of USWs. Finally, we measured the restorative benefits of USWs and explored the relationship between the key perceptual elements described above and environmental restorative benefits. Specifically, the study explored three questions:

1. What are the key perceptual elements of USWs?;
2. What are the restorative benefits of USWs?;
3. Do the key perceived elements of USWs influence its restorative benefits?

2. Materials and Methods

2.1. Study Design

As shown in Figure 1, the experimental design of this study was divided into three parts.

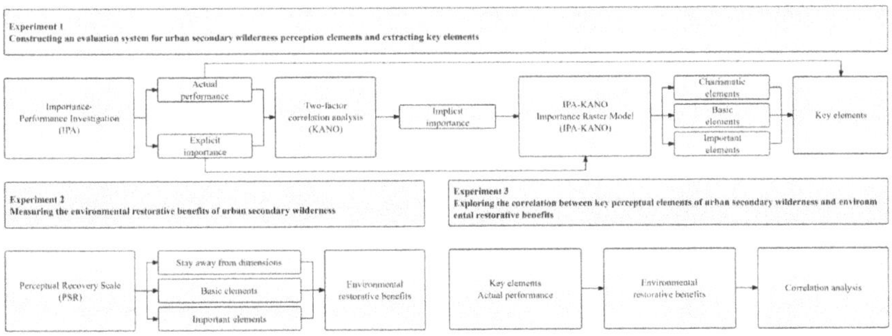

Figure 1. Experimental procedure.

The three parts are as follows:

1. Constructing a perceptual evaluation system for USWs and extracting key perceptual elements of USWs;
2. Measuring the environmental restorative benefits of USWs;
3. Analyzing the correlation between key perceptual elements of USWs and environmental restorative benefits.

The first part of the experiment includes six main steps. First, we summarized the perceived dimensions and evaluation elements of USWs, based on the results of related studies and field research. Second, we designed a questionnaire to evaluate the importance of and satisfaction with the perceived elements. The survey helps us to learn about residents' preferences for perceived elements and how they behave in reality, as urban green spaces can be seen as a service provided to residents [53]. Then, we counted the explicit importance and actual performance of USWs perception elements based on the importance and satisfaction evaluation results, respectively. Third, we conducted bivariate analysis of the importance of perceived elements and satisfaction, and extracted correlation coefficients to reflect the implied importance of these elements. Distinguishing from the subjects' direct evaluation of the importance of the elements, the implicit importance can reflect the implicit influence of individual perceptual element on the overall benefits. Fourth, we use the IPA-Kano model to analyze the quality attributes of the perceptual elements. The Importance–Performance Analysis (IPA) model evaluates the attributes of elements based on the importance–satisfaction matrix [54], and the Kano model categorizes element attributes based on user needs [55]. The IPA-Kano model builds on the similarities and differences between the IPA model and the Kano model to construct the quality measurement grid that can jointly assess importance and performance [56]. Origin is statistical and graphical software applied to statistics and the visualization of data, and we used it to map the IPA-Kano quality measurement grid of the perception elements of USWs. As shown in Figure 2, the horizontal axis used the importance data and the vertical axis used the implicit importance data, and the mean of the explicit and implicit importance of all the perceptual elements was used as the center coordinate, and each element was integrated into the coordinate system separately. Based on the IPA-Kano model, the important elements in Quadrant I have high explicit and implicit importance, the charismatic elements in Quadrant II have low explicit and high implicit importance, Quadrant III has unimportant elements with low explicit and implicit importance, and Quadrant IV has basic elements consisting of high explicit and low implicit importance. Fifth, we analyzed the impact of user needs on satisfaction and prioritize perceptual elements based on the Kano theory. Specifically, we focus on the important, charismatic, and basic perceptual elements and prioritize them according to their actual performance, filtering key perceptual elements ranked in the top one-third. The criteria for the key hierarchy of the perceptual elements are shown in Table 1.

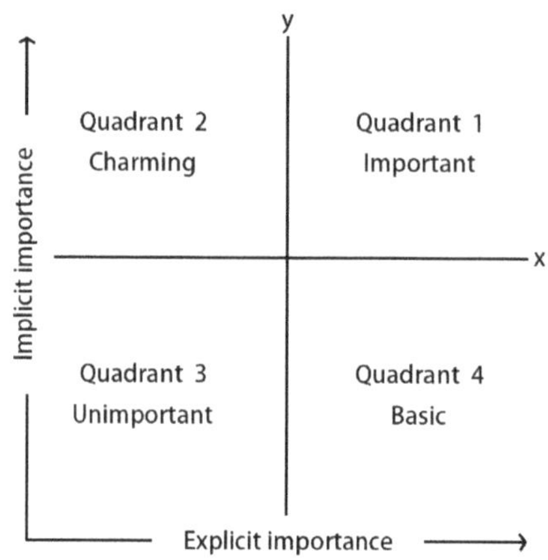

Figure 2. IPA-Kano quality measurement grid.

Table 1. Criteria for key hierarchy.

Key Hierarchy	Quality Attributes	Actual Performance Ranking
High	Important Charming Basic	1–6
Medium	Important Charming Basic	7–11
Low	Important Charming Basic	12–17

The second part of the experiment included two main steps. First, the Perceived Restoration Scale (PRS) was used to obtain subjects' perceptual evaluations of the restorative benefits of USWs. Second, SPSS software was used to calculate subjects' restorative benefits of the three cases.

In the third part of the experiment, Pearson correlation between satisfaction and the restorative benefits of USWs' key perceived elements was analyzed using SPSS software, and the correlation between USWs' key perceived elements and their restorative benefits was analyzed using the same method.

2.2. Indicator System

As shown in Table 2, based on the literature, the study categorized USWs perception into two types of dimensions, physiological and psychological, with a total of seventeen elements. Among them, the physiological perception element included twelve indicators, Plants (A1), Animals (A2), Landscape color (A3), Landscape seasonal phase (A4), Water (A5), Boundary (A6), Landform (A7), Climate (A8), Sound (A9), Air (A10), Service facilities (A11), and Cell phone signal (A12), which were used to obtain the subjects' perception evaluations through the means of vision, hearing, and smell. The psychological perception element includes five indicators, Comfort (B1), Freedom (B2), Stimulation (B3), Fun (B4), and Naturalness (B5), which were selected as representative indicators for the evaluation of USWs perception.

Table 2. Evaluation indicators for USWs perception elements.

Dimension	S/N	Index	Description	References
Physiological perception	A1	Plants	Visual perception of vegetation height, density, and richness	[57,58]
	A2	Animals	Visual perception of animal numbers, diversity	[57,59]
	A3	Landscape color	Visual perception of landscape color vibrancy and diversity	[60]
	A4	Landscape seasonal phase	Visual perception of landscape seasonal legibility	[61]
	A5	Water	Quality of the water environment, dynamic visual perception	[42]
	A6	Boundary	Visual perception of environmental boundary integrity, sense of shelter	[62]
	A7	Landform	Perception of terrain relief, diversity	[62]
	A8	Climate	Perception of solar radiation, temperature and humidity, and wind speed	[63]
	A9	Sound	Auditory perception of biological sounds, such as birdsong, and natural sounds, such as the wind	[64,65]
	A10	Air	Olfactory perception of air freshness	[66,67]
	A11	Facilities	Perceived completeness and cleanliness of manual services such as roads, streetlights, seats, toilets, etc.	[68,69]
	A12	Cell phone signal	Access to cell phone signal	[70,71]
psychological perception	B1	Comfort	Feeling of gentleness and harmony	[50]
	B2	Freedom	Feeling undisturbed within, freedom of movement, and relaxation of the mind	[72,73]
	B3	Stimulation	Experience a unique, stimulating, heart-pumping atmosphere	[62,74]
	B4	Fun	Enjoyable atmosphere and interest in the activity	[75,76]
	B5	Naturalness	Feeling close to nature and ecological integration	[59,77]

2.3. *Questionnaire Design*

The study was conducted to obtain data by administering a paper questionnaire in the field, which consisted of three parts. The first part of the questionnaire recorded the socio-demographic characteristics of the subjects, such as gender and age.

The second part of the questionnaire consisted of subjects' importance and satisfaction ratings of seventeen USWs perception elements. The importance questionnaire asked subjects to rate the USWs perception elements according to their subjective cognitive preferences, with questions such as "Do you think plant elements in the current environment are important?" The question was recorded on a 9-point Likert scale, with "1" indicating very unimportant, "5" indicating fairly unimportant, and "9" indicating very important. The satisfaction questionnaire asked subjects to rate USWs perception elements based on objective performance. For example, the questions included "Are you satisfied with the plant elements in your current environment?", and the answers were also recorded on a 9-point Likert scale, with "1" indicating very dissatisfied, "5" indicating fairly dissatisfied, and "9" indicating very satisfied. The arithmetic mean scores of the variables were used as importance and satisfaction scores, reflecting the apparent importance and actual performance of the perceived elements of the USWs [78]. The importance questionnaire scale had a Cronbach's alpha of 0.901, a KMO value of 0.910, and a significant Barlett's test of sphericity ($p < 0.001$). And the satisfaction questionnaire scale had a Cronbach's alpha of 0.911, a KMO value of 0.902, and a significant Barlett's test of sphericity ($p < 0.001$). The results showed that the scales have high reliability and validity.

The third part of the questionnaire was used to measure the restorative benefits of USWs. As shown in Table 3, this study utilized the Chinese version of the PRS developed by Ye et al. (2010) [79]. It is widely used in Chinese contexts and has been shown to have high reliability and validity [42,80,81]. Subjects were assessed using a 7-point Likert scale to evaluate the restorative benefits of three USWs cases. The questionnaire consisted of three dimensions and twenty-two questions, of which the first seventeen questions were positively scored, with "1" indicating strongly disagree, "4" indicating fair, and "7" indicating strongly agree, and the last five questions were scored in reverse, with "1" indicating strongly agree, "4" indicating fairly agree, and "7" indicating strongly disagree. The arithmetical mean of the scores for each question was used to reflect the overall environmental restoration benefits.

Table 3. Chinese version of the PRS.

Dimension	S/N	Description
Being away	Q1	Here I feel far away from what is expected of me.
	Q2	It's a place where I'm away from a lot of the pressures of real life.
	Q3	It's a lot different from the environment I'm exposed to in my normal life.
	Q4	When I'm here, I don't have to think about my responsibilities.
	Q5	Here, I can put aside my daily chores for a while.
Fascination and Compatibility	Q6	This environment is new to me.
	Q7	There is a certain monumental significance here.
	Q8	Here I can see, hear, feel and think about many things.
	Q9	When I stay here, I often make unexpected discoveries.
	Q10	There are attractive qualities here.
	Q11	My attention is drawn to many interesting things.
	Q12	I can do what I like here.
	Q13	I feel at one with this place.
	Q14	I want to stay here longer.
	Q15	I feel like I belong here.
	Q16	I can notice many things in this environment without trying hard.
	Q17	It brings back memories or imaginations.
Complexity	Q18	I find this environment boring.
	Q19	I find this environment monotonous.
	Q20	The landscape contains too many things.
	Q21	I feel like this place is a mess.
	Q22	I feel bored here.

The three sections of the questionnaire were analyzed for reliability and validity using SPSS 27 statistical software; the Cronbach values were basically above 0.9, proving that the questionnaire had good reliability. The significance value was 0 and the validity was good.

2.4. Study Area

As shown in Figure 3, the study area is Fuzhou, the capital city of Fujian Province in China. The city has one of the top five green coverage rates in China, and the large number of parks within this city limits not only attracts a large number of foreign tourists, but is also closely related to the well-being of local residents. However, in the context of decreasing local revenues, high maintenance costs have elevated the difficulty of managing parks, and some of the city's man-made green spaces have even entered an unmanaged state and are being transformed into natural wastelands [82,83].

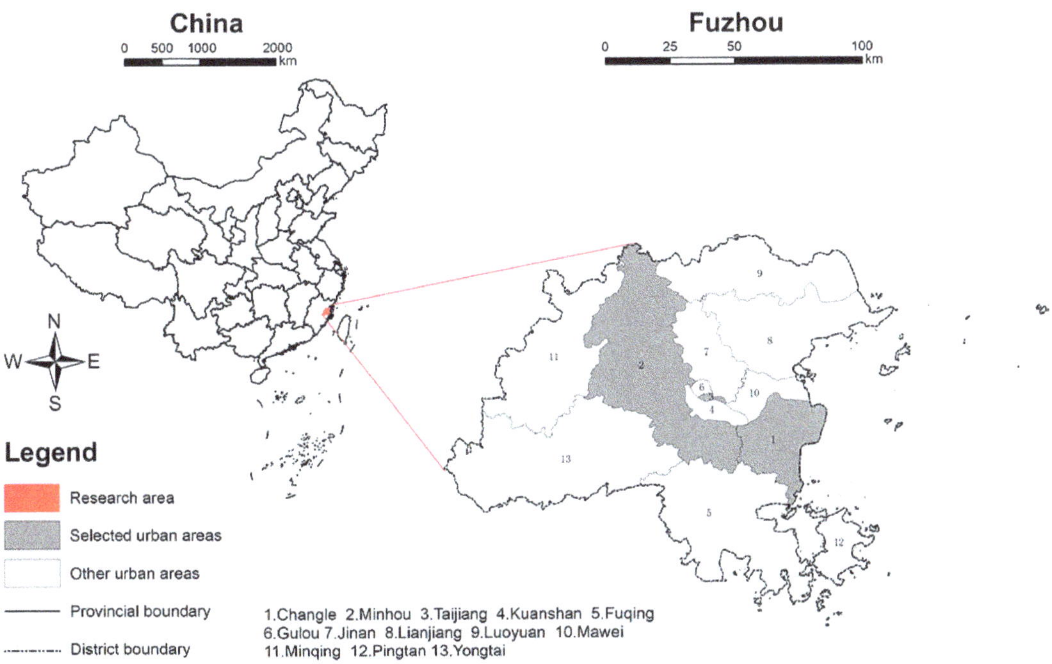

Figure 3. Study area.

In this study, Aofeng Park (Aofeng), Qishan Forest Park (Qishan), and Lujitan Forest Park (Lujitan) were selected as our study cases (Figure 4). Located in Taijiang District, Aofeng was once famous for the more than 700 crocodiles on display in the park, which has since been deserted for more than fifteen years due to poor management. Qishan, located in Minhou County, was once a 4A-class scenic spot in China, but subsequent poor performance led to the management's inability to afford the maintenance costs, and the park has been left unmanaged for the past five years. Located in Changle District, Lujitan features beautiful features such as lakes, waterfalls, and strange rocks, but a management incident that occurred more than ten years ago created a bad social impact and led to the park's abandonment. We assessed the feasibility of Aofeng, Qishan, and Lujitan as USWs study cases. First, the abandoned parks meet the definition of USWs, as urban green spaces that were once under artificial management and are now dominated by natural succession, and are capable of becoming USWs. Second, the three parks differ significantly in terms of length of abandonment, size, landscape type, and other characteristics. Lastly, the parks have sufficient traffic flow, and the main users include not only the residents of the neighborhood, but also external visitors such as hikers, fishermen, and wildlife enthusiasts. Combining these three factors, we believed that public surveys could be conducted in Aofeng, Qishan, and Lujitan.

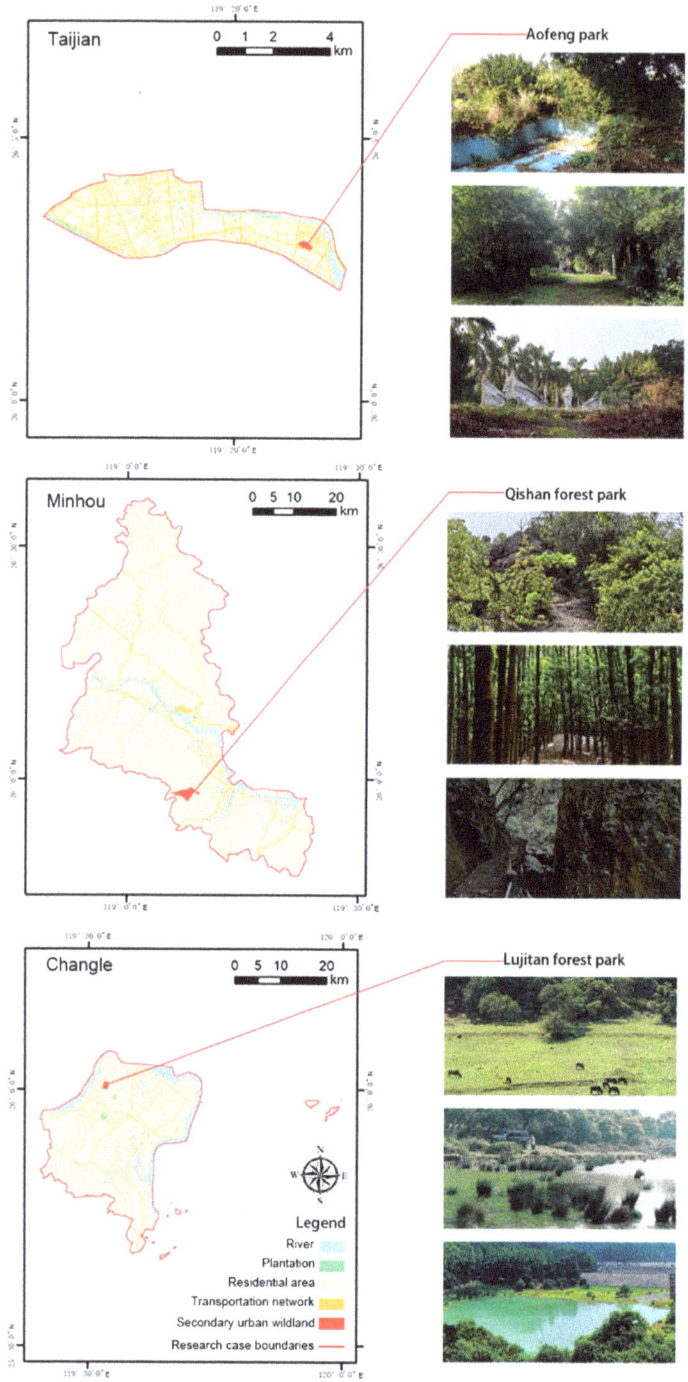

Figure 4. Three cases of USWs in Fuzhou.

2.5. Subjects

Data were collected over a seven-month period from 10 August 2023 to 7 March 2024, and the researchers were divided into three groups that visited each case site four times per month, on weekly rest days (Saturday, Sunday), with each visit lasting five hours (9:00–11:00, 13:00–16:00). The exact procedure of the investigation is as follows:

1. Researchers selected the main entrance to the USWs, randomly chose groups of visitors and used the "random number method" to invite visitors to participate in the survey;
2. Subjects were informed of the purpose and content of the survey, and could discontinue the ongoing survey at any time if they felt any discomfort;
3. Subjects were asked to visit the park for more than 15 min to ensure they had a basic understanding of their environment;
4. Respondents were then asked to fill out a questionnaire consisting of importance and satisfaction ratings of the perceived elements and the PRS, at the end of which each respondent was able to receive a small gift (valued at less than five dollars).

A total of 850 questionnaires were distributed in the study, and 797 valid questionnaires were returned after excluding invalid questionnaires, including 275 questionnaires from Aofeng, 280 questionnaires from Qishan, and 242 questionnaires from Lujitan, and the validity rate of questionnaire return was 93.7%. Information regarding the subjects is shown in Figure 5; the respondents mainly consisted of subjects under 35 years old (62%), of whom 54% were male and 46% were female.

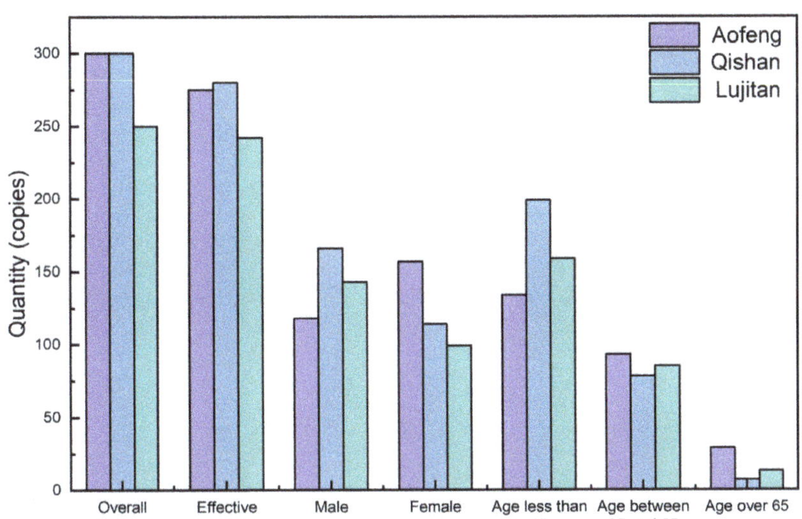

Figure 5. Subject statistics.

3. Results

3.1. Description of the Underlying Data

The results of the significance evaluation reflect the apparent significance of the perceived elements in the USWs cases (Appendix A). As shown in Figure 6, the perceptual elements that ranked high in the importance scores in all three cases included Landscape color (A3), Landscape seasonal phase (A4), Water (A5), Cell phone signal (A12), and Freedom (B2), indicating that the public recognized that these five perceptual elements were of high importance. In contrast, both Animals (A2) and Stimulation (B3) ranked at the bottom of the importance scores, suggesting that the public perceived these two perceptual elements as less important. Meanwhile, the perceptual elements that received attention in the different parks showed some similarity. Climate was regarded as the most important perceptual element by the subjects of Aofeng (7.473) and Lujitan (7.694), and Water was seen as the most important perceptual element of Qishan (7.050). It is also worth mentioning that none of the subjects cared about Stimulation, which ranked at the bottom of the scores for Aofeng (4.895), Qishan (5.425), and Lujitan (5.116).

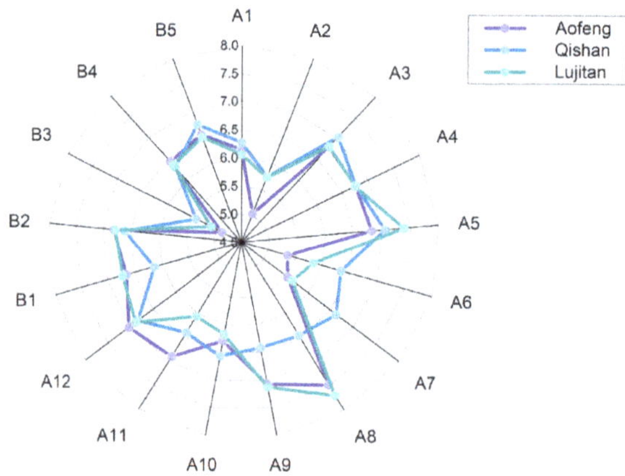

Figure 6. Importance of the perceptual elements of three cases.

The results of the satisfaction evaluation reflect the actual performance of the perceived elements in the USWs cases (Appendix B). As shown in Figure 7, the perceptual elements with the highest satisfaction scores in all three cases include Plants (A1), Landform (A7), and Climate (A8), indicating that the public recognized the actual performance of these three perceptual elements as good. In contrast, both Comfort (B1) and Stimulation (B3) were at the bottom of the list, suggesting that the public recognized these two perceptual elements as less important. Meanwhile, the actual performance of the perceptual elements also varies greatly from park to park. Climate (7.076) and Water (3.495) were the best- and worst-performing perceptual elements in Aofeng, while in Qishan the best- and worst-performing categories were Landform (6.953) and Comfort (4.689), and in Lujitan they were Water (7.198) and Comfort (4.231).

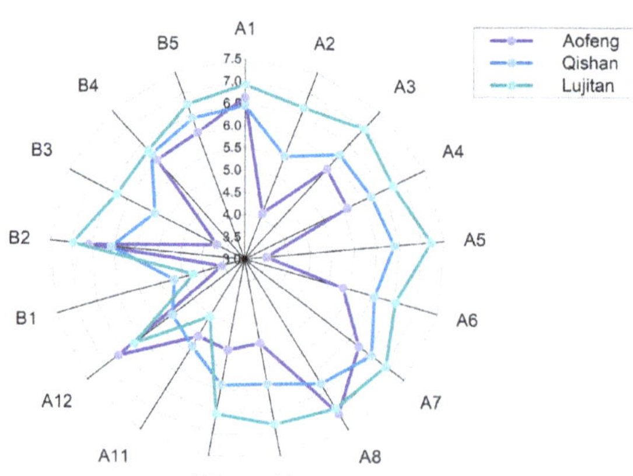

Figure 7. Satisfaction with the perceptual elements in three cases.

3.2. Extraction of Key Perceptual Elements

We found strong correlations in the evaluation data for some of the perceptual elements (r > 0.7), which means that there is multicollinearity in the data. Therefore, bivariate correlation analyses were chosen for the study to extract implied importance [84]. According to the data in Table 4, we extracted the importance of the perceived elements and satisfaction for correlation analysis; all the perceptual elements show implicit importance. In Aofeng, Animals (0.336 **) was the perceptual element with the highest implicit importance, followed by Sound (0.332 **) and Facilities (0.331 **), while Boundary (0.181 **) has the lowest implicit importance. In Qishan, the perceptual elements with the highest implicit importance were Fun (0.370 **), Cell phone signal (0.351 **), and Air (0.328 **), respectively, while Naturalness (0.162 **) had the lowest implicit importance. In Lujitan, Landscape color (0.372 **) was the perceptual element with the highest implicit importance, followed by Landform (0.370 **) and Climate (0.370 **), while Boundary (0.276**) had the lowest implicit importance.

Table 4. Implicit importance of the perceptual elements of three cases.

Factor	Aofeng		Qishan		Lujitan	
	Score	Ranking	Score	Ranking	Score	Ranking
A1	0.249 **	11	0.270 **	9	0.343 **	6
A2	0.336 **	1	0.291 **	5	0.335 **	8
A3	0.291 **	7	0.227 **	15	0.372 **	1
A4	0.228 **	13	0.229 **	14	0.356 **	5
A5	0.264 **	10	0.255 **	11	0.334 **	9
A6	0.181 **	17	0.285 **	7	0.276 **	17
A7	0.241 **	12	0.231 **	13	0.370 **	2
A8	0.198 **	16	0.282 **	8	0.370 **	3
A9	0.331 **	3	0.237 **	12	0.325 **	10
A10	0.226 **	14	0.328 **	3	0.281 **	16
A11	0.332 **	2	0.256 **	10	0.315 **	12
A12	0.294 **	6	0.351 **	2	0.287 **	15
B1	0.224 **	15	0.301 **	4	0.325 **	11
B2	0.283 **	8	0.203 **	16	0.359 **	4
B3	0.308 **	4	0.287 **	6	0.311 **	13
B4	0.297 **	5	0.370 **	1	0.339 **	7
B5	0.273 **	9	0.162 **	17	0.308 **	14

Note: ** indicates highly significant correlation at $p < 0.01$, the same below.

As shown in Figure 8, the quadrant in which the element is located delineates the quality of the perceptual elements. Regarding the important elements in Quadrant I, there are seven items in Aofeng (Landscape color, Sound, Facilities, Cell phone signal, Freedom, Fun, Naturalness), two items in Qishan (Air, Cell phone signal), and five items in Lujitan (Landscape color, Landscape seasonal phase, Water, Climate, Freedom). Regarding the charming elements in Quadrant II, there are two items in Aofeng (Animals, Stimulation), seven items in Qishan (Plants, Animals, Boundary, Climate, Comfort, Stimulation, Fun), and four items in Lujitan (Plants, Animals, Landform, Fun). Regarding the unimportant elements in Quadrant III, there are four items in both Aofeng (Plants, Boundary, Landform, Air) and Lujitan (Boundary, Air, Facilities, Stimulation), and Qishan has two items (Sound, Facilities). Regarding the basic elements in Quadrant IV, there were four items in Aofeng (Landscape seasonal phase, Water, Climate, Comfort) and Lujitan (Sound, Cell phone signal, Comfort, Naturalness), and Qishan has six items (Landscape color, Landscape seasonal phase, Water, Landform, Freedom, Naturalness). Similarities and differences coexisted in the quality measurements of the perceptual elements in the three USWs cases.

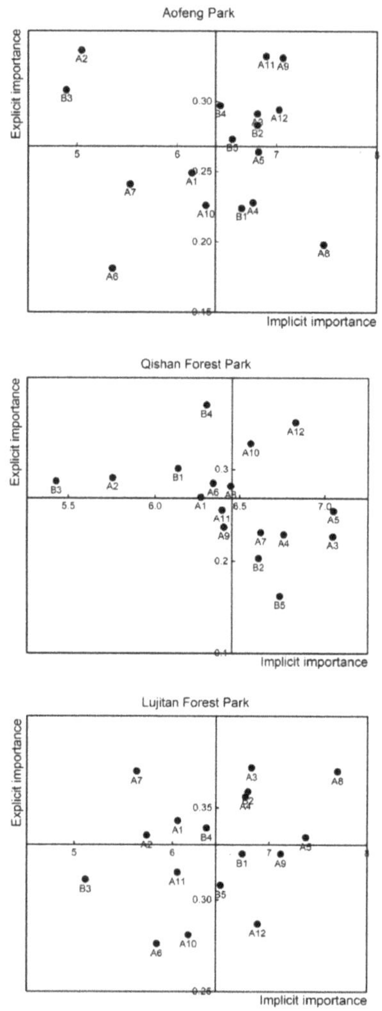

Figure 8. Quality attributes of the perceptual elements of three cases.

We combined the qualities of the perceptual elements and their actual performance to delineate the key hierarchy (Table 5). In the highest key hierarchy, the number of perceptual elements in Aofeng is the lowest, only four (Climate, Cell phone signal, Freedom, Naturalness); while the number in both Qishan (Plants, Landscape color, Water, Landform, Climate, Naturalness) and Lujitan (Plants, Landscape color, Water, Landform, Climate, Freedom) was six. Using repetition in at least two case categories as the extraction criterion, the study screened seven key perceptual elements based on USWs cases in Fuzhou, namely, Plants (A1), Landscape color (A3), Water (A5), Landform (A7), Climate (A8), Freedom (B2), and Naturalness (B5).

Table 5. Key hierarchy of perceptual elements in three cases.

Key Hierarchy	Aofeng	Qishan	Lujitan
High	A8 A12 B2 B5	A1 A3 A5 A7 A8 B5	A1 A3 A5 A7 A8 B2
Medium	A3 A4 B4	A4 A6 A10 B2 B4	A2 A4 A9 B5
Low	A2 A5 A9 A11 B1 B3	A2 A12 B1 B3	A12 B1 B4

3.3. Association of USWs Restorative Benefits with Key Perceptual Elements

The study used the PRS to assess the restorative benefits of the environment (Appendix C). Aofeng (4.972), Qishan (5.166), and Lujitan (4.979) all scored around five, suggesting good restorative benefits of USWs. In addition, Aofeng (5.440) had the highest scores in the being away dimension, suggesting that it is more differentiated from participants' everyday environment compared to Qishan (5.209) and Lujitan (5.170). In the dimension of fascination and compatibility, the performance of Qishan (5.234) was superior to both Aofeng (5.013) and Lujitan (4.905), indicating that it is more popular. In the complexity dimension, Qishan (4.957) was close to Lujitan (4.965) and higher than Aofeng (4.404), indicating that compared to the other two, the environment of Aofeng may be a bit cluttered.

The study used Pearson correlation analysis to explore the relationship between the key perceptual elements of USWs and its restorative benefits. According to the data in Table 6, all the key perceptual elements were significantly correlated with environmental restoration benefits ($p < 0.01$). Specifically, Climate (r = 0.325 **), Freedom (r = 0.477 **), and Naturalness (r = 0.522 **) in Aofeng showed moderate correlation. In Qishan, Plants (r = 0.481**), Landscape color (r = 0.342 **), Water (r = 0.415 **), Landform (r = 0.430 **), Climate (r = 0.392 **), and Naturalness (r = 0.363 **) showed moderate correlation. In Lujitan, Plants (r = 0.479 **), Landscape color (r = 0.464 **), Water (r = 0.517 **), Landform (r = 0.569 **), Climate (r = 0.567 **), and Freedom (r = 0.494 **) showed moderate correlations. It is noteworthy that all the correlation coefficients are positive (r > 0), so we conclude that the above seven key perceptual elements of USWs are positively driving its environmental restoration benefits.

Table 6. Pearson's correlation analysis of satisfaction and restorative benefits.

Factor	Aofeng	Qishan	Lujitan
A1		0.481 **	0.479 **
A3		0.342 **	0.464 **
A5		0.415 **	0.517 **
A7		0.430 **	0.569 **
A8	0.325 **	0.392 **	0.567 **
B2	0.477 **		0.494 **
B5	0.522 **	0.363 **	

Note: ** indicates highly significant correlation at $p < 0.01$.

4. Discussion

4.1. Key Perceptual Elements of USWs

In this study, three abandoned parks in Fuzhou were selected as the research cases to construct a perception evaluation system of USWs. Based on the results of the public perception evaluation, we selected seven key perceptual elements: Plants, Landscape color, Water, Landform, Climate, Freedom and Naturalness, combining the quality attributes of the elements and their real performance, and refined our perception of USWs as a special type of urban environment with both artificial green space and natural wilderness. This result completes our perception of USWs, which is a special type of urban environment that combines the characteristics of artificial green space and natural wilderness [85].

On the one hand, artificial green space perceptual elements such as Plants, Landscape color, Water and Climate are usually of high importance and preferred by the public. Palliwoda et al. (2021) investigated more than thirty urban parks and brownfields, and they emphasized that Plants are a key factor influencing the social, cultural, and ecological benefits of the environment [69], while plants are also positively correlated with perceptions of the ecological diversity of the environment and the complexity of the landscape, and are a valid predictor of how well the environment is experienced [86–89]. Ma et al. (2021) developed a landscape visual quality (LVQ) evaluation model, and they emphasized the importance of Landscape color as a perceptual dimension, which not only affects the aesthetic characteristics and overall presentation of urban green space landscapes [90], but also attracts the visual attention of the experiencers [91,92], as a "wow factor" that is appreciated by the public [93]. A study by Wang et al. (2021) concluded that Water is closely related to aesthetic and is an important factor in encouraging recreational activities for experiencers [94], as the public prefers the visual and auditory stimulation that come with the perception of Water [65]. Javadi et al. (2021) focused on the sustainable development of urban green spaces and generalized the relationship between Climate and social well-being [95], emphasizing that the willingness, duration, and benefits of outdoor activities of the experiencers were influenced by Climate factors such as air quality, temperature and humidity, and wind speed [96].

On the other hand, natural wilderness perception elements such as Landform, Freedom, and Naturalness also showed high criticality, and we believe that this result may indicate the desire of experiencers such as city dwellers to live in a natural environment. Nakarmi et al. (2023) surveyed public perceptions of a proposed geopark and found that wilderness features significantly influenced the visual quality of the environment, with the public showing a clear preference for undulating landforms such as rocks and cliffs [97]. Smith (2015) explored the significance of wilderness for youth [98], where Freedom has the value of enhancing the value of the recreational experience, boosting personal spirituality, and promoting the outdoors [99]. Ried et al. (2020) emphasized the recreational experience value of natural wilderness by studying nature reserves, where experiencers were able to recognize the beauty of wilderness during encounters with natural organisms, as well as their own desire to return to nature [100]. In summary, analyzing the results of the study, the diverse and natural landscapes in USWs areas reflect the aesthetic tendency of city dwellers to favor naturalization [101], and people are able to realize their own goals, get rid of the burden of responsibility, and act unrestrainedly in USWs experience [73], and this kind of transcendent experience will attract more and more city dwellers to try to get out of the city and explore the future of a nature-connected lifestyle [102].

4.2. Relationship Between Key Perceptual Elements of USWs and Its Restorative Benefits

While past studies have found that USWs have experiential value such as relieving daily stress for urban residents, our findings further suggest that USWs is the ideal place for residents to engage in restorative experiences. We also found significant positive correlations between key perceptions of USWs and its restorative properties, with perceptions of Plants, Landscape color, Water, Landform, Climate, Freedom, and Naturalness contributing to experiencers' access to the restorative benefits of USWs. This finding is not surprising as it is consistent with past research findings.

Physiological perceptual elements such as Plants, Landscape color, Water, Landform, and Climate can serve as positive predictors of environmental resilience [43,103–105]. According to Nghiem et al. (2021), increasing the biodiversity, landscape complexity, and perceived visual aesthetic quality of the environment helps experiencers to reap restorative benefits [65,106,107]. The rich landscape elements in USWs areas attract the attention of the experiencers, help people to escape from their daily lives, and satisfy them by enabling them to have a good rest [86,88]. Hoyle et al. (2017) emphasized that diverse and vibrant landscapes have greater perceptual appeal, and that aesthetically pleasing environments stimulate positive emotions and mobilize their supervisory energies to engage in perceptually restorative behaviors [93,108]. Rich vegetation communities and rare species in USWs areas are more preferred, encouraging people to increase the frequency of their visits to wilderness areas, which can help them reap restorative benefits [92]. A study by McAllister et al. (2017) concluded that environments with water features are more restorative than those without them [109]. The quality of Water in USWs areas, such as clear water quality, naturally sloping banks, and high levels of interactivity and safety, enhances the restorative benefits that people receive [110–112]. Rickard et al. (2021) found that exposure to the natural world is an effective way for people to escape from their daily stresses [113]. USWs provide spaces for diverse scenic experiences, such as flat man-made terrains and naturalized rolling landscapes, that can meet the restoration needs of different groups [114]. Sousa et al. (2024) emphasized that climate is an important determinant of public environmental perceptions [115], with factors such as light, temperature and humidity, and wind speed showing a significant relationship with environmental restoration potential [108]. USWs is effective in purifying the air and regulating the climate in a "refreshing", "comfortable" and "gentle" environment that can help experiencers obtain restorative benefits [58,116].

Similarly, there is an influence of psychological factors on environmental resilience [43,117]. Conradson et al. (2016) reviewed research on the psychological elements of environments and defined Freedom as the psychological perception of autonomy, emancipation, and transcendence [117]. The environmental features of USWs that differentiate them from everyday life have a special experiential potential to act as a refuge for the mind and help residents to self-realize, get rid of stress, regulate their moods, and increase their self-confidence [118,119]. In their study, Fisher et al. (2021) emphasized that more natural sites are more conducive to people's rejuvenation [59], whereas it is difficult for people to perceive ecological sensations in artificially maintained urban green spaces [120]. The natural successional character of USWs shapes the diversity of biomes and diverse landscape spaces, creates a link between ecological processes and people's perceptions, can facilitate the ecological perceptual experience, and can effectively enhance the restoration benefits of the environment [40,121].

5. Conclusions

In this study, we constructed a perception evaluation system of USWs, screened seventeen evaluation elements based on two types of perception dimensions, physiological and psychological. Meanwhile, we analyzed the quality of the perceptual elements based on the IPA-Kano model and extracted the key perceptual elements of USWs by combining their actual performance. In addition, we adopted the PRS to evaluate the restorative benefits of USWs. By investigating the case of USWs in Fuzhou, we identified Plants, Landscape color, Water, Landform, Climate, Freedom, and Naturalness as the key perceptual elements, and found that they contribute to the restoration benefits of the environment.

5.1. Theoretical and Practical Implications

The study focused on the urban and natural attributes of USWs, based on previous perceptual evaluation dimensions, integrating a larger number of evaluation indicators, and also focusing on perceptual elements such as Boundary, Cell phone signal, and Stimulation. In terms of methodology, the study focused on the influence of public perception and preference differences on the evaluation of the perceptual elements. According to the IPA-Kano model, bivariate correlation analyses were used to extract the correlation coefficients to express the implicit importance of the elements, and then key perceptual elements were extracted by combining this with the elements' actual performance, which is an experimental procedure that demonstrates higher reliability and efficiency. In addition, the study found that USWs has good restorative efficacy and this is associated with its key perceptual elements. Our findings suggested that physiological perceptual elements such as Plants, Landscape color, Water, Landform, and Climate, as well as psychological perceptual elements such as Freedom and Naturalness, are conducive to improving the restoration benefits of USWs. These findings can improve public awareness of USWs, which is not only conducive to the conservation and sustainable development of it, but also provides valuable evidence for the future harmonious development of humans and nature.

5.2. Limitations and Suggestions for Future Research

Several limitations of this study should be addressed. First, although perceptual evaluation facilitates the assessment of people's subjective perceptions and preferences for the environment in which they live, there is variability in landscape preferences among different groups of people, and the influencing elements include upbringing, cultural practices, and values [122]. Secondly, the IPA-Kano model used in this study has some flaws, and due to the multicollinearity of the study data, a bivariate correlation model was used instead of the commonly used partial correlation and regression analysis models to calculate the implicit importance of the elements [84]. Subsequent studies should consider using different analytical models and coefficient calculation methods to reduce the error according to the actual situation. In addition, the research object is limited to abandoned parks in the city, which makes it difficult to guide subsequent research on other types of USWs. Further expanding the scope of the study to include more diverse natural spaces in future research will not only enhance public awareness of USWs and protect precious natural experience resources, but also encourage people to pursue alternative modes of living and explore the realization of sustainable and prosperous living.

Author Contributions: Conceptualization, K.C. and X.X.; methodology, K.C.; software, K.C.; validation, K.C., X.X.; formal analysis, K.C.; investigation, K.C., Y.X., K.Z. and Y.G.; resources, X.X.; data curation, K.C., Y.X. and K.Z.; writing—original draft preparation, K.C.; writing—review and editing, X.X.; visualization, K.C.; supervision, X.X.; project administration, X.X.; funding acquisition, X.X. All authors have read and agreed to the published version of the manuscript.

Funding: This research was funded by Fujian Water Resources Department Project, grant number: KLh20015A; Fujian Water Conservancy Science and Technology Program, grant number: MSK202203.

Institutional Review Board Statement: The study was conducted in accordance with the Declaration of Helsinki, and approved by the Fujian Agriculture and Forestry University, College of Landscape Architecture and Art, Approval Code FAFUIRB202309052 Date of approval: 5 September 2023.

Informed Consent Statement: Informed consent was obtained from all subjects involved in the study.

Data Availability Statement: The data used to support the findings of this study are available from the authors upon request.

Conflicts of Interest: The authors declare no conflict of interest.

Appendix A

Table A1. Importance of perceived elements.

S/N	Aofeng		Qishan		Lujitan	
	Score	Ranking	Score	Ranking	Score	Ranking
A1	6.146	13	6.271	14	6.058	12
A2	5.044	16	5.754	16	5.731	15
A3	6.804	6	7.046	2	6.814	5
A4	6.760	8	6.764	4	6.752	7
A5	6.815	5	7.050	1	7.380	2
A6	5.349	15	6.343	12	5.835	14
A7	5.524	14	6.625	6	5.628	16
A8	7.473	1	6.446	9	7.694	1
A9	7.066	2	6.407	10	7.120	3
A10	6.287	12	6.564	8	6.169	11
A11	6.895	4	6.396	11	6.054	13
A12	7.029	3	6.832	3	6.880	4
B1	6.647	9	6.139	15	6.719	8
B2	6.804	7	6.611	7	6.777	6
B3	4.895	17	5.425	17	5.116	17
B4	6.429	11	6.304	13	6.351	10
B5	6.549	10	6.739	5	6.496	9

Appendix B

Table A2. Satisfaction with perceived elements.

S/N	Aofeng		Qishan		Lujitan	
	Score	Ranking	Score	Ranking	Score	Ranking
A1	6.636	2	6.454	2	6.926	6
A2	4.102	14	5.475	13	6.649	10
A3	5.749	7	6.186	6	6.979	3
A4	5.596	8	6.179	7	6.715	9
A5	3.495	17	6.379	4	7.198	1
A6	5.320	10	6.046	10	6.525	12
A7	6.247	5	6.593	1	6.988	2
A8	7.076	1	6.289	5	6.938	5
A9	4.920	13	5.854	12	6.769	7
A10	5.080	11	5.879	11	6.537	11
A11	5.040	12	5.311	14	4.517	16
A12	6.604	3	5.082	16	6.149	15
B1	3.546	16	4.689	17	4.231	17
B2	6.571	4	6.050	9	6.959	4
B3	3.720	15	5.304	15	6.260	14
B4	6.004	7	6.175	8	6.289	13
B5	6.051	6	6.396	3	6.731	8

Appendix C

Table A3. PRS score.

Dimension	S/N	Aofeng	Qishan	Lujitan
Being away	Q1	5.051	5.157	5.091
	Q2	5.647	5.396	5.298
	Q3	5.702	5.171	5.504
	Q4	4.996	4.804	4.769
	Q5	5.804	5.518	5.190
	Q1–5	5.440	5.209	5.170
Fascination and Compatibility	Q6	5.298	5.493	4.748
	Q7	4.415	5.089	4.335
	Q8	4.753	5.421	4.955
	Q9	5.120	5.350	5.095
	Q10	5.407	5.386	4.930
	Q11	5.313	5.450	5.236
	Q12	5.465	5.414	5.252
	Q13	4.520	5.146	5.033
	Q14	5.425	5.150	5.198
	Q15	4.185	4.711	4.620
	Q16	5.233	5.193	4.860
	Q17	5.025	5.007	4.603
	Q6–17	5.013	5.234	4.905
Complexity	Q18	4.887	5.061	5.227
	Q19	4.451	4.821	4.913
	Q20	3.949	4.539	4.517
	Q21	3.062	4.793	4.971
	Q22	5.673	5.571	5.198
	Q18–22	4.404	4.957	4.965
Restorative benefits	Q1–22	4.972	5.166	4.979

References

1. Soga, M.; Yamaura, Y.; Aikoh, T.; Shoji, Y.; Kubo, T.; Gaston, K.J. Reducing the extinction of experience: Association between urban form and recreational use of public greenspace. *Landsc. Urban Plann.* **2015**, *143*, 69–75. [CrossRef]
2. Colléony, A.; Cohen-Seffer, R.; Shwartz, A. Unpacking the causes and consequences of the extinction of experience. *Biol. Conserv.* **2020**, *251*, 108788. [CrossRef]
3. Awoyemi, A.G.; Ibáñez-Rueda, N.; Guardiola, J.; Ibáñez-Álamo, J.D. Human-nature interactions in the Afrotropics: Experiential and cognitive connections among urban residents in southern Nigeria. *Ecol. Econ.* **2024**, *218*, 108105. [CrossRef]
4. Pyle, R.M. *The Thunder Tree: Lessons from an Urban Wildland*; Houghton Mifflin: Boston, MA, USA, 1993.
5. Soga, M.; Gaston, K.J. Extinction of experience: The loss of human–nature interactions. *Front. Ecol. Environ.* **2016**, *14*, 94–101. [CrossRef]
6. Birch, J.; Rishbeth, C.; Payne, S.R. Nature doesn't judge you—How urban nature supports young people's mental health and wellbeing in a diverse UK city. *Health Place* **2020**, *62*, 102296. [CrossRef]
7. Jimenez, M.P.; DeVille, N.V.; Elliott, E.G.; Schiff, J.E.; Wilt, G.E.; Hart, J.E.; James, P. Associations between nature exposure and health: A review of the evidence. *Int. J. Environ. Res. Public Health* **2021**, *18*, 4790. [CrossRef]
8. Cameron, R.W.; Brindley, P.; Mears, M.; McEwan, K.; Ferguson, F.; Sheffield, D.; Jorgensen, A.; Riley, J.; Goodrick, J.; Ballard, L. Where the wild things are! Do urban green spaces with greater avian biodiversity promote more positive emotions in humans? *Urban Ecosyst.* **2020**, *23*, 301–317. [CrossRef]
9. Richardson, M.; Passmore, H.A.; Barbett, L.; Lumber, R.; Thomas, R.; Hunt, A. The green care code: How nature connectedness and simple activities help explain pro-nature conservation behaviours. *People Nat.* **2020**, *2*, 821–839. [CrossRef]
10. Martin, L.; White, M.P.; Hunt, A.; Richardson, M.; Pahl, S.; Burt, J. Nature contact, nature connectedness and associations with health, wellbeing and pro-environmental behaviours. *J. Environ. Psychol.* **2020**, *68*, 101389. [CrossRef]
11. Cleary, A.; Fielding, K.S.; Bell, S.L.; Murray, Z.; Roiko, A. Exploring potential mechanisms involved in the relationship between eudaimonic wellbeing and nature connection. *Landsc. Urban Plann.* **2017**, *158*, 119–128. [CrossRef]
12. Luo, Q.Y.; Bao, Y.; Wang, Z.T.; Chen, X.T. Potential recreation service efficiency of urban remnant mountain wilderness: A case study of Yunyan District of Guiyang city, China. *Ecol. Indic.* **2022**, *141*, 109081. [CrossRef]
13. Wang, X.; Wang, X. Urban wilderness based on the view of landscape architecture. *Chin. Landsc. Archit.* **2017**, *33*, 40–47. [CrossRef]

14. Kowarik, I. Urban wilderness: Supply, demand, and access. *Urban For. Urban Green.* **2018**, *29*, 336–347. [CrossRef]
15. Luo, S.; Patuano, A. Multiple ecosystem services of informal green spaces: A literature review. *Urban For. Urban Green.* **2023**, *81*, 127849. [CrossRef]
16. Wang, X. Urban interstitial space: Strategic research on the planning of wildscape in the context of living environment. *Urban Dev. Stud.* **2017**, *24*, 11–16.
17. Xiangron, W. Wilderness in the City. *Landsc. Archit.* **2019**, *26*, 4–5.
18. Kowarik, I. Working With Wilderness: A Promising Direction for Urban Green Spaces. *Landsc. Archit. Front.* **2021**, *9*, 92–104. [CrossRef]
19. Sikorska, D.; Ciężkowski, W.; Babańczyk, P.; Chormański, J.; Sikorski, P. Intended wilderness as a Nature-based Solution: Status, identification and management of urban spontaneous vegetation in cities. *Urban For. Urban Green.* **2021**, *62*, 127155. [CrossRef]
20. Drenthen, M. Rewilding in Layered Landscapes as a Challenge to Place Identity. *Environ. Values* **2018**, *27*, 405–425. [CrossRef]
21. Gammon, A.R. The Many Meanings of Rewilding: An Introduction and the Case for a Broad Conceptualisation. *Environ. Values* **2018**, *27*, 331–350. [CrossRef]
22. Yue, C.; Rui, Y. Urban wildness: Protection and creation of wild nature in urban areas. *Landsc. Archit.* **2019**, *26*, 20–24.
23. Martin, V.G.; Hill, M. Urban Wildness—A More Correct Term Than" Urban Wilderness". *Landsc. Archit. Front.* **2021**, *9*, 80–92. [CrossRef]
24. Fayet, C.M.J.; Reilly, K.H.; Van Ham, C.; Verburg, P.H. What is the future of abandoned agricultural lands? A systematic review of alternative trajectories in Europe. *Land Use Policy* **2022**, *112*, 105833. [CrossRef]
25. Ali, L.; Haase, A.; Heiland, S. Gentrification through green regeneration? Analyzing the interaction between inner-city green space development and neighborhood change in the context of regrowth: The case of Lene-Voigt-Park in Leipzig, Eastern Germany. *Land* **2020**, *9*, 24. [CrossRef]
26. Gao, Z.; Song, K.; Pan, Y.; Malkinson, D.; Zhang, X.; Jia, B.; Xia, T.; Guo, X.; Liang, H.; Huang, S.; et al. Drivers of spontaneous plant richness patterns in urban green space within a biodiversity hotspot. *Urban For. Urban Green.* **2021**, *61*, 127098. [CrossRef]
27. Zefferman, E.P.; McKinney, M.L.; Cianciolo, T.; Fritz, B.I. Knoxville's urban wilderness: Moving toward sustainable multifunctional management. *Urban For. Urban Green.* **2018**, *29*, 357–366. [CrossRef]
28. Yuan, J.; You, F.; Hou, C.; Ou, H.; Yin, Y. Reconstruction of urban wilderness habitats based on vegetation rewilding: Taking wildflower meadows as an example. *Landsc. Arch. Front.* **2021**, *9*, 26–39. [CrossRef]
29. Mahmoudi Farahani, L.; Maller, C. Investigating the benefits of 'leftover' places: Residents' use and perceptions of an informal greenspace in Melbourne. *Urban For. Urban Green.* **2019**, *41*, 292–302. [CrossRef]
30. Ulrich, R.S. Aesthetic and affective response to natural environment. In *Behavior and the Natural Environment*; Springer: Berlin/Heidelberg, Germany, 1983; pp. 85–125.
31. Kaplan, R.; Kaplan, S.; Brown, T. Environmental Preference: A Comparison of Four Domains of Predictors. *Environ. Behav.* **1989**, *21*, 509–530. [CrossRef]
32. Ulrich, R.S.; Simons, R.F.; Losito, B.D.; Fiorito, E.; Miles, M.A.; Zelson, M. Stress recovery during exposure to natural and urban environments. *J. Environ. Psychol.* **1991**, *11*, 201–230. [CrossRef]
33. Kaplan, S. The restorative benefits of nature: Toward an integrative framework. *J. Environ. Psychol.* **1995**, *15*, 169–182. [CrossRef]
34. Lev, E.; Kahn, P.H., Jr.; Chen, H.; Esperum, G. Relatively wild urban parks can promote human resilience and flourishing: A case study of Discovery Park, Seattle, Washington. *Front. Sustain. Cities* **2020**, *2*, 2. [CrossRef]
35. Lampinen, J.; Tuomi, M.; Fischer, L.K.; Neuenkamp, L.; Alday, J.G.; Bucharova, A.; Cancellieri, L.; Casado-Arzuaga, I.; Ceplova, N.; Cervero, L.; et al. Acceptance of near-natural greenspace management relates to ecological and socio-cultural assigned values among European urbanites. *Basic Appl. Ecol.* **2021**, *50*, 119–131. [CrossRef]
36. Yakınlar, N.; Akpınar, A. How perceived sensory dimensions of urban green spaces are associated with adults' perceived restoration, stress, and mental health? *Urban For. Urban Green.* **2022**, *72*, 127572. [CrossRef]
37. Janeczko, E.; Bielinis, E.; Wójcik, R.; Woźnicka, M.; Kędziora, W.; Łukowski, A.; Elsadek, M.; Szyc, K.; Janeczko, K. When urban environment is restorative: The effect of walking in suburbs and forests on psychological and physiological relaxation of young Polish adults. *Forests* **2020**, *11*, 591. [CrossRef]
38. Dobson, J.; Birch, J.; Brindley, P.; Henneberry, J.; McEwan, K.; Mears, M.; Richardson, M.; Jorgensen, A. The magic of the mundane: The vulnerable web of connections between urban nature and wellbeing. *Cities* **2021**, *108*, 102989. [CrossRef]
39. Kim, M.; Rupprecht, C.D.D.; Furuya, K. Residents' Perception of Informal Green SpaceA Case Study of Ichikawa City, Japan. *Land* **2018**, *7*, 102. [CrossRef]
40. Li, J.; Wang, Z.; Wu, S.; Xie, M. Progress of Research on Ecological and Cultural Values of Urban Wilderness. *Landsc. Archit.* **2024**, *31*, 89–96. [CrossRef]
41. Van den Berg, A.E. Individual Differences in the Aesthetic Evaluation of Natural Landscapes. 1999. Available online: https://research.rug.nl/en/publications/individual-differences-in-the-aesthetic-evaluation-of-natural-lan (accessed on 10 December 2023).
42. Liu, L.; Qu, H.; Ma, Y.; Wang, K.; Qu, H. Restorative benefits of urban green space: Physiological, psychological restoration and eye movement analysis. *J. Environ. Manag.* **2022**, *301*, 113930. [CrossRef] [PubMed]
43. Li, X.; Zhang, X.; Jia, T. Humanization of nature: Testing the influences of urban park characteristics and psychological factors on collegers' perceived restoration. *Urban For. Urban Green.* **2023**, *79*, 127806. [CrossRef]

44. Twedt, E.; Rainey, R.M.; Proffitt, D.R. Beyond nature: The roles of visual appeal and individual differences in perceived restorative potential. *J. Environ. Psychol.* **2019**, *65*, 101322. [CrossRef]
45. Menzel, C.; Reese, G. Seeing nature from low to high levels: Mechanisms underlying the restorative effects of viewing nature images. *J. Environ. Psychol.* **2022**, *81*, 101804. [CrossRef]
46. Wei, Y.; Hou, Y. Forest Visitors' Multisensory Perception and Restoration Effects: A Study of China's National Forest Parks by Introducing Generative Large Language Model. *Forests* **2023**, *14*, 2412. [CrossRef]
47. Sona, B.; Dietl, E.; Steidle, A. Recovery in sensory-enriched break environments: Integrating vision, sound and scent into simulated indoor and outdoor environments. *Ergo* **2019**, *62*, 521–536. [CrossRef]
48. Chiang, Y.-J. Multisensory Stimuli, Restorative Effect, and Satisfaction of Visits to Forest Recreation Destinations: A Case Study of the Jhihben National Forest Recreation Area in Taiwan. *Int. J. Environ. Res. Public Health* **2023**, *20*, 6768. [CrossRef] [PubMed]
49. Foellmer, J.; Kistemann, T.; Anthonj, C. Academic Greenspace and Well-Being—Can Campus Landscape be Therapeutic? Evidence from a German University. *Wellbeing Space Soc.* **2021**, *2*, 100003. [CrossRef]
50. DeLauer, V.; McGill-O'Rourke, A.; Hayes, T.; Haluch, A.; Gordon, C.; Crane, J.; Kossakowski, D.; Dillon, C.; Thibeault, N.; Schofield, D. The Impact of Natural Environments and Biophilic Design as Supportive and Nurturing Spaces on a Residential College Campus. *Cogent Soc. Sci.* **2022**, *8*, 2000570. [CrossRef]
51. Pichlerova, M.; Vybost'ok, J.; Onkal, D.; Lamatungga, K.E.; Tamatam, D.; Marcinekova, L.; Pichler, V. Increased appreciation of forests and their restorative effects during the COVID-19 pandemic. *Ambio* **2023**, *52*, 647–664. [CrossRef] [PubMed]
52. Gong, C.; Yang, R.; Li, S. The role of urban green space in promoting health and well-being is related to nature connectedness and biodiversity: Evidence from a two-factor mixed-design experiment. *Landsc. Urban Plann.* **2024**, *245*, 105020. [CrossRef]
53. Chen, K.-S.; Chen, H.-T. Applying Importance–Performance Analysis With Simple Regression Model and Priority Indices to Assess Hotels' Service Performance. *J. Test. Eval.* **2014**, *42*, 455–466. [CrossRef]
54. Martilla, J.A.; James, J.C. Importance-performance analysis. *J. Mark.* **1977**, *41*, 77–79. [CrossRef]
55. Berger, C.; Blauth, R.; Boger, D. Kano's methods for understanding customer-defined quality. *Cent. Qual. Manag. J.* **1993**, *2*, 3–36. Available online: https://sid.ir/paper/601079/en (accessed on 17 June 2023).
56. Matzler, K.; Sauerwein, E.; Heischmidt, K. Importance-performance analysis revisited: The role of the factor structure of customer satisfaction. *Serv. Ind. J.* **2003**, *23*, 112–129. [CrossRef]
57. Anderson, E.C.; Minor, E.S. Assessing social and biophysical drivers of spontaneous plant diversity and structure in urban vacant lots. *Sci. Total Environ.* **2019**, *653*, 1272–1281. [CrossRef]
58. Sikorski, P.; Gawryszewska, B.; Sikorska, D.; Chormanski, J.; Schwerk, A.; Jojczyk, A.; Ciezkowski, W.; Archicinski, P.; Lepkowski, M.; Dymitryszyn, I.; et al. The value of doing nothing—How informal green spaces can provide comparable ecosystem services to cultivated urban parks. *Ecosyst. Serv.* **2021**, *50*, 101339. [CrossRef]
59. Fisher, J.C.; Irvine, K.N.; Bicknell, J.E.; Hayes, W.M.; Fernandes, D.; Mistry, J.; Davies, Z.G. Perceived biodiversity, sound, naturalness and safety enhance the restorative quality and wellbeing benefits of green and blue space in a neotropical city. *Sci. Total Environ.* **2021**, *755*, 143095. [CrossRef]
60. Wei, F.; Chen, L.; Zan, P. Between Industrial Remains and the Green—Research on Public Perception of Rewilded Postindustrial Landscape. *Chin. Landsc. Archit.* **2022**, *38*, 36–41. [CrossRef]
61. Wang, R.; Zhao, J. Effects of evergreen trees on landscape preference and perceived restorativeness across seasons. *Landsc. Res.* **2020**, *45*, 649–661. [CrossRef]
62. Syed Othman Thani, S.K.; Hussein, H.; Siew Cheok, N. Landscape attributes as stimuli for cognitive restoration in the outdoor environment: A systematic review and thematic analysis. *J. Leis. Res.* **2024**, *55*, 441–461. [CrossRef]
63. Eldridge, D.J.; Cui, H.; Ding, J.; Berdugo, M.; Sáez-Sandino, T.; Duran, J.; Gaitan, J.; Blanco-Pastor, J.L.; Rodríguez, A.; Plaza, C.; et al. Urban greenspaces and nearby natural areas support similar levels of soil ecosystem services. *npj Urban Sustain.* **2024**, *4*, 15. [CrossRef]
64. Zhang, T.; Liu, J.; Li, H. Restorative effects of multi-sensory perception in urban green space: A case study of urban park in Guangzhou, China. *Int. J. Environ. Res. Public Health* **2019**, *16*, 4943. [CrossRef]
65. Deng, L.; Luo, H.; Ma, J.; Huang, Z.; Sun, L.-X.; Jiang, M.-Y.; Zhu, C.-Y.; Li, X. Effects of integration between visual stimuli and auditory stimuli on restorative potential and aesthetic preference in urban green spaces. *Urban For. Urban Green.* **2020**, *53*, 126702. [CrossRef]
66. Gillis, K. Nature-based restorative environments are needed now more than ever. *Cities Health* **2021**, *5*, S237–S240. [CrossRef]
67. Ríos-Rodríguez, M.L.; Rosales, C.; Lorenzo, M.; Muinos, G.; Hernández, B. Influence of perceived environmental quality on the perceived restorativeness of public spaces. *Front. Psychol.* **2021**, *12*, 644763. [CrossRef]
68. Manyani, A.; Shackleton, C.M.; Cocks, M.L. Attitudes and preferences towards elements of formal and informal public green spaces in two South African towns. *Landsc. Urban Plann.* **2021**, *214*, 104147. [CrossRef]
69. Palliwoda, J.; Priess, J.A. What do people value in urban green? Linking characteristics of urban green spaces to users' perceptions of nature benefits, disturbances, and disservices. *Ecol. Soc.* **2021**, *26*, 28. [CrossRef]
70. Peterson, B.A.; Fefer, J.P.; Sharp, R.L.; Brunson, M.M.; Skibins, J.C. To connect or not to connect: Visitor preferences for Wi-Fi and cellular network service at a national park. *Soc. Sci. Humanit. Open* **2021**, *4*, 100179. [CrossRef]
71. Ólafsdóttir, R.; Sæþórsdóttir, A.D. Public perception of wilderness in Iceland. *Land* **2020**, *9*, 99. [CrossRef]

72. Li, Y.; Zhang, J.; Jiang, B.; Li, H.; Zhao, B. Do all types of restorative environments in the urban park provide the same level of benefits for young adults? A field experiment in Nanjing, China. *Forests* **2023**, *14*, 1400. [CrossRef]
73. Dorman, J.D. Unconfined Wilderness Experiences: What Is Important to Feeling Unconfined While Visiting the Selway-Bitterroot Wilderness? 2019. Available online: https://scholarworks.umt.edu/etd/11341/ (accessed on 25 July 2023).
74. Tucker, A.R.; DeMille, S.; Newman, T.J.; Polachi Atanasova, C.; Bryan, P.; Keefe, M.; Smitherman, L. How adolescents view the role of the wilderness in wilderness therapy: "I am in the middle of nowhere and that is okay". *Child. Youth Serv. Rev.* **2023**, *153*, 107045. [CrossRef]
75. Johansson, M.; Hartig, T.; Frank, J.; Flykt, A. Wildlife and public perceptions of opportunities for psychological restoration in local natural settings. *People Nat.* **2024**, *6*, 800–817. [CrossRef]
76. Zhu, X.; Zhang, Y.; Luo, Y.Y.; Zhao, W. Natural or artificial? Exploring perceived restoration potential of community parks in Winter city. *Urban For. Urban Green.* **2023**, *79*, 127808. [CrossRef]
77. Menardo, E.; Brondino, M.; Hall, R.; Pasini, M. Restorativeness in natural and urban environments: A meta-analysis. *Psychol. Rep.* **2021**, *124*, 417–437. [CrossRef]
78. Du, Y.; Chen, J. Research on quality types classification of design elements of subway entrances in winter cities based on IPA-KANO Method. *Archit. J. S* **2018**, *1*, 144–148.
79. Ye, L.; Zhang, F.; Wu, J. Developing the restoration environment scale. *China J. Health Psychol.* **2010**, *12*, 15151518.
80. Wang, X.; Rodiek, S.; Wu, C.; Chen, Y.; Li, Y. Stress recovery and restorative effects of viewing different urban park scenes in Shanghai, China. *Urban For. Urban Green.* **2016**, *15*, 112–122. [CrossRef]
81. Li, H.; Du, X.; Ma, H.; Wang, Z.; Li, Y.; Wu, J. The effect of virtual-reality-based restorative environments on creativity. *Int. J. Environ. Res. Public Health* **2022**, *19*, 12083. [CrossRef] [PubMed]
82. Hu, X.; Xu, H. A new remote sensing index for assessing the spatial heterogeneity in urban ecological quality: A case from Fuzhou City, China. *Ecol. Indic.* **2018**, *89*, 11–21. [CrossRef]
83. Cai, Y.; Chen, Y.; Tong, C. Spatiotemporal evolution of urban green space and its impact on the urban thermal environment based on remote sensing data: A case study of Fuzhou City, China. *Urban For. Urban Green.* **2019**, *41*, 333–343. [CrossRef]
84. Lai, I.K.W.; Hitchcock, M. Importance–performance analysis in tourism: A framework for researchers. *Tour. Manag.* **2015**, *48*, 242–267. [CrossRef]
85. van der Jagt, S. Exploring How Urban Wilderness Connects Urban Residents with Nature. 2022. Available online: https://edepot.wur.nl/580395 (accessed on 5 March 2024).
86. Wang, R.; Zhao, J.; Meitner, M.J.; Hu, Y.; Xu, X. Characteristics of urban green spaces in relation to aesthetic preference and stress recovery. *Urban For. Urban Green.* **2019**, *41*, 6–13. [CrossRef]
87. Liu, Q.; Zhu, Z.; Zeng, X.; Zhuo, Z.; Ye, B.; Fang, L.; Huang, Q.; Lai, P. The impact of landscape complexity on preference ratings and eye fixation of various urban green space settings. *Urban For. Urban Green.* **2021**, *66*, 127411. [CrossRef]
88. Phillips, D.; Lindquist, M. Just weeds? Comparing assessed and perceived biodiversity of urban spontaneous vegetation in informal greenspaces in the context of two American legacy cities. *Urban For. Urban Green.* **2021**, *62*, 127151. [CrossRef]
89. Wu, Y.; Zhuo, Z.; Liu, Q.; Yu, K.; Huang, Q.; Liu, J. The Relationships between Perceived Design Intensity, Preference, Restorativeness and Eye Movements in Designed Urban Green Space. *Int. J. Environ. Res. Public Health* **2021**, *18*, 10944. [CrossRef]
90. Ma, B.; Hauer, R.J.; Xu, C.; Li, W. Visualizing evaluation model of human perceptions and characteristic indicators of landscape visual quality in urban green spaces by using nomograms. *Urban For. Urban Green.* **2021**, *65*, 127314. [CrossRef]
91. Gao, Y.; Zhang, T.; Zhang, W.; Meng, H.; Zhang, Z. Research on visual behavior characteristics and cognitive evaluation of different types of forest landscape spaces. *Urban For. Urban Green.* **2020**, *54*, 126788. [CrossRef]
92. Tomitaka, M.; Uchihara, S.; Goto, A.; Sasaki, T. Species richness and flower color diversity determine aesthetic preferences of natural-park and urban-park visitors for plant communities. *Environ. Sustain. Indic.* **2021**, *11*, 100130. [CrossRef]
93. Hoyle, H.; Hitchmough, J.; Jorgensen, A. All about the 'wow factor'? The relationships between aesthetics, restorative effect and perceived biodiversity in designed urban planting. *Landsc. Urban Plann.* **2017**, *164*, 109–123. [CrossRef]
94. Wang, R.; Jiang, W.; Lu, T. Landscape characteristics of university campus in relation to aesthetic quality and recreational preference. *Urban For. Urban Green.* **2021**, *66*, 127389. [CrossRef]
95. Javadi, R.; Nasrollahi, N. Urban green space and health: The role of thermal comfort on the health benefits from the urban green space; a review study. *Build. Environ.* **2021**, *202*, 108039. [CrossRef]
96. Niu, J.; Xiong, J.; Qin, H.; Hu, J.; Deng, J.; Han, G.; Yan, J. Influence of thermal comfort of green spaces on physical activity: Empirical study in an urban park in Chongqing, China. *Build. Environ.* **2022**, *219*, 109168. [CrossRef]
97. Nakarmi, G.; Strager, M.P.; Yuill, C.; Moreira, J.C.; Burns, R.C.; Butler, P. Assessing Public Preferences of Landscape and Landscape Attributes: A Case Study of the Proposed Appalachian Geopark Project in West Virginia, USA. *Geoheritage* **2023**, *15*, 85. [CrossRef]
98. Smith, K.; Kirby, M. Wilderness 2.0: What does wilderness mean to the Millennials? *J. Environ. Stud. Sci.* **2015**, *5*, 262–271. [CrossRef]
99. Boyes, M.; Mackenzie, S.H. Concepts of the wilderness experience and adventure mountaineering tourism. In *Mountaineering Tourism*; Routledge: London, UK, 2015; pp. 92–110.
100. Ried, A.; Monteagudo, M.J.; Benavides, P.; Le Bon, A.; Carmody, S.; Santos, R. Key aspects of leisure experiences in protected wilderness areas: Notions of nature, senses of place and perceived benefits. *Sustainability* **2020**, *12*, 3211. [CrossRef]

101. Van den Berg, A.E.; Hartig, T.; Staats, H. Preference for nature in urbanized societies: Stress, restoration, and the pursuit of sustainability. *J. Soc. Iss.* **2007**, *63*, 79–96. [CrossRef]
102. Venter, Z.S.; Barton, D.N.; Gundersen, V.; Figari, H.; Nowell, M.S. Back to nature: Norwegians sustain increased recreational use of urban green space months after the COVID-19 outbreak. *Landsc. Urban Plann.* **2021**, *214*, 104175. [CrossRef]
103. Li, C.; Yuan, Y.; Sun, C.; Sun, M. The perceived restorative quality of viewing various types of urban and rural scenes: Based on psychological and physiological responses. *Sustainability* **2022**, *14*, 3799. [CrossRef]
104. Zhu, Z.; Hassan, A.; Wang, W.; Chen, Q. Relationship between PSD of park green space and attention restoration in dense urban areas. *Brain Sci.* **2022**, *12*, 721. [CrossRef]
105. Dzhambov, A.M.; Lercher, P.; Browning, M.H.E.M.; Stoyanov, D.; Petrova, N.; Novakov, S.; Dimitrova, D.D. Does greenery experienced indoors and outdoors provide an escape and support mental health during the COVID-19 quarantine? *Environ. Res.* **2021**, *196*, 110420. [CrossRef]
106. Nghiem, T.; Wong, K.; Jeevanandam, L.; Chang, C.; Tan, L.; Goh, Y.; Carrasco, L. Biodiverse urban forests, happy people: Experimental evidence linking perceived biodiversity, restoration, and emotional wellbeing. *Urban For. Urban Green.* **2021**, *59*, 127030. [CrossRef]
107. Kang, Y.; Kim, E.J. Differences of restorative effects while viewing urban landscapes and green landscapes. *Sustainability* **2019**, *11*, 2129. [CrossRef]
108. Lu, S.; Oh, W.; Ooka, R.; Wang, L. Effects of environmental features in small public urban green spaces on older adults' mental restoration: Evidence from Tokyo. *Int. J. Environ. Res. Public Health* **2022**, *19*, 5477. [CrossRef]
109. McAllister, E.; Bhullar, N.; Schutte, N.S. Into the woods or a stroll in the park: How virtual contact with nature impacts positive and negative affect. *Int. J. Environ. Res. Public Health* **2017**, *14*, 786. [CrossRef]
110. Zhao, J.; Xu, W.; Ye, L. Effects of auditory-visual combinations on perceived restorative potential of urban green space. *Appl. Acoust.* **2018**, *141*, 169–177. [CrossRef]
111. Liu, F.; Liu, P.; Kang, J.; Meng, Q.; Wu, Y.; Yang, D. Relationships between landscape characteristics and the restorative quality of soundscapes in urban blue spaces. *Appl. Acoust.* **2022**, *189*, 108600. [CrossRef]
112. Poulsen, M.N.; Nordberg, C.M.; Fiedler, A.; DeWalle, J.; Mercer, D.; Schwartz, B.S. Factors associated with visiting freshwater blue space: The role of restoration and relations with mental health and well-being. *Landsc. Urban Plann.* **2022**, *217*, 104282. [CrossRef]
113. Rickard, S.C.; White, M.P. Barefoot walking, nature connectedness and psychological restoration: The importance of stimulating the sense of touch for feeling closer to the natural world. *Landsc. Res.* **2021**, *46*, 975–991. [CrossRef]
114. Choi, S.; Kim, I. Role of restorative natural environments in predicting hikers' pro-environmental behavior in a nature trail context. *J. Travel Tour. Mark.* **2024**, *41*, 596–613. [CrossRef]
115. Sousa-Silva, R.; Zanocco, C. Assessing public attitudes towards urban green spaces as a heat adaptation strategy: Insights from Germany. *Landsc. Urban Plann.* **2024**, *245*, 105013. [CrossRef]
116. Vieira, J.; Matos, P.; Mexia, T.; Silva, P.; Lopes, N.; Freitas, C.; Correia, O.; Santos-Reis, M.; Branquinho, C.; Pinho, P. Green spaces are not all the same for the provision of air purification and climate regulation services: The case of urban parks. *Environ. Res.* **2018**, *160*, 306–313. [CrossRef]
117. Conradson, D. Freedom, space and perspective: Moving encounters with other ecologies. In *Emotional Geographies*; Routledge: London, UK, 2016; pp. 103–116.
118. Hinners, S.J.; Rose, J.; Choi, D.-A.; Park, K. Geographically evaluating urban-wildland juxtapositions across 36 urban areas in the United States. *Geogr. Sustain.* **2022**, *3*, 139–151. [CrossRef]
119. Wang, Y.; Luo, F.; Gazal, K.A.; Wen, Y.; Lei, H.; Xiao, Z. Exploring the Impact of Psychological Accessibility on the Restorative Perception in Urban Forests: A Case Study of Yuelu Mountain, Central China. *Forests* **2023**, *14*, 721. [CrossRef]
120. Wakulich, K.A. Connecting People to Community and Nature through Ecological Sense of Place: Case Study the Missouri Flat Creek Restoration. Ph.D. Thesis, Washington State University, Pullman, WA, USA, 2017.
121. Shen, H.; He, X.; He, J.; Li, D.; Liang, M.; Xie, X. Back to the Village: Assessing the Effects of Naturalness, Landscape Types, and Landscape Elements on the Restorative Potential of Rural Landscapes. *Land* **2024**, *13*, 910. [CrossRef]
122. Liu, Q.; Wu, Y.; Xiao, Y.; Fu, W.; Zhuo, Z.; van den Bosch, C.C.K.; Huang, Q.; Lan, S. More meaningful, more restorative? Linking local landscape characteristics and place attachment to restorative perceptions of urban park visitors. *Landsc. Urban Plann.* **2020**, *197*, 103763. [CrossRef]

sustainability

MDPI

Article

Eco-Physiological Vulnerability of *Quararibea funebris* in Peri-Urban Landscapes: Integrating Gender and Nature-Based Solutions in the Central Valleys of Oaxaca, Mexico

Yolanda Donají Ortiz-Hernández [1,*], Marco Aurelio Acevedo-Ortiz [2,*], Gema Lugo-Espinosa [2], Fernando Elí Ortiz-Hernández [3], Edgar García-Sánchez [2] and Salatiel Velasco-Pérez [2]

1 Instituto Politécnico Nacional, CIIDIR Oaxaca, Santa Cruz Xoxocotlán 71230, Oaxaca, Mexico
2 Secretaría de Ciencia, Humanidades, Tecnología e Innovación (SECIHTI)—Instituto Politécnico Nacional, CIIDIR Oaxaca, Santa Cruz Xoxocotlán 71230, Oaxaca, Mexico; glugoe@ipn.mx (G.L.-E.); egarcias@ipn.mx (E.G.-S.); svelascop2500@alumno.ipn.mx (S.V.-P.)
3 Instituto Politécnico Nacional, ESIME Culhuacán, Coyoacán, Ciudad de Mexico 04440, Mexico; fortizh@ipn.mx
* Correspondence: yortiz@ipn.mx (Y.D.O.-H.); macevedoo@ipn.mx (M.A.A.-O.); Tel.: +52-9515170610 (ext. 82723) (Y.D.O.-H.); +52-9511726274 (M.A.A.-O.)

Abstract

Nature-based Solutions (NbS) are essential for peri-urban resilience; however, a critical research gap exists regarding the lack of species-specific eco-physiological validation for interventions within complex biocultural systems. This study addresses this gap by assessing the vulnerability of *Quararibea funebris*, a shade-tolerant tree and biocultural keystone for the tejate economy in Oaxaca, Mexico, currently caught in an anthropogenic ecological trap. A mixed-methods approach was employed, integrating a geospatial analysis of land-use change (1992–2021), microclimatic monitoring, and ethnographic assessment of gendered management. Results reveal the loss of 1552 ha of forest buffer, which has degraded the thermal niche below the species optimum. Urban specimens are subjected to a Daily Light Integral exceeding 38 mol m^{-2} d^{-1}, triggering biometric stunting and oxidative stress. Furthermore, given that seed recalcitrance limits ex situ conservation, the species' persistence relies strictly on a domestic monopoly of irrigation managed by women, who effectively subsidize the environmental deficit. The study concludes that the current backyard conservation model has hit its ecological ceiling; sustainability requires a transition toward landscape-scale NbS—specifically biocultural corridors governed by local female knowledge—to restore the multi-strata canopy required to regulate the species' eco-physiological limits.

Keywords: biocultural conservation; peri-urban agriculture; tejate; ecological trap; gender environmental subsidy; daily light integral; shade tolerance

1. Introduction

The current context of climate crisis and biodiversity loss has driven the global emergence of Nature-based Solutions (NbS) as an essential paradigm for resource management [1–3]. These strategies are fundamental for ensuring critical ecosystem services such as food security and human health [4–6]; however, their implementation in urban and peri-urban environments often faces the challenge of reconciling biological conservation with local development dynamics [1,3,5–7].

A paradigmatic example of this tension occurs in the Central Valleys of Oaxaca, Mexico, the home of *Quararibea funebris*. This tree, native to the tropical rainforests of Mesoamerica [8,9], is not only a symbol of identity in the municipality of San Andrés Huayapam, where it is locally known as "Rosital", but also holds immense biocultural value as the source of the "rosita de cacao". This flower is a fundamental ingredient of "tejate", an ancestral beverage made with maize dough, cacao (*Theobroma cacao*), "pixtle" (toasted seed of *Pouteria sapota*), and oak ash (*Quercus* sp.) [8,10,11]. Beyond its cultural importance, recent investigations have validated the nutraceutical potential of its flowers, which are rich in bioactive compounds such as salicylic acid, kaempferol-3-O-glucoside, and linoleic acid, conferring significant antioxidant and antimicrobial properties against human pathogens [12].

Ecologically, *Q. funebris* is native to the mid-canopy stratum of tropical rainforests [8,13]. In this natural habitat, the tree functions under a protective vegetative ceiling that filters direct solar radiation and maintains high humidity. Consequently, its photosynthetic machinery is evolutionarily adapted to low-light environments, when exposed to full sun conditions [14–16] typical of urban heat islands, the plant creates an energy imbalance. Instead of fueling biomass accumulation, excess energy triggers oxidative stress, leading to photo-inhibition and tissue necrosis. This physiological constraint forces a trade-off: the tree reallocates resources from vertical growth to cellular survival mechanisms, often resulting in biometric stunting [17].

However, this system faces a critical socio-ecological paradox: the commercial success of "tejate" drives gastronomic tourism and accelerated urbanization. Ironically, these very forces degrade the microclimate and habitat of the tree that sustains such an economy [10,11,18], transforming the peri-urban landscape into an ecological trap [19–21] where the species persists due to its cultural value, but under environmental conditions that compromise its survival. Women producers (tejateras), who act as custodians of ecological knowledge and germplasm through intensive backyard management, directly face this vulnerability, which puts both the biological resource and their financial autonomy at risk [10,22–24].

This study addresses a critical information gap regarding these dynamics, as NbS interventions often lack validation of the specific eco-physiological limits of species in altered environments [5–7]. In the case of *Q. funebris*, it remains unclear how urban abiotic stressors—specifically irradiance and ambient humidity—compromise the physiology of this tropical forest species isolated within heat islands [25,26]. To address this gap, our research is guided by three fundamental questions: (1) how has the peri-urban landscape structure changed and fragmented over the last three decades; (2) what are the specific physiological stress indicators on *Q. funebris* within this altered environment; and (3) how do traditional gendered management practices respond to these biological constraints?

Based on the theoretical framework of urban ecology [19,20,27,28], our central hypothesis posits that the peri-urban expansion of San Andrés Huayapam acts as an ecological trap for *Q. funebris*. Specifically, we argue that while anthropogenic maintenance (irrigation) acts as a positive cue promoting species persistence, the degradation of the microclimatic niche exceeds the species physiological plasticity, compromising its vegetative structure and reproductive viability.

Given that *Q. funebris* seeds are recalcitrant and lose viability rapidly [29], traditional ex situ conservation is unfeasible, making it essential to understand how the empirical knowledge of women producers (tejateras) compensates for these biological deficiencies. In this context, analyzing the role of gender is a methodological requirement to assess eco-physiological vulnerability, as the intensive care provided by the tejateras effectively subsidizes the environmental deficit, potentially masking the true extent of abiotic stress

https://doi.org/10.3390/su18031630

through supplementary irrigation. Therefore, the objective of this work is to assess the eco-physiological vulnerability of *Q. funebris* through a multi-scale diagnosis that integrates geospatial analysis of land-use changes, monitoring of physiological stress, and an assessment of gender-based management. This comprehensive approach aims to establish the scientific and social foundations required to transition toward landscape-scale NbS, such as biocultural corridors, capable of restoring ecosystem functionality.

To articulate this comprehensive diagnosis, the subsequent sections follow an integrated narrative of our findings and methodology. Section 2 describes the mixed-methods and geospatial approach used for this territorial analysis, Section 3 presents results regarding landscape transformation, the physiological stress documented in the *Q. funebris* trees, and the gender dynamics that sustain the current management system, while Section 4 discusses the main results and Section 5 synthesizes the conclusions and outlines the proposed NbS to ensure the future of this biocultural heritage.

2. Materials and Methods

2.1. Geographic Context and Hydrological Importance

The research focused on San Andrés Huayapam, a Zapotec community located in the Central Valleys region of Oaxaca [30]. Huayapam is known as the "cradle of tejate", and its identity is indissolubly linked to this ancestral beverage [31,32]. Geographically, the municipality is located at coordinates 96°40′ West longitude and 17°06′ North latitude, at an altitude of 1710 masl. The climate in the urban area is semi-warm subhumid with summer rains, characterized by a mean annual precipitation of 687.6 mm and a mean annual temperature of 25.9 °C, with fluctuations ranging from 0 °C (November to December) to 39 °C (April to July) [33–35].

Historically, this territory held critical importance due to its environmental suitability for forest conservation and its role in regional hydrological regulation. The municipality is situated within a strategic recharge section of the Río Atoyac RH-20-A hydrological region. The "Río Grande", which originates in the spring zone (at 2750 masl), has traditionally represented a primary source of potable water for both the local community and the broader Oaxaca City metropolitan area [30]. However, as analyzed in this study, landcover transformation challenges this function. Furthermore, the land tenure regime is predominantly communal [33–35], which acts as a determining factor for the collective management and governance of these biocultural resources.

2.2. The Rosital Tree: Study System and Biological Characteristics

This member of the Malvaceae family, native to the tropical rainforests of Mesoamerica, serves as the biocultural keystone of San Andrés Huayapam [8,9,36]. The *Q. funebris*, is locally known as "Rosital", while its fragrant white flowers, which are edible and possess antioxidant and antimicrobial properties [12], are called "Rosita de cacao" or "Cacahuaxochitl". It is essential to clarify that, despite these common names, the species is not botanically related to the cacao bean (*Theobroma cacao*); however, they converge in the ancestral "tejate" beverage [10,32].

While *Q. funebris* is unusual in other regions of Oaxaca, it was adopted and protected in Huayapam (Figure 1), becoming an emblematic element of the community [36]. In Huayapam, the "rositales" are cultivated as ornamental trees within the domestic courtyards of the urban area (Figure 1), comprising a population that includes both young specimens and mature individuals older than 50 years. The biometric inventory conducted within the urban matrix of San Andrés Huayapam (Figure 1) established the current structural status of the *Q. funebris* population. To assess the environmental pressures acting on these individuals, an eco-physiological monitoring protocol was implemented. Due to the strict

 https://doi.org/10.3390/su18031630

regime of private land tenure (domestic backyards), which limits unrestricted access for continuous instrumentation, a "sentinel tree" approach was adopted. This representative specimen was selected based on its spatial isolation from architectural or vegetative obstructions, ensuring free exposure to solar radiation without shading interference.

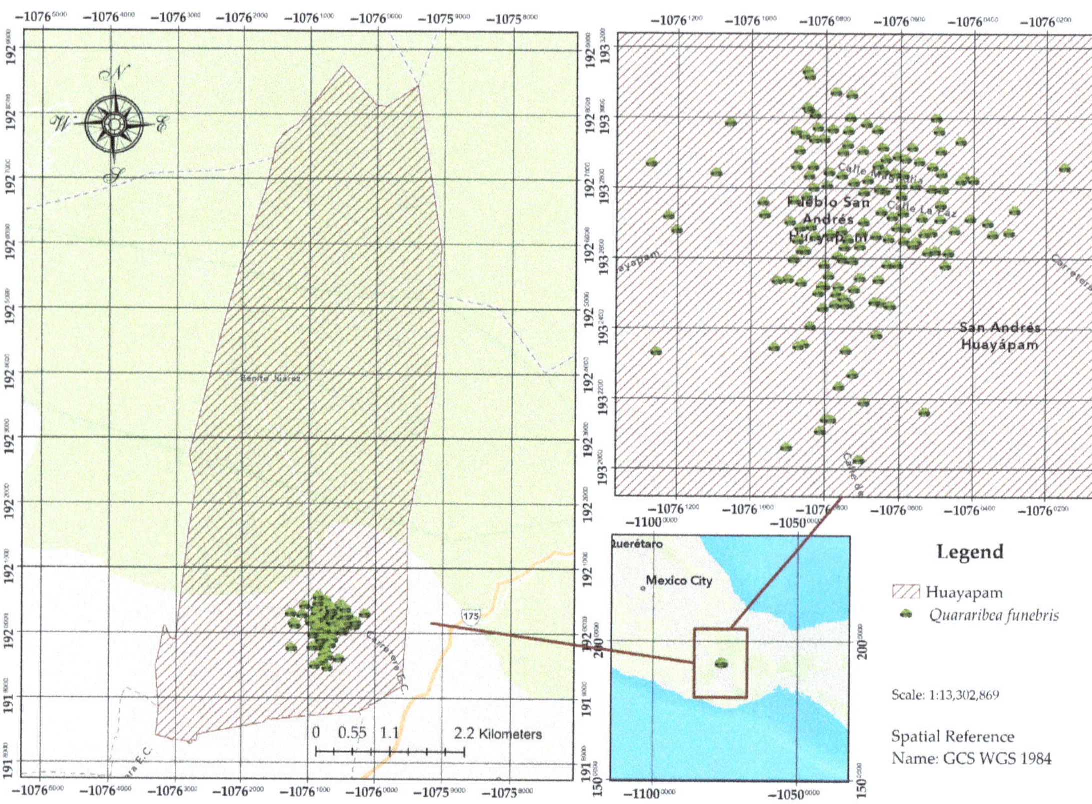

Figure 1. Location of the sampled *Q. funebris* specimens within the urban matrix of San Andrés Huayapam, Oaxaca. The map highlights the fragmented distribution of the "Rosital" trees (green icons) embedded within the municipality layer.

On sunny days, Photosynthetically Active Radiation (PAR, μmol m^{-2} s^{-1}) was recorded over four consecutive weeks in March 2025 on both the east and west sides of the tree (Figure 2). Readings were taken hourly from 07:00 to 18:00 h using a LI-191SA® quantum sensor (LI-COR, Lincoln, NE, USA) with a 127 cm^2 sensitive area, calibrated for the visible light band (400–700 nm), and connected to a LI-1400® Data Logger (LI-COR, Lincoln, NE, USA). The sensor was positioned at a height of 1 m above the ground and at a distance of 2 m from the trunk. Simultaneously, ambient temperature and relative humidity were monitored using LI-1400-101 and LI-1400-104 sensors, respectively.

While the quantum sensor measures the instantaneous photon flux (μmol m^{-2} s^{-1}), the use of units in mol m^{-2} d^{-1} is recommended [16,37] as plant growth is determined by the daily integrated light flux. Therefore, once hourly average PAR values were obtained, the Daily Light Integral (DLI) was calculated [16] using the following equation: DLI (mol m^{-2} d^{-1}) = (hourly readings in mmol m^{-2} s^{-1} $\times$ 0.0036), where 0.0036 is derived from the conversion factor: (60 s/min $\times$ 60 min/h)/1,000,000 mmol mol^{-1}. Data visualization (mean values and standard error) was performed using Microsoft Excel®, and a mean comparison test (Tukey, $p \leq 0.05$) was conducted using SAS software version 9.0 [38].

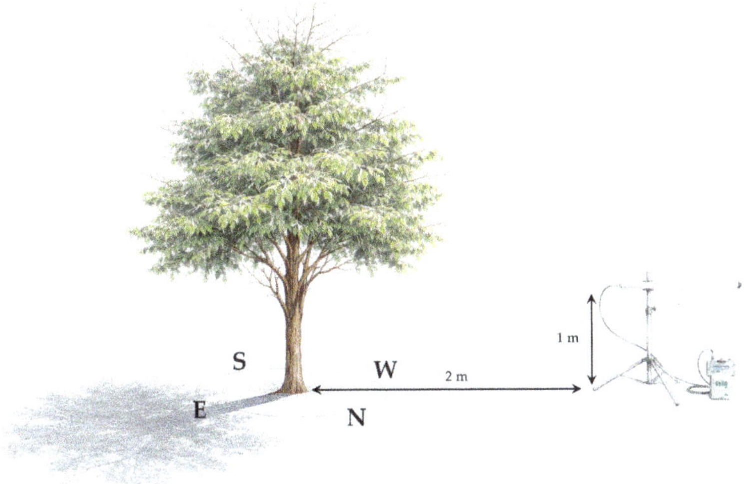

Figure 2. Schematic representation of the tree's cardinal orientation used for the weekly monitoring of Photosynthetically Active Radiation (PAR) and microclimatic variables.

Finally, phenological monitoring focused on flowering and fruiting cycles, which are driven by these climatic factors and plant age. Floral buds were sampled on branches located at a height of 1.20 to 1.5 m, specifically selected with an eastern orientation due to the higher abundance of flowers (Figure 3). Data recording spanned from floral bud emergence to full bloom and subsequent fruiting.

2.3. Spatio-Temporal Analysis of Landscape Transformation

Landscape transformation was analyzed using vector data of Land Use and Vegetation from the Instituto Nacional de Estadística y Geografía (INEGI) for the years 1992 (Series I) [39], 2001 (Series II) [40], 2005 (Series III) [41], 2009 (Series IV) [42], 2013 (Series V) [43], 2017 (Series VI) [44] and 2021 (Series VII) [45]. All spatial data were processed using ArcGIS PRO (v.3.5.4) software [46]. To ensure temporal comparability across the 30-year period, a standardization of projections (WGS 84/UTM Zone 14N) and cartographic categories was implemented [47].

To address the heterogeneous nomenclature used by INEGI across three decades, a harmonization process was implemented. This involved grouping 55 specific original labels into four standardized functional categories (Table 1) based on their structural role in the peri-urban landscape of Huayapam. This reclassification was essential to quantify the transition from mature forest cover to secondary stages and human expansion [47]. The quantitative analysis utilized transition matrices via the Tabulate Intersection tool to identify tipping points in habitat degradation, particularly the structural loss of mature ecosystems.

Complementarily, to contextualize the prevailing abiotic conditions during this period of transformation, historical records from the Comisión Nacional del Agua (CONAGUA) were processed [48]. Climatological Normals (1991–2020) from meteorological station 20,079 (Oaxaca de Juárez) were analyzed, extracting specific variables: monthly mean, maximum, and minimum temperatures (°C) and cumulative precipitation (mm) [48]. This station was selected due to its geographic contiguity with San Andrés Huayapam and its temporal correspondence with the land-use change data. The analysis of these

https://doi.org/10.3390/su18031630

datasets served as the selection criterion for the field monitoring period, allowing for the identification of the window of maximum environmental pressure—defined by the historical intersection of the prolonged drought season (November–April) and the highest annual thermal peaks. This ensures that the eco-physiological assessment of the Rosital tree is conducted during the most critical stage of the annual cycle, providing a robust baseline for the design of Nature-based Solutions (NbS).

(a) (b) (c)

Figure 3. (**a**) "Rosital" in full bloom; (**b**) different flower development phases, from buds to opening, also showing "jarritos"; (**c**) flower ready for harvest.

Table 1. Harmonization of INEGI Land Use and Vegetation labels (1992–2021).

Standardized Category	Original INEGI Labels (as Identified in Series I to VII)	Functional Criteria
Primary Forest	Pine Forest, Oak Forest, and Pine-Oak Forest	Mature forest ecosystems with closed canopy and high microclimatic buffering
Secondary Vegetation	Secondary Arboreal and Shrub Vegetation of Pine, Oak, Pine-Oak, and Low Deciduous Forest.	Successional stages resulting from disturbance; fragmented cover with reduced capacity to regulate irradiance.
Agriculture & Grassland	Rainfed Agriculture and Induced Grassland.	Areas of complete canopy removal and high soil exposure, forcing dependency on manual irrigation.
Human Settlements	Human Settlements.	Urban expansion and soil sealing that creates urban heat islands.

2.4. Ethnographic Approach and Biocultural Triangulation

A mixed-methods approach was adopted to bridge ecological data with the social dimensions of *Q. funebris* management [49–51]. The social component involved semi-structured interviews with 30 tejate producers (tejateras), who were selected based on their status as primary custodians of Traditional Ecological Knowledge (TEK) and their active participation in local production unions [10,24,31,32,52]. The interviews were designed

https://doi.org/10.3390/su18031630

to document local perceptions regarding tree health, flowering patterns, and the socio-economic impacts of environmental change (Figures A1 and A2).

The qualitative data from interview transcripts were analyzed using thematic analysis [49–51]. This process involved a manual coding system to categorize narratives into four analytical nodes: (1) perceived changes in phenology, (2) management strategies against water scarcity, (3) barriers to natural and domestic propagation, and (4) the role of gender in resource governance.

This social information was then subject to a biocultural triangulation process [53], acting as a participatory monitoring system. Methodologically, this involved cross-referencing the categorized narratives of the tejateras with the quantitative metrics obtained from the physiological (DLI) and geospatial analysis (transition matrices). The objective of this convergence was to perform a multidimensional diagnosis of the territory, ensuring that the proposed Nature-based Solutions (NbS) are grounded in a participatory monitoring system that validates scientific evidence through local social reality [54–57].

3. Results

3.1. Spatio-Temporal Landscape Dynamics (1992–2021)

The geospatial analysis reveals a continuous structural transformation of the territory, involving dynamic shifts among the four functional categories identified (Figure 4, Table 2). The temporal assessment segregates this degradation process into three distinct chronological phases, each characterized by specific drivers of change ranging from natural succession to urban competition:

Table 2. Land use changes in San Andrés Huayapam (hectares).

Land Type Use	1992	2001	2005	2009	2013	2017	2021
Pine Forest	353	432	0	0	0	0	
Secondary Arboreal Vegetation of Pine Forest	0	0	432	432	429	429	429
Oak Forest	665	572	0	0	0	0	
Secondary Shrub Vegetation of Oak Forest	0	0	533	533	533	533	533
Pine-Oak Forest	534	858	0	0	0	0	
Secondary Arboreal Vegetation of Pine-Oak Forest	0	0	419	419	419	419	419
Secondary Shrub Vegetation of Pine-Oak Forest	0	0	439	439	439	439	439
Agriculture	804	886	926	612	567	567	536
Secondary Shrub Vegetation of Low Deciduous Forest	402	9	9	91	91	91	91
Induced Grassland	0	0	0	199	160	160	160
Human Settlements	0	0	0	33	119	119	150

3.1.1. Reconfiguration and Successional Recovery (1992–2001)

During the first decade, the landscape was dominated by mature structures, totaling 1552 ha of Primary Forest (Pine, Oak, and Pine-Oak). Between 1992 and 2001, the data reveal a positive trend of ecological succession: the Pine-Oak Forest category expanded significantly from 534 ha to 858 ha. This increase coincides with the reduction in Secondary Shrub Vegetation of Low Deciduous Forest, which decreased from 402 ha in 1992 to just 9 ha in 2001. This suggests that the shrub layer matured and integrated into the mixed forest canopy. In this period, human settlements remained below detectable cartographic thresholds (Figure 5).

https://doi.org/10.3390/su18031630

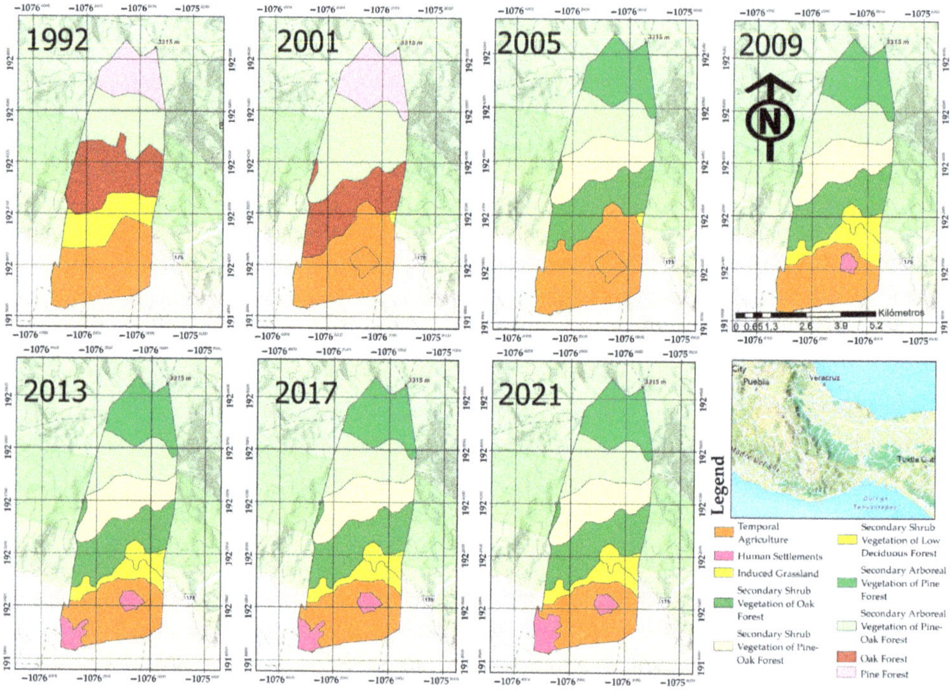

Figure 4. Spatio-temporal evolution of land use in San Andrés Huayapam (1992–2021). The map series, derived from INEGI vector data [39–45].

3.1.2. Structural Collapse and Successional Transition (2001–2005)

A critical tipping point occurred between 2001 and 2005. During this interval, the three primary forest categories disappeared from the record, being quantitatively replaced by secondary vegetation categories. The data reveal a direct transition pathway: the 858 ha of mature Pine-Oak Forest recorded in 2001 were fragmented into exactly 858 ha of secondary pine-oak vegetation (419 ha arboreal and 439 ha shrub) by 2005. Similarly, mature Pine and Oak forests were replaced by their respective secondary stages, marking the replacement of complex canopy structures with simplified successional stages (Figure 6).

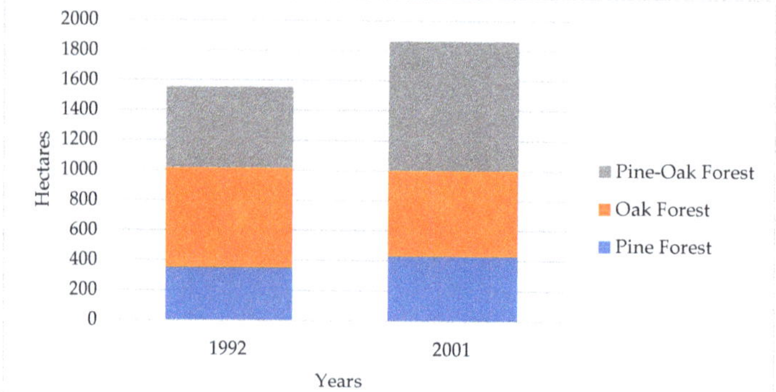

Figure 5. Ecological succession and canopy integration (1992–2001). The spatial analysis identifies a recovery period where Secondary Shrub Vegetation (Low Deciduous Forest) transitioned into mature Pine-Oak Forest structures, increasing the area's biological density prior to urbanization.

https://doi.org/10.3390/su18031630

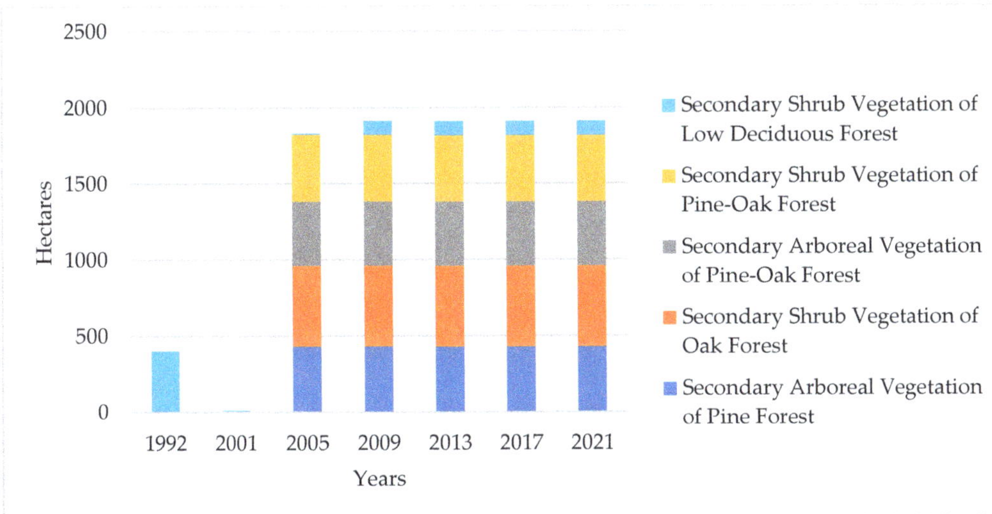

Figure 6. The structural tipping point (2001–2005). Visual representation of the critical transition where 1552 ha of primary forest were fragmented and quantitatively replaced by secondary vegetation and agricultural zones, marking the loss of the microclimatic buffer.

3.1.3. Urban Consolidation and Fragmentation (2009–2021)

The most recent phase is defined by the sustained growth of Human Settlements, which emerged in 2009 with 33 ha and reached 150 ha by 2021. This expansion coincided with a steady decline in active productive land: Temporal Agriculture fell from its 2005 peak of 926 ha to 536 ha by the end of the study period. Concurrently, Induced Grassland appeared in the record in 2009 (199 ha) and stabilized around 160 ha by 2021. This trend indicates a process of soil sealing and the conversion of productive zones into induced pastures within the peri-urban matrix (Figure 7).

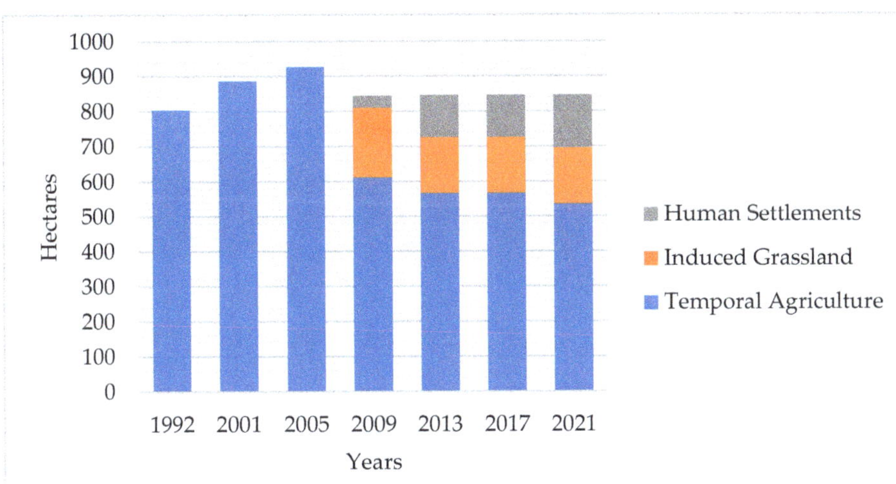

Figure 7. Peri-urban land-use competition (2009–2021). The map highlights the "sealing" of the territory, where the expansion of Human Settlements (reaching 150 ha) and Induced Grassland competes directly with productive agricultural land.

https://doi.org/10.3390/su18031630

3.2. Eco-Physiological Stress and Structural Vulnerability

The biological performance of the *Q. funebris* population is strictly conditioned by the local abiotic baseline. Analysis of climatological normals (1991–2020) indicates a defined drought season from November to April, coinciding with maximum annual temperatures ranging between 33.3 °C and 34.9 °C (Table 3). Under these limiting conditions and lacking the original forest buffer identified in Section 3.1, the urban specimens exhibit specific morphological adjustments.

Table 3. Climatological Normals (1991–2020) Oaxaca de Juárez Station (20079) [48].

Normal Climate Parameter	Jan	Feb	Mar	Apr	May	Jun	Jul	Aug	Sep	Oct	Nov	Dec	Annual
Mean Temperature (°C)	19.6	21.4	23.5	25.4	25.3	23.7	23.0	23.0	22.6	21.9	20.6	20.0	22.5
Maximum Temperature (°C)	28.9	31.1	33.3	34.9	33.8	30.9	30.1	30.0	29.1	29.3	29.0	28.9	30.8
Minimum Temperature (°C)	10.3	11.6	13.6	15.8	16.7	16.7	15.9	15.9	16.1	14.5	12.3	11.0	14.2
Precipitation (mm)	2.4	4.6	17.6	47.3	97.9	188.3	118.6	131.8	163.6	63.1	9.1	9.0	853.3

3.2.1. Biometric Stunting and Radiative Stress

The biometric inventory established the structural status of the population within the urban matrix (Figure 1). Measurements indicate that adult trees (>50 years) average a maximum height of 12.5 m. This recorded value corresponds to 50% of the maximum potential height (25 m) described for the species in its native tropical rainforest range [8].

Regarding canopy health, the inventory revealed that trees exposed to the West (afternoon sun) present higher incidences of chlorosis and apical desiccation compared to protected specimens (Figure 8).

(**a**) (**b**) (**c**)

Figure 8. *Q. funebris.* (**a**) Specimen showing apical chlorosis and foliage reduction, symptoms consistent with the "urban ecological trap" effect; (**b**,**c**) Comparative view of specimens with healthy foliage density.

This aligns with the physiological monitoring of the Daily Light Integral (DLI). The data reveal a significant differential in radiative load: the West-facing canopy receives 38.81 $\pm$ 0.77 mol m^{-2} d^{-1}, significantly higher than the East-facing side (29.33 $\pm$ 0.83 mol m^{-2} d^{-1}) (Table 4) (Figures 9 and 10).

https://doi.org/10.3390/su18031630

Table 4. Daily Light Integral (DLI) and microclimatic parameters recorded on *Q. funebris* canopy.

Location	DLI (mol m^{-2} d^{-1})	Temperature (°C)	Relative Humidity (%)
East	29.33 ± 0.83 b	28.26 ± 0.38 b	30.31 ± 1.36 a
West	38.81 ± 0.77 a	31.19 ± 0.26 a	31.52 ± 1.45 a

Means with different letters within a column indicate significant differences (Tukey, $p \leq 0.05$).

(a) (b) (c)

Figure 9. Condition of mature *Q. funebris* (>50 years). (**a**) General view exhibiting structural stunted growth and canopy thinning; (**b**) Detail of leaf sunscald and oxidative stress on the West-facing canopy; (**c**) Contrast with healthy foliage density.

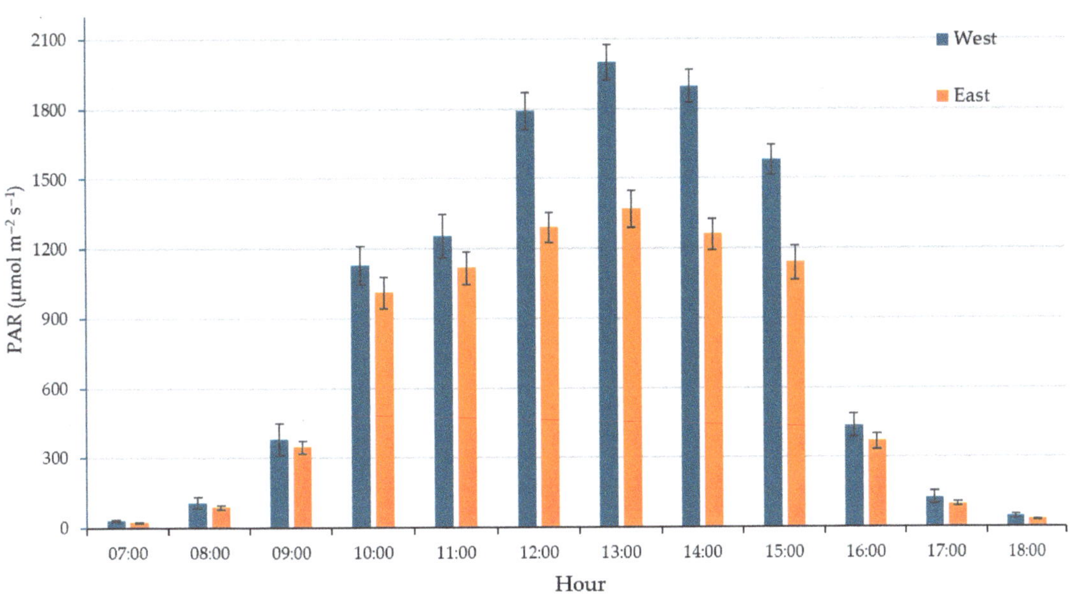

Figure 10. Photosynthetically Active Radiation (PAR, μmol m^{-2} s^{-1}) received by the *Q. funebris* tree in full sun on the East and West sides. Data represents the average of five measurements ± standard error per hour of the day.

https://doi.org/10.3390/su18031630

This high irradiance coincides spatially with the necrotic lesions (sunscald) observed on the foliage (Figure 11).

(a) (b) (c)

Figure 11. Details of foliar damage. (**a**,**b**) Chlorotic leaves showing marginal necrosis (sunscald), typical of exposure to high evapotranspiration demand; (**c**) Healthy leaf morphology from a protected canopy section.

3.2.2. Phenological Dynamics and Productive Yield

Phenological monitoring indicates that "Rosital" trees begin their reproductive phase approximately five years after transplantation, with vigor increasing with age. Although the species has the physiological capacity to flower year-round, field records show a marked decrease in floral production during the dry season (November–April), followed by a blooming peak with the onset of rains (Figure 12).

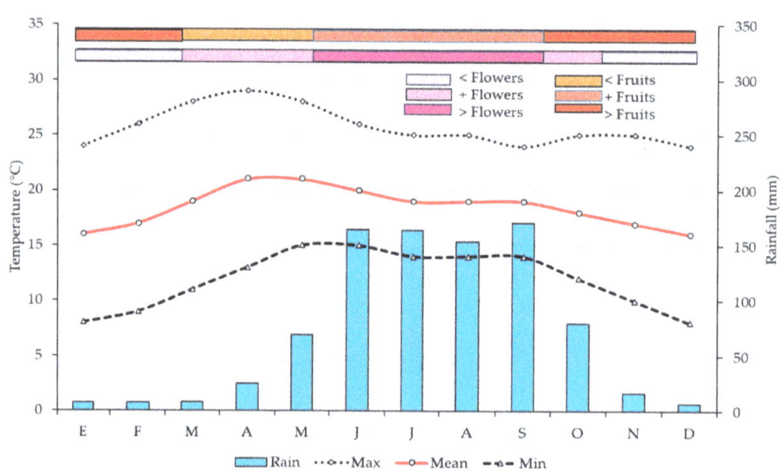

Figure 12. Phenological behavior of *Quararibea funebris* in relation to local climatic variables in San Andrés Huayapam. The ombrothermic diagram displays monthly precipitation (blue bars) and temperature profiles (minimum, mean, and maximum). The horizontal bars at the top illustrate the intensity of reproductive phases: flowering (pink scale) and fruiting (orange/brown scale) throughout the annual cycle, classified by abundance (<, +, >).

https://doi.org/10.3390/su18031630

The reproductive cycle, from floral bud emergence to full anthesis, spans a period of 45 to 65 days (Figure 13). Following pollination, the transition from petal fall to the formation of the immature fruit (locally known as "jarrito") takes approximately 60 to 75 days.

Figure 13. *Q. funebris*. (**a**) Full anthesis of the "Rosita de cacao" marking the start of the harvest window (45–65 days); (**b**) Transition phase showing tepal dehiscence and initial fruit formation ("jarrito"); (**c**) Immature fruit development prior to harvest.

Following pollination, the transition to the immature fruit stage ("jarrito") takes approximately 60 to 75 days. Regarding productivity, ethnobotanical data indicate that mature trees (>50 years) yield an estimated average of 10 to 12 kg of dry flower annually, despite the high dehydration rate of the biomass (94% loss) (Table 5).

Table 5. Phenological dynamics and productive performance of *Q. funebris* under peri-urban management.

Category	Parameter	Quantitative Value	Condition/Context
Reproductive Cycle	Anthesis development	45–65 days	From bud appearance to full opening (in situ)
	Fruit set duration	60–75 days	From petal fall to "jarrito" stage (immature fruit)
Post-Harvest Dynamics	Biomass loss rate	94% reduction	Total weight loss (Fresh to Dry)
	Drying time (Traditional)	5 days	Shade drying on petate/metal
	Drying time (Accelerated)	2 days	Direct solar exposure
Productive Yield	Annual harvest estimate	10–12 kg (dry weight)	Mature trees (>50 years) with supplementary irrigation

3.3. Reproductive Biology and Seed Recalcitrance

Following the reproductive phenology described in Table 5, the cycle concludes with the maturation of the fruit (Figure 14), an indehiscent berry. Post-harvest monitoring indicates that the fruit structure lacks mechanisms to retain moisture once detached from the parent tree (Figure 14a). Quantitative observations show that fresh fruits lose 43% of their biomass within six days under ambient conditions (Figure 14b).

Regarding propagation potential, the species exhibits characteristics of seed recalcitrance. The seeds are encased in a thick testa (Figure 14c), but the embryo is highly sensitive to desiccation. While germination in controlled nursery conditions occurs between 15 and 20 days, natural regeneration in the peri-urban understory is virtually absent. This field observation aligns with the physiological constraints documented for the species, which reports a rapid decay in viability, dropping from 99.3% at collection to 71.3% after

https://doi.org/10.3390/su18031630

12 months, and significantly lower in non-controlled humidity environments [29]. This biological trait prevents the formation of persistent soil seed banks in the study area.

Figure 14. Mature Fruit. (**a**) Mature indehiscent fruits transitioning to yellow; (**b**) Structural tearing of the berry due to rapid water loss (43% biomass loss in 6 days); (**c**) Exposed seed showing the thick testa, which fails to protect the recalcitrant embryo from dehydration.

3.4. Gender-Based Management and Local Ecological Knowledge (TEK)

The ethnographic assessment conducted with the "Union of tejate Producers" (n = 30) characterized the sociotechnical system sustaining the *Q. funebris* population. Results indicate that 100% of the active germplasm is managed within the domestic sphere, specifically in backyards owned or managed by women (Figure 15). This structure constitutes a "domestic monopoly" of the resource, where access to the tree is strictly linked to family lineage and membership in the producer's union.

Figure 15. The Domestic Monopoly system. The image illustrates how *Q. funebris* management is restricted to private backyards, where trees compete for space with housing infrastructure but survive due to the direct care and gray-water recycling provided by tejateras.

3.4.1. Management Intensity and Environmental Subsidy

Regarding management practices, the primary strategy recorded is supplementary irrigation, which acts as an environmental subsidy. Interviews revealed that irrigation regimes vary significantly based on water availability: while some producers apply approximately 500 L per tree to individuals >15 years old, others are limited to applying 100 L every two days. Notably, producers report that trees lacking this supplementary input during the dry season suffer high mortality rates.

3.4.2. Harvest Dynamics and Economic Valorization

Harvesting is driven by phenological and market constraints. While the onset of the rainy season triggers a peak in blooming, it introduces a phytosanitary trade-off: wet-season flowers require immediate harvesting to prevent fungal decay or staining caused by excess moisture, which compromises their quality.

https://doi.org/10.3390/su18031630

This intensive management is justified by the high market value of the resource. Market analysis indicates that fresh flowers are sold for 0.057 to 0.29 USD per unit, while the processed dry flower reaches a value of approximately 91.48 USD per kilogram. Given that a mature tree (>50 years) yields an estimated average of 10 to 12 kg annually (Table 5), the tree represents a significant pillar of the household economy.

3.4.3. Post-Harvest Technology and Perception

In terms of processing, a technological duality was documented: while traditional drying in the shade (on palm mats or "petates") preserves quality but requires 5 days, producers increasingly resort to accelerated sun-drying on metal surfaces (2 days) to meet market demand, despite the trade-off in organoleptic quality (Figure 16).

(a) (b)

Figure 16. Harvest and post-harvest technological duality [58]. (**a**) Volumetric quantification of floral yield using standard containers, the traditional unit of measure; (**b**) Accelerated sun-drying process on metal surfaces (2 days), a modern adaptation to meet market demand despite the trade-off in organoleptic quality compared to traditional shade drying.

The Local Ecological Knowledge (TEK) held by the tejateras has also identified a decline in ecosystem services. A substantial majority of interviewees (90%) reported a perceived decrease in floral yield per tree compared to previous decades, attributing this loss to "excess heat" and "lack of root space" (Table 6).

Table 6. Summary of Management Practices and Local Ecological Knowledge (TEK) regarding *Q. funebris*.

Category	Indicator/Practice	Consensus Result
Resource Tenure	Access Regime	100% Domestic/Private (Backyards). No communal forest extraction reported.
	Gender Role	Women control harvest, processing, and sale. Men assist in pruning and planting.
Water Management	Irrigation Frequency (Dry Season)	Daily (Trees > 50 years); 2–3 times/week (Young trees).
	Water Source	Potable municipal network and private wells.
Post-harvest Handling	Harvesting Technique	Manual collection with "carrizo" (reed) poles to avoid damaging buds.
	Drying Method	Dual practice: Shade drying (5 days) for quality vs. Sun drying on metal (2 days) for speed.
Ecological Perception	Threat Identification	90% identify "Heat/Drought" as the main threat.
	Replacement Strategy	60% have planted replacement saplings but report high mortality rates.
Economic Value	Market Price	High value: ~91.48 USD/kg (Dry) driving intensive care.
	Rainy Season Constraint	Immediate harvest is required to prevent fungal decay/staining.

https://doi.org/10.3390/su18031630

4. Discussion

4.1. Landscape Transformation and the Mechanism of the Ecological Trap

The integration of geospatial analysis with climatological normals confirms the hypothesis that the *Q. funebris* population in San Andrés Huayapam is caught in an anthropogenic ecological trap [19–21]. In ecological theory, a trap occurs when an organism prefers or persists in a habitat due to attractive cues, even though the habitat quality lowers its survival or reproductive success. In Huayapam, the attractive cue is the supplementary irrigation provided by the tejateras, which masks the hydrological deficit. However, our results show that the structural quality of the habitat has degraded below the physiological optimum of the species [8,9].

The disappearance of 1552 ha of primary forest buffer, documented in Section 3.1.2, removed the regional thermal regulator [18,25,30,33–35]. Consequently, the current urban matrix (150 ha) functions as a heat island, exposing a species evolutionarily adapted to the tropical rainforest understory [26,59,60] to maximum temperatures exceeding 33 °C. This creates a functional mismatch: the tree receives water (anthropogenic subsidy) that allows it to survive, but is exposed to an atmospheric demand (DLI > 38 mol m^{-2} d^{-1}) that limits its development [8,26]. This validates that unplanned urbanization has compromised the functional niche of the "Rosital", transforming the territory into a sink habitat sustained only by human intervention [26,59,60].

4.2. Biometric Constraints and Physiological Stress Impact

The biometric stunting observed in the urban population, averaging 12.5 m compared to the potential 25 m documented in natural habitats [8,29,61], should be interpreted as an adaptive response to the restrictive urban environment, rather than simple variability [27,62,63]. This phenotypic plasticity reflects the evolutionary constraints of a species native to the rainforest mid-canopy [8,13,14,64]. Lacking the protective vegetative ceiling of its original habitat, the urban specimens face a metabolic conflict: they must prioritize cellular repair over biomass accumulation.

As noted in the literature regarding shade-tolerant tropical trees [4,14,64], when the Daily Light Integral (DLI) exceeds the species saturation point (confirmed here by values > 38 mol m^{-2} d^{-1} on the West face), the plant limits its vertical growth to reduce the surface area exposed to dehydration [15–17]. This energy trade-off explains the structural simplification observed: the tree stays small to minimize the radiative load it cannot physiologically process [14,17,65].

This restriction comes with a physiological cost. The chlorosis and apical necrosis observed in the upper canopy (Section 3.2.1) are physical manifestations of this oxidative stress [17,62,63,66]. This has critical implications for the tejate economy. Since the flowers of *Q. funebris* are valued for their antioxidant metabolites and specific flavor profile [10,12,32], the chronic stress observed suggests that the tree is diverting energy towards survival (cellular repair) rather than optimal phytochemical synthesis [12,67,68]. Therefore, the environmental degradation identified not only affects the tree's survival but also potentially compromises the organoleptic quality of the harvest [29,68,69], threatening the heritage value managed by the tejateras [11,32,69].

4.3. Reproductive Vulnerability and the Impossibility of Ex Situ Conservation

Beyond individual stress, our results confirm that the system faces a reproductive lock-in [8,9,29]. The rapid dehydration of fruits (43% biomass loss in 6 days) and the absence of natural regeneration align with the recalcitrant nature of *Q. funebris* seeds [29]. The rapid decay in viability (<12 months) renders the formation of persistent soil seed banks impossible in the paved peri-urban environment [19,63,64].

https://doi.org/10.3390/su18031630

This biological constraint has a profound management implication: traditional ex situ conservation strategies (seed banks) are unfeasible for this species. Consequently, the survival of germplasm relies exclusively on the continuous maintenance of living individuals [15,64]. This validates the critical role of the domestic monopoly identified in Section 3.4; without the immediate propagation and care provided in backyards, the species lacks the biological mechanisms to persist in this landscape naturally [64,70].

4.4. The Gendered Environmental Subsidy

The survival of *Q. funebris* in this hostile abiotic context is actively subsidized by the labor of women producers. The daily irrigation regimes (up to 500 L/tree) documented in the ethnographic results effectively mask the hydrological deficit of the territory [25,33–35,48]. From a political ecology perspective, this constitutes a gendered environmental subsidy, where women's unpaid care work compensates for the loss of ecosystem services caused by the deforestation analyzed in Section 3.1.

However, this model faces a growth paradox [22,27,71]. The economic success of tejate drives tourism and urbanization [10,11,18,23,58,72], which in turn degrades the microclimate required by the tree. The reported decline in floral yield by 90% of producers suggests that the "domestic gardening" strategy has hit an ecological ceiling. Backyards alone can no longer buffer the extreme DLI and temperature. Therefore, sustainability requires a shift toward Nature-based Solutions (NbS) at a landscape scale [1–3,71]. Specifically, the implementation of Biocultural Biological Corridors [19,20] is required to restore the multi-strata canopy along hydrological tributaries. Unlike isolated domestic planting, restoring the forest buffer would regulate DLI and ambient humidity, dismantling the ecological trap while validating the governance and TEK of the local women.

5. Conclusions

The comprehensive diagnosis presented in this study confirms that the unplanned urbanization of San Andrés Huayapam has generated a functional ecological trap for the *Quararibea funebris* population. By integrating geospatial analysis with eco-physiological monitoring, we conclude that the loss of the forest buffer has degraded the microclimatic niche below the species physiological optimum, subjecting the remnant specimens to a chronic radiative load that exceeds their biological limits. This environmental mismatch has forced a negative plastic response in the trees, manifested as stunting and oxidative stress, which currently compromises the material base of the tejate cultural heritage. While this diagnosis establishes a robust baseline, we recognize that our sentinel-tree approach and reliance on regional seed viability data represent specific methodological limitations; therefore, future research must expand sampling across the canopy gradient and conduct in situ germination trials to refine these eco-physiological thresholds.

Biologically, the study validates that preserving this resource cannot depend on ex situ germplasm banks due to the recalcitrant nature of the seeds. Consequently, the persistence of the species relies strictly on in situ conservation. However, our findings suggest that the current strategy of domestic gardening has hit its ecological ceiling: the supplementary irrigation provided by families is no longer sufficient to buffer the evapotranspirative demand of the urban heat island. Furthermore, given the oxidative stress documented, there is an urgent need to investigate the phytochemical profile of urban flowers in future studies to determine if this chronic abiotic stress is altering the organoleptic quality—specifically flavor and antioxidant content—valued by the market, thereby linking environmental health directly to product quality.

Socially, this research confirms that the system's sustainability relies on a gendered environmental subsidy, where women tejateras compensate for the territory's environmental

https://doi.org/10.3390/su18031630

deficit through unpaid care work. However, this model faces a critical growth paradox, where the economic success of the beverage drives the very urbanization that degrades the tree's habitat. To resolve this contradiction, sustainability requires a paradigm shift from private conservation to landscape-scale Nature-based Solutions (NbS). Since our diagnosis proves that the core vulnerability is the loss of the shade-tolerant niche, simple reforestation is insufficient; we must reconstruct the forest architecture. Therefore, we propose the implementation of Biocultural Biological Corridors to restore the multi-strata canopy along hydrological tributaries. This intervention implies a major governance challenge: moving from a domestic monopoly to a collective management model led by the Union of tejate Producers, ensuring that the restoration of the landscape validates their Traditional Ecological Knowledge (TEK) and reinforces their economic sovereignty.

Author Contributions: Conceptualization, Y.D.O.-H., M.A.A.-O. and G.L.-E.; methodology, Y.D.O.-H., M.A.A.-O. and G.L.-E.; software, Y.D.O.-H. and M.A.A.-O.; validation, Y.D.O.-H., M.A.A.-O. and G.L.-E.; formal analysis, Y.D.O.-H., M.A.A.-O. and E.G.-S.; investigation, Y.D.O.-H., M.A.A.-O., G.L.-E., F.E.O.-H. and S.V.-P.; resources, Y.D.O.-H., M.A.A.-O., G.L.-E., F.E.O.-H. and S.V.-P.; data curation, Y.D.O.-H., M.A.A.-O. and G.L.-E.; writing—original draft preparation, Y.D.O.-H., M.A.A.-O. and G.L.-E.; writing—review and editing, Y.D.O.-H., M.A.A.-O. and G.L.-E.; visualization, Y.D.O.-H., M.A.A.-O. and G.L.-E.; supervision, Y.D.O.-H., M.A.A.-O. and E.G.-S.; project administration, Y.D.O.-H. and E.G.-S.; funding acquisition, Y.D.O.-H., F.E.O.-H. and E.G.-S. All authors have read and agreed to the published version of the manuscript.

Funding: This research was funded by the Instituto Politécnico Nacional (IPN), project SIP-20251329.

Institutional Review Board Statement: Ethical review and approval were waived for this study by Institution Committee of Instituto Politécnico Nacional due to ethical guidelines and data regulations established by the Instituto Politécnico Nacional and current Mexican legislation.

Informed Consent Statement: Informed consent for publication was obtained from all identifiable human participants.

Data Availability Statement: The data presented in this study is available on request from the corresponding author.

Acknowledgments: The authors are grateful to the Instituto Politécnico Nacional (IPN) México, Secretaría de Investigación y Posgrado, and CIIDIR Oaxaca for their support, as well as Secretaría de Ciencia, Humanidades, Tecnología e Innovación (SECIHTI), and Sistema Nacional de Investigadoras e Investigadores (SNII-SECIHTI).

Conflicts of Interest: The authors declare no conflicts of interest.

Abbreviations

The following abbreviations are used in this manuscript:

NbS	Nature-based Solutions
DLI	Daily Light Integral
PAR	Photosynthetically Active Radiation
TEK	Traditional Ecological Knowledge

https://doi.org/10.3390/su18031630

Appendix A

Semi-Structured Interview Script: Tejate and Rosita de Cacao Producers

I. Introduction and Informed Consent

Good morning/afternoon. We are conducting academic research on the importance of ***Tejate*** *and the current status of* ***Rosita de Cacao*** *trees in our region.*

The objective is to gather your experience and expert opinion as a producer regarding the changes you have noticed in the environment and in the preparation of this beverage.

Before we begin, I want to assure you that:

1. *All information you share will be used* ***strictly for educational and scientific purposes****.*
2. *This interview is* ***anonymous****; we will NOT collect personal data (such as your full name or home address).*
3. *Your participation is entirely voluntary, and you may stop the interview at any time you decide.*
4. *We will* ***take a picture*** *during the interview for validation purposes and* ***could be*** *used in our reports or articles.*

Do you have any questions about this? ***Do you agree to participate and share your experience with us?"***

[ ] ***YES, consent granted*** *(Check if the person agrees to proceed)*

II. General Context Data (No personal identification)

- ***Approximate age:_____***
- ***Years engaged in the activity: _____***
- ***Member of a producer organization?*** *Yes / No*
- ***Source of the flower:***

  [ ] Own harvest (backyard)

  [ ] Market purchase

  [ ] Wild collection (forest)

Figure A1. Consent agreement for structured interview.

https://doi.org/10.3390/su18031630

III. Assessment Dimensions

A. Environmental Dimension (The Tree and the Climate)

Objective: To understand the perception of ecological changes.

1. ***Climate Perception:*** *"In the years you have been working, have you noticed changes in the weather (heat, rainfall patterns) that affect the Rosita trees? Does the tree yield the same amount of flowers as before?"*
2. ***Tree Health:*** *"Have you noticed if the trees get sick more often now? Do you observe any pests or 'bugs' that were not present before?"*
3. ***Perceived Causes:*** *"From your experience, why do you think it is becoming harder for the flower to grow or bloom? (e.g., lack of water, intense sun, mishandling?)"*

B. Socio-Cultural Dimension (Knowledge and Voice)

Objective: To identify the role of transmission and participation.

4. ***Knowledge Transmission:*** *"Who taught you how to work with the Rosita flower and Tejate? Are you teaching your daughters or granddaughters, or are they no longer interested?"*
5. ***Social Organization:*** *"Do the women dedicated to this activity meet or organize in any way? Do you support each other within the community?"*
6. ***Decision Making:*** *"When there are community issues regarding trees or water, is the opinion of women taken into account in the assemblies? Do you feel that your voice is heard?"*

C. Economic Dimension (Income and Labor)

Objective: To evaluate the impact of scarcity on the family economy.

7. ***Costs and Scarcity:*** *"If there are fewer flowers or the yield decreases, how does this affect your sales? Have you had to raise prices, or do you earn less profit?"*
8. ***Workload:*** *"To obtain the flower, do you now have to work harder, travel further, or dedicate more time than before?"*
9. ***Economic Dependence:*** *"How significant is the sale of Tejate for your household expenses? If the Rosita tree were to disappear, how would it impact your family's economy?"*

Figure A2. Structured Interview Format.

References

1. Dunlop, T.; Khojasteh, D.; Cohen-Shacham, E.; Glamore, W.; Haghani, M.; van den Bosch, M.; Rizzi, D.; Greve, P.; Felder, S. The evolution and future of research on Nature-based Solutions to address societal challenges. *Commun. Earth Environ.* **2024**, *5*, 132. [CrossRef]
2. Anderegg, W.R.L.; Blanchard, L.; Anderson, C.; Badgley, G.; Cullenward, D.; Gao, P.; Goulden, M.L.; Haya, B.; Holm, J.A.; Hurteau, M.D.; et al. Towards more effective nature-based climate solutions in global forests. *Nature* **2025**, *643*, 1214–1222. [CrossRef]
3. Olivadese, M. Rethinking Nature-Based Solutions: Unintended Consequences, Ancient Wisdom, and the Limits of Nature. *Land* **2025**, *14*, 1272. [CrossRef]
4. Pauleit, S.; Zölch, T.; Hansen, R.; Randrup, T.B.; Konijnendijk van den Bosch, C. Nature-Based Solutions and Climate Change—Four Shades of Green. In *Nature-Based Solutions to Climate Change Adaptation in Urban Areas: Linkages Between Science, Policy and Practice*; Kabisch, N., Korn, H., Stadler, J., Bonn, A., Eds.; Springer International Publishing: Cham, Switzerland, 2017; pp. 29–49.
5. International Union for Conservation of Nature and Natural Resources. *Nature-Based Solutions to Address Global Societal Challenges*; IUCN: Gland, Switzerland, 2016; p. 97.
6. Sowińska-Świerkosz, B.; García, J. What are Nature-based solutions (NBS)? Setting core ideas for concept clarification. *Nat.-Based Solut.* **2022**, *2*, 100009. [CrossRef]
7. Hock, L.; Chua, C.; Kirkwood, N.; Keesstra, S.; Veraart, J.; Verhagen, J.; Visser, S.; Kragt, M.; Linderhof, V.; Appelman, W.; et al. Nature-based solutions as building blocks for the transition towards sustainable climate-resilient food systems. *Sustainability* **2023**, *15*, 4475. [CrossRef]

8. Pennington, T.D.; Sarukhán, J. *Arboles Tropicales de México: Manual para la Identificación de las Principales Especies*; Universidad Autónoma de México, Fondo de Cultura Económica: Ciudad de Mexico, Mexico, 2005; p. 523.
9. Alverson, W.S. *Matisia* and *Quararibea* (Bombacaceae) should be retained as separete genera. *TAXON* **1989**, *38*, 377–388. [CrossRef]
10. Rothman, L. Tejate, Oaxaca's Favorite Drink Then and Now—Culinary Backstreets. Available online: https://culinarybackstreets.com/stories/oaxaca/liquid-assets-tejate-oaxacas-drink-of-the-gods (accessed on 5 December 2025).
11. Soleri, D.; Cleveland, D.A.; Aragón Cuevas, F.; Jimenez, V.; Wang, M.C. Traditional Foods, Globalization, Migration, and Public and Planetary Health: The Case of Tejate, a Maize and Cacao Beverage in Oaxacalifornia. *Challenges* **2023**, *14*, 9. [CrossRef]
12. Jiménez-Fernández, M.; Juárez Trujillo, N.; Mendoza, M.; Monribot-Villanueva, J.; Guerrero-Analco, J. Nutraceutical potential, and antioxidant and antibacterial properties of *Quararibea funebris* flowers. *Food Chem.* **2023**, *411*, 135529. [CrossRef]
13. Ibarra-Manríquez, G.; Cornejo-Tenorio, G. Tree fruit traits diversity in the tropical rain forest of Mexico. *Acta Bot. Mex.* **2010**, *90*, 51–104. [CrossRef]
14. Krause, G.; Winter, K.; Matsubara, S.; Krause, B.; Jahns, P.; Virgo, A.; Aranda, J.; Garcia, M. Photosynthesis, photoprotection, and growth of shade-tolerant tropical tree seedlings under full sunlight. *Photosynth. Res.* **2012**, *113*, 273–285. [CrossRef]
15. Faust, J.; Holcombe, V.; Rajapakse, N.; Layne, D. The Effect of Daily Light Integral on Bedding Plant Growth and Flowering. *HortScience* **2005**, *40*, 645–649. [CrossRef]
16. Chang, X.; Alderson, P.G.; Wright, C.J. Solar irradiance level alters the growth of basil (*Ocimum basilicum* L.) and its content of volatile oils. *Environ. Exp. Bot.* **2008**, *63*, 216–223. [CrossRef]
17. Chauhan, J.; Prathibha, M.D.; Singh, P.; Choyal, P.; Mishra, U.N.; Saha, D.; Kumar, R.; Anuragi, H.; Pandey, S.; Bose, B.; et al. Plant photosynthesis under abiotic stresses: Damages, adaptive, and signaling mechanisms. *Plant Stress* **2023**, *10*, 100296. [CrossRef]
18. SADER. Orgullo Oaxaca. Tejate: La Bebida de Los Dioses. Available online: https://www.oaxaca.gob.mx/sedeco/orgullo-oaxaca/ (accessed on 13 June 2025).
19. Wang, Z.; Fu, B.; Wu, X.; Li, Y.; Wang, S.; Lu, N. Escaping social–ecological traps through ecological restoration and socioeconomic development in China's Loess Plateau. *People Nat.* **2023**, *5*, 1364–1379. [CrossRef]
20. Zuñiga-Palacios, J.; Zuria, I.; Castellanos, I.; Lara, C.; Sánchez-Rojas, G. What do we know (and need to know) about the role of urban habitats as ecological traps? Systematic review and meta-analysis. *Sci. Total Environ.* **2021**, *780*, 146559. [CrossRef]
21. Colléony, A.; Shwartz, A. Beyond Assuming Co-Benefits in Nature-Based Solutions: A Human-Centered Approach to Optimize Social and Ecological Outcomes for Advancing Sustainable Urban Planning. *Sustainability* **2019**, *11*, 4924. [CrossRef]
22. Edwards, M.G. The growth paradox, sustainable development, and business strategy. *Bus. Strategy Environ.* **2021**, *30*, 3079–3094. [CrossRef]
23. Mace, G.M. Whose conservation? *Science* **2014**, *345*, 1558–1560. [CrossRef]
24. Hernández, R. Las Guardianas del Tejate. Available online: https://www.eluniversal.com.mx/articulo/menu/2017/03/7/las-guardianas-del-tejate/ (accessed on 27 December 2025).
25. SISPLADE. Sistema de Planeación para el Desarrollo—San Andrés Huayápam. Available online: https://sisplade.oaxaca.gob.mx/sisplade/smPublicacionesMunicipio.aspx?idMunicipio=91#divPlanesT (accessed on 16 October 2025).
26. Vázquez-Negrín, I.; Castillo-Acosta, O.; Valdez-Hernández, J. Estructura y composición florística de la selva alta perennifolia en el ejido Niños Héroes Tenosique, Tabasco, México. *Polibotánica* **2011**, *32*, 41–61.
27. Boutreux, T.; Bourgeois, M.; Bellec, A.; Commeaux, F.; Kaufmann, B. Addressing the sustainable urbanism paradox: Tipping points for the operational reconciliation of dense and green morphologies. *npj Urban Sustain.* **2024**, *4*, 38. [CrossRef]
28. Perevochtchikova, M.; Hernández Flores, J. How can social-ecological system trajectories be understood? An investigation of conceptualization through a systematic literature review. *Estud. Demográficos Urbanos* **2025**, *40*, 2025. [CrossRef]
29. Ríos-García, C.A.; Orantes-García, C.; Moreno-Moreno, R.A.; Farrera-Sarmiento, Ó. Efecto del almacenamiento sobre la viabilidad y germinación de dos especies arbóreas tropicales. *Ecosistemas Recur. Agropecu.* **2018**, *5*, 103–109. [CrossRef]
30. SEIE. Análisis Municipal. Consulta de Información Municipal. Available online: https://www.economia.gob.mx/datamexico/es/profile/geo/san-andres-huayapam (accessed on 2 February 2026).
31. Miranda, F. Tejate, Sabores, Conocimiento y Una Forma de Vida Que se Hereda Entre Mujeres de Oaxaca. Available online: https://oaxaca.eluniversal.com.mx/mas-de-oaxaca/tejate-sabores-conocimiento-y-una-forma-de-vida-que-se-hereda-entre-mujeres-de-oaxaca/ (accessed on 5 December 2025).
32. Mulato, A. El Tejate, Detrás de la Bebida Mexicana Que Mantiene Vivo el Mundo Prehispánico en Cada Sorbo. Available online: https://www.trtespanol.com/article/ebce73060577 (accessed on 5 December 2025).
33. PMD. Plan Municipal de Desarrollo 2020–2022. San Andrés Huayapam. Available online: https://sisplade.oaxaca.gob.mx/bm_sim_services/PlanesMunicipales/2020_2022_/091.pdf (accessed on 5 December 2025).
34. PMD. Plan Municipal de Desarrollo 2023–2025. San Andrés Huayapam. Available online: https://huayapam.gob.mx/wp-content/uploads/2025/05/PMD-HUAYAPAM-06.MAY_.25.pdf (accessed on 5 December 2025).
35. PMD. Plan Municipal de Desarrollo 2011–2013. San Andrés Huayapam. Available online: https://sisplade.oaxaca.gob.mx/bm_sim_services/PlanesMunicipales/2011_2013/091.pdf (accessed on 5 December 2025).

https://doi.org/10.3390/su18031630

36. Congreso de Oaxaca Dictamen Num. 56. Tejate como Patrimonio Cultural del Estado de Oaxaca. Available online: https://www.congresooaxaca.gob.mx/docs65.congresooaxaca.gob.mx/dictamen/1186.pdf (accessed on 2 February 2026).
37. Korczynski, P.C.; Logan, J.; Faust, J.E. Mapping Monthly Distribution of Daily Light Integrals across the Contiguous United States. *HortTechnology* **2001**, *12*, 12–16. [CrossRef]
38. SAS Institute Inc. *SAS/STAT Software*, 9.4; SAS Institute Inc.: Cary, NC, USA, 2023.
39. INEGI. Conjunto de Datos Vectoriales de la Carta de Uso del Suelo y Vegetación. Escala 1:250 000. Serie I. Continuo Nacional. Instituto Nacional de Estadística y Geografía. Available online: https://www.inegi.org.mx/app/biblioteca/ficha.html?upc=702825007020 (accessed on 24 April 2025).
40. INEGI. Conjunto de Datos Vectoriales de la Carta de Uso del Suelo y Vegetación. Escala 1:250 000. Serie II. Continuo Nacional. Instituto Nacional de Estadística y Geografía. Available online: https://www.inegi.org.mx/app/biblioteca/ficha.html?upc=702825007021 (accessed on 24 April 2025).
41. INEGI. Conjunto de Datos Vectoriales de la Carta de Uso del Suelo y Vegetación. Escala 1:250 000. Serie III. Continuo Nacional. Instituto Nacional de Estadística y Geografía. Available online: https://www.inegi.org.mx/app/biblioteca/ficha.html?upc=702825007022 (accessed on 24 April 2025).
42. INEGI. Conjunto de Datos Vectoriales de la Carta de Uso del Suelo y Vegetación. Escala 1:250 000. Serie IV. Conjunto Nacional. Instituto Nacional de Estadística y Geografía. Available online: https://www.inegi.org.mx/app/biblioteca/ficha.html?upc=702825007023 (accessed on 24 April 2025).
43. INEGI. Conjunto de Datos Vectoriales de la Carta de Uso del Suelo y Vegetación. Escala 1:250 000. Serie V. Conjunto Nacional. Instituto Nacional de Estadística y Geografía. Available online: https://www.inegi.org.mx/app/biblioteca/ficha.html?upc=702825007024 (accessed on 24 April 2025).
44. INEGI. Conjunto de Datos Vectoriales de la Carta de Uso del Suelo y Vegetación. Escala 1:250 000. Serie VI. Conjunto Nacional. Instituto Nacional de Estadística y Geografía. Available online: https://www.inegi.org.mx/app/biblioteca/ficha.html?upc=889463598459 (accessed on 24 April 2025).
45. INEGI. Conjunto de Datos Vectoriales de la Carta de Uso del Suelo y Vegetación. Escala 1:250 000. Serie VII. Conjunto Nacional. Instituto Nacional de Estadística y Geografía. Available online: https://www.inegi.org.mx/app/biblioteca/ficha.html?upc=889463842781 (accessed on 24 April 2025).
46. ESRI. *ArcGIS PRO*, 3.5.4; Environmental Systems Research Institute (ESRI): Redlands, CA, USA, 2025.
47. INEGI. *Diccionario de Datos de Uso del Suelo y Vegetación. (Vectorial) Escala 1:250 000*, 1st ed.; Instituto Nacional de Estadística, Geografía e Informática, INEGI.: Mexico City, Mexico, 1998.
48. Servicio Meteorológico Nacional. CONAGUA. Normales Climatológica por Estado. Available online: https://smn.conagua.gob.mx/tools/RESOURCES/Normales_Climatologicas/Normales9120/oax/nor9120_20079.txt (accessed on 10 December 2025).
49. Foster, L.; Sloan, L.; Clark, T.; Bryman, A. *Bryman's Social Research Methods*, 6th ed.; Oxford University Press: New York, NY, USA, 2022.
50. Lim, W.M. What Is Qualitative Research? An Overview and Guidelines. *Australas. Mark. J.* **2025**, *33*, 199–229. [CrossRef]
51. Rapport, F. Combining methodological approaches in research: Ethnography and interpretive phenomenology. *J. Adv. Nurs.* **2000**, *31*, 219–225. [CrossRef]
52. Coordinación de Comunicación Social. Gobierno del Estado de Oaxaca. Recibe Certificación Hecho en Oaxaca la Unión de Productores de Tejate Sabor a Huayápam A.C. Available online: https://www.oaxaca.gob.mx/comunicacion/recibe-certificacion-hecho-en-oaxaca-la-union-de-productores-de-tejate-sabor-a-huayapam-a-c/ (accessed on 5 December 2025).
53. Arias-Valencia, M.M. Principles, Scope, and Limitations of the Methodological Triangulation. *Investig. Educ. Enfermería* **2022**, *40*, e03.
54. López-Sánchez, M.P.; Alberich, T.; Aviñó, D.; Francés García, F.; Ruiz-Azarola, A.; Villasante, T. Herramientas y métodos participativos para la acción comunitaria. Informe SESPAS 2018. *Gac. Sanit.* **2018**, *32*, 32–40. [CrossRef]
55. de Almeida, D.R.A.; Stark, S.C.; Valbuena, R.; Broadbent, E.N.; Silva, T.S.F.; de Resende, A.F.; Ferreira, M.P.; Cardil, A.; Silva, C.A.; Amazonas, N.; et al. A new era in forest restoration monitoring. *Restor. Ecol.* **2020**, *28*, 8–11. [CrossRef]
56. Gardner, T. *Monitoring Forest Biodiversity: Improving Conservation Through Ecologically Responsible Management*; Routledge: London, UK, 2010; pp. 1–360. [CrossRef]
57. Perz, S.G.; Arteaga, M.; Baudoin Farah, A.; Brown, I.F.; Mendoza, E.R.H.; De Paula, Y.A.P.; Perales Yabar, L.M.; Pimentel, A.D.S.; Ribeiro, S.C.; Rioja-Ballivián, G.; et al. Participatory action research for conservation and development: Experiences from the amazon. *Sustainability* **2021**, *14*, 233. [CrossRef]
58. Lugo-Espinosa, G.; Acevedo-Ortiz, M.A.; Ortiz-Hernández, F.E. Análisis del turismo en huayapam, desde la perspectiva de género y en el contexto de pandemia. In *Innovación, Turismo y Perspectiva de Genero en el Desarrollo Regional*; Rózga-Luter, R.E., Serrano-Oswald, S.E., Mota-Flores, V.E., Eds.; Recuperación transformadora de los territorios con equidad y sostenibilidad; UNAM-AMECIDER: Ciudad de Mexico, Mexico, 2021; Volume V, pp. 517–534.

https://doi.org/10.3390/su18031630

59. Velasco-Murguía, A.; Medina, E.; Garcia, R.; Bray, D. Cambios en la cobertura arbolada de comunidades indígenas con y sin iniciativas de conservación, en Oaxaca, México. *Investig. Geográficas* **2014**, *22*, 55–73. [CrossRef]
60. Wolff, N.H.; Masuda, Y.J.; Meijaard, E.; Wells, J.A.; Game, E.T. Impacts of tropical deforestation on local temperature and human well-being perceptions. *Glob. Environ. Change* **2018**, *52*, 181–189. [CrossRef]
61. Geist, H.J.; Lambin, E.F. Proximate Causes and Underlying Driving Forces of Tropical Deforestation: Tropical forests are disappearing as the result of many pressures, both local and regional, acting in various combinations in different geographical locations. *BioScience* **2002**, *52*, 143–150. [CrossRef]
62. Arsenov, D.; Borišev, M.; Nikolić, N.; Horak, R.; Pajević, S. Physiological and Biochemical Indicators of Urban Environmental Stress in Tilia, Celtis, and Platanus: A Functional Trait-Based Approach. *Plants* **2025**, *14*, 3451. [CrossRef] [PubMed]
63. Czaja, M.; Kołton, A.; Muras, P. The Complex Issue of Urban Trees—Stress Factor Accumulation and Ecological Service Possibilities. *Forests* **2020**, *11*, 932. [CrossRef]
64. Lopez-Toledo, L.; Martínez, M.; van Breugel, M.; Sterck, F.J. Soil and light effects on the sapling performance of the shade-tolerant species *Brosimum alicastrum* (Moraceae) in a Mexican tropical rain forest. *J. Trop. Ecol.* **2008**, *24*, 629–637. [CrossRef]
65. Jangid, R.; Kumar, A.; Masu, M.M.; Kanade, N.; Pant, D. Alternate Bearing in Fruit Crops: Causes and Control Measures. *Asian J. Agric. Hortic. Res.* **2023**, *10*, 10–19. [CrossRef]
66. Coetzee, B.W.T.; van Zyl, L. The Environmental Light Characteristics of Forest Under Different Logging Regimes. *Ecol. Evol.* **2024**, *14*, e70623. [CrossRef]
67. Dueñas-Deyá, A.; González-Anduaga, G.M.; Medina-Torres, L.; Balderas-López, J.L.; Sandoval-Flores, S.D.; Gutiérrez-Rodelo, C.; Manero, O.; Navarrete, A. Fundamental understanding of *Quararibea funebris* flowers mucilage: An evaluation of chemical composition, rheological properties, and cytotoxic estimation. *Food Hydrocoll.* **2025**, *167*, 111431. [CrossRef]
68. Pérez-Ramírez, I.F.; Cariño-Sarabia, A.; Castaño-Tostado, E.; Vázquez-Landaverde, P.A.; Ramos-Gómez, M.; Reynoso-Camacho, R.; Amaya-Llano, S.L. Chemical and sensorial characterization of Tejate, a Mexican traditional maize-cocoa beverage, and improvement of its nutritional value by protein addition. *J. Food Sci. Technol.* **2021**, *58*, 3548–3560. [CrossRef]
69. Sotelo, A.; Soleri, D.; Wacher, C.; Sánchez-Chinchillas, A.; Argote, R.M. Chemical and nutritional composition of tejate, a traditional maize and cacao beverage from the Central Valleys of Oaxaca, Mexico. *Plant Foods Hum. Nutr.* **2012**, *67*, 148–155. [CrossRef]
70. Kundathil, A.; Varga, Z.; Szalay, K.; Sipos, L.; Jung, A. Daily Light Integral (DLI) Mapping Challenges in a Central European Country (Slovakia). *Appl. Sci.* **2025**, *15*, 12254. [CrossRef]
71. Acevedo-Ortiz, M.A.; Lugo-Espinosa, G.; Ortiz-Hernández, Y.D.; Pérez-Pacheco, R.; Ortiz-Hernández, F.E.; Martínez-Tomás, S.H.; Tavera-Cortés, M.E. Nature-Based Solutions for Conservation and Food Sovereignty in Indigenous Communities of Oaxaca. *Sustainability* **2024**, *16*, 8151. [CrossRef]
72. Secretaría de Desarrollo Económico Oaxaca. La Unión de Productores de Tejate "Sabor a Huayapam A.C." Presentó su Marca Colectiva. Available online: https://www.youtube.com/watch?v=Lf3J4_ttBAI (accessed on 5 December 2025).

 https://doi.org/10.3390/su18031630

sustainability

MDPI

Article

Diversity and Geographic Distribution Patterns of Wild Magnoliaceae Species in China

Weihao Yao [1], Yu Fan [1], Zenghui Wang [1], Danyang Liu [1], Zeyang Ding [2] and Jing Ou [1,*]

[1] College of Forestry, Guizhou University, Guiyang 550025, China
[2] College of Agriculture, Guizhou University, Guiyang 550025, China
* Correspondence: coloroj@126.com; Tel.: +86-139-8418-4857

Abstract: Magnoliaceae, an ancient lineage of relict plants, were once extensively distributed throughout geological history. Understanding their species diversity, geographic distribution patterns, and hotspots is essential for providing a theoretical foundation for germplasm resource assessment, conservation, and utilization. This study presents a comprehensive catalog of wild Magnoliaceae species in China, derived from an extensive literature review integrating geographic distribution and endemism data. Utilizing GIS 10.2 software, we analyzed data from up to July 2024, revealing that China hosts 157 wild Magnoliaceae species across 13 genera, including 11 varieties, two subspecies, and three forms. These species are distributed across 24 provincial-level administrative regions, including special administrative regions and municipalities. Yunnan Province exhibits the highest species richness, with 12 genera and 94 species, followed by Guizhou, with 9 genera and 65 species, and Guangxi, with 9 genera and 62 species. At the county level, species are found in 609 regions, with Malipo County in Yunnan recording the highest count (32 species), followed by Leishan County in Guizhou (28 species). Vertically, Magnoliaceae are most abundant in the H3 elevation band ($1000 < H3 \leq 1500$ m), while the H7 elevation band ($H7 > 3000$ m) shows the lowest species diversity. China has two endemic genera, *Pachylarnax* and *Woonyoungia*, along with 102 endemic species, of which *Michelia* accounts for the highest number of endemics at 33. The species composition in Guizhou shows significant similarity to that in Guangxi, Sichuan, and Yunnan, indicating close phylogenetic relationships among these floras. The Yunnan–Guizhou–Guangxi region is identified as a "modern distribution center", "diversity preservation center", "hotspot", and "refuge" for wild Magnoliaceae in China. This study highlights the importance of understanding the diversity and distribution patterns of wild Magnoliaceae for the effective conservation, development, and sustainable utilization of plant resources.

Keywords: Magnoliaceae; species diversity; geographic distribution patterns; endemic species; distribution center

1. Introduction

Magnoliaceae can be categorized using broad and narrow definitions. In the Engler system, Magnoliaceae are defined broadly, encompassing the subfamilies Magnolioideae, Illicioideae, Schisandraceae, and Tetracentraceae [1,2]. Conversely, the Hutchinson system [3] adopts a narrower definition, focusing specifically on the subfamilies Magnolioideae and Liriodendroideae [1,4]. This study concentrates on the narrowly defined Magnoliaceae to provide a more targeted understanding. Magnoliaceae, which originated during the Mesozoic Era, specifically in the Cretaceous or Jurassic periods, are considered ancient relict plants [5]. As pioneers of fruit-bearing flowering plants [6,7], they are among the most representative examples of primitive angiosperms and their evolutionary trajectories [8]. These plants are integral to both evergreen and deciduous broadleaf forests, spanning tropical to temperate regions, and they hold substantial scientific, ecological, ornamental, medicinal, and economic importance [9]. Originating in tropical mountainous regions [10],

Magnoliaceae were once widely distributed but are now predominantly found in the tropical and subtropical zones of Asia and the Americas, encompassing approximately 18 genera and over 300 species [11,12] (Figure 1). They are also significant in landscaping, serving as sources of spices, medicinal resources, and industrial timber [13].

Figure 1. Magnoliaceae plants (partial); (**a**) Yulania biondii. (**b**) Manglietia dandyi. (**c**) Yulania amoena. (**d**) Manglietia conifera. (**e**) Michelia odora. (**f**) Michelia wilsonii subsp. szechuanica. (**g**) Manglietia yuyuanensis. (**h**) Manglietia calcarea. (**i**) Michelia maudiae. (Image provided by Scholar Jiang Guobin, Guizhou Academy of Forestry).

China serves as a significant hub for the diverse Magnoliaceae family, with about 70% of its species listed by the IUCN, many of which are rare, endangered, or endemic. In 1995, Liu Yuhu pioneered a systematic study on the origins and distribution of Magnoliaceae [14]. He compiled extensive data on the family in China in 1996 [15], and, in 2004, published Magnolias of China, covering 11 genera and 178 species, proposing Liu's Taxonomic System of Magnoliaceae. During the compilation of the Flora of China, Xia Nianhe and colleagues revised the classification of some Magnoliaceae species [16]. Subsequently, in 2007, they proposed Xia's Taxonomic System of Magnoliaceae, categorizing them into 17 genera. In the same year, Sima Yongkang introduced the Sima–Lu Taxonomic System of Magnoliaceae. Based on DNA data and morphological characteristics, they proposed new combinations and classifications [17], categorizing the family into 15 genera. Current research primarily focuses on taxonomic revisions [18], phylogenetics [19], and reproductive biology [20] of Magnoliaceae, along with investigations into the diversity and geographic distribution patterns within specific provinces [21–23]. However, studies on the geographic distribution of Magnoliaceae across China remain limited.

Species diversity is the core of biodiversity research [24], reflecting the composition, distribution, community structure, and adaptability of species to their environments [25]. The distribution patterns of species result from adaptation and coordination with the environment. Studies indicate that species diversity decreases with increasing latitude [26,27] and elevation [28] on a large scale. Scholars have proposed that with climate warming, the Magnoliaceae will experience geographical migration, altering their distribution patterns [29]. The spatial distribution patterns of species diversity are fundamental to conservation biology and represent key areas of focus within both ecology and biogeography [26,30]. Understanding these patterns is crucial for identifying hotspots of species diversity. Numerous scholars have investigated the diversity and geographic distribution of Orchidaceae [31], *Ilex* [32], and *Chimonobambusa* [33], thereby providing a scientific basis for their conservation efforts. The distribution of Magnoliaceae plants is influenced by contemporary geographic and ecological factors as well as historical ecological conditions, resulting in complex distribution patterns. Existing studies, which draw from botanical monographs and relevant literature, reveal high species richness in southwestern, central, and southern China, reflecting the overall richness pattern of the Magnoliaceae family. Despite the continuous increase in new distribution records and species discoveries, along with advancements in species classification and taxonomic revisions, systematic organization of wild Magnoliaceae diversity data in China remains lacking. This gap hinders the effective conservation and sustainable utilization of Magnoliaceae species.

This study synthesizes a substantial body of literature and incorporates the latest research on newly described Magnoliaceae taxa and their distributions, creating a comprehensive catalog detailing species and their distribution patterns. It examines hotspot regions and patterns of similarity in Magnoliaceae distribution, elucidating their horizontal and vertical distribution patterns. The findings provide crucial data support for understanding the origins, adaptive mechanisms, and distribution predictions of Magnoliaceae, offering a scientific foundation for the survey, conservation, and utilization of germplasm resources.

2. Materials and Methods

2.1. Data Sources

Currently, a unified classification system for Magnoliaceae plants has not been established. This study employs the APG IV classification system to assign species to their respective families and genera. The data sources include the following: (1) 53 references from Flora of China (Vol. 7), regional floras, and plant lists, including online resources; (2) 59 monographs and related books, such as Magnoliaceae of Guizhou and Magnolias of China; (3) 42 academic papers, including published theses and dissertations up to July 2024, as well as literature on new Magnoliaceae taxa and records in China; (4) specimen data from the National Specimen Resource Sharing Platform (NSII, http://www.nsii.org.cn/2017/home.php (accessed on 1 August 2024)) and the Chinese Virtual Herbarium (CVH, https://www.cvh.ac.cn/ (accessed on 25 August 2024)), totaling 852 records, with questionable and cultivated specimens excluded; and (5) distribution information from scientific surveys of wild Magnoliaceae in various natural reserves across China, comprising 84 documents.

Data standardization was performed using Flora of China (Vol. 7) and the Catalogue of Life China 2024 Annual Checklist (http://www.sp2000.org.cn/CoLChina (accessed on 10 June 2024)), with meticulous verification. Refer to the IUCN Red List for species conservation assessments and check the conservation status of species using the Information System of Chinese Rare and Endangered Plants (ISCREP, https://www.iplant.cn/rep/protlist?page=8 (accessed on 15 April 2024)). Place names were cross-checked using standard maps from the Ministry of Natural Resources (http://bzdt.ch.mnr.gov.cn/ (accessed on 7 May 2024)), the NSII's regional change query tool, and place name conversion tools from the Chinese Virtual Herbarium. The inventory of wild Magnoliaceae in China was used to correlate species distribution data with administrative boundaries using ArcGIS 10.2. The natural breaks classification method (Jenks) was then applied to produce a map depicting the current distribution of wild Magnoliaceae in China.

2.2. Screening Algorithm

Utilizing the inventory of wild Magnoliaceae species in China, this study applies the Dobson exclusion algorithm to delineate hotspots of Magnoliaceae distribution based on species richness [34]. The methodology is as follows: (1) provincial species richness is ranked according to the species inventory, identifying the province with the highest richness, and subsequently removing its species from the list; (2) the remaining species and provinces are re-ranked by species richness, with iterative selections and exclusions continuing until all species are accounted for, thereby pinpointing provincial hotspots for Magnoliaceae; (3) provincial hotspots are established by evaluating species richness rankings and analyzing the distribution patterns to determine the central areas of these hotspots; (4) for county-level hotspots, the province with the highest species richness is selected, applying a similar methodology as used for provincial hotspots; (5) in instances where regions exhibit identical species richness, priority is given to the smaller area [35].

2.3. Data Analysis

(1) For the assessment by county-level administrative regions, the study counts the number of wild Magnoliaceae species in China. To address the impact of varying sizes of administrative regions on species density analysis, we applied the species–area relationship theory [36], using the following formula:

$$\rho s = \log Ns / \log A \quad (1)$$

In this formula, Ns represents the number of wild Magnolia species within a given county-level administrative region, A denotes the area of that region, and ρs indicates the species distribution density for the county-level administrative region.

(2) The similarity in plant composition between different regions reflects the phylogenetic relationships among their floras. Typically, the Coefficient of Similarity is used as a fundamental quantitative measure to assess similarity. For these calculations, the Czekanowski coefficient is employed [37,38]. The formula is as follows:

$$\text{Similarity Coefficient} = 2w/(a + b) \quad (2)$$

In this formula, a represents the total number of species in region A, b denotes the total number of species in region B, and w is the number of species common to both regions. By selecting regions at the provincial and county levels with $\geq$17 species of Magnoliaceae, we compared the species composition and abundance to calculate the similarity coefficient.

(3) Analysis of Vertical Spatial Distribution Similarity: The numbers of species and genera across different elevation ranges were quantified, and the Jaccard similarity coefficient was used for calculation [37,38]. The formula is as follows:

$$Cj = j/(a + b - j) \quad (3)$$

In this formula, Cj denotes the similarity coefficient; a and b represent the numbers of Magnoliaceae species in elevation ranges A and B, respectively; and j indicates the number of Magnoliaceae species common to both elevation ranges. Based on the recorded altitude distribution data for wild Magnoliaceae, the elevation range of Magnoliaceae in China spans from 0 to 3500 m. The elevation range of a species is determined by its recorded maximum and minimum altitudes. Divisions were made at 500 m intervals, resulting in a total of seven elevation gradients, labeled H1 through H7.

3. Results

3.1. Species Diversity of Wild Magnoliaceae in China

As of July 2024, the wild Magnoliaceae in China comprise 13 genera and 157 species, including 11 varieties, two subspecies, and three forms, with 26 new taxa described over the past 27 years. According to Wu Zhengyi's classification of seed plant genera in China [37], these plants are categorized into five distribution types and four forms. As detailed in Table 1, the predominant distribution type spans from Tropical Southeast Asia to Indo-Malaya and the Tropical Southwest Pacific Islands, accounting for 46.1% of the total genera. This is followed by disjunct distribution types in East Asia and North America disjuncted and East Asia (Tropical and Subtropical) and Tropical South America disjuncted, representing 23.1% and 15.4%, respectively. The East Asia and Endemic to China distribution types represent the lowest proportions, each at 7.7%.

Table 1. Types of distribution areas of wild Magnoliaceae genera in China.

Areal—Type	Areal—Subtype	Genus	Total Genus	Rate (%)
East Asia (Tropical and Subtropical) and Tropical South America disjuncted	--	Talauma	2	15.4
	Tropical and Subtropical Central to South America (including central Mexico and West Indies)	Magnolia		
Tropical Southeast Asia to Indo-Malaya and the Tropical Southwest Pacific Islands	--	Michelia; Manglietia; Pachylarnax	6	46.1
	Tropical India to South China (especially southern Yunnan)	Alcimandra		
	Vietnam (or Indochinese Peninsula) to South or Southwest China	Woonyoungia; Lirianthe		
East Asia and North America disjuncted	--	Liriodendron; Houpoea; Yulania	3	23.1
Endemic to China	--	Parakmeria	1	7.7
East Asia	Sino-Japan	Oyama	1	7.7

Wild Magnoliaceae in China are primarily evergreen trees (Table 2) and are mainly distributed in the southwestern regions of the country. The Magnoliaceae in China demonstrate significant genus and species diversity, representing approximately 50% of the global wild Magnoliaceae resources. Of these, 102 endemic species, including regional endemic species and China's endemic species, constitute 65% of China's Magnoliaceae resources. At the genus level, *Michelia* exhibits the most diversity, with 57 species (33 endemic species), accounting for 36% of the total. It is followed by *Manglietia*, with 40 species (24 endemic species), making up 25%, and *Yulania*, with 28 species (25 endemic species), representing 18% of the total. These three genera are the principal components of wild Magnoliaceae in China. Genera with fewer than 20 species include *Parakmeria*, *Woonyoungia*, and *Lirianthe*, totaling 10 genera and 32 species, which account together for 21% of the total species. Notably, *Talauma* and *Pachylarnax* each have only a single species distributed in China.

Table 2. Status table of wild Magnoliaceae resources in China.

Genus	Total Species	Rate (%)	Endemic Species	Rate (%)	Endangered Status	Protected Status
Magnolia	1	0.64	1	0.64	--	--
Liriodendron	1	0.64	1	0.64	--	1
Talauma	1	0.64	0	0.00	1	1
Pachylarnax	1	0.64	1	0.64	1	1
Woonyoungia	1	0.64	1	0.64	1	1
Alcimandra	1	0.64	0	0.00	1	1
Lirianthe	13	8.28	9	5.73	7	3
Oyama	5	3.18	2	1.27	3	2
Yulania	28	17.83	25	15.92	9	8
Manglietia	40	25.48	24	15.29	26	11
Michelia	57	36.31	33	21.02	27	14
Houpoea	3	1.91	1	0.64	2	2
Parakmeria	5	3.18	4	2.55	4	4

Situation of risk: species included in the IUCN Red List. Refer to the IUCN Red List of Threatened Species. Protected status: species included in national or provincial protection in China receive policy-based conservation. List of National Key Protected Wild Plants in China, Information System of Chinese Rare and Endangered Plants ((ISCREP) https://www.iplant.cn/rep/protlist/4 (accessed on 1 September 2024)).

3.2. Horizontal Distribution of Diversity

3.2.1. Provincial Scale

Species diversity across China's provincial regions was analyzed and categorized (Figure 2). The wild Magnoliaceae in China exhibit a broad distribution range from 22° to 28° N and 98° to 112° E, spanning 24 provincial-level regions. Yunnan Province leads with 94 species (59.87% of the total), followed by Guizhou and Guangxi with 65 and

62 species, respectively. Sichuan, Jiangxi, and five other provinces are classified as Level II; Hainan, Zhejiang, and five additional provinces as Level III; Henan, Shaanxi, and ten other provinces as Level IV; no species are found in Level V areas (Figure 2a). Endemic species are most numerous in Yunnan, Guizhou, and Guangxi, with 46, 40, and 35 species, respectively. Sichuan and Hunan follow with 24 and 23 species, while Xinjiang, Jilin, and Liaoning lack endemic species (Figure 2b). Endemism generally contrasts with species richness, being higher in Shanxi, Jiangsu, and Gansu, and lower in the Dian–Qian–Gui region, with Tibet exhibiting the lowest endemism (Figure 2c). The species-to-genus ratio, reflecting floristic differentiation, is highest in Yunnan, followed by Guizhou and Guangxi, with Jilin showing the lowest ratio (Figure 2d). Overall, the Dian–Qian–Gui region is notable for its high species richness. The Magnoliaceae developmental system is comprehensive, demonstrating significant representativeness at each stage and complex differentiation, marking it as a key area for Magnoliaceae in China.

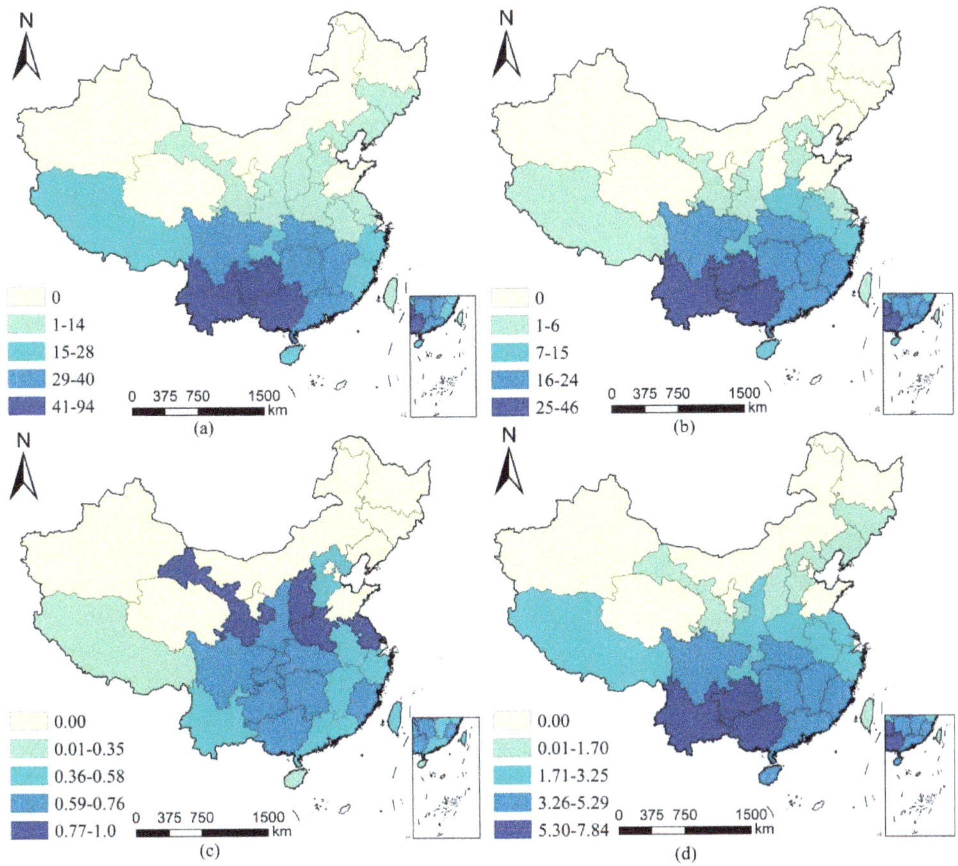

Figure 2. Provincial distribution of Magnoliaceae species. (**a**) Species richness. (**b**) Abundance of endemic species. (**c**) Ratio of endemic/all species. (**d**) Ratio of species/genus.

3.2.2. County Scale

Wild Magnoliaceae are distributed across 609 county-level regions, showcasing extensive coverage (Figure 3). Twelve counties host over 20 species each, with Malipo County, Yunnan, having the highest richness at 32 species, followed by Leishan County, Guizhou, and Jinping County, Yunnan, each with 28 species (Figure 3a). Endemic species are found in 496 counties, with Leishan County (17 species), Rongjiang County (16 species), Liping County (16 species), Libo County (15 species), Huanjiang County (15 species), and Malipo County (15 species) having the highest numbers of endemic species (Figure 3b). Endemism contrasts with species richness, often showing a ring-like scattered pattern. There are 158 counties with 100% endemism, predominantly in Sichuan, Shaanxi, and Gansu (Figure 3c). The species-to-genus ratio, reflecting species richness, is highest in Libo County, Guizhou; Leishan County, Guizhou; Huanjiang County, Guangxi; Xichou County, Yunnan; and Maguan County, Yunnan (Figure 3d). Southeastern Yunnan, southwestern Guizhou, and northern Guangxi are crucial for Magnoliaceae differentiation, representing multiple evolutionary stages and offering key insights into their development. High species density is noted at the Guizhou–Hunan–Guangxi junction and southeastern Yunnan, particularly in Leishan, Libo, Liping, Rongjiang, Xichou, Malipo, and Emei Mountain (Figure 4). In summary, wild Magnoliaceae in China exhibit a distribution pattern centered on the Dian–Qian–Gui region, with high-density hubs at the Guizhou–Hunan–Guangxi junction and southeastern Yunnan, creating a multi-point, radiating pattern.

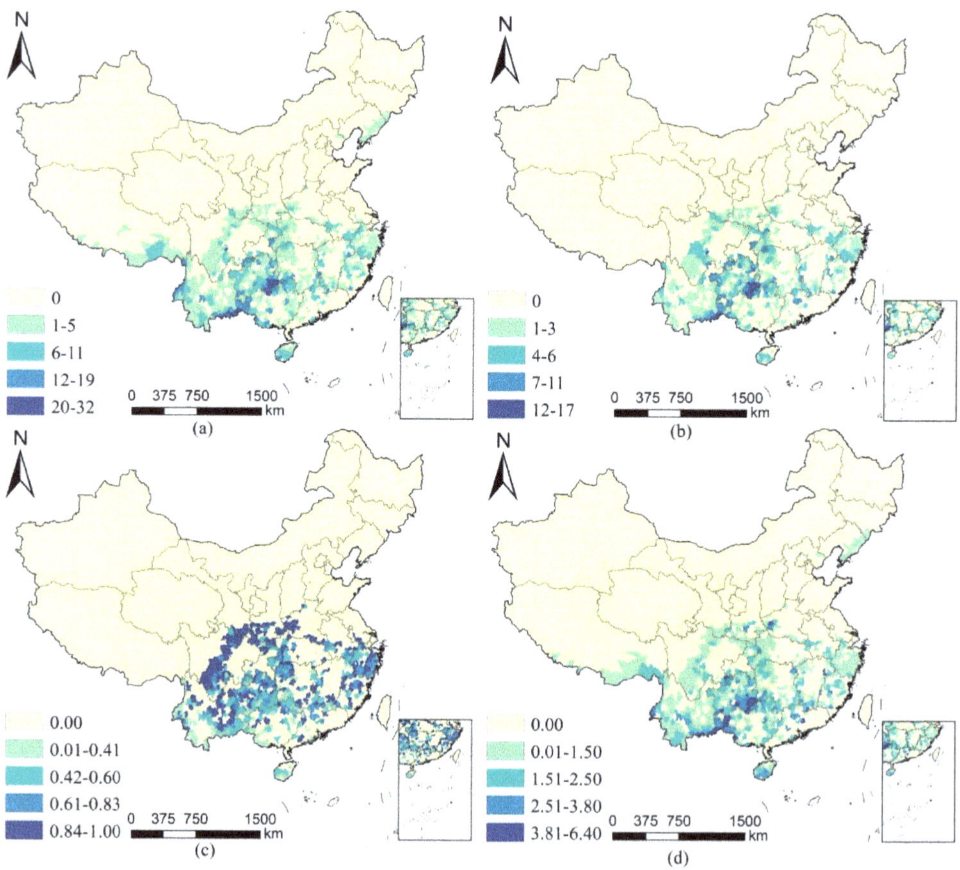

Figure 3. County distribution of Magnoliaceae species. (**a**) Species richness. (**b**) Abundance of endemic species. (**c**) Ratio of endemic/all species. (**d**) Ratio of species/genus.

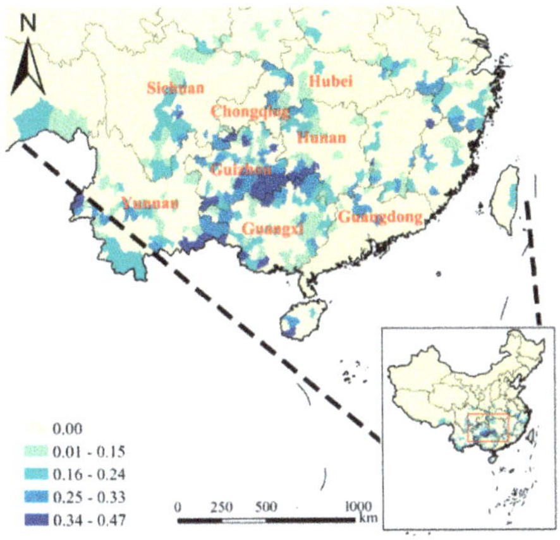

Figure 4. Distribution map of species density of wild Magnoliaceae at county scale.

3.2.3. Hotspots

Using the Dobson exclusion algorithm, we identified distribution areas of species hotspots. In the provincial hotspot selection, a 100% species selection rate yielded 157 species and 16 hotspots (Figure 5a). Yunnan is the primary provincial hotspot for Magnoliaceae in China, hosting 94 species, followed by the Guangdong–Guangxi–Jiangxi region. The overall distribution pattern centers on Yunnan as the primary hotspot, with Guangdong–Guangxi–Jiangxi as a secondary hotspot, exhibiting a decreasing gradient to the north and resulting in a dual hotspot distribution pattern centered in the southwestern region, with the South China region as secondary. In the selection of county-level hotspots, we focused on Yunnan and Guizhou provinces, which have high species richness. With a 100% species selection rate, 100 species and 19 hotspots were identified (Figure 5b). Major hotspots include Malipo County and Jinping County in Yunnan, and Leishan County in Guizhou, while secondary hotspots include Libo County in Guizhou and Maguan County in Yunnan, covering a total of 73 species, with Malipo County having the highest species count of 32.

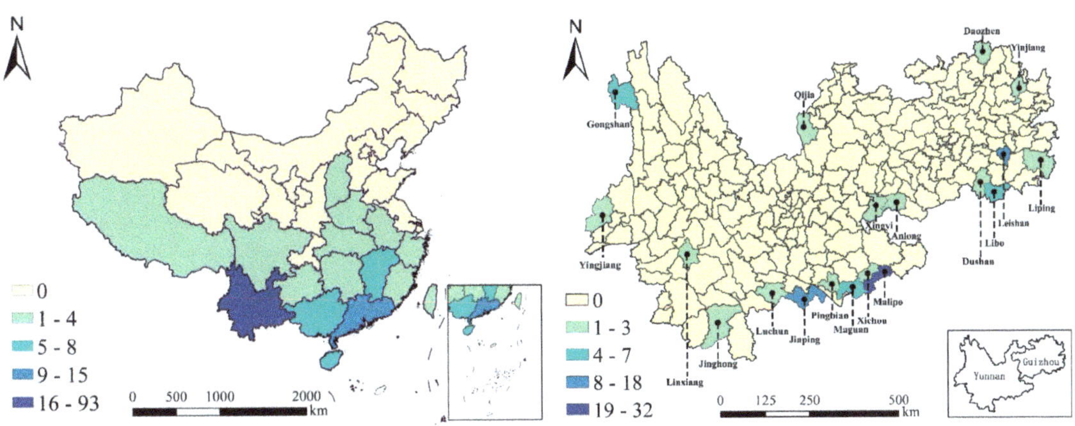

Figure 5. Distribution hotspots of wild Magnoliaceae species in China established by sieve removal algorithm at provincial scale. (**a**) Provincial hotspot distribution. (**b**) Distribution of hotspots at county level.

3.3. Vertical Distribution Pattern

Based on the vertical distribution of Magnoliaceae plants (Figure 6), *Michelia macclurei* is found at an elevation of 0 m (sea level), while *Oyama sieboldii* reaches the highest elevation of 3500 m. Other species close to sea level include *Michelia retusa*, which is found at a minimum elevation of 20 m. In the altitude gradients from H1 to H7, the H3 elevation zone has the highest species richness, encompassing 12 genera and 96 species, while the H7 zone has the lowest richness, with only 2 genera and 5 species. *Oyama* is predominantly distributed above 500 m, while *Magnolia*, *Talauma*, and *Woonyoungia* are primarily found below 1000 m. *Michelia* and *Manglietia* are mainly distributed between 500 and 2000 m, while *Yulania* is present across all elevation zones.

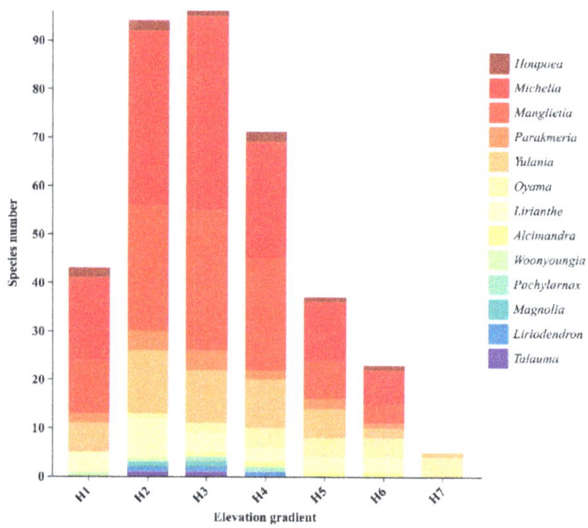

Figure 6. Vertical distribution analysis of Magnoliaceae: $0 < H1 \leq 500$ m, $500 < H2 \leq 1000$ m, $1000 < H3 \leq 1500$ m, $1500 < H4 \leq 2000$ m, $2000 < H5 \leq 2500$ m, $2500 < H6 \leq 3000$ m, and $H7 > 3000$ m. For example, if a species has an elevation range of 600–1200 m, it is recorded in both the H1 and H2 elevation gradients.

3.4. Similarity in Species Distribution

Analyzing shared species and distribution similarity is crucial to understanding the Magnoliaceae family's flora and phylogenetic relationships. Regions with a species richness of ≥17 were included in the analysis. The results indicate that in the provincial-level analysis, there is a high similarity in species distribution between Guizhou and the provinces of Guangxi, Yunnan, Sichuan, and Hunan (Figure 7a). The highest similarity of 0.68 is observed between Guizhou and Guangxi, while the lowest, 0.02, is between Hainan and Chongqing. In the county-level analysis, Guizhou's Rongjiang, Leishan, Congjiang, and Liping counties exhibit high species similarity. Guangxi's Lingui, Longsheng, and Huanjiang counties, along with Yunnan's Malipo, Maguan, and Xichou counties, also show high similarity (Figure 7b). The highest similarity, 0.86, is between Rongjiang and Congjiang in Guizhou, followed by 0.84 between Liping in Guizhou and Huanjiang in Guangxi. The lowest similarity (0) is found between Ledong in Hainan and Jiangkou in Guizhou. Higher species distribution similarity suggests closer phylogenetic relationships. Vertical distribution similarity (Figure 7c) averages 0.24, with a minimum value of 0 (H1–H7) and a maximum of 0.62 (H5–H6). Greater similarity between adjacent elevation gradients suggests that many Magnoliaceae species are adaptable to varying altitudes.

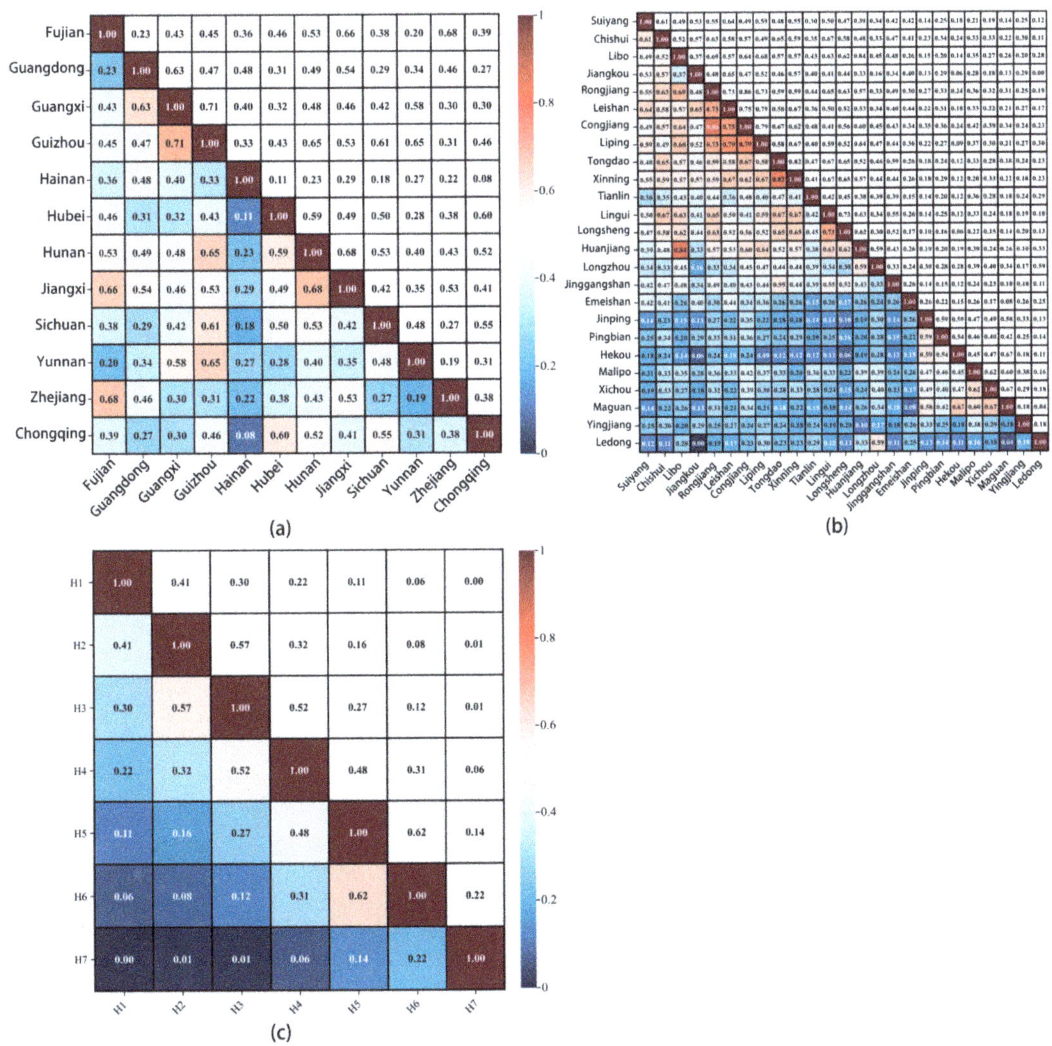

Figure 7. Distributional similarity analysis of Magnoliaceae species. (**a**) Species distribution similarity at the provincial scale. (**b**) Species distribution similarity at the county scale. (**c**) Species similarity across altitudinal gradients.

4. Discussion

4.1. Diversity Characteristics of Wild Magnoliaceae Plants in China

The wild Magnoliaceae plants of China exhibit remarkable diversity, consisting of 13 genera and 157 species. *Michelia* and *Manglietia* are the most species-rich genera, with 57 and 40 species, respectively. In contrast, *Woonyoungia*, *Alcimandra*, and *Pachylarnax* are monospecific genera, each represented by a single species. Similarly, *Liriodendron*, and *Talauma* also contain only one species each. *Pachylarnax* is an endemic genus, and among the 102 endemic species, such as *Manglietia glaucifolia*, 59 are exclusive to China while 43 are regionally endemic. The distribution types of wild Magnoliaceae in China are categorized into five primary types and four variants, primarily ranging from Tropical

Southeast Asia to Indo-Malaya and the Tropical Southwest Pacific Islands and East Asia and North America disjuncted.

4.2. Distribution Centers and Patterns of Magnoliaceae Plants

This study employed Geographic Information Systems (ArcGIS10.2) to analyze the spatial distribution characteristics of Magnoliaceae species at both the provincial and county levels with enhanced precision and effectiveness. The distribution of wild Magnoliaceae species in China spans 24 provincial-level administrative regions (including municipalities directly under the central government) and 582 county-level administrative regions. Statistical analysis reveals a concentration of Magnoliaceae species in southwestern China, with the highest species richness in Yunnan Province, followed by Guizhou and Guangxi provinces. According to Wang Hesheng's [39] definition of distribution centers, the Yunnan–Guizhou–Guangxi region is identified as a primary distribution center (or diversity center) for Magnoliaceae plants in China. At the provincial level, Yunnan Province, hosting 94 species, is classified as a provincial hotspot. At the county level, 73 species are concentrated in Malipo County (Yunnan) and Leishan County and Libo County (Guizhou), forming a county-level hotspot. Overall, this indicates that the Yunnan–Guizhou–Guangxi region is a significant hotspot for Magnoliaceae plants. Based on phytogeographical theories [39,40], Wu Lufu's [41] criteria for distribution centers, and the differentiation rates and intensities reflected in the species-to-genus ratio [42], this region demonstrates high differentiation rates. It encompasses the key stages of evolutionary development in the Magnoliaceae family, exhibiting high developmental integrity and strong representativeness. Consequently, the Yunnan–Guizhou–Guangxi region is identified as both a differentiation center [40] and a diversification center (evolutionary center) [43,44] for Magnoliaceae plants in China. These findings are consistent with research indicating that the southwestern region of China, as a biodiversity hotspot, serves as a differentiation center for many taxa [45].

Based on the discovery of Magnoliaceae fossils and the completeness of their developmental system [46,47], along with the niche conservatism hypothesis [48], *Michelia* and *Manglietia*, as the most primitive genera in the Magnoliaceae family [49], are densely and extensively distributed in southwestern China. This region is consequently recognized as one of the primary origins of Magnoliaceae plants. Recent studies increasingly employ ecological niche models to predict suitable habitats [40,50–54], suggesting that the southwestern mountains of China, including the Himalayan region, Hengduan Mountains, and Sichuan basin, acted as significant "refuges" for Magnoliaceae during glacial periods. Liu Yuhu et al. [14] have identified the East Asia–Southeast Asia region as the modern distribution center for Magnoliaceae, reinforcing the critical role of southwestern China in their distribution. Species composition varies across different altitudinal ranges, as elevation changes directly influence water and thermal conditions. These changes affect physiological processes and limit distribution, thereby impacting species diversity [55]. Magnoliaceae species are predominantly concentrated between 1000 and 1500 m above sea level, encompassing 12 genera and 96 species. This elevation range is identified as a key vertical distribution center for Magnoliaceae.

4.3. The Similarity in the Distribution of Magnoliaceae Species

At the horizontal scale, Guizhou Province and its surrounding areas exhibit a high similarity, attributed to several factors: they are located in the transition zone between the Yunnan–Guizhou Plateau and the Hunan–Guangxi Hills, share the same Hongshui River basin and are characterized by a similar karst topography [55] and subtropical monsoon climate. Additionally, the complex terrain, including mountainous and plateau regions, contributes to geographical isolation from other areas. The adjacent regions of Yunnan, Guizhou, and Guangxi are rich in plant resources, with floristic compositions that are distinctly transitional and intermingled. Furthermore, the protected areas maintain relatively pristine habitat conditions, with contiguous distributions forming biological corridors that facilitate species exchange. At the vertical scale, species distribution similarity is higher between the H2 and H3, as well as the H5 and H6, altitudinal gradient, while it is lower between the H1 and H7 altitudinal gradient. This pattern is primarily due to the broad span of the altitudinal gradient and increased habitat variability, which decrease species distribution similarity and enhance the prominence of vertical distribution patterns. Additionally, Magnoliaceae plants, predominantly trees and shrubs, exhibit notable environmental adaptability.

5. Conclusions

To date, wild Magnoliaceae in China encompass 13 genera and 157 species. This study provides a comprehensive analysis of their distribution patterns, focusing on factors such as species richness, endemic species richness, endemism, differentiation rates, similarity, and distribution hotspots. The findings reveal that wild Magnoliaceae are primarily concentrated in the Yunnan–Guizhou–Guangxi region, which demonstrates a radial distribution pattern with southeastern Yunnan as the core area and the Guizhou–Hunan–Guangxi border as a key hotspot. This region is identified as a "modern distribution center", "evolutionary center", "diversity conservation center", "distribution hotspot", and "origination center" for wild Magnoliaceae in China. Consequently, the Yunnan–Guizhou–Guangxi region emerges as a crucial area for the conservation of these diverse plants. According to the species records, only 49 Magnoliaceae species receive protection at the provincial level or higher, while 82 species are classified as threatened (NT or higher) on the IUCN Red List, highlighting a significant gap in conservation efforts. To address this, it is imperative to accelerate the protection of rare and endangered species, enhance in situ conservation measures, implement ex situ conservation strategies, boost public awareness, and prioritize conservation actions in their distribution areas.

Author Contributions: Conceptualization, W.Y. and Y.F.; methodology, W.Y.; software, W.Y.; validation, Y.F., Z.W. and D.L.; formal analysis, Z.D.; investigation, Z.W., D.L. and Z.D.; resources, Y.F.; data curation, D.L. and Z.D.; writing—original draft preparation, W.Y.; writing—review and editing, W.Y., Y.F., Z.W. and J.O.; visualization, W.Y.; supervision, J.O.; project administration, J.O.; funding acquisition, J.O. All authors have read and agreed to the published version of the manuscript.

Funding: This research was funded by the National Key Wildlife and Plant Conservation Project for Central Forestry and Grassland Transfer Payments (MCHC-SC20232028).

Institutional Review Board Statement: Not applicable.

Informed Consent Statement: Not applicable.

Data Availability Statement: The datasets presented in this article are not readily available because the data are part of an ongoing study. Access to the dataset can be obtained through the following private link: (https://www.scidb.cn/s/iYni2a (accessed on 10 September 2024)).

Conflicts of Interest: The authors declare no conflicts of interest.

References

1. Liu, Y. A Preliminary Study on the Taxonomy of the Family Magnoliaceae. *J. Syst. Evol.* **1984**, *22*, 89–109.
2. Li, Y. *National Protected Wild Plants in Yunnan Province, China*; Yunnan Science and Technology Publishing House: Kunming, China, 2005.
3. Hutchinson, J. The Family Winteraceae. *Bull. Misc. Inf. (R. Bot. Gard. Kew)* **1921**, *5*, 185–191. [CrossRef]
4. Sima, Y.; Lu, S.; Han, M.; Ma, H.; Xu, T.; Xu, L.; Hao, J.; Li, D.; Chen, S. Advances in Magnoliaceae Systematic Research. *J. West China For. Sci.* **2012**, *41*, 116–127.
5. Li, J. Analysis of Genetic Relationship of Magnoliaceae Based on ISSR. Master's Thesis, Central South University of Forestry and Technology, Changsha, China, 2021.
6. Gong, X.; Shi, S.; Pan, Y.; Huang, Y.; Yin, Q. An Observation on the Main Taxonomic Characters of Subfamiliy Magnolioideae in China. *Acta Bot. Yunnanica* **2003**, *25*, 447–456.
7. Mu, N.; Zhu, K.; Li, S.; Dong, J.; Dong, Z.; Xiao, N. Discovery of Magnoliaceae Taxa and Its Research Progress in China. *J. Anhui Agric. Sci.* **2022**, *50*, 222–226.
8. Zhang, B. Floral analysis of Magnoliaceae. *Guihaia* **2001**, *21*, 315–320.

9. Wu, Z.; Lu, A.; Tang, Y.; Chen, Z.; Li, D. *The Families and Genera of Angiosperms in China a Comprehensive Analysis;* Science Press: Beijing, China, 2003; pp. 57–68.
10. Li, W.; Mao, P. The distribution of Magnoliaceae and continental drift. *Trop. Geogr.* **1990**, *10*, 138–142.
11. Xie, H.; Tang, Y.; Fu, J.; Chi, X.; Du, W.; Dimitrov, D.; Liu, J.; Xi, Z.; Wu, J.; Xu, X. Diversity patterns and conservation gaps of Magnoliaceae species in China. *Sci. Total Environ.* **2022**, *813*, 152665. [CrossRef]
12. Editorial Committee of Flora of China; Chinese Academy of Sciences (Eds.) *Flora of China: Vol.30;* Science Press: Beijing, China, 1996; pp. 82–85.
13. Sima, Y.; Wang, J.; Cao, L.; Wang, B.; Wang, Y. Prefoliation Features of the Magnoliaceae and their Systematic Significance. *J. Yunnan Univ. Nat. Sci. Ed.* **2001**, *23*, 71–78.
14. Liu, Y.; Xia, N.; Yang, H. The origin, evolution and phytogeography of Magnoliaceae. *J. Trop. Subtrop. Bot.* **1995**, *3*, 1–12.
15. Liu, Y.; Wu, R. Materials for Chinese Magnoliaceae. *J. Syst. Evol.* **1996**, *34*, 87–91.
16. Xia, N.; Deng, Y. Notes on Magnoliaceae. *J. Trop. Subtrop. Bot.* **2002**, *2*, 128–132.
17. Sima, Y.; Yu, H.; Ma, H.; Hao, J.; Chen, S.; Li, S.; Fu, Y. New Combinations in Magnoliaceae. *J. West China For. Sci.* **2020**, *49*, 29–40.
18. Wang, Y.; Liu, B.; Nie, Z.; Chen, H.; Chen, F.; Figlar, R.; Wen, J. Major clades and a revised classification of *Magnolia* and Magnoliaceae based on whole plastid genome sequences via genome skimming. *J. Syst. Evol.* **2020**, *58*, 673–695. [CrossRef]
19. Fu, D.; Fu, H.; Duan, R.; Qin, Y. Evolutionary System of Magnoliaceae Based on Chloroplast Genomic and Morphological Evolutionomy. *Am. J. Agric. For.* **2024**, *12*, 22–50. [CrossRef]
20. Lin, L.; Cai, L.; Fan, L.; Ma, J.; Yang, X.; Hu, X. Seed dormancy, germination and storage behavior of Magnolia sinica, a plant species with extremely small populations of Magnoliaceae. *Plant Divers.* **2022**, *44*, 94–100. [CrossRef]
21. He, F.; Ma, Q.; Liu, X.; Wang, J.; He, D. Characteristics of biology and geographical distribution of wild Magnoliaceae plants in Sichuan Province. *Sci. Soil Water Conserv.* **2008**, *06*, 48–53.
22. Li, R.; Jin, X.; Ning, Y.; Chen, J.; Cai, M. Studies on the Resources and Geographical Distribution of the Magnoliaceae in Hunan Province. *J. Fudan Univ. (Nat. Sci.)* **2016**, *55*, 267–274.
23. Yuan, C.; Yang, Y.; Dai, X.; Li, H.; Yang, C.; Wu, Y.; Wang, J. Magnoliaceae Plants and Its Distribution Pattern in Guizhou Province. *J. West China For. Sci.* **2017**, *46*, 68–75.
24. Duffy, J.; Godwin, C.; Cardinale, B. Biodiversity effects in the wild are common and as strong as key drivers of productivity. *Nature* **2017**, *549*, 261–264. [CrossRef]
25. Flynn, D.; Mirotchnick, N.; Jain, M.; Palmer, M.; Naeem, S. Functional and phylogenetic diversity as predictors of biodiversity-ecosystem-function relationships. *Ecology* **2011**, *92*, 1573–1581. [CrossRef]
26. Wu, A.; Deng, X.; Ren, X.; Xiang, W.; Zhang, L.; Ge, R.; Niu, Z.; He, H.; He, L. Biogeographic patterns and influencing factors of the species diversity of tree layer community in typical forest ecosystems in China. *Acta Ecol. Sin.* **2018**, *38*, 7727–7738.
27. Xu, M.; Ma, L.; Jia, Y.; Liu, M. Integrating the effects of latitude and altitude on the spatial differentiation of plant community diversity in a mountainous ecosystem in China. *PLoS ONE* **2017**, *12*, e0174231. [CrossRef] [PubMed]
28. Vázquez-García, J.A.; Givnish, T. Altitudinal gradients in tropical forest composition, structure, and diversity in the Sierra de Manantlán. *J. Ecol.* **1998**, *86*, 999–1020.
29. Wu, H.; Liu, Y.; He, Q.; Ye, J.; Tian, B. Differential distribution shifts in two subregions of East Asian subtropical evergreen broadleaved forests-a case of Magnoliaceae. *Front. Plant Sci.* **2023**, *14*, 1326207. [CrossRef]
30. Wang, S. Spatial distribution pattern and diversity of plant resources in Northeast China. *Nat. Resour. Informatiz.* **2023**, *01*, 27–31.
31. Chang, M.; Yu, J.; Ma, L.; Chen, S.; Lan, S.; Wang, S. Species diversity and geographical pattern of wild Orchidaceae in Fujian Province. *J. Fujian Agric. For. Univ. (Nat. Sci. Ed.)* **2024**, *53*, 622–628.
32. Zhou, P.; Zhu, Y.; Liu, B.; Li, F.; Huang, J.; Li, M. Geographical distribution pattern of species diversity of the genus *llex* in China. *J. Cent. South Univ. For. Technol.* **2022**, *42*, 126–132.
33. Zhang, Y.; Xia, N.; Lin, S.; Ding, Y. Spatial distribution characteristics and influencing factors of Chimonobambusa in China. *J. Nanjing For. Univ. (Nat. Sci. Ed.)* **2024**, *42*, 1–10.
34. Dobson, A.P.; Rodriguez, J.P.; Roberts, W.M.; Wilcove, D.S. Geographic Distribution of Endangered Species in the United States. *Science* **1997**, *275*, 550–553. [CrossRef]
35. Huang, M.; Li, M.; Huang, H.; Zhang, J.; Bai, X. Species diversity and geographic distribution of wild Gesneriaceae in Guizhou. *Guihaia* **2022**, *42*, 210–219.
36. Tian, L.; An, M.; Yang, Y.; Li, F. Composition characteristics and geographical distribution pattern of nationakey protected wild plants distributed in Guizhou Province. *J. Plant Resour. Environ.* **2023**, *32*, 83–91.
37. Wu, Z.; Zhou, Z.; Sun, H.; Li, D.; Peng, H. *The Areal-Types of Seed Plants and Their Origin and Differentiation;* Yunnan Science and Technology Press: Kunming, China, 2006; pp. 146–452.
38. Yu, S.; Li, M.; Yuan, T.; Ren, L.; Bai, X. Species Diversity and Geographic Distribution of Wild *Impatiens* in Guizhou Province. *Acta Bot. Boreali-Occident. Sin.* **2021**, *41*, 863–871.
39. Wang, H. *Floristic Plant Geography;* Science Press: Beijing, China, 1992; pp. 1–149.
40. Takhtajan, A. *Floristic Regions of the World;* Huang, G., Translator; Science Press: Beijing, China, 1988; pp. 1–522.
41. Wulff, E.V. *Historical Geography of Plants: History of Floras of the Earth;* Zhong, C., Translator; Science Press: Beijing, China, 1964; pp. 1–632.

42. Zhang, Y.; Du, H.; Jin, X.; Ma, K. Species diversity and geographic distribution of wild Orchidaceae in China. *Chin. Sci. Bull.* **2015**, *60*, 179–188.
43. Lin, L.; Liu, Q. Geographical Distribution of *Avena* L. (Poaceae). *J. Trop. Subtrop. Bot.* **2015**, *23*, 111–122.
44. Liu, H.; Zeng, F.; Liu, Q. Geographical Distribution of *Sorghum* Moench (Poaceae). *J. Trop. Subtrop. Bot.* **2014**, *22*, 1–11.
45. Liu, Q.; Wu, X.; Xing, H.; Chi, K.; Wang, W.; Song, L.; Xing, X. Orchid diversity and distribution pattern in karst forests in eastern Yunnan Province, China. *For. Ecosyst.* **2023**, *10*, 348–356. [CrossRef]
46. Zhou, A.; Cheng, Y.; Chen, H.; Xu, C.; Zhang, Y. Prediction of potential suitable areas of *Michelia figo* in China based on MaxEnt model. *J. Anhui Sci. Technol. Univ.* **2023**, *37*, 19–27.
47. Huan, Z.; Geng, X.; Xu, X.; Liu, W.; Zhu, Z.; Tang, M. Potential Geographical Distribution of *Michelia martinii* under Different Climate Change Scenarios Based on MaxEnt Model. *J. Ecol. Rural. Environ.* **2023**, *39*, 1277–1287.
48. Anoop, N.R.; Babu, S.; Nagarajan, R.; Sen, S. Identifying suitable reintroduction sites for the White-rumped Vulture (*Gyps bengalensis*) in India's Western Ghats using niche models and habitat requirements. *Ecol. Eng.* **2020**, *158*, 106034. [CrossRef]
49. Lin, X.; Yu, Z. Floristic analysis of Magnoliaceae in Jiangxi. *Acta Bot. Boreali-Occident. Sin.* **2004**, *3*, 272–273.
50. Liang, J.; Jiang, Q.; Jiang, L.; Zhang, M.; Wu, Y.; Yang, J.; Yang, Y.; Lan, Y.; Tang, G. The MaxEnt modeling for predicting the potential geographical distribution of *Manglietia pachyphylla* in China under climate change. *Guangxi Sci.* **2023**, *30*, 1133–1143.
51. Zhuang, H.; Zhang, Y.; Wang, W.; Ren, Y. Optimized hot spot analysis for probability of species distribution under different spatial scales based on MaxEnt model: *Manglietia insignis* case. *Biodivers. Sci.* **2018**, *26*, 931–940. [CrossRef]
52. Zou, W.; Wang, L.; Lai, T.; Zhu, X.; Chen, F.; Zeng, J. Distribution, influencing factors and gaps analysis of *Manglietia fordiana*: Implications of conservation to mountain broad-leaved forests. *Acta Ecol. Sin.* **2022**, *42*, 1889–1900.
53. He, Q.; Wu, H.; Ye, J.; Tian, B. Distributional Shift of East Asian Endemic Genus *Oyama* Based on MaxEnt. *J. Yunnan Agric. Univ. Nat. Sci.* **2023**, *38*, 846–855.
54. Lan, X.; Wang, J.; Fu, C.; Li, L.; Yuan, M.; Tan, T.; Du, F. Prediction of Suitable Distribution Area of *Magnolia sieboldii* in China Based on the Optimized MaxEnt Model. *J. Northwest For. Univ.* **2022**, *37*, 100–106.
55. Zou, D.; Wang, Q.; Luo, A.; Wang, Z. Species richness patterns and resource plant conservation assessments of Rosaceae in China. *Chin. J. Plant Ecol.* **2019**, *43*, 1–15. [CrossRef]

MDPI AG
Grosspeteranlage 5
4052 Basel
Switzerland
Tel.: +41 61 683 77 34

MDPI Books Editorial Office
E-mail: books@mdpi.com
www.mdpi.com/books

www.ingramcontent.com/pod-product-compliance
Lightning Source LLC
Chambersburg PA
CBHW041942260926
51850CB00018B/466
* 9 7 8 3 7 2 5 8 7 9 2 7 4 *